Developments in Geochemistry

2

DEVELOPMENTS IN GEOCHEMISTRY
Advisory Editor: W.S. Fyfe

1. W.S. FYFE, N.J. PRICE and A.B. THOMPSON
FLUIDS IN THE EARTH'S CRUST

Developments
in
Geochemistry
2

RARE EARTH ELEMENT GEOCHEMISTRY

edited by

P. HENDERSON

Department of Mineralogy, British Museum (Natural History), London, U.K.

ELSEVIER
Amsterdam – Oxford – New York – Tokyo 1984

ELSEVIER SCIENCE PUBLISHERS B.V.
1 Molenwerf
P.O. Box 211, 1000 AE Amsterdam, The Netherlands

Distributors for the United States and Canada:

ELSEVIER SCIENCE PUBLISHING COMPANY INC.
52, Vanderbilt Avenue
New York, N.Y. 10017

Library of Congress Cataloging in Publication Data
Main entry under title:

Rare earth element geochemistry.

(Developments in geochemistry ; 2)
Includes bibliographical references and index.
1. Earths, Rare. 2. Geochemistry. I. Henderson, Paul. II. Series.
QE516.R2R37 1983 551.9 83-1536
ISBN 0-444-42148-3 (U.S.)

ISBN 0-444-42148-3 (Vol. 2)
ISBN 0-444-41635-8 (Series)

Printed in The Netherlands

PREFACE

Nearly two centuries ago, in 1787, Carl Alex Arrhenius collected an unusual black mineral from a feldspar quarry at Ytterby, near Stockholm. From this mineral, later to be named gadolinite, Gadolin in 1794 extracted the earth "yttria" — a mixture of several rare earth oxides. Thus began the discovery and separation of the rare earths, but it was not until 1907 that Urbain completed the separation of all the naturally-occurring rare earths by extracting lutetium. The problem facing the early chemists was the extreme chemical coherence of the rare earth element group, which makes separation of the individual members so difficult. This coherence is shown in their geochemical behaviour but sufficiently significant fractionations do occur, which have proved to be of great importance in the interpretation of petrogenetic processes.

This book reflects the remarkable developments that have come about in our knowledge of the chemistry and geochemistry of the rare earth elements since the beginning of the century. The geochemical advances stem from the introduction of new analytical techniques, together with the recognition that rare earth fractionation occurs naturally in various ways and is capable of interpretation. These advances were aided by improvements in data presentation, especially that of chondrite-normalization, which allows a clear visual appreciation of the type and degree of fractionation. Much of the geochemical work has been in the field of igneous rock petrogenesis; this book, therefore, also has that emphasis. More information on metamorphic rocks and processes would have been welcomed but this proved to be unforthcoming. For the future, it seems clear that the application of rare earth element geochemistry to the interpretation of the processes of metasomatism, ore formation, rock alteration, and mineral authigenesis in marine and fresh waters is likely to yield beneficial results.

The following chapters have been written by recognized authorities in their chosen field; they reflect the current state of knowledge in an area of science undergoing rapid development. They help to show the value of the rare earths in geochemistry but illustrate also how interpretations obtained from rare earth element data often need to be supported by other geochemical or geological evidence. The book is addressed to all those who study geochemistry and petrology, whether they be undergraduates, lecturers, or researchers. Some of the chapters emphasizing general principles (e.g., Chapter 4: "Petrogenetic modelling — use of rare earth elements") will be of especial use to the student while others (e.g., Chapter 2: "Mineralogy of the rare earth elements") will be of most value to those needing a source of reference. In all, the book attempts to provide a much-needed synthesis of a diverse geochemical field at a time of its continuing growth.

I wish to thank the many people who acted as reviewers for the following chapters or who made many useful suggestions in discussion.

I wish to thank the many people who acted as reviewers for the following chapters or who made many useful suggestions in discussion.

PAUL HENDERSON
October, 1982.

LIST OF CONTRIBUTORS

W.V. BOYNTON	*Department of Planetary Sciences and Lunar and Planetary Laboratory, University of Arizona, Tucson, AZ 85721, U.S.A.*
A.M. CLARK	*Department of Mineralogy, British Museum (Natural History), London SW7 5BD, U.K.*
R.L. CULLERS	*Department of Geology, Kansas State University, Manhattan, KS 66506, U.S.A.*
A.J. FLEET	*Exploration and Production Division, BP Research Centre, Chertsey Road, Sunbury-on-Thames, Middlesex TW16 7LN, U.K.*
F.A. FREY	*Department of Earth and Planetary Sciences, Massachusetts Institute of Technology, Cambridge, MA 02139, U.S.A.*
J.L. GRAF	*Department of Geology, Kansas State University, Manhattan, KS 66506, U.S.A.*
L.A. HASKIN	*Department of Earth and Planetary Sciences and McDonnell Center for the Space Sciences, Washington University, St. Louis, MO 63130, U.S.A.*
C.J. HAWKESWORTH	*Department of Earth Sciences, Open University, Walton Hall, Milton Keynes, MK7 6AA, U.K.*
P. HENDERSON	*Department of Mineralogy, British Museum (Natural History), London SW7 5BD, U.K.*
D.E. HIGHLY	*Minerals Strategy and Economics Research Unit, Institute of Geological Sciences, London SW7 2DE, U.K.*
S.E. HUMPHRIS	*Sea Education Association and Woods Hole Oceanographic Institution, Woods Hole, MA 02543, U.S.A.*
C.R. NEARY	*Minerals Strategy and Economics Research Unit, Institute of Geological Sciences, London SW7 2DE, U.K.*
R.J. PANKHURST	*British Antarctic Survey, Natural Environment Research Council, c/o Institute of Geological Sciences, 64-78 Gray's Inn Road, London WC1X 8NG, U.K.*
A.D. SAUNDERS	*Department of Geology, Bedford College, London NW1 4NS, U.K.*
P.W.C. van CALSTEREN	*Department of Earth Sciences, Open University, Walton Hall, Milton Keynes, MK7 6AA, U.K.*

SYMBOLS

Most of the symbols, abbreviations and acronyms used in this book are standard ones, and many are defined within the body of the text. The more commonly-used, and less usual, symbols are listed below.

D	Nernst distribution coefficient
D_{WR}	distribution coefficient (solid/liquid) summed for all minerals in a rock
f_{O_2}	oxygen fugacity
K	equilibrium constant
R	gas constant
$T_{1/2}$	half-life
a	year
t	tonne
cn	chondrite normalized
l	liquid
s	solid
HREE	heavy rare earth elements, Gd—Lu
LREE	light rare earth elements, La—Eu
M	a chemical element
RE	rare earth(s)
REE	rare earth element(s)
REO	rare earth oxide (no specific stoichiometry is implied)
EPR	East Pacific Rise
IOR	Indian Ocean Ridge
MAR	Mid-Atlantic Ridge
MORB	mid-oceanic ridge basalt(s)

The rare earth elements, in alphabetical order, are:

Ce	cerium	Nd	neodymium
Dy	dysprosium	Pm	promethium
Er	erbium	Pr	praseodymium
Eu	europium	Sm	samarium
Gd	gadolinium	Tb	terbium
Ho	holmium	Tm	thulium
La	lanthanum	Yb	ytterbium
Lu	lutetium		

CONTENTS

Preface V
List of Contributors VII
Symbols VIII

CHAPTER 1. GENERAL GEOCHEMICAL PROPERTIES AND ABUNDANCES OF THE RARE EARTH ELEMENTS
P. Henderson

1.1. Introduction 1
1.2. Abundances in the solar system, Sun and Earth 3
1.3. Data presentation 6
1.4. Oxidation states 10
1.5. Element co-ordination and ionic radii 14
1.6. Element substitution and partition 17
1.7. Geochemical reference literature 28
References 29

CHAPTER 2. MINERALOGY OF THE RARE EARTH ELEMENTS
A.M. Clark

2.1. Introduction 33
2.2. Halides 38
2.3. Carbonates 39
2.4. Carbonates with fluoride 40
2.5. Oxides 41
2.6. Borate 45
2.7. Silicates 45
2.8. Silicates with other anions 51
2.9. Phosphates 51
2.10. Arsenates 54
2.11. Sulphate 54
References 54

CHAPTER 3. COSMOCHEMISTRY OF THE RARE EARTH ELEMENTS: METEORITE STUDIES
W.V. Boynton

3.1. Introduction 63
3.2. Condensation theory 65
3.3. REE abundances in Ca, Al-rich inclusions 71
3.4. REE in other components of chondrites 86
3.5. Whole-rock chondrite analyses 89
3.6. Analyses of achondrites 94

3.7. REE in other differentiated meteorites . . . 103
3.8. Summary of REE cosmochemistry and future directions . . . 106
References . . . 107

CHAPTER 4. PETROGENETIC MODELLING — USE OF RARE EARTH ELEMENTS
L.A. Haskin
4.1. Introduction . . . 115
4.2. Modelling of separation processes . . . 117
4.3. Simple chemical partitioning — the Nernst D . . . 118
4.4. Equilibrium melting and crystallization . . . 122
4.5. Fractional crystallization: continuous removal of crystals . . . 128
4.6. Fractional crystallization: crystal zoning-surface equilibrium . . . 133
4.7. Incremental partial melting . . . 138
4.8. Zone melting . . . 140
4.9. Extraction . . . 142
4.10. Mixing . . . 145
4.11. References to combined models, and other comments . . . 145
References . . . 148

CHAPTER 5. RARE EARTH ELEMENT ABUNDANCES IN UPPER MANTLE ROCKS
F.A. Frey
5.1. Introduction . . . 153
5.2. Alpine peridotites . . . 154
5.3. Ultramafics associated with ophiolites . . . 166
5.4. Ocean floor peridotites . . . 170
5.5. Ultramafic inclusions in basalts and kimberlites . . . 176
5.6. Implications of REE data for upper mantle composition and upper mantle processes . . . 192
References . . . 196

CHAPTER 6. THE RARE EARTH ELEMENT CHARACTERISTICS OF IGNEOUS ROCKS FROM THE OCEAN BASINS
A.D. Saunders
6.1. Introduction . . . 205
6.2. History of research . . . 206
6.3. Ocean ridge basalts . . . 208
6.4. Ocean islands . . . 214
6.5. Ocean plateau volcanism . . . 220
6.6. Factors controlling the distribution of REE in oceanic basalts . . . 222
6.7. Summary and conclusions . . . 229
References . . . 231

CHAPTER 7. RARE EARTH ELEMENTS IN IGNEOUS ROCKS OF THE CONTINENTAL CRUST: PREDOMINANTLY BASIC AND ULTRABASIC ROCKS
R.L. Cullers and J.L. Graf
7.1. Introduction . . . 237
7.2. Kimberlite . . . 239
7.3. Carbonatites and associated alkalic silicate rocks . . . 243
7.4. Lamprophyres . . . 249

7.5. Komatiites and associated basic rocks 251
7.6. Alkaline mafic rocks and associated intermediate and felsic rocks 254
7.7. Tholeiitic basalts — continental, back-arc basins and island arcs 262
7.8. Summary 267
References 268

CHAPTER 8. RARE EARTH ELEMENTS IN IGNEOUS ROCKS OF THE CONTINENTAL CRUST: INTERMEDIATE AND SILICIC ROCKS — ORE PETROGENESIS
R.L. Cullers and J.L. Graf

8.1. Introduction 275
8.2. Andesites and associated rocks 276
8.3. Anorthosites and associated jotunites, mangerites, and charnockites 281
8.4. Granitic rocks — quartz diorite, tonalite, granodiorite, and trondhjemite 285
8.5. Granitic rocks — monzogranites and syenogranites 293
8.6. REE in hydrothermal systems 301
8.7. Summary 306
References 308

CHAPTER 9. THE MOBILITY OF THE RARE EARTH ELEMENTS IN THE CRUST
Susan E. Humphris

9.1. Introduction 317
9.2. Factors affecting the degree of mobility of the REE 318
9.3. Igneous crystallization history — the distribution of REE in the unaltered rock and their susceptibility to weathering and alteration 319
9.4. The fluid phase 322
9.5. Uptake of the REE by secondary minerals 334
9.6. Conclusions 337
References 340

CHAPTER 10. AQUEOUS AND SEDIMENTARY GEOCHEMISTRY OF THE RARE EARTH ELEMENTS
A.J. Fleet

10.1. Introduction 343
10.2. Weathering and transport 346
10.3. The marine environment 350
10.4. REE contents of sediments and sedimentary rocks 365
10.5. Diagenesis 367
References 369

CHAPTER 11. RADIOGENIC ISOTOPES — SOME GEOLOGICAL APPLICATIONS
C.J. Hawkesworth and P.W.C. van Calsteren

11.1. Introduction 375
11.2. Radioactive decay 376
11.3. Geochronology — the dating of rocks and minerals 379
11.4. Model Nd, or T^{Nd}_{CHUR}, ages 383
11.5. Isotopes as tracers in petrogenesis 388
11.6. Isotope variations in seawater 405
11.7. Models for the evolution of the Earth's crust and mantle 408
11.8. Lu-Hf isotopes 411
11.9. Concluding remarks 413

Appendix 414
References 416

CHAPTER 12. THE ECONOMIC IMPORTANCE OF THE RARE EARTH ELEMENTS
C.R. Neary and D.E. Highley

12.1. Introduction. 423
12.2. Abundance and ore mineralogy 424
12.3. Geological characteristics of the deposits 424
12.4. World reserves and resources 436
12.5. Mining, mineral processing, extraction and separation. 438
12.6. Historical production and demand 443
12.7. Current uses 447
12.8. Use development. 460
References 461

CHAPTER 13 ANALYTICAL CHEMISTRY
P. Henderson and R.J. Pankhurst

13.1. Introduction. 467
13.2. Neutron activation analysis 468
13.3. Mass-spectrometric isotope dilution analysis 478
13.4. Inductively coupled plasma emission spectrometry 488
13.5. Other methods 491
13.6. Concluding statement. 494
References 495

Subject Index. 501

Chapter 1

GENERAL GEOCHEMICAL PROPERTIES AND ABUNDANCES OF THE RARE EARTH ELEMENTS

PAUL HENDERSON

1.1. Introduction

The rare earth elements, lanthanum to lutetium (atomic numbers 57—71), are members of Group IIIA in the periodic table (Fig. 1.1) and all have very similar chemical and physical properties. This uniformity arises from the nature of their electronic configurations, leading to a particularly stable 3+ oxidation state and a small but steady decrease in ionic radius with increase in atomic number for a given co-ordination number. Despite the similarity in their chemical behaviour, these elements can be partially fractionated, one from the other, by several petrological and mineralogical processes. The wide variety of types and sizes of the cation co-ordination polyhedra in rock-forming minerals provides the means for this chemical fractionation: it is this phenomenon which has important consequences in geochemistry.

The significant growth of interest in the geochemistry of the rare earth elements (REE) has come about because of the realization that the observed (measured) degree of REE fractionation in a rock or mineral can be a pointer

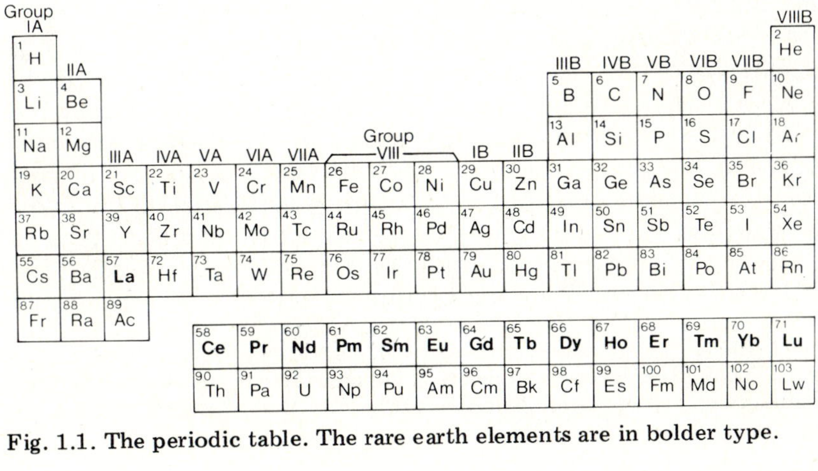

Fig. 1.1. The periodic table. The rare earth elements are in bolder type.

to its genesis, and also because accurate quantitative analysis for the REE, both as a group and individually, is now possible on a routine basis even when the elements occur at very low concentrations (Chapter 13). The application of REE abundances to petrogenetic problems has centred on the evolution of igneous rocks where such processes as partial melting of crustal or mantle materials, fractional crystallization, and/or mixing of magmas are involved (Chapters 5 to 8). In these studies the matching of observed REE abundance with those provided by the theoretical modelling of petrogenetic processes (Chapter 4) has helped considerably to restrict the number of possible hypotheses on the genesis of a rock or mineral suite.

Yttrium (Y, atomic number 39) is also a member of Group IIIA and shows a similar chemistry to that of the REE, and is sometimes included with them in descriptive accounts. The term "lanthanons" (abbreviated Ln) is applied to the sixteen elements in the group La to Lu plus Y. The lightest element in Group IIIA, scandium, shows a sufficiently distinct chemistry, owing to the relatively small radius of its 3+ ion, to warrant separate description. The term lanthanides is sometimes used as a name for the fourteen elements following lanthanum in the periodic table (i.e. Ce to Lu). This book is concerned with the fifteen elements La to Lu where, to conform with past and current geochemical usage, they are referred to as *rare earth elements*†. Chapters 2 and 12 also include Y in their discussion.

It has been found convenient to divide the REE into two sub-groups: those from La to Sm (i.e. lower atomic numbers and masses) being referred to as the light rare earth elements (abbreviated LREE) and those from Gd to Lu (higher atomic numbers and masses) being referred to as the heavy rare earth elements (abbreviated HREE). Very occasionally the term middle rare earth elements (abbreviated MREE) is loosely applied to the elements from about Pm to about Ho. Atomic weights are given in Table 1.1.

The electronic configurations of the REE are given in Table 1.1 for the ground state and for three different oxidation states. Lanthanum has an outer electronic configuration in the ground state of $5d^1 6s^2$, but the next element, cerium, has an electron in the $4f$ sub-shell (see Table 1.1). The following elements have the electrons entering the $4f$ sub-shell, until at ytterbium the $4f$ sub-shell is filled. The $4f$ electrons are well shielded by the eight electrons in the $5s^2$ and $5p^6$ sub-shells, so that they are not significantly involved in chemical interactions. Hence, any difference in the number of electrons in the $4f$ sub-shell does not lead to much difference in chemical behaviour, nor to significant ligand field affects. The REE, therefore, tend to occur in any natural occurrence as a group rather than singly or as a combination of a few of their number. They are lithophile, in that they concentrate

†In 1968 the International Union for Pure and Applied Chemistry recommended that "rare earth elements" should refer to the elements scandium, yttrium, lanthanum and the lanthanides, but this usage has not been generally adopted in the geochemical literature and is not followed here.

TABLE 1.1

Atomic weights (1973) and ground state electronic configurations

Atomic No.	Atomic weight	Symbol	Configuration 0	+1	+2	+3
57	138.9055	La	[Xe] $5d^1 6s^2$	[Xe] $5d^2$	[Xe] $5d^1$	[Xe] $4f^0$
58	140.12	Ce	[Xe] $4f^1 5d^1 6s^2$	[Xe] $4f^1 5d^1 6s^1$	[Xe] $4f^2$	[Xe] $4f^1$
59	140.9077	Pr	[Xe] $4f^3 6s^2$	[Xe] $4f^3 6s^1$	[Xe] $4f^3$	[Xe] $4f^2$
60	144.24	Nd	[Xe] $4f^4 6s^2$	[Xe] $4f^4 6s^1$	[Xe] $4f^4$	[Xe] $4f^3$
61	(145)	Pm	[Xe] $4f^5 6s^2$	[Xe] $4f^5 6s^1$	[Xe] $4f^5$	[Xe] $4f^4$
62	150.4	Sm	[Xe] $4f^6 6s^2$	[Xe] $4f^6 6s^1$	[Xe] $4f^6$	[Xe] $4f^5$
63	151.96	Eu	[Xe] $4f^7 6s^2$	[Xe] $4f^7 6s^1$	[Xe] $4f^7$	[Xe] $4f^6$
64	157.25	Gd	[Xe] $4f^7 5d^1 6s^2$	[Xe] $4f^7 5d^1 6s^1$	[Xe] $4f^7 5d^1$	[Xe] $4f^7$
65	158.9254	Tb	[Xe] $4f^9 6s^2$	[Xe] $4f^9 6s^1$	[Xe] $4f^9$	[Xe] $4f^8$
66	162.50	Dy	[Xe] $4f^{10} 6s^2$	[Xe] $4f^{10} 6s^1$	[Xe] $4f^{10}$	[Xe] $4f^9$
67	164.9304	Ho	[Xe] $4f^{11} 6s^2$	[Xe] $4f^{11} 6s^1$	[Xe] $4f^{11}$	[Xe] $4f^{10}$
68	167.26	Er	[Xe] $4f^{12} 6s^2$	[Xe] $4f^{12} 6s^1$	[Xe] $4f^{12}$	[Xe] $4f^{11}$
69	168.9342	Tm	[Xe] $4f^{13} 6s^2$	[Xe] $4f^{13} 6s^1$	[Xe] $4f^{13}$	[Xe] $4f^{12}$
70	173.04	Yb	[Xe] $4f^{14} 6s^2$	[Xe] $4f^{14} 6s^1$	[Xe] $4f^{14}$	[Xe] $4f^{13}$
71	174.97	Lu	[Xe] $4f^{14} 5d^1 6s^2$	[Xe] $4f^{14} 6s^2$	[Xe] $4f^{14} 6s^1$	[Xe] $4f^{14}$

[Xe] = configuration of xenon: $1s^2 2s^2 2p^6 3s^2 3p^6 3d^{10} 4s^2 4p^6 4d^{10} 5s^2 5p^6$.

predominantly in the silicate rather than the metal or sulphide phases when these coexist. They also tend to be "dispersed" elements since they are present in trace quantities in many minerals, only rarely occurring at high concentrations (see Chapter 2).

This chapter reviews some of the fundamental aspects of REE geochemistry and gives data on abundances in the solar system, the bulk Earth and the Earth's crust. It describes the state of knowledge on the partitioning of the REE, especially in igneous rock systems, and cites reference works concerned with the REE.

1.2. Abundances in the solar system, Sun and Earth

Estimates of the composition of the solar system based on the concentrations of elements in carbonaceous chondrites and in young stars have provided data on the relative abundances of the REE. Part of a recent compilation by Cameron (1973) is given in Table 1.2, together with an estimate of the relative abundances (to Si = 10^6 atoms) in the solar atmosphere determined from spectral analysis (Ross and Aller, 1976). The solar-system abundances are plotted in Fig. 1.2 where the rhythmic alternation in abundance between elements of even and odd atomic number can be seen. This alternation arises from variations in the binding energy, and hence

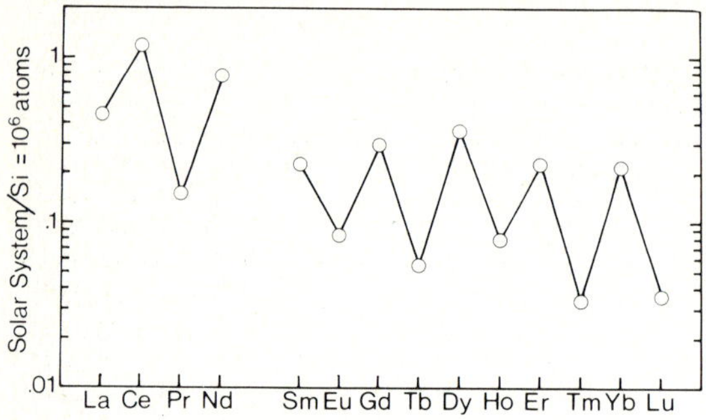

Fig. 1.2. Rare earth element abundances (log scale), in the solar system, relative to 10^6 atoms of silicon, plotted against atomic number. Data from Cameron (1973).

stability, of a nucleus being dependent on whether the neutron number (N) and the proton number (Z) are odd or even. Those nuclei with both N and Z even are the stablest while those with both N and Z odd are the least stable. Table 1.2 shows that the low relative abundance values of the REE follow the general trend of decreasing abundance with increasing atomic number.

Perhaps not surprisingly it is more difficult to determine element concentrations in the bulk Earth than in the Sun's atmosphere. In the lack of direct evidence, it is necessary to make assumptions as to the nature of proportions of the different components that aggregated to form the bulk Earth, or to make assumptions about the available materials which could be representative of the interior parts of our planet. Ganapathy and Anders (1974) in their attempt to estimate the composition of the bulk Earth, used the former approach with the assumption that the inner planets were produced by the same process that gave chondritic meteorites, and also used theoretical condensation sequences of nebular gases. They concluded that the early condensate material from the solar nebula was the sole contributor of REE to the bulk Earth. Smith (1977), using the latter approach, summed feasible contributions from the hydrosphere, atmosphere, crust, mantle and core to obtain estimates for most elements including the REE (Table 1.3). For the REE, both estimates are subject to large error, because of problems relating to element condensation sequences, and the existence of varied REE concentrations in chondritic meteorites as well as in material that is likely to be representative of early condensates (see discussion by Boynton, Chapter 3). There is also uncertainty of the REE concentrations in the major shells that constitute the Earth. Hence, the values in Table 1.3 should be treated with caution.

There have been several attempts to establish the composition of the Earth's crust but most of these have not included the REE. One important exception is in the work of Taylor (1964) who used the very different abundances of the REE in granitic and basaltic rocks as a basis for estimating the composition of the continental crust. He showed that a mix of 1:1 mafic

TABLE 1.2

Relative abundances of the REE, and other selected elements, in the Sun and in the solar system (normalized to Si = 10^6 atoms)

Atomic No.	Element	Abundances	
		solar[a]	solar system[b]
1	H	2.24×10^{10}	3.18×10^{10}
11	Na	4.27×10^4	6.0×10^4
12	Mg	8.91×10^5	1.016×10^6
20	Ca	5.01×10^4	7.21×10^4
21	Sc	24.5	35
26	Fe	7.08×10^5	8.3×10^5
39	Y	2.82	4.8
40	Zr	12.6	28
50	Sn	2.2	3.6
57	La	0.302	0.445
58	Ce	0.794	1.18
59	Pr	0.102	0.149
60	Nd	0.380	0.78
62	Sm	0.12	0.226
63	Eu	0.01	0.085
64	Gd	0.295	0.297
65	Tb	n.a.	0.055
66	Dy	0.257	0.36
67	Ho	n.a.	0.079
68	Er	0.13	0.225
69	Tm	0.041	0.034
70	Yb	0.2	0.216
71	Lu	0.13	0.036
72	Hf	0.14	0.21
82	Pb	1.91	4
92	U	<0.09	0.0262

n.a. = not available.
[a]Based on data in Ross and Aller (1976).
[b]From Cameron (1973).

to silicic igneous rock would produce the observed REE distribution in "average sediment". This igneous rock mix was considered, therefore, to represent the composition of average continental crust. The figures for the REE are given in Table 1.3 and in a chondrite-normalized plot in Fig. 1.3 (see next section for explanation of these plots). The fact that the REE are sometimes mobilized during rock weathering (see Chapters 9 and 10) throws doubt on the accuracy of Taylor's method, but the general nature of the abundance pattern is probably correct, except that the apparently anomalous abundance of Dy is almost certainly wrong. Fig. 1.3 shows that the LREE, relative to chondrites, are more abundant in the crust than are the HREE.

TABLE 1.3

Estimated abundances (ppm) of REE in the bulk Earth according to Ganapathy and Anders (1974) and Smith (1977), and in the Earth's continental crust, according to Taylor (1964)

	Bulk Earth		Crust
	Ganapathy and Anders (1974)	Smith (1977)	Taylor (1964)
La	0.48	0.78	30
Ce	1.28	2.2	60
Pr	0.162		8.2
Nd	0.87	1.2	28
Sm	0.26	0.22	6
Eu	0.100	0.066	1.2
Gd	0.37	0.35	5.4
Tb	0.067		0.9
Dy	0.45	0.21	3
Ho	0.101		1.2
Er	0.29	0.093	2.8
Tm	0.044		0.48
Yb	0.29		3
Lu	0.049	0.015	0.5

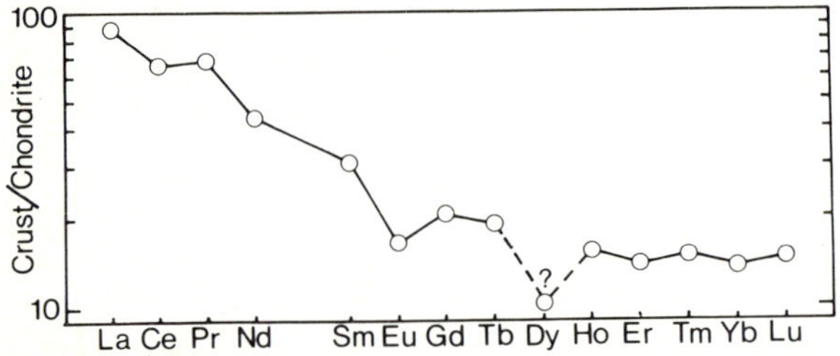

Fig. 1.3. Chondrite-normalized abundances (log scale) of the REE in the continental crust plotted against atomic number. Data from Taylor (1964). The value for dysprosium is probably incorrect.

The relative natural abundances of the REE isotopes, together with the half-lives of those which are radioactive, are given in Table 1.4. Frey (Chapter 5, section 5.1) discusses the probable REE contents of the mantle.

1.3. Data presentation

Rare earth element concentrations can be presented in several other ways than simple tabulation. The two most common methods are graphical and both involve the normalization of the concentrations in the sample to those

TABLE 1.4

Natural abundances (atomic percent) and half-lives (years) of the REE isotopes

	Abundance	$T_{1/2}$		Abundance	$T_{1/2}$
^{138}La	0.089	1.05×10^{11}	^{159}Tb	100	
^{139}La	99.91				
			^{156}Dy	0.052	2×10^{14}
^{136}Ce	0.193		^{158}Dy	0.090	
^{138}Ce	0.250		^{160}Dy	2.29	
^{140}Ce	88.48		^{161}Dy	18.88	
^{142}Ce	11.07	$>5 \times 10^{16}$	^{162}Dy	25.53	
			^{163}Dy	24.97	
^{141}Pr	100		^{164}Dy	28.18	
^{142}Nd	27.17		^{165}Ho	100	
^{143}Nd	12.20				
^{144}Nd	23.79	2.1×10^{15}	^{162}Er	0.136	
^{145}Nd	8.29		^{164}Er	1.56	
^{146}Nd	17.18		^{166}Er	33.41	
^{148}Nd	5.748		^{167}Er	22.94	
^{150}Nd	5.625		^{168}Er	27.07	
			^{170}Er	14.88	
^{144}Sm	3.075				
^{147}Sm	15.00	1.06×10^{11}	^{169}Tm	100	
^{148}Sm	11.24	8×10^{15}			
^{149}Sm	13.82	$>1 \times 10^{16}$	^{168}Yb	0.135	
^{150}Sm	7.38		^{170}Yb	3.03	
^{152}Sm	26.74		^{171}Yb	14.31	
^{154}Sm	22.75		^{172}Yb	21.82	
			^{173}Yb	16.13	
^{151}Eu	47.82		^{174}Yb	31.84	
^{153}Eu	52.18		^{176}Yb	12.73	
^{152}Gd	0.20	1.1×10^{14}	^{175}Lu	97.41	
^{154}Gd	2.15		^{176}Lu	2.59	2.6×10^{10}
^{155}Gd	14.73				
^{156}Gd	20.47				
^{157}Gd	15.68				
^{158}Gd	24.87				
^{160}Gd	21.90				

in a chosen reference material (i.e. the concentration of each REE in the sample is divided by the concentration of the same REE in the reference material). The plot is usually given as the logarithm of the normalized abundance versus atomic number (or ionic radius, but see below).

Method 1: Normalization to a reference that is external to the system under investigation

The reference concentrations are those either in chondritic meteorites, in a sedimentary "average rock", or those estimated for the Earth's crust. Normalization to abundances in chondritic meteorites is the commonest choice and the resultant graph — see Fig. 1.4 — is sometimes referred to as a Masuda-Coryell diagram after some of the people who proposed this procedure (Masuda, 1962; Coryell et al., 1963).

The advantages of this method are that the abundance variation between REE of odd and even atomic number is eliminated, and the extent of any fractionation amongst the various REE in the specimen is discernible because there is considered to have been no fractionation between the light and heavy REE in chondrites. The fractionation of one REE from another can also be expressed in the form of a chondrite-normalized element ratio, e.g.

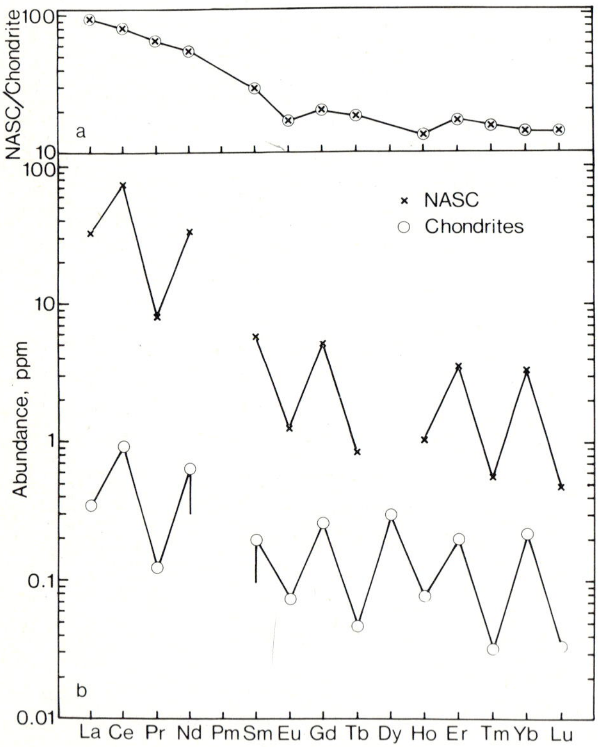

Fig. 1.4. (a) Chondrite-normalized abundances of the REE in "North American shales composite", abbreviated NASC. (b) Actual abundances in NASC given by Haskin et al. (1968) and in the composite for ordinary chondrites by Wakita et al. (1971). The values plotted in (b) were used to construct Fig. 1.4a.

$(La/Lu)_{cn}$. The $(La/Lu)_{cn}$ ratio gives a measure of the overall slope of the normalized plot if this approximates to a straight line, and thereby the extent of fractionation of the LREE from the HREE. $(La/Sm)_{cn}$ and $(Gd/Lu)_{cn}$ ratios give comparable information for the LREE and HREE, respectively. These and similar ratios are useful in defining the nature of a chondrite-normalized plot of REE abundances, but care must be exercised in their use. For example, the $(La/Lu)_{cn}$ ratio of some peridotites and dunites is close to 1, but this value gives no indication of the markedly concave nature of some of their chondrite-normalized plots (see Chapter 5).

The plotted position for Eu sometimes lies away from the trend defined by the other REE on a chondrite-normalized abundance diagram (see plot for feldspar in Fig. 1.5a). This departure is called a *europium anomaly* and is additionally referred to as being *positive* if Eu_{cn} is greater than each of Sm_{cn} and Gd_{cn}, or *negative* if vice versa. The extent of the anomaly is defined by the ratio [Eu]/[Eu*] where [Eu] is the europium concentration and [Eu*] is the value obtained at the europium position by straight line interpolation between the plotted points for Sm and Gd†. Similar anomalies are occasionally recorded for a few other REE, especially Ce.

Unfortunately, adoption of only one agreed set of chondrite-normalizing concentrations has not occurred. Instead, several sets have been used (Table 1.5) although some much more than others. Those by Wakita et al. (1971) and the average of nine chondrites given by Haskin et al. (1968) are very commonly used and are similar to each other. The values by Wakita et al. are used throughout this book unless stated otherwise. A further discussion on chondrite normalization is given by Boynton (Chapter 3, section 3.5, and see Fig. 3.7).

Any new "recommended" set of normalizing values (such as those suggested by Boynton, Chapter 3, section 3.5) should be agreed internationally but, as yet, there is no satisfactory mechanism by which this may be done. Insofar as the normalization procedure is primarily to clarify the presentation of REE concentration data and to facilitate data comparison, it is suggested that the established normalization values by Wakita et al. (1971) or Haskin et al. (1968) be used until a new consensus is reached. In the case of sediments it is often more appropriate, as well as being common practice, to use a set of REE concentrations in a suitable sediment or average sediment; e.g. North American shales composite, abbreviated NASC (see Table 1.5 and Chapter 10).

---

†Hence, positive europium anomalies have [Eu]/[Eu*] ratios >1, while negative anomalies have the ratio <1. The [Eu]/[Eu*] ratio is frequently written as Eu/Eu*.

TABLE 1.5

Rare earth element abundances in chondritic meteorites, and in North American shales composite (NASC), used for normalization

	Composite of 12 chondrites[a]	Composite of 9 chondrites[b]	Leedey chondrite[c]	Average of 10 chondrites[d]	Average of C1 chondrites[e]	NASC[b]
La	0.34	0.330	0.378	0.329	0.2446	32
Ce	0.91	0.88	0.976	0.865	0.6379	73
Pr	0.121	0.112	—	—	0.09637	7.9
Nd	0.64	0.60	0.716	0.630	0.4738	33
Sm	0.195	0.181	0.230	0.203	0.1540	5.7
Eu	0.073	0.069	0.0866	0.0770	0.05802	1.24
Gd	0.26	0.249	0.311	0.276	0.2043	5.2
Tb	0.047	0.047	—	—	0.03745	0.85
Dy	0.30	—	0.390	0.343	0.2541	—
Ho	0.078	0.070	—	—	0.05670	1.04
Er	0.20	0.200	0.255	0.225	0.1660	3.4
Tm	0.032	0.030	—	—	0.02561	0.50
Yb	0.22	0.200	0.249	0.220	0.1651	3.1
Lu	0.034	0.034	0.0387	0.0339	0.02539	0.48

[a]H. Wakita and D. Zellmar (unpublished), quoted in Wakita et al. (1971).
[b]Haskin et al. (1968).
[c]Masuda et al. (1973).
[d]Nakamura (1974).
[e]Evensen et al. (1978).

Method 2: Normalization using a reference that is part of the system under investigation

In this method the reference material could be a specific rock or mineral. For example, concentrations in minerals can be normalized to the REE concentrations of the rock they constitute. This procedure can show clearly the amount by which the different minerals fractionate the REE from one another. An example is given in Fig. 1.5b, which can be compared with the chondrite-normalized plot in Fig. 1.5a.

The use of ionic radius as a parameter in either method is to be avoided since the radius of any ion is a function of co-ordination number and ionic charge, the former varying from mineral to mineral and the latter for some REE ions occurring in more than one state within the same rock system.

1.4. Oxidation states

The REE are strongly electropositive and so most of their chemistry is characteristic of ionic bonding, with only a minimal covalent contribution.

The general sequence for the ionization of the REE is considered to be

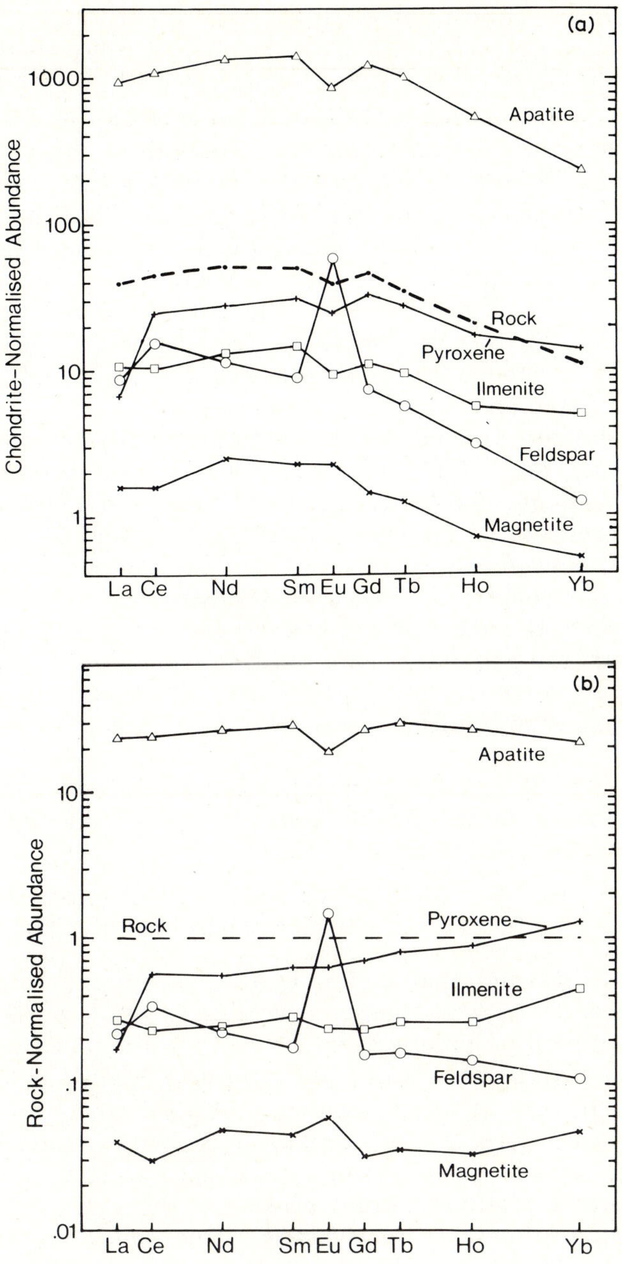

Fig. 1.5. (a) Plot of REE abundances in a rock and its constituent minerals, normalized to REE abundances in chondrites, versus atomic number. (The rock is from the Skaergaard intrusion.) Data from Paster et al. (1974). (b) As (a) but with abundances normalized to those in the rock.

the removal first of the two 6*s* electrons, then either a 5*d* or a 4*f* electron which is relatively close in energy to that of a 6*s* electron. A 4*f* electron would be the fourth to be removed but the ionization energy is too high for the +4 state to be common. In fact all the REE show a remarkably constant valency of three in their chemistry and geochemistry. Oxidation states of +2 may be shown by Eu and Yb, and of +4 by Ce and Tb. The existence of these states can be explained partly on the basis of the enhanced stability of half-filled (Eu^{2+} and Tb^{4+}) and completely-filled (Yb^{2+}) 4*f* sub-shells, while Ce^{4+} has the electronic configuration of the noble gas xenon. Very occasionally the other REE may exhibit 2+ or 4+ oxidation states in their chemistry but not, apparently, in their geochemistry. (Goldschmidt, 1954, suggested that there was some evidence for the presence of Sm^{2+} in minerals but this has not been substantiated and is unlikely.)

Evidence for the occurrence of 2+ and 4+ states in natural systems is substantial only for Eu^{2+} and Ce^{4+}. Tb^{4+} has not been recorded in any mineral or natural aqueous medium. The existence of Yb^{2+} has been invoked as an explanation for the presence of Eu and Yb negative anomalies, and the coherence of Eu and Yb concentrations in some inclusions in a carbonaceous chondrite (Allende, see Chapter 3). However, the existence of Yb^{2+} would require extremely reducing conditions (more so, for example, than during the genesis of lunar rocks) and so alternative explanations for the Yb anomalies need to be sought (Mason and Martin 1977; Boynton, Chapter 3). Under the usual conditions prevailing in the crust, Yb is trivalent.

The proportions of the different states of variable-valence ions could be dependent on composition, temperature and pressure of a given chemical system. These effects are now discussed in turn.

Compositional control

Morris and Haskin (1974) showed that in Ca,Mg,Al-silicate liquids the Eu^{2+}/Eu^{3+} ratio rises significantly with increasing atomic ratio (Al + Si)/O, i.e. with increase in degree of melt polymerization. The compositional variations studied by Morris and Haskin were wide, and have only limited relevance to the smaller variations seen in natural systems. This led Drake and Weill (1975) to suggest that variations in the Eu^{2+}/Eu^{3+} ratio can be used to establish redox conditions, as discussed below, since natural compositional variations would have a relatively small effect on the ratio, but more research on the effect of composition is needed. The behaviour of Eu as described by Morris and Haskin has been interpreted in terms of the amphoteric character of europium oxides in silicate melts (Fraser, 1975).

Redox conditions

The determination of redox conditions in rock or water systems from measurement of M^{2+}/M^{3+} or M^{4+}/M^{3+} ratios is feasible provided that the effects of temperature, pressure and composition are known and can be allowed for. Most attention has been given to the Eu^{2+}/Eu^{3+} ratio in minerals as an indicator of the oxygen fugacity prevailing during mineral formation (Towell et al., 1965; Philpotts, 1970; Drake, 1975). Philpotts (1970) suggested that the close similarity in the ionic radii and charge of Sr^{2+} and Eu^{2+} would lead to almost identical partitioning behaviour of the two ions in rock systems. This similarity could be used to calculate the Eu^{2+}/Eu^{3+} ratio, and hence the oxygen fugacity, in a system where the partitioning of Eu^{2+} is significantly different from that of Eu^{3+} as, for example, between silicate melt and plagioclase feldspar. The general principle is as follows. The europium concentration relationships for two coexisting phases, α and β at equilibrium are:

$$[Eu]_\alpha = [Eu^{2+}]_\alpha + [Eu^{3+}]_\alpha$$

$$[Eu]_\beta = [Eu^{2+}]_\beta + [Eu^{3+}]_\beta$$

Therefore:

$$[Eu]_\beta = D^{2+}_{\beta/\alpha}\,([Eu]_\alpha - [Eu^{3+}]_\alpha) + D^{3+}_{\beta/\alpha}\,[Eu^{3+}]_\alpha \tag{1.1}$$

or:

$$[Eu^{3+}]_\alpha = ([Eu]_\beta - D^{2+}_{\beta/\alpha}\,[Eu]_\alpha)/(D^{3+}_{\beta/\alpha} - D^{2+}_{\beta/\alpha}) \tag{1.2}$$

Also:

$$[Eu^{2+}]_\alpha = [Eu]_\alpha - [Eu^{3+}]_\alpha \tag{1.3}$$

where $D^{n+}_{\beta/\alpha}$ is the distribution coefficient of Eu in the nth valency state between phase β and phase α. $D^{2+}_{\beta/\alpha}$ may be substituted by the distribution coefficient value for Sr between β and α. Hence, the following information needs to be established for application of the method:

(a) The Sr distribution coefficient between phases β and α.

(b) The Eu concentrations in phases α and β.

(c) The value of $D^{3+}_{\beta/\alpha}$. This can be obtained from a plot of distribution coefficients for the other trivalent REE between phases β and α against atomic number, by interpolation at the Eu position between Sm and Gd.

The method has been applied successfully to europium distribution amongst feldspar, pyroxene and matrix phases of basalts, andesites and dacites

(Philpotts, 1970). It can be seen from equation (1.2), however, that unless the distribution coefficient for Eu^{2+} is very different from that for Eu^{3+}, errors in the determined $D_{\beta/\alpha}$ values could give rise to very large errors in the $[Eu^{3+}]_{\alpha}$.

To be able to use the $[Eu^{2+}]/[Eu^{3+}]$ ratio to determine the oxygen fugacity (f_{O_2}) of the system it is necessary to make a calibration which relates the two parameters for a given composition. Drake (1975) did this experimentally for plagioclase feldspar in melts of basaltic and andesitic compositions, and showed the relationship to be of the form:

$$\log f_{O_2} = -4 \log ([Eu^{2+}]/[Eu^{3+}]) + A \qquad (1.4)$$

where A is a constant. This expression was obtained with the assumption that the $EuO/EuO_{1.5}$ activity ratio in the melt, and the $EuAl_2Si_2O_8/EuAl_3SiO_8$ activity ratio in the plagioclase are directly proportional to the $[Eu^{2+}]/[Eu^{3+}]$ ratios in both phases; such an assumption is not strictly valid as there is a compositional control on the oxidation state ratio, but this effect is probably small in naturally occurring systems (Drake, 1975).

Further aspects of the control of oxygen fugacity on europium partition are discussed in section 1.6.

Temperature and pressure controls

A limited study (Morris et al., 1974) of a variation in the Eu^{2+}/Eu^{3+} ratio in quenched Ca,Mg-silicate and Ca,Al-silicate liquids as a function of temperature showed that the reduction of Eu^{3+} to Eu^{2+} is endothermic; the Eu^{2+}/Eu^{3+} ratio increasing with rising temperature (over the studied range of 1415—1580°C). It would be imprudent to extend this result to other silicate systems principally because of the compositional dependence of the oxidation state ratio (see above). More research is needed on the effect of temperature.

There are no experimental data on the effect of pressure on the oxidation state of europium in magmas.

1.5. Element co-ordination and ionic radii

The REE occupy a wide variety of co-ordination polyhedra in minerals, from six-fold to twelve-fold or even higher co-ordinations. The smaller REE ions can occupy six-fold co-ordination sites but do so only rarely in minerals. Normally the co-ordination is greater; most co-ordinations from seven to twelve being represented, e.g. 7-fold in sphene, 8-fold in zircon, 9-fold in monazite, 11-fold in allanite, and 12-fold in perovskite. The diversity of sites occupied by REE ions undoubtedly leads to complexity in REE mineral chemistry, much of which still remains to be elucidated. Felsche (1978) gives a detailed account of the co-ordination chemistry of REE in a wide range of compounds, and some details of co-ordination in REE-bearing minerals are given in Chapter 2 of this book.

A correlation between co-ordination and ionic radius is to be expected; the larger ions will tend to occupy larger sites and vice versa. The REE exhibit a gradual and steady decrease in their atomic volumes with increase in atomic number as a result of imperfect shielding of one electron by another in the same sub-shell, so that the effective nuclear charge acting on each $4f$ electron increases with increasing atomic number, thereby leading to a reduction in the size of the $4f$ sub-shell. This reduction is referred to as the *lanthanide contraction* and is reflected by a steady decrease in ionic radius of the REE with increase in atomic number (Fig. 1.6). Ionic radius is also a function of charge and co-ordination number.

Shannon and Prewitt (1969, 1970) compiled an empirical set of ionic radii in oxides and fluorides, which was revised by Shannon (1976). The revised values for the REE are given in Table 1.6, for an oxygen ionic radius of 1.40 Å. Comparison of the relative sizes of the REE ions with those of other cations shows that there are few ions of similar sizes. Na^+ and Ca^{2+} have similar sizes to the lightest REE in their trivalent state, (e.g. in six-fold co-ordination: Na^+ 1.02; Ca^{2+} 1.00 Å). Eu^{2+} has a similar radius to that of Sr^{2+} (e.g. in six-fold co-ordination: Sr^{2+} 1.18; Eu^{2+} 1.17 Å). K^+, Rb^+, Cs^+ and Ba^{2+} are larger than any trivalent REE ion, while most of the transition element ions are smaller, but with Mn^{2+}, Y^{3+}, Th^{4+} and U^{4+} as important exceptions. The substitutional behaviour of the REE ions is discussed in the next section.

The relatively large size of the REE, especially of the LREE ions, reduces covalent and electrostatic interactions and, thereby, is one of the principal factors tending to prevent complexing of the REE. In solution the trivalent REE can form ion-pair associations with CO^-, Br^-, I^-, NO_3^-, and SO_4^{2-} amongst others. Geochemically, the probable existence of carbonate, sulphate,

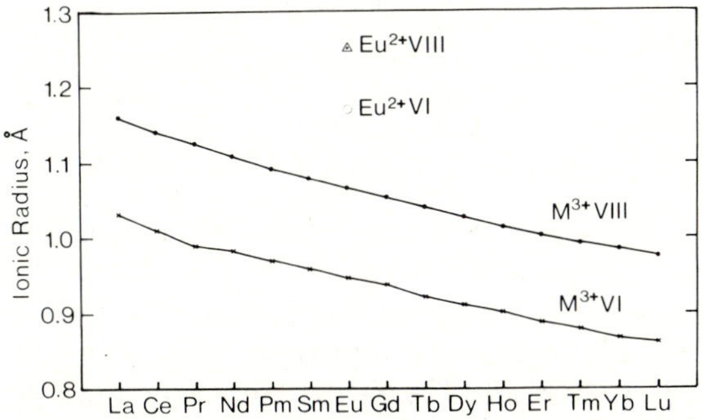

Fig. 1.6. Ionic radius variation with atomic number, for trivalent REE, and for Eu^{2+}, in six-fold and eight-fold co-ordination. Radii from Shannon (1976).

TABLE 1.6

Ionic radii (Å) of REE ions and other selected ions in different co-ordination numbers (CN) and based on a radius for O^{2-} of 1.40 Å (from Shannon, 1976)

Ion	CN	Radius	Ion	CN	Radius	Ion	CN	Radius	Ion	CN	Radius
La^{3+}	VI	1.032	Sm^{3+}	VIII	1.079	Er^{3+}	VI	0.890	Sr^{2+}	VI	1.18
	VII	1.10		IX	1.132		VII	0.945		VII	1.21
	VIII	1.160		XII	1.24		VIII	1.004		VIII	1.26
	IX	1.216	Eu^{2+}	VI	1.17		IX	1.062		IX	1.31
	X	1.27		VII	1.20	Tm^{2+}	VI	1.03		X	1.36
	XII	1.36		VIII	1.25		VII	1.09		XII	1.44
Ce^{3+}	VI	1.01		IX	1.30	Tm^{3+}	VI	0.880	Mn^{2+}	IV[a]	0.66
	VII	1.07		X	1.35		VIII	0.994		V[a]	0.75
	VIII	1.143	Eu^{3+}	VI	0.947		IX	1.052		VI[a]	0.830
	IX	1.196		VII	1.01	Yb^{2+}	VI	1.02		VII[a]	0.90
	X	1.25		VIII	1.066		VII	1.08		VIII	0.96
	XII	1.34		IX	1.120		VIII	1.14	Pb^{2+}	IV	0.98
Ce^{4+}	VI	0.87	Gd^{3+}	VI	0.938	Yb^{3+}	VI	0.868		VI	1.19
	VIII	0.97		VII	1.00		VII	0.925		VII	1.23
	X	1.07		VIII	1.053		VIII	0.985		VIII	1.29
	XII	1.14		IX	1.107		IX	1.042		IX	1.35
Pr^{3+}	VI	0.99	Tb^{3+}	VI	0.923	Lu^{3+}	VI	0.861		X	1.40
	VIII	1.126		VII	0.98		VIII	0.977		XI	1.45
	IX	1.179		VIII	1.040		IX	1.032		XII	1.49
Pr^{4+}	VI	0.85		IX	1.095	Na^{+}	IV	0.99	Y^{3+}	VI	0.900
	VIII	0.96	Tb^{4+}	VI	0.76		V	1.00		VII	0.96
Nd^{2+}	VIII	1.29		VIII	0.88		VI	1.02		VIII	1.019
	IX	1.35	Dy^{2+}	VI	1.07		VII	1.12		IX	1.075
Nd^{3+}	VI	0.983		VII	1.13		VIII	1.18	Th^{4+}	VI	0.94
	VIII	1.109		VIII	1.19		IX	1.24		VIII	1.05
	IX	1.163	Dy^{3+}	VI	0.912		XII	1.39		IX	1.09
	XII	1.27		VII	0.97	Ca^{2+}	VI	1.00		X	1.13
Sm^{2+}	VII	1.22		VIII	1.027		VII	1.06		XI	1.18
	VIII	1.27		IX	1.083		VIII	1.12		XII	1.21
	IX	1.32	Ho^{3+}	VI	0.901		IX	1.18	U^{4+}	VI	0.89
Sm^{3+}	VI	0.958		VIII	1.015		X	1.23		VII	0.95
	VII	1.02		IX	1.072		XII	1.34		VIII	1.00
				X	1.12					IX	1.05
										XII	1.17

[a] High-spin state only.

chloride, and fluoride complexes seems to be important. It is known, for example, that the REE are extremely mobile in CO_2-rich solutions. One interesting experimental study (Wendlandt and Harrison, 1979) showed that the REE, especially the HREE, were partitioned preferentially into the carbonate melt of an immiscible silicate melt/carbonate melt system, and that the co-existing CO_2-rich vapour was enriched further in the REE. Flynn and Burnham (1978) have shown also that the REE were complexed by chlorine in an experimental acid silicate melt-vapour phase couple, since the vapour/melt distribution coefficients of the REE increased with increase in vapour chloride molality. Other aspects of complexing are discussed in Chapter 9.

In the absence of a suitable fluid phase, the REE are relatively immobile in silicate rock systems, even under metamorphic conditions. Thus it is possible in such circumstances to use the REE patterns of a metamorphic rock as evidence for the nature of the parent rock (e.g. Schubert, 1979), but this approach must be backed up with other geochemical and mineralogical evidence if there is the slightest possibility of a fluid phase having been involved (see Chapter 9).

1.6. Element substitution and partition

The large ionic radii of the REE preclude significant substitution of these elements into minerals, except where the substituted cation is also large. Substitutions of trivalent REE are observed for Ca^{2+}, Y^{3+}, Th^{4+}, U^{4+}, Mn^{2+} and Zr^{4+} (ionic radius in six-fold co-ordination: 0.72 Å). These substitutions are expected on the criterion of ionic radius except perhaps in the case of Zr^{4+} which has a relatively small radius. The wide range of radii of trivalent REE ions means that some minerals are selective in their uptake of particular REE. This phenomenon is shown below where, in some cases, there is a very strong dependence of the distribution coefficient on ionic radius.

The substitution of a trivalent REE for a cation of different charge (i.e. altervalent substitution) requires the operation of a charge-compensating mechanism. This can be either by an additional substitution as, for example, in the exchange of a trivalent REE (symbol R^{3+}) for Ca^{2+} in anorthite involving the exchange of Al^{3+} for Si^{4+}:

$$R^{3+} + Al^{3+} + CaAl_2Si_2O_8 \rightarrow RAl_3SiO_8 + Ca^{2+} + Si^{4+},$$

or by the following:

$$R^{3+} + Na^{+} + 2CaAl_2Si_2O_8 \rightarrow 2(R,Na)\ Al_2Si_2O_8 + 2Ca^{2+},$$

or by production of a vacancy (symbol □), e.g.:

$$2R^{3+} + \underset{\text{apatite}}{Ca_5(PO_4)_3OH} \rightarrow (R_2Ca_2\square)(PO_4)_3OH + 3Ca^{2+},$$

or possibly by addition of an *anion* in an interstitial position in the crystal structure.

Eu^{2+} can substitute for Pb^{2+}, Ca^{2+}, Sr^{2+} and Na^{+}. The radii of these ions are given in Table 1.6.

Partition

The partition of an element between two phases, *A* and *B*, whether they be mineral/liquid, mineral/mineral, or melt/melt can be described conveniently in terms of the distribution coefficient *D*:

D = concentration in phase *A*/concentration in phase *B*.

The constant *D* will be independent of the concentration of the element only over the concentration range in which Henry's law (see below) is obeyed. Despite this limitation the distribution coefficient has been widely and usefully adopted in discussions of the partitioning behaviour of the REE. Some evidence on the concentration ranges over which the REE obey Henry's law is discussed below.

The value of *D* depends on temperature, pressure and composition of the phases but before these variables are discussed it is convenient to describe the control on partitioning shown by some mineral structures. Onuma et al. (1968) and later Jensen (1973) demonstrated the influence that ionic radii and charge have on element partition in mineral/melt systems by constructing graphs of log *D* versus radius, on which ions of the same charge can be seen to define fairly smooth curves. The REE, with their similar chemical properties and their steady change in ionic radius with changing atomic number, are particularly amenable to this treatment (Fig. 1.7). The departure of Eu from the smooth curves defined by the other REE in Fig. 1.7 is a result of the presence of some Eu^{2+} as well as Eu^{3+}. Those minerals with large co-ordination polyhedra for cations, such as allanite, favour the larger LREE ions, while those with relatively small polyhedra, such as zircon, favour the smaller HREE. Minerals with intermediate-sized polyhedra (e.g. apatite, sphene) have a more uniform set of associated *D* values or favour the middle REE. Even when the substituted cation is the same, the nature of the partitioning curve for different minerals can be very different (see curves in Fig. 1.7a, c, and d, which involve the substitution of Ca^{2+}).

Temperature and pressure dependence

It is to be expected from consideration of the integrated form of the van't Hoff equation that the relationship between the distribution coefficient *D* and temperature will be of the form:

$$\ln D = -A/RT + B$$

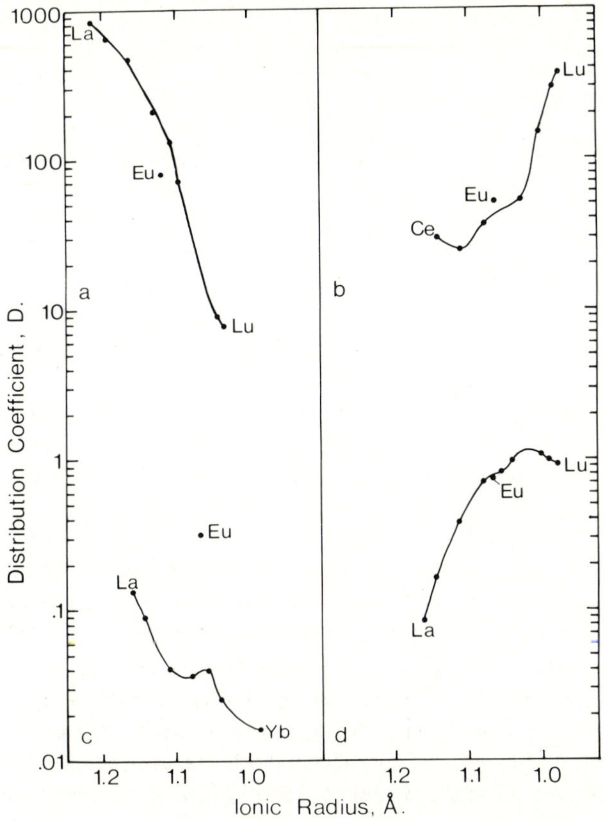

Fig. 1.7. Relationship between the distribution coefficient and ionic radius for REE partition between specific minerals and their coexisting melt. (a) Allanite (data from Brooks et al., 1981). (b) Zircon (data from Nagasawa, 1970). (c) Plagioclase feldspar. (d) Clinopyroxene. Data for (c) and (d) from Matsui et al., 1977. The curves were constructed without considering the point plotted for Eu.

where T is in degrees kelvin and A and B are constants. It is possible for A and B to be either negative or positive such that D can correlate positively or negatively with inverse temperature. A linear relationship between ln D and inverse temperature was observed by Sun et al. (1974) in their experimental investigation of Eu distribution in plagioclase/liquid and clinopyroxene/liquid pairs, using an oceanic ridge basalt which they had enriched in Eu. The distribution coefficients showed a marked variation with temperature: from 0.26 to 0.11 (at $f_{O_2} = 10^{-10}$) for plagioclase (Fig. 1.8) and from 0.39 to 0.29 (at $f_{O_2} = 10^{-11}$) for clinopyroxene over the studied temperature range of 1140—1190°C.

One of the principal difficulties, however, in the experimental investigation of the relationship between partition and temperature is that since many

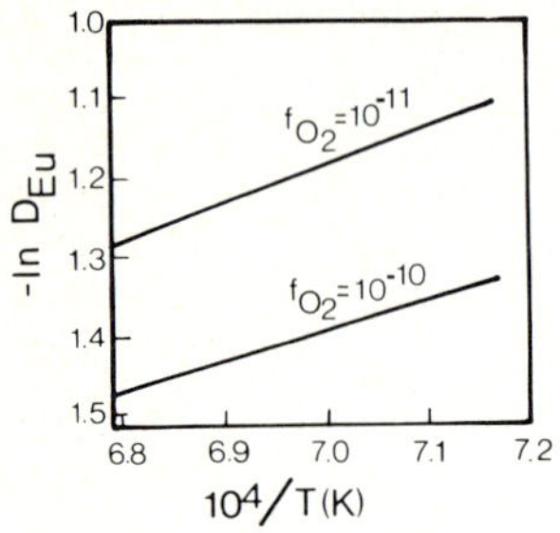

Fig. 1.8. Variation of the natural logarithm of the distribution coefficient for Eu between plagioclase feldspar and silicate melt as a function of inverse absolute temperature at two oxygen fugacities. The lines are linear least square fits to the experimental data points. Based on Sun et al. (1974).

minerals belong to solid solution series, a variation in temperature causes a change in their major element composition which, in turn, can affect trace element partition. The interdependence of the variables temperature, and composition, as well as pressure, have been stressed by Apted and Boettcher (1981) in their study of REE distribution between garnet and silicate melt.

There are no adequate data yet that allow one to evaluate the general effect of pressure on mineral/magma partition in anhydrous systems. Apted and Boettcher (1981) record an increase in $D_{\text{garnet/melt}}$ for REE in an andesitic melt with increasing pressure, but Watson and Green (1981) record no difference in REE partition between apatite and melt at 8 or 20 kbar. Data exist for REE distribution between a water-rich vapour and minerals as a function of pressure (e.g. Mysen, 1979a); these show that there is a strong positive pressure dependence for $D_{\text{vapour/mineral}}$.

Oxygen fugacity

Since the proportion of variable-valence ions, such as Eu^{2+}-Eu^{3+}, is affected by the prevailing oxygen fugacity, so also is the distribution coefficient of the element between two phases. The principles of this phenomenon are discussed in section 1.4 above.

A few experimental studies have been performed on the effect of partitioning of Eu as a function of oxygen fugacity (e.g. Sun et al., 1974; Weill and McKay, 1975). The Eu distribution coefficient is not a linear function of f_{O_2}; an experimentally determined variation of D versus log f_{O_2} is shown in Fig. 1.9 (Weill and McKay, 1975) for partition between plagioclase and a silicate melt. As the fugacity decreases so the distribution coefficient increases because the Eu^{2+}/Eu^{3+} ratio in the melt increases. In the limits of high oxygen fugacity and very low oxygen fugacity, the partitioning represents that of only Eu^{3+} and of only Eu^{2+}, respectively. Sun et al. (1974), however, were able to obtain straight lines which define well the relationships

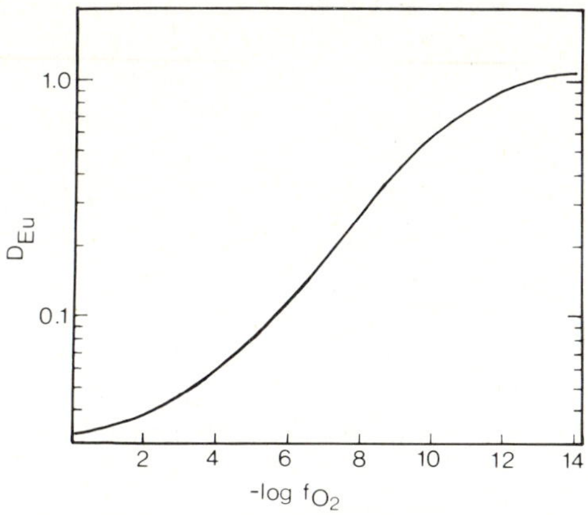

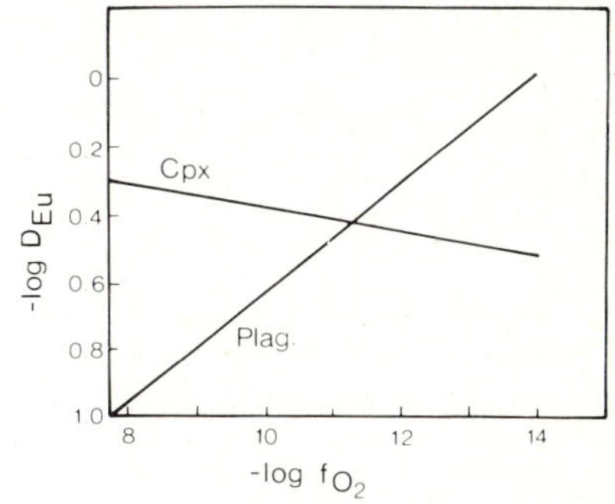

Fig. 1.9 (left). Nature of the variation of the distribution coefficient for Eu between plagioclase feldspar and melt as a function of the logarithm of the oxygen fugacity, at 1200°C in the system $(Mg,Fe)_2SiO_4$-$CaAl_2Si_2O_8$. Based on Weill and McKay (1975).

Fig. 1.10 (right). Nature of the variation of the distribution coefficient of Eu with oxygen fugacity (log-log plot) for clinopyroxene/melt and plagioclase/melt pairs. The lines are linear least squares fits of the data points obtained experimentally for a basaltic composition at 1150°C by Sun et al. (1974).

between europium distribution coefficients determined experimentally for clinopyroxene/melt and plagioclase/melt pairs and oxygen fugacity, in a plot of log D versus log f_{O_2} (Fig. 1.10).

Compositional dependence

There are two other aspects of the effect of compositional variation that concern us here. One is the effect that changes in the major constituents of a silicate melt have on REE partition between the melt and a given mineral; the other is the REE concentration ranges, in the crystal and in the coexisting melt, over which Henry's law is obeyed.

The effect of major element composition is well seen in the work of Watson (1976) on element distribution between two immiscible silicate melts in the system SiO_2-FeO-Al_2O_3-K_2O, at 1180°C and 1 atmosphere pressure. Among other elements three REE – La, Sm and Lu – were studied. All three were quite strongly enriched in the basic melt (SiO_2 55.8 wt%, Al_2O_3 3.6%, FeO 18.4%, K_2O 4.0%) relative to the acidic melt (SiO_2 74.0%, Al_2O_3 2.3%, FeO 40.1%, K_2O 1.8%) by factors of 4–5. It is to be expected, therefore, that increasing silica content of a magma will tend to increase mineral/melt REE distribution coefficients but it should be remembered that the chosen experimental system has only a limited comparability to natural magmas, especially since calcium is absent. Similar results, however, were obtained by Ryerson and Hess (1978, 1980) in an experimental study using Ca-bearing immiscible silicate liquids. They showed also that the presence of phosphorus (4.5 wt.% P_2O_5) increased the REE distribution coefficients to a value of about 15.

There have been several studies aimed at establishing the concentration ranges over which trace element partition obeys Henry's law (e.g. Mysen, 1976; 1978; Harrison and Wood, 1980; Harrison 1981a; Drake and Holloway, 1978), but because the reproducibility of results by different workers has been bad, a controversy has arisen which has not yet been resolved satisfactorily, (e.g. Mysen, 1979b, 1982; Drake and Holloway, 1981, 1982). Although this controversy has centred on the study of Ni partition, it could have important implications in the interpretation of REE distributions.

Henry's law states that when a solution is dilute, the activity of a solute component, a_i, is proportional to the solute's mole fraction, x_i:

$$a_i = kx_i \quad \text{as } x_i \to 0$$

where k is the Henry's law constant. This means that the distribution coefficient of a trace element will be constant since the activity coefficient of the element in each phase is constant. Deviations from Henry's law may, however, be observed at very low concentrations. For example, Harrison (1981a) has shown that when the concentration in garnet of a REE is just a few parts per million, the partition coefficient between the garnet and coexisting equilibrium liquid (hydrous synthetic silicate melt) can be dependent on the REE concentration. The dependence, for Sm, is of the form shown in Fig. 1.11, where it can be seen that the partition coefficient is not constant for Sm concentrations in pyrope of less than 12 ppm. Harrison and Wood (1980) have interpreted this type of behaviour as resulting from two mechanisms of altervalent substitution. At low REE concentrations, substitution of the divalent cations (Mg^{2+}, Ca^{2+}) for trivalent REE may be charge-balanced by the generation of a vacancy (see above), but as the

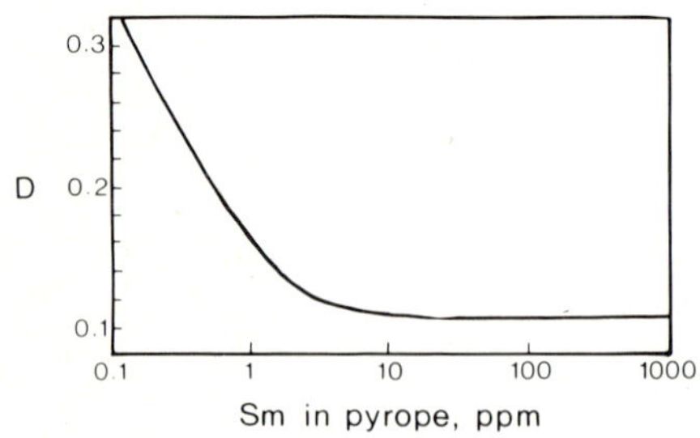

Fig. 1.11. Nature of the variation of the distribution coefficient of Sm between pyrope and hydrous liquid as a function of the Sm content of the garnet. Based on Harrison (1981a).

proportion of vacancies increases the mechanism becomes energetically less favourable and hence the partition is dependent on the concentration of vacancies which, in turn, is dependent on the REE concentration in the crystal. At higher REE concentrations, partitioning involves a coupled substitution, e.g. (REE^{3+}, Al^{3+}) $\leftrightarrows$ (Ca^{2+}, Si^{4+}) that obeys Henry's law.

The results of the work of Harrison (1981a) and Harrison and Wood (1980) have not been brought into question; they form part of a useful beginning to a large field of study that will require several years of research effort. Until more of this work has been done it is not possible to make a realistic assessment of the general significance that deviations from Henry's law are likely to have in petrogenetic modelling and interpretations. Harrison (1981a, b) discusses some implications for models of basalt petrogeneses involving garnet.

Distribution coefficients

Mineral/melt distribution coefficients may be determined either in natural systems by analysing the crystals and host matrix of extrusive igneous rocks or experimentally in the laboratory by the crystallization of specific minerals, and the analysis of these and their coexisting melts by conventional analytical methods or with the use of radiotracers. The determination of meaningful distribution coefficients in natural systems can be fraught with difficulties, principally uncertainty over the attainment of chemical equilibrium, the clean separation of phases, the possible presence of inclusions in the minerals, and lack of knowledge of the temperature and pressure conditions of crystallization. Provided the analyses are made on crystals and matrix which have been cleanly separated, the results give a good indication of the distribution behaviour of elements actually in rock systems, and in this sense they can be, and have been, used in the theoretical modelling of petrogenetic processes. Experimental determinations do not suffer from the same associated uncertainties as do those from natural systems, and they can provide a quantitative measure of the effects of temperature, pressure, composition and structure on element distribution. The experiments, however, do not reproduce all the conditions prevailing during rock formation, so that the use of the data in modelling is also subject to limitations.

One of the earliest studies of REE partition in natural systems was by Schnetzler and Philpotts (1970) who carefully separated and analysed the phenocrysts and matrices of several lavas. A plot of some of their data is given in Fig. 1.12, which shows clearly the different extents to which different minerals fractionate the REE. Since then, there have been several studies of REE partitioning in igneous rocks. A compilation of mineral/melt distribution coefficient data for natural systems has been made by Henderson (1982) which has provided the basis for the values listed in Table 1.7. Some specific features revealed by the data in Table 1.7 are:

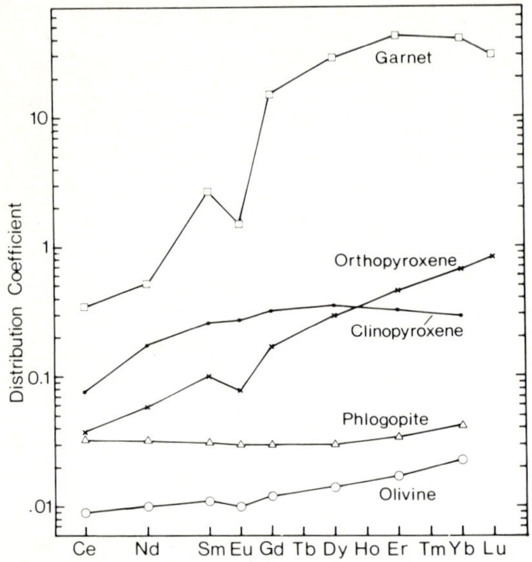

Fig. 1.12. Examples of the variation in REE distribution coefficients for different minerals as a function of atomic number. The values were obtained by analysis of separated phenocrysts and matrices of extrusive igneous rocks. Based on data by Schnetzler and Philpotts (1970).

(a) There is often a wide range in the values of distribution coefficients given for any one REE and mineral/melt pair (see Fig. 1.13). This variation is sometimes as much as one order of magnitude or greater (Fig. 1.13b, especially Yb); it results from the effects of variable temperature, pressure and composition, as well as from mineral impurities, etc. (see above).

(b) Average D values for the REE, excepting Eu, are commonly less than 1 for many rock-forming minerals. However, in acid igneous rocks, the D values are often greater than 1 for clinopyroxene and amphibole. The HREE are strongly partitioned, relative to the LREE, into garnet in both basic and silica-rich systems (e.g. Fig. 1.12).

(c) Accessory minerals can play a significant role in REE distribution. Distribution coefficients can be very large (Table 1.7c) and the REE can be strongly fractionated from each other. For example, D_{La} for allanite can be about two orders of magnitude greater than D_{Lu} (~800 cf. 8). Some accessory minerals favour the LREE (e.g. allanite), others the HREE (e.g. zircon; Fig. 1.14).

(d) The D values show that europium anomalies can be generated by the fractional crystallization or fusion of plagioclase feldspar, garnet, apatite, allanite, magnetite, and possibly clinopyroxene and amphibole.

(e) Mineral/melt distribution coefficients for the REE tend to be higher in silica-rich systems than in basic ones (Table 1.7a, b).

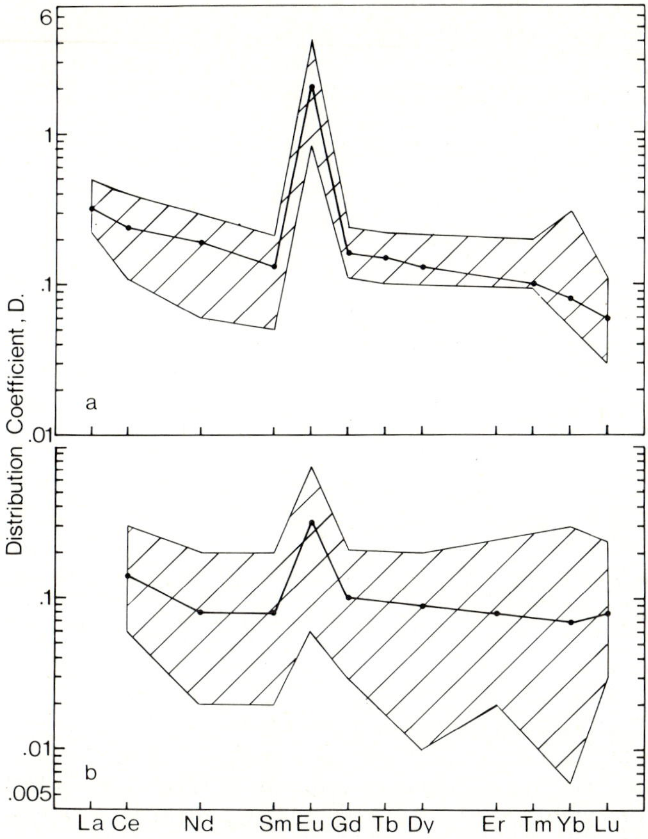

Fig. 1.13. Range and averages (thick line) of REE distribution coefficients for plagioclase/melt pairs plotted against atomic numbers, for: (a) acid igneous rocks, and (b) basaltic and andesitic rocks. Based on Henderson (1982).

Results from experimental studies of REE partitioning between minerals and synthetic melts are consistent with the results obtained directly from analysis of the phases from igneous rocks. Irving (1978) has given an excellent review of these studies for both REE and other element partitioning and the interested reader is referred to that paper for the details. Since that time there have been further experimental studies involving the partitioning of the REE, including the minerals olivine, pyroxene, garnet, and amphibole (Mysen, 1978; Nicholls and Harris, 1980; Mysen and Virgo, 1980; Apted and Boettcher, 1981; Harrison, 1981a). Watson and Green (1981) have determined experimentally the partitioning behaviour of REE and Sr between apatite and melt for a variety of compositions ranging from basanitic to granitic.

Although average D values such as those given in Table 1.7 are used in the

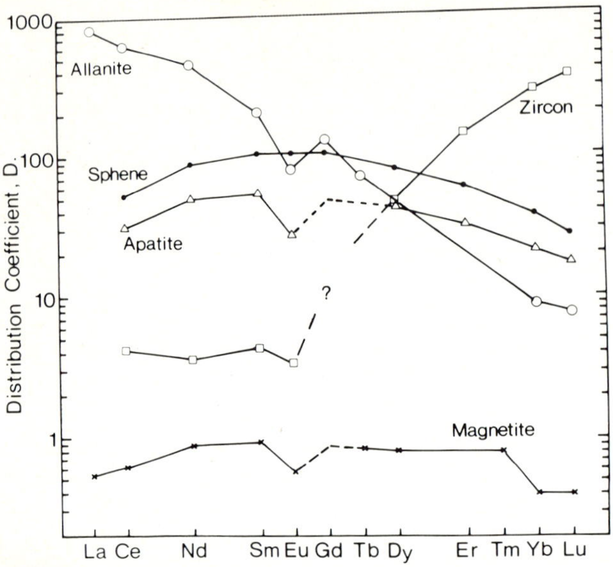

Fig. 1.14. Variation of REE distribution coefficients (mineral/melt) with atomic number for some accessory minerals in acid igneous rocks. Based on data in Table 1.7.

TABLE 1.7a

Average mineral/melt distribution coefficients of REE in basaltic and andesitic rocks

	Olivine	Ortho-pyroxene	Clino-pyroxene	Amphibole	Plagioclase	Phlogopite[a]	Garnet
La			0.08[a]	0.27	0.14[a]		0.05[a]
Ce	0.009[a]	0.02	0.34	0.34	0.14[a]	0.03	0.05[a]
Nd	0.009[a]	0.05[a]	0.6	0.19[a]	0.08	0.03	
Sm	0.009	0.05	0.9	0.91	0.08	0.03	0.6
Eu	0.008[a]	0.05[a]	0.9	1.0	0.32	0.03	0.9
Gd	0.012[a]		0.9[a]	1.1	0.10		3.7[a]
Tb		0.05[a]	1.0[a]	1.4			5.6[a]
Dy	0.012[a]	0.20[a]	1.1	0.64[a]	0.09	0.03	
Ho							19
Er	0.013[a]	0.31[a]	1.0	0.48[a]	0.08		
Tm			1.1[a]			0.03	
Yb		0.34	1.0	0.97	0.07		30
Lu		0.11[a]	0.8[a]	0.89	0.08	0.04	35

[a]Only one or two determinations available.

TABLE 1.7b

Average mineral/melt distribution coefficients of REE in dacitic and rhyolitic rocks

	Ortho-pyroxene	Clino-pyroxene	Amphibole	Plagioclase	Alkali feldspar[a]	Garnet	Magnetite
La	0.65	0.6	0.9[a]	0.32		0.39	0.5
Ce	0.5	0.9	1.2	0.24	0.04	0.62	0.6
Nd	0.6	2.1	3.2	0.19	0.03	0.63	0.9
Sm	0.7	2.7	5.4	0.13	0.02	2.2	0.9
Eu	0.5	1.9	3.6	2.0	1.13	0.7	0.6
Gd	1.1	3.1		0.16		7.7	0.8
Tb	1.2	3.0	3.0[a]	0.15		12	0.8
Dy	0.7	3.3	9	0.13	0.006	29[a]	
Ho						28	
Er	0.61		8	0.05	0.006	43[a]	
Tm	1.4		5[a]	0.1			0.8
Yb	1.0	2.1	6.2	0.08	0.012	43	0.4
Lu	1.1	2.3	4.5	0.06		38	0.4

[a]Only one or two determinations available.

TABLE 1.7c

Average mineral/melt distribution coefficients of REE for some accessory minerals in dacitic and rhyolitic rocks

	Allanite[a]	Apatite	Sphene[a]	Zircon
La	820			
Ce	635	31	53.3	4.2
Nd	460	50	88.3	3.6
Sm	205	54	102	4.3
Eu	80	27	101	3.4
Gd	130	22[a]	102	
Tb	71			
Dy		42	80.6	48
Er		31	58.7	140
Yb	8.9	21	37.4	280
Lu	7.7	17	26.9	345

[a]Only one determination is available.

Tables 1.7a to 1.7c are based on a compilation given by Henderson (1982), in which ranges of coefficient values are given, and original sources of data are cited. Data for sphene are from Simmons and Hedge (1978).

theoretical modelling of petrogenetic processes despite uncertainty over their true applicability, nevertheless the general nature of REE distribution coefficient patterns (i.e. *D* versus atomic number) for many important minerals are sufficiently well defined to allow significant constraints to be placed on such models (see later chapters).

There have been very few studies on REE partitioning in metamorphic rocks (e.g. Orsa et al., 1967; Reitan and Roelandts, 1973). Pride and Muecke (1981) determined the REE distribution amongst the minerals of some granulite-facies gneisses from Scotland and concluded that since regular and consistent distribution patterns exist for each mineral, the mineral/mineral distribution coefficients are equilibrium values. These values are similar to those recorded for the same mineral pairs of dacitic rocks. Metamorphism will cause local redistribution of REE if new minerals are formed or if the temperature and pressure conditions are different from those prevailing during the formation of the parent rock. Large-scale redistribution of the REE is to be expected only when a metamorphic or metasomatic fluid is generated or introduced; even then, the nature of the fluid is important (see section 1.5). Muecke et al. (1979) report convincing evidence of REE immobility during the regional metamorphism of the sedimentary and volcanic rocks of the White Rock Formation, southwestern Nova Scotia, Canada, despite the development of some local metasomatic zones, and the movement of other elements (notably the alkalis). The conditions under which the REE are likely to be mobile or immobile are poorly known (see also Chapter 9).

Information is also lacking on the distribution of REE between a fluid and coexisting minerals in natural aqueous systems. Marchand et al. (1976) give semi-quantitave data obtained experimentally on REE partition between fluorite and a solution of calcium chloride at 120°C. The REE distribution coefficients strongly favour fluorite (values are 100—1000) and those for the HREE are greater than those for the LREE. Many more studies are needed on natural and experimental aqueous systems.

1.7. Geochemical reference literature

The principal compilations or reviews of REE data pertaining to geochemistry are as follows. For data on abundances in rocks, minerals, natural waters, etc., together with short accounts of geochemical behaviour:

Felsche, J., 1978, and Herrmann, A.G., 1970. Yttrium and the lanthanides. In: K.H. Wedepohl (Executive Editor), *Handbook of Geochemistry, Vol. II/5*. Springer-Verlag, Berlin.

Hoffman, W., Kubach, I. and Topper, W., 1980. Y, La und Lanthanide. Kosmochemie. In: I. Kubach and W. Topper (Editors), *Gmelin Handbuch der Anorganischen Chemie: Sc, Y, La—Lu Seltenerdelemente*. Springer-Verlag, Berlin, 8th ed., A3.

Topper, W., 1979. Y, La und Lanthanide. Kristallchemische Grundlagen. In: I. Kubach (Editor), *Gmelin Handbuch der Anorganischen Chemie: Sc, Y, La—Lu Seltenerdelemente.* Springer-Verlag, Berlin, 8th ed., A4.

Kubach, I., Schubert, P. and Topper, W., 1981. Y, La und Lanthanide. Geochemie: Gesamterde. Magmatische Abfolge. In: I. Kubach and W. Topper (Editors), *Gmelin Handbuch der Anorganischen Chemie: Sc, Y, La—Lu Seltenerdelemente.* Springer-Verlag, Berlin, 8th ed., A5.

Data on the thermodynamic properties of some REE and some of their oxides may be found in:

Robie, R.A., Hemingway, B.S. and Fisher, J.R., 1978 (reprinted with corrections in 1979). Thermodynamic properties of minerals and related substances at 298.15 K and 1 bar (10^5 pascals) pressure and at higher temperatures. *Bull., U.S. Geol. Surv.*, 1452: 456 pp.

A bibliography to 1971 is:

Adams, J.A. and Iberall, E.R., 1973. Bibliography of the geology and mineralogy of the rare earths and scandium to 1971. *Bull., U.S. Geol. Surv.*, 1366: 195 pp.

References

Apted, M.J. and Boettcher, A.L., 1981. Partitioning of rare earth elements between garnet and andesite melt: an autoradiographic study of *P-T-X* effects. *Geochim. Cosmochim. Acta*, 45: 827—837.

Brooks, C.K., Henderson, P. and Rønsbo, J.G., 1981. Rare-earth partition between allanite and glass in the obsidian of Sandy Braes, Northern Ireland. *Mineral. Mag.*, 44: 157—160.

Cameron, A.G.W., 1973. Abundances of the elements in the solar system. *Space Sci. Rev.*, 15: 121—146.

Coryell, C.G., Chase, J.W. and Winchester, J.W., 1963. A procedure for geochemical interpretation of terrestrial rare-earth abundance patterns. *J. Geophys. Res.*, 68: 559—566.

Drake, M.J., 1975. The oxidation state of europium as an indicator of oxygen fugacity. *Geochim. Cosmochim. Acta*, 39: 55—64.

Drake, M.J. and Holloway, J.R., 1978. "Henry's Law" behavior of Sm in a natural plagioclase/melt system: importance of experimental procedure. *Geochim. Cosmochim. Acta*, 42: 679—683.

Drake, M.J. and Holloway, J.R., 1981. Partitioning of Ni between olivine and silicate melt: the "Henry's Law problem" re-examined. *Geochim. Cosmochim. Acta*, 45: 431—437.

Drake, M.J. and Holloway, J.R., 1982. Partitioning of Ni between olivine and silicate melt: the "Henry's law problem" re-examined — reply to discussion by B. Mysen. *Geochim. Cosmochim. Acta*, 46: 299.

Drake, M.J. and Weill, D.F., 1975. Partition of Sr, Ba, Ca, Y, Eu^{2+}, Eu^{3+}, and other REE between plagioclase feldspar and magmatic liquid: an experimental study. *Geochim. Cosmochim. Acta*, 39: 669—712.

Evensen, N.M., Hamilton, P.J. and O'Nions, R.K., 1978. Rare earth abundances in chondritic meteorites. *Geochim. Cosmochim. Acta*, 42: 1199—1212.

Felsche, J., 1978. Yttrium and lanthanides. Crystal chemistry. In: K.H. Wedepohl (Editor), *Handbook of Geochemistry, II/5.* Springer-Verlag, Berlin, pp. 39,57-71-A-1 to 39,57-71A-42.

Flynn, R.T. and Burnham, C.W., 1978. An experimental determination of rare earth partition coefficients between a chloride containing vapor phase and silicate melts. *Geochim. Cosmochim. Acta*, 42: 685—701.

Fraser, D.G., 1975. Activities of trace elements in silicate melts. *Geochim. Cosmochim. Acta*, 39: 1525—1530.

Ganapathy, R. and Anders, E., 1974. Bulk compositions of the moon and earth, estimated from meteorites. *Proc. 5th. Lunar Sci. Conf.*, pp. 1181—1206.

Goldschmidt, V.M., 1954. *Geochemistry*, Clarendon Press, Oxford, 730 pp.

Harrison, W.J., 1981a. Partition coefficients for REE between garnets and liquids: implications of non-Henry's Law behaviour for models of basalt origin and evolution. *Geochim. Cosmochim. Acta*, 45: 1529—1544.

Harrison, W.J., 1981b. Partitioning of REE between minerals and coexisting melts during partial melting of a garnet lherzolite. *Am. Mineral.*, 66: 242—259.

Harrison, W.J. and Wood, B.J., 1980. An experimental investigation of the partitioning of REE between garnet and liquid with reference to the role of defect equilibria. *Contrib. Mineral. Petrol.*, 72: 145—155.

Haskin, L.A., Haskin, M.A., Frey, F.A. and Wildeman, T.R., 1968. Relative and absolute terrestrial abundances of the rare earths. In: L.H. Ahrens (Editor), *Origin and Distribution of the Elements, 1.* Pergamon, Oxford, pp. 889—911.

Henderson, P., 1982. *Inorganic Geochemistry*. Pergamon, Oxford, 353 pp.

Irving, A.J., 1978. A review of experimental studies of crystal/liquid trace element partitioning. *Geochim. Cosmochim. Acta*, 42: 743—770.

Jensen, B.B., 1973. Patterns of trace element partitioning. *Geochim. Cosmochim. Acta*, 37: 2227—2242.

Marchand, L., Joseph, D., Touray, J.C. and Treuil, M., 1976. Critères d'analyse géochimique des gisements de fluorine basés sur l'étude de la distribution des lanthanides — application au gîte du Main (71-Cordesse, France). *Miner. Deposita*, 11: 357—359.

Mason, B. and Martin, P.M., 1977. Geochemical differences among components of the Allende meteorite. *Smithsonian Contrib. Earth Sci.*, 19: 84—95.

Masuda, A., 1962. Regularities in variation of relative abundances of lanthanide elements and an attempt to analyse separation-index patterns of some minerals. *J. Earth Sci. Nagoya Univ.*, 10: 173—187.

Masuda, A., Nakamura, N. and Tanaka, T., 1973. Fine structures of mutually normalized rare-earth patterns of chondrites. *Geochim. Cosmochim. Acta*, 37: 239—248.

Matsui, Y., Onuma, N. Nagasawa, H., Higuchi, H. and Banno, S., 1977. Crystal structure control in trace element partition between crystal and magma. *Bull. Soc. Fr. Mineral. Cristallogr.*, 100: 315—324.

Morris, R.V. and Haskin, L.A., 1974. EPR measurement of the effect of glass composition on the oxidation state of europium. *Geochim. Cosmochim. Acta*, 38: 1435—1445.

Morris, R.V., Haskin, L.A., Biggar, G.M. and O'Hara, M.J., 1974. Measurement of the effects of temperature and partial pressure of oxygen on the oxidation states of europium in silicate glasses. *Geochim. Cosmochim. Acta*, 38: 1447—1459.

Muecke, G.K., Pride, C. and Sarkar, P., 1979. Rare earth element geochemistry of regional metamorphic rocks. In: L.H. Ahrens (Editor), *Origin and Distribution of the Elements, 2.* Pergamon, Oxford, pp. 449—464.

Mysen, B.O., 1976. Partitioning of samarium and nickel between olivine, orthopyroxene, and liquid, preliminary data at 20 kbar and 1025°C. *Earth Planet. Sci. Lett.*, 31: 1—7.

Mysen, B.O., 1978. Partitioning of nickel between liquid, pargasite, and garnet peridotite minerals and concentration limit of behavior according to Henry's Law at high pressure and temperature. *Am. J. Sci.*, 278: 217—243.

Mysen, B.O., 1979a. Trace element partitioning between garnet peridotite minerals and water-rich vapor: experimental data from 5 to 30 kbar. *Am. Mineral.*, 64: 274—287.

Mysen, B.O., 1979b. Nickel partitioning between olivine and silicate melt: Henry's law revisited. *Am. Mineral.*, 64: 1107—1114.

Mysen, B., 1982. Partitioning of Ni between olivine and silicate melt: the "Henry's law problem" re-examined: discussion. *Geochim. Cosmochim. Acta*, 46: 297—298.

Mysen, B.O. and Virgo, D., 1980. Trace element partitioning and melt structure: an experimental study at 1 atm pressure. *Geochim. Cosmochim. Acta*, 44: 1917—1930.

Nagasawa, H., 1970. Rare earth concentrations in zircons and apatites and their host dacites and granites. *Earth Planet. Sci. Lett.*, 9: 359—364.

Nakamura, N., 1974. Determination of REE, Ba, Fe, Mg, Na and K in carbonaceous and ordinary chondrites. *Geochim. Cosmochim. Acta*, 38: 757—775.

Nicholls, I.A. and Harris, K.L., 1980. Experimental rare earth element partition coefficients for garnet, clinopyroxene and amphibole coexisting with andesitic and basaltic liquids. *Geochim. Cosmochim. Acta*, 44: 287—308.

Onuma, N., Higuchi, H., Wakita, H. and Nagasawa, H., 1968. Trace element partition between two pyroxenes and the host lava. *Earth Planet. Sci. Lett.*, 5: 47—51.

Orsa, V.I., Yeliseyeva, G.D. and Kazantseva, A.I., 1967. Rare earth assemblages in the accessory minerals of the ancient crystalline rocks of the Middle Dnepr region. *Geochem. Int.*, 44: 170—173.

Paster, T.P., Schauwecker, D.S. and Haskin, L.A., 1974. The behavior of some trace elements during solidification of the Skaergaard layered series. *Geochim. Cosmochim. Acta*, 38: 1549—1577.

Philpotts, J.A., 1970. Redox estimation from a calculation of Eu^{2+} and Eu^{3+} concentrations in natural phases. *Earth Planet. Sci. Lett.*, 9: 257—268.

Pride, C. and Muecke, G.K., 1981. Rare earth element distributions among coexisting granulite facies minerals, Scourian complex, NW Scotland. *Contrib. Mineral. Petrol.*, 76: 463—471.

Reitan, P.H. and Roelandts, I., 1973. Rare earth partitioning: coexisting metamorphic pyroxenes. *Geol. Soc. Am., Abstr.*, 5: 778—779.

Ross, J.E. and Aller, L.H., 1976. The chemical composition of the sun. *Science*, 191: 1223—1229.

Ryerson, F.J. and Hess, P.C., 1978. Implications of liquid-liquid distribution coefficients to mineral-liquid partitioning. *Geochim. Cosmochim. Acta*, 42: 921—932.

Ryerson, F.J. and Hess, P.C., 1980. The role of P_2O_5 in silicate melts. *Geochim. Cosmochim. Acta*, 44: 611—624.

Schnetzler, C.C. and Philpotts, J.A., 1970. Partition coefficients of rare-earth elements between igneous matrix material and rock-forming mineral phenocrysts, II. *Geochim. Cosmochim. Acta*, 34: 331—340.

Schubert, K., 1979. Amphibolites from the central Alps: REE patterns and their relation to the possible parent rocks. *Schweiz. Mineral. Petrogr. Mitt.*, 59: 299—308.

Shannon, R.D., 1976. Revised effective ionic radii and systematic studies of interatomic distances in halides and chalcogenides. *Acta Crystallogr., Sect. A*, 32: 751—767.

Shannon, R.D. and Prewitt, C.T., 1969. Effective ionic radii in oxides and fluorides. *Acta Crystallogr., Sect. B*, 25: 925—946.

Shannon, R.D. and Prewitt, C.T., 1970. Revised values of effective ionic radii. *Acta Crystallogr., Sect. B*, 26: 1046—1048.

Simmons, E.C. and Hedge, C.E., 1978. Minor-element and Sr-isotope geochemistry of Tertiary stocks, Colorado mineral belt. *Contrib. Mineral. Petrol.*, 67: 379—396.

Smith, J.V., 1977. Possible controls on the bulk composition of the earth: implications for the origin of the earth and moon. *Proc. 8th. Lunar Sci. Conf.*, pp. 333—369.

Sun, C.-O., Williams, R.J. and Sun, S.-S., 1974. Distribution coefficients of Eu and Sr for plagioclase-liquid and clinopyroxene-liquid equilibria in oceanic ridge basalt: an experimental study. *Geochim. Cosmochim. Acta*, 38: 1415—1433.

Taylor, S.R., 1964. The abundance of chemical elements in the continental crust — a new table. *Geochim. Cosmochim. Acta*, 28: 1273—1285.

Towell, D.G., Winchester, J.W. and Spirn, R.V., 1965. Rare earth distributions in some rocks and associated minerals of the batholith of southern California. *J. Geophys. Res.*, 70: 3485—3496.

Wakita, H., Rey, P. and Schmitt, R.A., 1971. Abundances of the 14 rare-earth elements and 12 other trace-elements in Apollo 12 samples: five igneous and one breccia rocks and four soils. *Proc. 2nd Lunar Sci. Conf.*, pp. 1319—1329.

Watson, E.B., 1976. Two-liquid partition coefficients: experimental data and geochemical implications. *Contrib. Mineral. Petrol.*, 56: 119—134.

Watson, E.B. and Green, T.H., 1981. Apatite/liquid partition coefficients for the rare earth elements and strontium. *Earth Planet. Sci. Lett.*, 56: 405—421.

Weill, D.F. and McKay G.A., 1975. The partitioning of Mg, Fe, Sr, Ce, Sm, Eu and Yb in lunar igneous systems and a possible origin of KREEP by equilibrium partial melting. *Proc. 6th Lunar Sci. Conf.*, pp. 1143—1158.

Wendlandt, R.F. and Harrison, W.J., 1979. Rare earth partitioning between immiscible carbonate and silicate liquids and CO_2 vapor: results and implications for the formation of light rare earth-enriched rocks. *Contrib. Mineral. Petrol.*, 69: 409—419.

Chapter 2

MINERALOGY OF THE RARE EARTH ELEMENTS

ANDREW M. CLARK

2.1. Introduction

The behaviour and properties of the rare earth elements in the natural environment arise from the group's closely related chemical properties. There are four series of transition elements in the periodic table. The first two, beginning at scandium and yttrium, consist of elements in which electrons are successively added to unfilled *d*-orbitals surrounded by already complete sub-shells. The other two series, beginning at lanthanum (the lanthanides) and actinium (the actinides) have unfilled *f*-orbitals. In the lanthanide series the electrons are added to the 4*f* sub-shell which lies deep within the 6*s* sub-shell. As the latter shell is completely filled in all the REE, the 4*f* electrons are so well shielded that the chemical properties of these elements are almost identical. For this reason the REE are always found as associated groups in minerals and rocks; in no case is one of the REE found in complete isolation. The other significant feature of the lanthanide series is that, due to the imperfect electronic screening between the nuclear charge and unfilled 4*f* sub-shell, a regular contraction of ionic radii occurs with increasing atomic number and is known as the lanthanide contraction. The importance of ionic radius in the geochemical behaviour of the rare earths was discussed in Chapter 1. For the purpose of the present discussion it is necessary to mention that the heavy REE Gd—Lu are frequently referred to as the yttrium group; the ionic radius of yttrium and its electronic configuration ally it with the heavier REE with which it is invariably associated in minerals and rocks. The light REE La—Eu are known as the cerium group.

In most rock-forming processes the REE are dispersed as minor or trace constituents of phases in which they are not essential components. Nevertheless, all minerals may be placed in one of three groups according to their total REE content:

(1) Minerals usually with very low REE concentrations. These include many of the common rock-forming minerals. Relative REE concentration levels may be inferred from distribution coefficient values, examples of which are given in Chapter 1. The distribution patterns of light and heavy REE in these minerals show wide variation.

(2) Minerals containing minor amounts of the REE, but not as essential constituents. Around 200 minerals are known to contain more than 0.01 wt.% of the REE (Herrmann, 1970). With these minerals it is frequently possible to recognize characteristic trends in the REE distribution.

(3) Minerals with major and usually essential contents of the REE. Over 70 minerals fall in this category and include all the REE species together with a few minerals which are lanthanide-rich equivalents of low-REE minerals, such as allanite and yttrofluorite (REE varieties of epidote and fluorite, respectively).

The causes of the variations in REE distributions in minerals have been debated for a number of years. One group of workers, notably Semenov (1957, 1958), considers that the structure of the mineral plays the predominant role in admitting particular REE ions. Thus minerals with high co-ordination numbers for their REE sites (10—12) are Ce-selective; those with the low co-ordination number of 6 are Y-selective; and those with intermediate numbers (7—9) have complex compositions, with both light and heavy REE present. Semenov (1963) has defined a REE distribution as selective if the ratio between the maximum and minimum concentration of the even-numbered REE (including Y) exceeds 50. Below 50 the distribution is complex. Other workers (e.g. Murata et al., 1953, 1957, 1959) consider the paragenetic conditions under which the minerals form to be of major importance in determining the REE distributions. Neumann et al. (1966) have concluded that both processes operate and give the factors controlling the REE distributions in minerals as (a) the availability of elements of suitable ionic radius, and (b) the appropriate bonding forces, charge, and optimum ionic radius for the given structural position. Ionic radius is, therefore, now regarded as a major contribution to fractionation within the REE but some minerals have strong structural constraints on the range of acceptable REE. Monazite $CePO_4$ has a monoclinic unit cell, but xenotime, the corresponding yttrium phosphate, is tetragonal; this structural modification is required to accommodate the heavier but smaller atoms of the yttrium group. Other minerals, however, are representative in their REE distributions of the medium from which they form.

Igneous rocks can contain several hundred parts per million of the lanthanides, distributed across both the major and accessory minerals. Schnetzler and Philpotts (1968, 1970) have provided data on REE distribution coefficients between mineral phenocrysts and matrix material for a number of igneous rocks. Of the common minerals, calcic clinopyroxenes and amphiboles proved to have the largest coefficients. The former, in some instances, gave values close to, or slightly greater than 1, showing that clinopyroxenes may act marginally as concentrators of the REE. Generally, values for these two minerals lie between 0.1 and 1. Other minerals with distribution coefficients around 0.1 include feldspars, pigeonite, mica, and orthopyroxene. Values for olivine of around 0.01 were the lowest found among the minerals analysed (see Chapter 1).

Feldspars invariably show a pronounced positive europium anomaly. This element appears to be the only rare earth that will reduce to the divalent state in nature and the feldspar structure is such that Eu^{2+} is readily accepted, resulting in excess Eu in the mineral relative to the REE of adjacent atomic number and a disruption of the straightforward pattern of REE fractionation in igneous rocks in which it occurs. Cerium can occur as Ce^{4+} under oxidizing conditions, evidence being provided by the existence of the mineral cerianite CeO_2, although here the change in ionic radius is relatively small so no change in the pattern of REE fractionation is seen in this case.

In granitic rocks the REE are mainly concentrated in accessory minerals such as sphene, apatite, and monazite. These minerals tend to concentrate the light REE and, consequently, whole-rock samples of these rocks are frequently enriched in the light REE. Of the major rock-forming minerals, plagioclase, K-feldspar, and biotite, in that order of abundance, act as hosts for the remaining REE (Condie and Lo, 1971). A large increase in mineral/melt distribution coefficients has been observed as melts become more silicic (see Chapter 1).

In sedimentary rocks, such as carbonates and sandstones, clay minerals are usually present in abundance. The latter have much higher REE and yttrium concentrations than carbonate minerals or quartz and may therefore contain the bulk of the REE in the rock as a whole. Roaldset (1975) showed that phyllosilicate minerals (the clay and mica groups) do not concentrate the REE in igneous rocks, but acquire them through surface adsorption as clays develop during weathering. Clay minerals, as products of weathering of igneous minerals, tend to inherit and average the REE distributions of their sources. Oceanic sediments, both biogenic and authigenic, show REE distributions similar to that in seawater, evidently deriving from that source. Bender et al. (1971) demonstrate this in their analysis of a calcite core from the East Pacific Rise.

Among the metamorphic minerals garnet is a very efficient concentrator of the REE, fractionating the heavier lanthanides. Schnetzler and Philpotts (1970) gave distribution coefficients of 0.35 for Ce rising to nearly 43 for Er in a garnet from a Japaneses dacite. Several authors (e.g. White et al., 1972) have analysed garnet and pyroxene separated from eclogites, showing that the heavier REE concentrate in the garnet and the lighter lanthanides in the pyroxene.

As mentioned above, major concentrations of the REE occur in accessory minerals of some rocks, either as essential constituents (e.g. monazite) or concentrated in certain minerals (e.g. apatite). Another common situation for REE-bearing minerals is in pegmatites, since these elements are frequently concentrated in the residual magmatic fluids from which these rocks form. Here much variation in their total REE contents and distribution patterns is found, arising from the tendency of the minerals to reflect the REE abundances in the final, often highly differentiated magmatic fluids.

In Table 2.1 the major lanthanide-containing minerals are listed and in the

TABLE 2.1

REE-containing minerals (the section numbers refer to the accompanying text)

Mineral	Formula	Crystal system	Section
Aeschynite	$(Ce,Ca,Fe,Th)(Ti,Nb)_2(O,OH)_6$	orthorhombic (metamict)	2.5
Agardite	$(Y,Ca)Cu_6(AsO_4)_3(OH)_6\cdot 3H_2O$	hexagonal	2.10
Agrellite	$NaCa_2Si_4O_{10}F$	triclinic	2.8
Allanite	$(Ce,Ca,Y)_2(Al,Fe^{2+},Fe^{3+})_3(SiO_4)_3OH$	monoclinic	2.7
Ancylite	$SrCe(CO_3)_2OH\cdot H_2O$	orthorhombic	2.3
Apatite	$Ca_5(PO_4)_3F$	hexagonal	2.9
Ashcroftine	$KNaCaY_2Si_6O_{12}(OH)_{10}\cdot 4H_2O$	tetragonal	2.7
Bastnäsite	$(Ce,La)(CO_3)F$	hexagonal	2.4
Belovite	$(Sr,Ce,Na,Ca)_5(PO_4)_3OH$	hexagonal	2.3
Braitschite	$(Ca,Na_2)_7(Ce,La)_2B_{22}O_{43}\cdot 7H_2O$	hexagonal	2.6
Brannerite	$(U,Ca,Ce)(Ti,Fe)_2O_6$	monoclinic (metamict)	2.5
Britholite	$(Ce,Ca)_5(SiO_4,PO_4)_3(OH,F)$	hexagonal	2.7
Brockite	$(Ca,Th,Ce)PO_4\cdot H_2O$	hexagonal	2.9
Burbankite	$(Na,Ca,Sr,Ba,Ce)_6(CO_3)_5$	hexagonal	2.3
Calkinsite	$(Ce,La)_2(CO_3)_3\cdot 4H_2O$	orthorhombic	2.3
Cappelenite	$BaY_6B_6Si_3O_{25}$	hexagonal	2.7
Carbocernaite	$(Ca,Ce,Na,Sr)CO_3$	orthorhombic	2.3
Caysichite	$(Y,Ca)_4Si_4O_{10}(CO_3)_3\cdot 4H_2O$	orthorhombic	2.8
Cerianite	$(Ce^{4+},Th)O_2$	cubic	2.5
Cerite	$(Ce,Ca)_9(Mg,Fe^{2+})Si_7(O,OH,F)_{28}$	trigonal	2.7
Cerotungstite	$CeW_2O_6(OH)_3$	monoclinic	2.5
Chevkinite	$(Ca,Ce,Th)_4(Fe^{2+},Mg)_2(Ti,Fe^{3+})_3Si_4O_{22}$	monoclinic	2.7
Chukhrovite	$Ca_3(Y,Ce)Al_2(SO_4)F_{13}\cdot 10H_2O$	cubic	2.11
Churchite	$YPO_4\cdot 2H_2O$	monoclinic	2.9
Cordylite	$(Ce,La)_2Ba(CO_3)_3F_2$	hexagonal	2.4
Davidite	$(La,Ce)(Y,U,Fe^{2+})(Ti,Fe^{3+})_{20}(O,OH)_{38}$	trigonal (metamict)	2.5
Donnayite	$Sr_3NaCaY(CO_3)_6\cdot 3H_2O$	triclinic	2.3
Dysanalyte	$(Ca,Ce,Na)(Ti,Nb,Ta)O_3$	cubic	2.5
Eudialyte	$(Ca,Na,Ce)_5(Zr,Fe)_2Si_6(O,OH,Cl)_{20}$	trigonal	2.7
Euxenite	$(Y,Ca,Ce,U,Th)(Nb,Ta,Ti)_2O_6$	orthorhombic (metamict)	2.5
Ewaldite	$Ba(Ca,Y,Na,K)(CO_3)_2$	hexagonal	2.3
Fergusonite	$(Y,Er)(Nb,Ta)O_4$	tetragonal (metamict)	2.5
Fersmite	$(Ca,Ce,Na)(Nb,Ti,Fe,Al)_2(O,OH,F)_6$	orthorhombic	2.5
Florencite	$CeAl_3(PO_4)_2(OH)_6$	trigonal	2.9
Fluocerite	$(Ce,La)F_3$	hexagonal	2.2
Formanite	$(Y,Er)(Ta,Nb)O_4$	tetragonal (metamict)	2.5
Gadolinite	$(Y,Ce)_2Fe^{2+}Be_2Si_2O_{10}$	monoclinic	2.7
Gagarinite	$NaCaY(F,Cl)_6$	trigonal	2.2
Hellandite	$(Ca,Y)_6(Al,Fe^{3+})Si_4B_4O_{20}(OH)_4$	monoclinic	2.7
Hibonite	$(Ca,Ce)(Al,Ti,Mg)_{12}O_{19}$	hexagonal	2.5
Huanghoite	$CeBa(CO_3)_2F$	hexagonal	2.4
Iimoriite	$Y_5(SiO_4)_3(OH)_3$	triclinic	2.7
Ilimaussite	$Ba_2Na_4CeFeNb_2Si_8O_{28}\cdot 5H_2O$	monoclinic	2.7
Ilmajokite	$(Na,Ce,Ba)_2TiSi_3O_5(OH)_{10}\cdot nH_2O$	monoclinic	2.7
Iraqite	$(K,La,Ce,Th)(Ca,Na,La)_2Si_8O_{20}$	hexagonal	2.7

TABLE 2.1 (*continued*)

Mineral	Formula	Crystal system	Section
Joaquinite	$Ba_2NaCe_2Fe^{2+}(Ti,Nb)_2Si_8O_{26}(OH,F)\cdot H_2O$	monoclinic	2.7
Kainosite	$Ca_2(Y,RE)_2(Si_4O_{12})CO_3\cdot H_2O$	orthorhombic	2.7
Karnasurtite	$(Ce,La,Th)(Ti,Nb)(Al,Fe^{3+})(Si,P)_2O_7(OH)_4\cdot 3H_2O$	hexagonal (?)	2.7
Keilhauite	$(Ca,Y,Ce)(Ti,Al,Fe^{3+})SiO_5$	monoclinic	2.7
Knopite	$(Ca,Ce)TiO_3$	cubic	2.5
Lanthanite	$(La,Ce)_2(CO_3)_3\cdot 8H_2O$	orthorhombic	2.3
Laplandite	$Na_4CeTiPSi_7O_{22}\cdot 5H_2O$	orthorhombic	2.7
Loparite	$(Ce,Na,Ca)(Ti,Nb)O_3$	orthorhombic (?)	2.5
Loveringite	$(Ca,Ce)(Ti,Fe^{3+},Cr,Mg)_{21}O_{38}$	trigonal (metamict)	2.5
Mckelveyite	$Na_2Ba_4(Y,Ca,Sr,U)_3(CO_3)_9\cdot 5H_2O$	triclinic	2.3
Melanocerite	$Ce_4CaBSi_2O_{12}(OH)$	hexagonal (metamict)	2.7
Monazite	$(Ce,La)PO_4$	monoclinic	2.9
Mosandrite	$(Na,Ca,Ce)_3Ti(SiO_4)_2F$	monoclinic (metamict)	2.7
Nordite	$Na_3Ce(Sr,Ca)(Mn,Mg.Fe,Zn)_2Si_6O_{18}$	orthorhombic	2.7
Okanoganite	$(Na,Ca)_3(Y,Ce,Nd,La)_{12}Si_{12}B_2O_{27}F_{14}$	hexagonal	2.8
Parisite	$(Ce,La)_2Ca(CO_3)_3F_2$	hexagonal	2.4
Perrierite	$(Ca,Ce,Th)_4(Mg,Fe^{2+})_2(Ti,Fe^{3+})_3Si_4O_{22}$	monoclinic	2.7
Phosinaite	$H_2Na_3(Ca,Ce)(SiO_4)(PO_4)$	orthorhombic	2.8
Polycrase	$(Y,Ca,Ce,U,Th)(Ti,Nb,Ta)_2O_6$	orthorhombic (metamict)	2.5
Polymignite	$(Ca,Fe,Y,Th)(Nb,Ti,Ta)O_4$	orthorhombic (metamict)	2.5
Priorite	$(Y,Ca,Fe,Th)(Ti,Nb)_2(O,OH)_6$	orthorhombic (metamict)	2.5
Pyrochlore	$(Na,Ca,Ce)_2Nb_2O_6(OH,F)$	cubic	2.5
Retzian	$Mn_2Y(AsO_4)(OH)_4$	orthorhombic	2.10
Rhabdophane	$(Ce,La)PO_4\cdot H_2O$	hexagonal	2.9
Röntgenite	$(Ce,La)_3Ca_2(CO_3)_5F_3$	hexagonal	2.4
Sahamalite	$(Mg,Fe)(Ce,La)_2(CO_3)_4$	monoclinic	2.3
Samarskite	$(Y,Ce,U,Fe)(Nb,Ta,Ti)_2(O,OH)_6$	orthorhombic (metamict)	2.5
Saryarkite	$Ca(Y,Th)Al_5(SiO_4)_2(PO_4,SO_4)_2(OH)_7\cdot 6H_2O$	hexagonal	2.8
Sazhinite	$Na_3CeSi_6O_{15}\cdot 6H_2O$	orthorhombic	2.7
Scheteligite	$(Ca,Fe,Mn,Bi,Y)_2(Ti,Ta,Nb,W)_2(O,OH)_7$	orthorhombic	2.5
Semenovite	$(Ca,Ce,La,Na)_{10-12}(Fe^{2+},Mn)(Si,Be)_{20}(O,OH,F)_{48}$	orthorhombic	2.7
Spencite	$(Y,Ca,La,Fe)_5(Si,B,Al)_3(O,OH,F)_{13}$	hexagonal (metamict)	2.7
Steenstrupine	$(Ce,La,Na,Mn)_6(Si,P)_6O_{18}OH$	trigonal	2.7
Stillwellite	$(Ce,La,Ca)BSiO_5$	trigonal	2.7
Synchysite	$(Ce,La)Ca(CO_3)_2F$	hexagonal	2.4
Tengerite	$CaY_3(CO_3)_4(OH)_3\cdot 3H_2O$	monoclinic	2.3
Thalenite	$Y_3Si_3O_{10}(OH)$	monoclinic	2.7
Thortveitite	$(Sc,Y)_2Si_2O_7$	monoclinic	2.7
Tombarthite	$Y_4(Si,H_4)_4O_{12-x}(OH)_{4+2x}$	monoclinic	2.7
Törnebohmite	$(Ce,La)_3Si_2O_8OH$	hexagonal	2.7
Tranquillityite	$Fe_8(Zr,Y)_2Ti_3Si_3O_{24}$	monoclinic	2.7
Tritomite	$(Ce,La,Y,Th)_5(Si,B)_3(O,OH,F)_{13}$	hexagonal (metamict)	2.7

TABLE 2.1 (*continued*)

Mineral	Formula	Crystal system	Secti
Tundrite	$Na_3(Ce,La)_4(Ti,Nb)_2(SiO_4)_2(CO_3)_3O_4(OH)\cdot 2H_2O$	triclinic	2.7
Tveitite	$Ca_{1-x}Y_xF_{2+x}$	monoclinic	2.2
Vitusite	$Na_3(Ce,La,Nd)(PO_4)_2$	orthorhombic	2.9
Xenotime	YPO_4	tetragonal	2.9
Yttrocerite	$(Ca,Ce)F_2$	cubic	2.2
Yttrocrasite	$(Y,Th,Ca,U)(Ti,Fe^{3+})_2(O,OH)_6$	orthorhombic (metamict)	2.5
Yttrofluorite	$(Ca,Y)F_2$	cubic	2.2
Yttrotantalite	$(Y,U,Fe)(Ta,Nb)O_4$	monoclinic (metamict)	2.5
Yttrotungstite	$YW_2O_6(OH)_3$	monoclinic	2.5
Zhonghuacerite	$Ba_2Ce(CO_3)_3F$	trigonal	2.4
Zircon	$(Zr,Y)(Si,P)O_4$	tetragonal	2.7

following pages notes are given on their total REE abundances, distribution patterns, and modes of occurrence. References are confined as far as possible to papers containing usable REE data. The space available does not allow the inclusion of full optical, physical, and crystallographic data, but these are readily available from other sources. The minerals are grouped alphabetically within each major cationic group.

2.2. Halides

Fluocerite (tysonite). $(Ce,La)F_3$. Hexagonal. Quite common in rare earth pegmatites, often partially altered to bastnäsite. Data give a total of 70.6 wt.% REE for fluocerite from the northwestern Ukranian Shield where it occurs as inclusions in fluorite and yttrofluorite (Gurov et al., 1975). This occurrence is strongly Ce-selective, with a La/Lu ratio of near 1000:1.

Fluorite (yttrofluorite and yttrocerite). CaF_2. Cubic. Fluorite commonly contains small amounts of the REE and Y; occasionally the amounts are high, up to 13.7 wt.% Y and 14.1 wt.% Ce having been reported. The REE replace Ca in the structure, which is in 8-fold coordination, indicating that both heavy and light REE enrichment can occur. Fleischer (1969) notes a concentration of the HREE in fluorites from granitic pegmatites, and of light lanthanides in fluorites from alkaline rocks. Those from granites and hydrothermal deposits showed distributions with the whole series of REE present. Jacob (1974) and others have demonstrated that REE fractionation occurs in fluorite and that, in particular, the Tb/Ca vs. Tb/La ratios can be used to differentiate between sedimentary and hydrothermal fluorites.

Gagarinite. $NaCa(F,Cl)_6$. Hexagonal. Originally found in quartz-microcline veins in highly albitized granites in Kazakhstan but since at several other Russian localities, notably at Tuva in an albitite from a quartz syenite (Akelin and Kazakova, 1963). The mineral alters easily and is replaced by tengerite, synchysite, and yttrofluorite. The rare earth data in

Stepanov and Severov (1961) show a slight relative enrichment of the Y group in a crystalline sample of the mineral (46.2 wt.% RE_2O_3) compared with the more common cryptocrystalline material at the type locality (45.2 wt.%). Khomyakov (1970), in assessing the potentiality of gagarinite as a geothermometer, notes that, at the Tarbagatay granitic pluton, the transition from granite to outer-contact metasomatite is accompanied by an increase in the proportion of heavier REE in the mineral, whereas the REE in the associated pyrochlore behave in the opposite way.

Tveitite. $Ca_{1-x}Y_xF_{2+x}$ with $x \sim 0.3$. Monoclinic, pseudo-cubic. Found in a cleavelandite pegmatite at Høydalen, South Norway, as an irregular lump altering on the surface to kainosite. It contains 30.26 wt.% (Y + REE) with the lanthanide maximum at Ce (Bergstøl et al., 1977).

2.3. Carbonates

Ancylite. $SrCe(CO_3)_2OH{\cdot}H_2O$. Orthorhombic. Ca can replace Sr (calcioancylite) and the Sr/REE ratio varies. A full structural formulation has been given by Dal Negro et al. (1975) for a crystal from Mont St. Hilaire, Quebec. The isostructural end-member $RE(OH)CO_3$ has been synthesized by Sawyer et al. (1973). Ancylite occurs in pegmatites of some alkaline rocks and in carbonatites. The REE distribution of calcioancylite (CaO 8.35, SrO 3.07, RE_2O_3 50.75 wt.%) from Cornog, Pennsylvania (Keidel et al., 1971) shows marked LREE enrichment.

Burbankite. $(Na,Ca,Sr,Ba,Ce)_6(CO_3)_5$. Hexagonal. The REE substitute for Ca in the structure, RE_2O_3 varying between 2 and 22 wt.% in specimens from different localities. It was first recorded, intergrown with ancylite, in carbonate veins in shonkinite from Montana; later in lacustrine deposits of the Green River Formation in Wyoming; in an altered alkaline rock in the Zeerust District, Transvaal; from the Vuorijärvi pluton, U.S.S.R.; and in a nepheline syenite at Mont St. Hilaire, Quebec. A comparison between burbankite and associated ewaldite and mckelveyite in the Green River Formation showed that the Ce and Y earths are strongly partitioned — the former concentrating in burbankite (Fitzpatrick and Pabst, 1977).

Calkinsite. $(Ce,La)_2(CO_3)_3{\cdot}4H_2O$. Orthorhombic. Occurs at Big Sandy Creek, Bearpaw Mountains, Montana, with lanthanite, as an alteration product of burbankite. The REE data of Pecora and Kerr (1953) show that calkinsite concentrates the REE in forming from burbankite, being some seven times richer in these elements (RE_2O_3 62.73 wt.%). The REE distributions for the two associated minerals are, however, quite similar, both showing LREE enrichment.

Carbocernaite. $(Ca,Ce,Na,Sr)CO_3$. Orthorhombic. Occurs in dolomite-calcite carbonatite veins from the Vuorijärvi Massif, Kola Peninsula; in a metasomatite from North Vietnam; and in an albite-biotite-calcite carbonatite from Sturgeon Lake, Ontario. Rare earth data are recorded for the mineral from the Kola Peninsula (RE_2O_3 26.1 wt.%; Bulakh et al., 1961) and from Ontario (RE_2O_3 21.47 wt.%; Harris, 1972). Both are Ce-rich but the former has an anomalously high Sm value.

Donnayite. $Sr_3NaCaY(CO_3)_6{\cdot}3H_2O$. Triclinic, pseudo-rhombohedral. Occurs in pegmatite dyke, miarolitic cavities, and interstices in the nepheline syenite at Mont St. Hilaire, Quebec. The $(Y + RE)_2O_3$ content is given as 15.38 wt.% (Chao et al., 1978) with Nd and La as the principal lanthanides. Donnayite is isomorphous with welagonite and mckelveyite.

Ewaldite. $Ba(Ca,Y,Na,K)(CO_3)_2$. Hexagonal. Mckelveyite from the Green River Formation, Sweetwater Co., Wyoming was found to be a mixture of two phases in syntactic overgrowth. Ewaldite is the name given to the second phase. They occur in association with burbankite. Rare earth data in Donnay and Donnay (1971) give 7.7 wt.% $(Y + RE)_2O_3$ with the Y group dominant, but with notable enrichment in Gd and Er.

Lanthanite. $(La,Ce)_2(CO_3)_3 \cdot 8H_2O$. Orthorhombic. It was originally found coating cerite at the Bastnäs mine, Västmanland, Sweden (Lindström, 1910); now known in oxidized zinc ores in Lehigh Co., Pennsylvania; as an alteration product of allanite in a pegmatite at Baringer Hill, Llano Co., Texas, and at several other localities.

Mckelveyite. $Na_2Ba_4(Y,Ca,Sr,U)_3(CO_3)_9 \cdot 5H_2O$. Triclinic. Found at four localities in Sweetwater Co., Wyoming, from the Green River Formation. Occurs in syntactic overgrowth with ewaldite and associated with burbankite. The latter is the concentrator of the Ce-group rare earths while mckelveyite, with 13.39 wt.% $(Y + RE)_2O_3$, and ewaldite show similar HREE-enriched distributions. Mckelveyite is notable for the high Tb_2O_3 value recorded (C. Milton et al., 1965).

Sahamalite. $(Mg,Fe)(Ce,La)_2(CO_3)_4$. Monoclinic. Found in a baryte-dolomite rock at Mountain Pass, San Bernardino Co., California, associated with bastnäsite and parisite (Jaffe et al., 1953). It is thought to form from dolomite and bastnäsite under hydrothermal conditions.

Tengerite. $CaY_3(CO_3)_4(OH)_3 \cdot 3H_2O$. Monoclinic. Occurs at Ytterby in Sweden, Barringer Hill in Texas, and several other localities. It is formed by the alteration of gadolinite. Tengerite from Kazakhstan (Shipovalov and Stepanov, 1971) has 55.61 wt.% RE_2O_3, the REE distribution showing enrichment in Sm and other medium-weight REE. Tengerite from the Pyörönmaa pegmatite, Finland, however, shows marked enrichment in the heavier REE (Siivola, 1975).

2.4. Carbonates with fluoride

These minerals can be grouped, following Fleischer (1978a), in the *bastnäsite* group of hexagonal (or trigonal) minerals:

bastnäsite	$(Ce,La)(CO_3)F$
var. hydroxyl-bastnäsite	$Ce(CO_3)(OH,F)$
bastnäsite-(La)	$(La,Ce)(CO_3)F$
bastnäsite-(Y)	$(Y,Ce,Dy)(CO_3)F$
thorbastnäsite	$(Ca,RE)Th(CO_3)_2F_2 \cdot 3H_2O$
parisite	$(Ce,La)_2Ca(CO_3)_3F_2$
cordylite	$(Ce,La)_2Ba(CO_3)_3F_2$
röntgenite	$(Ce,La)_3Ca_2(CO_3)_5F_3$
synchysite	$(Ce,La)Ca(CO_3)_2F$
var. synchysite-(Y)	$(Y,Ce)Ca(CO_3)_2F$
synchysite-(Nd)	$(Nd,La)Ca(CO_3)_2F$
huanghoite	$CeBa(CO_3)_2F$

Bastnäsite. (Ce,La)(CO_3)F. Hexagonal. It is one of the more widespread rare earth minerals, occurring in many pegmatites, usually associated with allanite, cerite and fluocerite, sometimes as an alteration product of the latter. Fleischer (1965) lists the minerals as strongly Ce-selective, in keeping with a mineral having the lanthanides in 11-fold co-ordination. Bastnäsite-(Y) is an exception to this general rule. Fleischer's (1978a) compilation of 100 analyses of minerals in the group shows an increase in LREE in bastnäsite-group minerals from granites and granite pegmatites through to hydrothermal, alkaline rock, and carbonatitic occurrences. Bastnäsite-(Y) occurs as pseudomorphs after gagarinite in a microcline-quartz pegmatite in Kazakhstan. It contains 60 wt.% RE_2O_3 and the distribution matches exactly that of cryptocrystalline gagarinite from the same locality (Mineev et al., 1970). Thorbastnäsite (Pavlenko et al., 1965) is a variety occurring in albitites from alkaline metasomatic rocks in eastern Siberia. The REE are partially replaced by Ca and Th, and the 7.46 wt.% RE_2O_3 shows an even distribution of heavy and light REE.

Cordylite. $(Ce,La)_2Ba(CO_3)_3F_2$. Hexagonal isomorphous with parisite (hence sometimes called Ba-parisite). It occurs sparsely in pegmatite veins in nepheline syenite, associated with aegirine, synchysite, ancylite, and neptunite at Narsarsuk, Greenland (Flink, 1901).

Huanghoite. $CeBa(CO_3)_2F$. Hexagonal. Originally found in hydrothermal deposits near the Huang-Ho river, Inner Mongolia; later in vugs in an ankerite carbonatite from Siberia. Both minerals have similar rare earth contents: RE_2O_3 38.4 wt.%, Huang-Ho river (Semenov and Chang, 1961) and 37.90 wt.%, Siberia (Kapustin, 1972). The distributions of the constituent REE are also similar.

Parisite. $(Ce,La)_2Ca(CO_3)_3F_2$. Hexagonal. Originally described from carbonaceous shale beds in the emerald deposits of Muzo District, Colombia. Now known from many other localities, particularly in pegmatite pipes in riebeckite-aegirine granites and in altered trachyte. Several occurrences rich in Y have been recorded (e.g. Gerasimovskii, 1964).

Röntgenite. $(Ce,La)_3Ca_2(CO_3)_5F_3$. Trigonal. Occurs intergrown with synchysite, parisite and bastnäsite at Narsarsuk, Greenland (Donnay, 1953; Donnay and Donnay, 1953).

Synchysite. $(Ce,La)Ca(CO_3)_2F$. Hexagonal. A fairly widespread rare earth fluocarbonate, occurring in a number of pegmatites and carbonatites, usually associated with other members of the bastnäsite group. Fleischer (1978a) notes that, of the minerals in the group, synchysite shows a rather lower increase in LREE content in occurrences ranging from granitic through to carbonatitic. The mineral also occurs in varieties enriched in Nd, synchysite-(Nd) (Maksimovic and Panto, 1978); and enriched in the HREE, synchysite-(Y). The latter was formerly named doverite (Smith et al., 1960) but has been renamed following acceptance of Levinson's (1966) system for the nomenclature of rare earth minerals.

Zhonghuacerite. $Ba_2Ce(CO_3)_3F$. Trigonal. Found as a fine-grained granular aggregate in Proterozoic metamorphic dolomite from Bayan Obo in North China. The rare earth content, principally Ce and La, is near 24 wt.% RE_2O_3 (Peishan and Kejie, 1981).

2.5. Oxides

Aeschynite-priorite series. $(Ce,Ca,Fe,Th)(Ti,Nb)_2(O,OH)_6$-$(Y,Ca,Fe,Th)(Ti,Nb)_2(O,OH)_6$. Orthorhombic but usually metamict. This series is unusual in grading from Ce-dominant aeschynite through to Y-dominant priorite. Fleischer (1966) has compiled 26 analyses from the literature and shows that the REE compositions in the group are regularly distributed throughout the series, but with aeschynite somewhat commoner than priorite. A correlation of composition with environment was also found — those from granite pegmatites are enriched in Y while those from alkaline rocks are enriched in the light lanthanides. The total rare earth content of minerals in the series is rather variable. Fleischer (1966) quotes values from 16 to 36 wt.% RE_2O_3. This arises, in the main, from the chemical substitutions which occur in the group. In particular at the aeschynite end of the series niobo-, alumo-, thoro-, titan-, and tantal-aeschynite have been recorded. *Vigezzite* (Graeser et al., 1979) is a Ca-rich member of the aeschynite group with Ce_2O_3 = 10.5 wt.%.

Betafite — see pyrochlore group.

Brannerite. $(U,Ca,Ce)(Ti,Fe)_2O_6$. Monoclinic but invariably metamict. RE_2O_3 concentrations of up to 11.6 wt.% have been recorded but are usually considerably lower. The formula is usually given as shown but the available analyses suggest that the Y group rather than the Ce group are the dominant REE in the minerals. Brannerite occurs chiefly as a primary mineral in pegmatites or as a detrital mineral derived from pegmatities or vein deposits and is the major ore mineral in several uranium deposits. At several localities it is found in association with gold deposits. A review of occurrences has been given by Ferris and Ruud (1971).

Cerianite. $(Ce^{4+},Th)O_2$. Cubic. Found originally as minute radioactive octahedra separated from carbonate rock at Sudbury, Ontario. Van Wambeke (1977) has also reported the mineral from the Karonge rare earth deposits, Burundi. He ascribes the very low concentration of the other REE relative to Ce (Ce/La ~16:1) to the limitations on replacement of Ce^{4+} by other rare earths owing to the differences in ionic radii. The limitation imposed by the composition on charge balance for $RE^{3+} \rightleftharpoons Ce^{4+}$ substitution could also be a contributory factor. At Karonge, cerianite and goethite have formed by the alteration of bastnäsite, the rare earths becoming concentrated in the cerianite. The Canadian occurrence has ~80 wt.% CeO_2 and ~4.5 wt.% RE_2O_3 for the remaining rare earths (Graham, 1955).

Cerotungstite — see yttrotungstite.

Davidite. $(La,Ce)(Y,U,Fe^{2+})(Ti,Fe^{3+})_{20}(O,OH)_{38}$. Trigonal (Gatehouse et al., 1979). Often found in a metamict form, occurring as a primary mineral at a number of localities, particularly in pegmatities and hydrothermal veins. Fleischer (1965) reports that the rare earth composition of the mineral (up to 10 wt.% RE_2O_3) is remarkable in showing a high concentration of the LREE, a high La/Nd ratio with La normally exceeding Ce, and a surprisingly constant lanthanide composition regardless of paragenesis. Dixon and Wylie (1951) and Hayton (1960) report enrichment at both ends of the lanthanide series in davidites from several localities.

Dysanalyte — see perovskite group.

Euxenite-polycrase series. $(Y,Ca,Ce,U,Th)(Nb,Ta,Ti)_2O_6$-$(Y,Ca,Ce,U,Th)(Ti,Nb,Ta)_2O_6$. Orthorhombic, generally metamict. Euxenite and priorite are dimorphous forms of $YNbTiO_6$ with euxenite forming at higher temperatures than priorite. In the aeschynite-priorite series the REE distribution changes from Ce- to Y-rich, whereas in the euxenite-polycrase series the Y group predominates throughout. Komkov (1973) has found that if the LREE are dominant the euxenite structure is metastable. The substitution in the series occurs in the B-cation site, where Ti replaces Nb and Ta. Blomstrandine (distinct from blomstrandite, a variety of betafite) is a synonym for polycrase. The REE distributions in the group do not serve to distinguish the minerals from priorite, both showing Y-group selectivity. Butler (1958) found the minerals contain between 20 and 30 wt.% $(Y + RE)_2O_3$ with no correlation between the rare earths and ThO_2 or TiO_2 contents. Smellie et al. (1978) report a member of the group from Ampangabé, Malagasy Republic unusually rich in thulium. Minerals of the euxenite-polycrase series are widespread in granitic pegmatites; also as an accessory mineral in granites and sometimes in placers.

Fergusonite-formanite series. $(Y,Er)(Nb,Ta)O_4$-$(Y,Er)(Ta,Nb)O_4$. Tetragonal but usually metamict. Minerals at the fergusonite end of the series are more common. They occur at a number of localities, in pegmatites associated with euxenite, monazite, gadolinite and other rare earth minerals. A monoclinic modification, β-fergusonite is known, and a Ce-rich variety (brocenite) has been recorded. Butler and Hall (1960) carried out complete analyses of 11 members of the group and showed the overall dominance of the Y-group REE. The individual distributions, however, showed considerable variation; the majority had Yb as the most abundant REE but others, with maxima of Gd and Dy, suggest that the structure is capable of accommodating a range of HREE substitution.

Fersmite. $(Ca,Ce,Na)(Nb,Ti,Fe,Al)_2(O,OH,F)_6$. Orthorhombic. A Ca niobate structurally analogous to euxenite, but differs from the latter in that fersmite is generally enriched in the Ce-group REE (e.g. Hess and Trumpour, 1959) with RE_2O_3 contents of over 6 wt.% having been recorded. However, from a granite pegmatite in the Ilmen Mountains, U.S.S.R., a Y-rich variety with 16 wt.% RE_2O_3 has been recorded (Makarochkin et al., 1963) with Dy as the maximum REE and with a similar REE distribution to euxenite from Arendal, Norway. Fersmite occurs in carbonatites, nepheline syenites, carbonate veins, and pegmatites. In several instances it is found replacing columbite.

Formanite — see fergusonite.

Hibonite. $(Ca,Ce)(Al,Ti,Mg)_{12}O_{19}$. Hexagonal. A rather rare mineral occurring terrestrially in a metamorphosed limestone in the Malagasy Republic, in siliceous marbles in Siberia, and in granulite-facies rocks from Tanzania. In addition, it is common in Ca,Al-rich inclusions in carbonaceous chondritic meteorites. The rare earths, however, are confined to the terrestrial occurrences, occurring in amounts up to 4.1 wt.% in substitution for Ca, with the Ce group dominant (Maaskant et al., 1980).

Knopite — see perovskite group.

Loparite — see perovskite group.

Loveringite. $(Ca,Ce)(Ti,Fe^{3+},Cr,Mg)_{21}O_{38}$. Trigonal but metamict; isostructural with crichtonite and senaite (Gatehouse et al., 1978). Found in bronzite cumulate layers in the Jimberlana layered intrusion, Norseman, Western Australia. The published analyses show up to 6 wt.% RE_2O_3.

Microlite — see pyrochlore group.

Perovskite group. $CaTiO_3$. Cubic. *Knopite* is a perovskite-group mineral with Ce-group REE substituting for Ca; up to 8 wt.% RE_2O_3 has been reported (Shilin and Yanchenko, 1964). It is found as an accessory mineral in alkaline rocks, pyroxenites, apatite-nepheline ores, and basic pegmatites. *Dysanalyte*, $(Ca,Ce,Na)(Ti,Nb,Ta)O_3$, is a perovskite-group mineral containing Nb and Ta, usually with Nb > Ta (Murdoch, 1951). The REE and Na are non-essential. It is a rare mineral, occurring in some carbonatites and also in lunar material. *Loparite*, (Ce,Na,Ca)(Ti,Nb)O_3, may contain up to 27 wt.% RE_2O_3. Sitnin and Leonova (1961) confirm the Ce-selective nature of the mineral, but record an unusually high value for Tb. Loparite occurs in nepheline syenites and occasionally in albitized and greisenized granites.

Polycrase — see euxenite-polycrase series.

Polymignite. (Ca,Fe,Y,Th)(Nb,Ti,Ta)O_4. Orthorhombic but generally metamict. Chemically similar to aeschynite with which it may be identical. Early analyses suggest rare earth contents of near 13 wt.% $(Y + RE)_2O_3$ with substantial contents of both heavy and light REE (Brögger, 1890). It is found in several Norwegian pegmatites and at a few other localities.

Priorite — see aeschynite-priorite series.

Pyrochlore group. $A_2B_2O_6$(O,OH,F), where A sites are occupied by atoms of Na, Ce, REE, K, U, etc., and B sites by Nb, Ta, and Ti. Cubic. Three sub-groups (see IMA Commission nomenclature report; Hogarth, 1977) are defined according to the relative contents of Nb, Ta, and Ti; pyrochlore is the Nb-dominant subgroup, *microlite* is Ta-dominant, and *betafite* Ti-dominant. Species within these subgroups are designated according to the various substitutions in the A site. Pyrochlore and betafite are the principal carriers of the rare earths, although REE-bearing microlite is known. The designated rare earth species are *yttropyrochlore*, *ceriopyrochlore*, and *yttrobetafite*, the criteria being that the REE are the dominant atoms in the A site. apart from Na and Ca. *Plumbopyrochlore*, *uranpyrochlore* and *uranmicrolite* are also known to be REE-bearing species. The available analyses suggest that rare earths of the Ce group are more common in pyrochlore (RE_2O_3 up to 13 wt.%) and the Y group in betafite (up to 16 wt.%). Vlasov (1966) reports a number of geochemical characteristics of the REE in pyrochlore. Those from nepheline-syenite and alkali-syenite massifs have high contents of CaO and TiO_2 and marked enrichment in the lighter Ce earths. In a formation from the Vishnevye Mountains, Urals, showing a progression from high- to low-temperature conditions, the content of CaO, RE_2O_3, and U_3O_8 in pyrochlore is reported to decrease, while Nb_2O_5 and Na_2O increase (Es'kova and Nazarenko, 1960). Pyrochlore from carbonatites does not generally contain more than 2—3 wt.% of the Ce earths and Nd was found to be invariably greater than La. A betafite from a Precambrian pegmatite (Marchenko et al., 1969) also showed Ce-Nd enrichment. The mineral was formed by the replacement of yttrotitanite (keilhauite) and the REE distributions of the two phases are practically identical. Khomyakov (1970) has noted that, at the Tarbagatay granitic pluton, the transition from granite to outer-contact metasomatite is accompanied by an increase in the proportion of lighter REE in pyrochlore, while REE in the associated gagarinite behave in the opposite way.

Samarskite. (Y,Ce,U,Fe)$(Nb,Ta,Ti)_2(O,OH)_6$. Orthorhombic, invariably metamict. A complex niobate-tantalate characterized by the predominance of Nb over Ti and Ta. *Ishikawaite* is a variety containing up to 23% UO_2. Y and the Y earths are dominant in the REE distributions of samarskites with Sm as the principal cerium earth. Butler (1958) measured the rare earth compositions of four minerals from different localities ($(Y + RE)_2O_3$ 16—25 wt.%). He found among them fairly strong selective-assemblage tendencies with Gd, Tb, and Dy preferentially concentrated. Gd was the major REE in two specimens and Dy in the other two. Nilssen (1970) examined four Norwegian samarskites and obtained a distribution for one of them almost identical with the one Norwegian specimen in Butler's work, with a very pronounced Gd maximum. Nilssen's other specimens all showed marked Dy peaks but two had Yb maxima. The Bulgarian samarskite of Pashov and Kussovski (1971) also has Dy as its maximum lanthanide. Samarskite is widespread in granite pegmatites, often associated with columbite; sometimes as a detrital mineral in placer deposits.

Scheteligite. $(Ca,Fe,Mn,Bi,Y)_2(Ti,Ta,Nb,W)_2(O,OH)_7$. Orthorhombic. Found as small black crystals embedded in a pegmatite at Torvelona, Iveland, Norway. The total $(Y + RE)_2O_3$ content is reported as 6 wt.% (Bjørlykke, 1937).

Vigezzite — see aeschynite-priorite series.

Yttrocrasite. $(Y,Th,Ca,U)(Ti,Fe^{3+})_2(O,OH)_6$. Orthorhombic, often metamict (Hidden and Warren, 1906). Thought to be the Ti-rich end-member of the euxenite group. It occurs in granite pegmatites at several localities in Texas and California.

Yttrotantalite. $(Y,U,Fe)(Ta,Nb)O_4$. Monoclinic but invariably metamict. Probably the Ta-rich analogue of samarskite. Analyses show pronounced Y-group selectivity with $(Y + RE)_2O_3$ contents up to 18 wt.%. Found in several pegmatites in Sweden and Norway and more recently in Bulgaria (Yb/La ~ 17:1; Aleksiev and Ivanov, 1970).

Yttrotungstite and *cerotungstite.* $YW_2O_6(OH)_3$ and $CeW_2O_6(OH)_3$. Monoclinic. Yttrotungstite has been reported from the Kramat Pulai tin workings in Perak (Malaysia); cerotungstite as a secondary mineral in tungsten deposits in northern Rwanda and in the Kigezi District of southwestern Uganda (Sahama et al., 1970). Their respective $(Y + RE)_2O_3$ concentrations are 20.74 and 24.12 wt.%.

2.6. Borate

Braitschite. $(Ca,Na_2)_7(Ce,La)_2B_{22}O_{43}\cdot 7H_2O$. Hexagonal. Found in nodules in anhydrite rock from Grand Co., Utah. It contains 18.3 wt.% RE_2O_3 and a complex REE distribution with the maximum (6.81 wt.%) at Ce but also 1.34 wt.% of Y_2O_3 and measurable quantities of the heavy lanthanides (Raup et al., 1968).

2.7. Silicates

Allanite (orthite). $(Ce,Ca,Y)_2(Al,Fe^{2+},Fe^{3+})_3(SiO_4)_3OH$. Monoclinic but frequently metamict. Allanite is a Ce-rich member of the epidote group, its composition being governed by the coupled substitution Ca ⇌ RE and Al ⇌ Fe^{2+}. It is a characteristic accessory mineral in many granites, granodiorites, monzonites, and syenites. The REE are located in the A(2) structural site and the fact that this site will accept rare earth ions of any size led Adams (1969) to conclude that allanite is one of those minerals whose REE distribution closely reflects the available REE assemblage at the time of crystallization. He cited the similarity in distribution patterns of composite North American shales (Haskin and Haskin, 1968) with allanite from Nevada (Lee and Bastron, 1967). Jensen (1967) quotes several examples of allanite in closely related associations, also concluding that the mineral forms without appreciable REE fractionation. More recently, however, Brooks et al. (1981) have shown that REE partition between phenocrystal allanite and glass of the Sandy Braes perlitic obsidian of Northern Ireland strongly favours the LREE in allanite. Exley (1980) also found variations in the degree of REE fractionation in allanites from several rocks in Skye.

Ashcroftine. $KNaCaY_2Si_6O_{12}(OH)_{10}\cdot 4H_2O$. Tetragonal. Found in vugs in augite syenite at Narsarsuk, Greenland. It was formerly thought to be a zeolite but it is now established as an independent species with major Y and minor amounts of the other REE (Moore et al., 1969).

Britholite. $(Ce,Ca)_5(SiO_4,PO_4)_3(OH,F)$. Hexagonal. Britholite is a rare earth mineral related to apatite through the substitution of $(RE^{3+} + Si^{4+})$ for $(Ca^{2+} + P^{5+})$. It occurs chiefly in nepheline syenites and contact metasomatic deposits related to alkali syenites and granites. In the Shonkin Sag laccolith, Montana, it occurs instead of, or replacing, apatite in the more highly evolved rocks (Nash, 1972). Its rare earth chemistry at this locality shows marked Ce enrichment (RE_2O_3 60.54 wt.% with Ce_2O_3 22.5 wt.%). In britholite the REE are located in 7-fold co-ordination accounting for the complex REE compositions generally observed (Fleischer, 1965, p. 767). Various REE distributions have been recorded, and a Y-rich variety, known as britholite-(Y) or abukumalite, recorded from several localities.

Cappelenite. $BaY_6B_6Si_3O_{25}$. Hexagonal. A borosilicate of Y and Ba with over 40 wt.% $(Y + RE)_2O_3$. It was found originally in nepheline-syenite pegmatites in Norway (Brögger, 1890) and later in metasomatic rocks associated with quartz-feldspar veins in Kazakhstan (Shipovalov and Stepanov, 1971).

Caryocerite — see melanocerite.

Cerite. $(Ce,Ca)_9(Mg,Fe^{2+})Si_7(O,OH,F)_{28}$. Trigonal. A Ce silicate containing up to 70 wt.% of the Ce earths. Found in pegmatites and hydrothermal veins in alkali syenites and granites. The REE data of Glass et al. (1958) demonstrate the Ce-selective composition of the mineral.

Chevkinite. $(Ca,Ce,Th)_4(Fe^{2+},Mg)_2(Ti,Fe^{3+})_3Si_4O_{22}$. Monoclinic. It is closely related, both structurally and chemically, to perrierite. Chevkinite is usually found in pegmatites but has also been reported in igneous rocks and volcanic ash. It is strongly Ce-selective and most analyses show Ce > La. However, at the Norwegian occurrences detailed by Segalstad and Larsen (1978a) La is consistently greater than Ce. In one pegmatite (Bjørkedalen, Stokkøya) chevkinite and perrierite both occur and, although not in co-existence, give nearly identical REE distributions.

Eudialyte. $(Ca,Na,Ce)_5(Zr,Fe)_2Si_6(O,OH,Cl)_{20}$. Trigonal. A silicate of Na, Ca, and Zr, in which isomorphous replacement of Na by Ca or the REE is accompanied by simultaneous replacement of Zr by Fe or partial replacement of oxygen by OH and Cl. It has a complex rare earth composition, up to 6.4 wt.% RE_2O_3 having been recorded, usually with the Ce groups as the major REE, but invariably associated with significant Y (e.g. Edgar and Blackburn, 1972). Eudialyte occurs chiefly in nepheline syenites and their pegmatites in association with microcline, nepheline, aegirine, and other minerals.

Gadolinite. $(Y,RE)_2Fe^{2+}Be_2Si_2O_{10}$. Monoclinic. Gadolinite has been shown by Ito and Hafner (1974) to be part of the ternary system $Fe^{2+}RE_2^{3+}Be_2Si_2O_{10}$ (gadolinite)-$Fe^{3+}RE^{3+}CaBe_2Si_2O_{10}$ (calciogadolinite)-$H_2RE_2^{3+}Be_2Si_2O_{10}$. Most analyses fall in the gadolinite area of the system. It occurs chiefly in granites and granite pegmatites, often associated with fluorite and allanite. In naturally occurring gadolinites Y is usually the dominant element of those in 8-fold co-ordination, but gadolinites enriched in the larger REE have been reported. Gadolinite-(Ce) from Bjϕrkedalen, Norway, contains up to 24.5 wt.% Ce_2O_3 (Segalstad and Larsen, 1978b). Vainshtein et al. (1960) showed that the REE distribution in gadolinite from deposits of different genetic types depends on the conditions of their formation. Those most enriched in Y are associated with granitic and alkalic pegmatites and the richest in Ce are the accessory gadolinites. It is therefore one of the minerals showing complex REE compositions. Plots of the chondrite-normalized REE contents of the gadolinite from Hundholmen, North Norway (Nilssen, 1973) show near linearity, from a low La ratio ($\sim 1.2 \times 10^3$) to a high Lu ratio ($\sim 2.5 \times 10^5$) but with a distinct negative Eu anomaly. The data of Gibson and Ehlmann (1970) on gadolinite from Rode Ranch, Texas, are also characterized by a negative Eu anomaly but with higher ratios for the larger REE and lower ratios for the smaller REE, and a maximum ratio at Tb. Segalstad and Larsen's gadolinite-(Ce) shows the reverse composition trend, but again a negative Eu anomaly. This anomaly is thought to result from the incorporation of Eu^{2+}, which has a similar ionic radius to Ca^{2+}, in the early-crystallizing plagioclase.

Hellandite. $(Ca,Y)_6(Al,Fe^{3+})Si_4B_4O_{20}(OH)_4$. Monoclinic. The mineral is a borosilicate of Ca and the REE. It has been recorded from three localities: originally in much altered crystals from granite pegmatites at Kragerϕ, Norway; from the Predazzo granite, Italy; and from a rare earth pegmatite near Wakefield Lake, Quebec. Available analyses suggest a variable degree of Ca-REE substitution but $(Y + RE)_2O_3$ concentrations of up to 46 wt.% have been recorded. Hellandite from all three localities is enriched in Y, but the REE distributions are somewhat different (Hogarth et al., 1972). The Canadian and Norwegian occurrences show Yb maxima and fairly regular Y-group selectivity, while the most abundant REE in the Italian occurrence is Gd. *Tadzhikite*, $Ca_3(Ce,Y)_2(Ti,Al,Fe^{3+})B_4Si_4O_{22}$, from aegirine-quartz-microcline pegmatite veins in Tadzhikstan, U.S.S.R., is structurally similar, and chemically related, to hellandite but shows an intermediate REE composition with a high concentration of Y as well as major Ce-group REE (Yefimov et al., 1970).

Iimoriite. $Y_5(SiO_4)_3(OH)_3$. Triclinic. Occurs in quartz-microcline pegmatites in Fukushima Prefecture, Japan, and is believed to form through the alteration of thalenite. Strongly Y-selective with $(Y + RE)_2O_3$ near 66 wt.% (Kato and Nagashima, 1970).

Ilimaussite. $Ba_2Na_4CeFeNb_2Si_8O_{28} \cdot 5H_2O$. Hexagonal. A RE-Nb-Ba silicate found in a hydrothermal vein in a nepheline-sodalite syenite of the Ilimaussaq alkaline massif, Greenland (Semenov et al., 1968: RE_2O_3 10.6 wt.%), and in a nepheline-aegirine-feldspar section of a pegmatite from the Khibiny pluton, U.S.S.R. Both occurrences give highly ceric REE distributions.

Ilmajokite. $(Na,Ce,Ba)_2TiSi_3O_5(OH)_{10} \cdot nH_2O$. Monoclinic. Forms bright yellow granular crusts in cavities in pegmatites of the Lovozero tundra, Kola Peninsula, U.S.S.R. (Bussen et al., 1972). The RE_2O_3 content approaches 5 wt.%. with the Ce group dominant.

Iraqite. $(K,La,Ce,Th)(Ca,Na,La)_2Si_8O_{20}$. Hexagonal. A rare earth member of the ekanite group occurring in northern Iraq in a coarsely crystalline granite in contact with dolomitic marble (Livingstone et al., 1976). The REE are principally of the Ce group, with RE_2O_3 15.06 wt.% and La_2O_3 6.78% as the maximum REE.

Joaquinite. $Ba_2NaCe_2Fe^{2+}(Ti,Nb)_2Si_8O_{26}(OH,F) \cdot H_2O$. Monoclinic. A rather rare mineral, occurring at three known localities: in natrolite veins with neptunite and benitoite in San Benito Co., California; in fenitized gneisses and alkaline syenites at Seal Lake, Quebec; and in nepheline-sodalite syenitic pegmatites from the Ilimaussaq alkaline massif, South Greenland. Analyses of specimens from the three localities (Semenov et al., 1967a) show a variation in rare earth content from 11.5 to 22.59 wt.% RE_2O_3 with the Ce group dominant.

Johnstrupite — see mosandrite.

Kainosite (cenosite). $Ca_2(Y,RE)_2(Si_4O_{12})CO_3 \cdot H_2O$. Orthorhombic. A rare earth silicate-carbonate with around 38 wt.% $(Y + RE)_2O_3$, found in a few granite pegmatites. The several available analyses confirm it as an Y-selective mineral (e.g. Pouliot et al., 1964).

Karnasurtite. $(Ce,La,Th)(Ti,Nb)(Al,Fe^{3+})(Si,P)_2O_7(OH)_4 \cdot 3H_2O$. Hexagonal (?) but invariably metamict. It occurs in a zoned pegmatite stock in the Lovozero alkaline massif (Kuzmenko and Kozhanov, 1959). Crystalline after heating to 900°C giving an X-ray pattern close to huttonite. The available analyses give Ce-rich compositions with RE_2O_3 contents of near 17 wt.%.

Laplandite. $Na_4CeTiPSi_7O_{22} \cdot 5H_2O$. Orthorhombic. Found as radiating fibrous deposits in the natrolite zone of Jubilee pegmatite, Mt. Karnasurt, Kola Peninsula, U.S.S.R. (Es'kova et al., 1974a). The RE_2O_3 content of 16.79 wt.% is dominated by REE of the Ce group.

Lovchorrite — see mosandrite.

Melanocerite. $Ce_4CaBSi_2O_{12}(OH)$. Hexagonal, but invariably metamict. Caryocerite is thought to be a thorian melanocerite and both minerals have been tentatively allied with the apatite group. Melanocerite occurs in alkali pegmatites. Rare earth analyses are given for material from a fenite of the Burpala alkaline intrusion by Portnov et al. (1969) (43.5 wt.% RE_2O_3) and for caryocerite from Siberia (56.4 wt.%) by Vasil'yeva and Kupriyanova (1972). Both minerals show a preponderance of the Ce earths.

Mosandrite (rinkite, rinkolite, lovchorrite, johnstrupite). $(Na,Ca,Ce)_3Ti(SiO_4)_2F$. Triclinic but frequently metamict (Sahama and Hytönen, 1957). Mosandrite and related minerals constitute a group characteristic of nepheline-syenite pegmatites. The REE are generally of the Ce group but several occurrences have shown significant Y.

Nordite. $Na_3Ce(Sr,Ca)(Mn,Mg,Fe,Zn)_2Si_6O_{18}$. Orthorhombic. A highly selective Ce mineral with RE_2O_3 near 20 wt.%. Occurs in sodalite syenites and pegmatites in the Lovozero alkaline massif, Kola Peninsula (Bakakin et al., 1970). Highly unstable under surface conditions and invariably found coated with rare earth hydroxides and Sr carbonate-ancylite.

Perrierite. $(Ca,Ce,Th)_4(Mg,Fe^{2+})_2(Ti,Fe^{3+})_3Si_4O_{22}$. Monoclinic. Closely related structurally and chemically to chevkinite and is usually found in pegmatites but has also been reported in igneous rocks and volcanic ash. Strongly Ce-selective, usually with Ce > La but a Norwegian occurrence (Bjørkedalen, Stokkøya) described by Segalstad and Larsen (1978a) has La greater than Ce. In this pegmatite both perrierite and chevkinite occur, but not together; the two minerals at this occurrence have nearly identical REE distributions.

Rinkite and rinkolite — see mosandrite.

Rowlandite — see thalenite.

Sazhinite. $Na_3CeSi_6O_{15} \cdot 6H_2O$. Orthorhombic. Ce-rich with RE_2O_3 near 21 wt.%. It is found in the Lovozero alkaline massif; in fine-grained aggregates, rims around steenstrupine, and intergrown with neptunite (Es'kova et al., 1974b). Formed under hydrothermal conditions from solutions saturated with Na.

Semenovite. $(Ca,Ce,La,Na)_{10-12}(Fe^{2+},Mn)(Si,Be)_{20}(O,OH,F)_{48}$. Orthorhombic. Found in cavities and fractures in albitites of the Ilimaussaq alkaline intrusive, South Greenland. The available analyses suggest a complex rare earth distribution (RE_2O_3 up to 19 wt.%) with substantial Ce earths but also 2.3% Y (Mazzi et al., 1979).

Spencite — see tritomite-spencite

Sphene (titanite). $CaTiSiO_5$. Monoclinic. A widespread mineral occurring chiefly as an accessory component in igneous rocks; also in schists, gneisses, and other metamorphic rocks. It is an important concentrator of the rare earths. For instance, Staatz et al. (1977) found that sphenes from four different areas in west-central Alaska contained from 20,350 to 39,180 ppm total REE. They also concluded, as have several others, that the sphene structure will accommodate the REE available when the mineral crystallized, without appreciable fractionation (see also Chapter 1). As well as occurring in minor concentrations in most sphenes, the REE occur in major quantities in keilhauite (yttrotitanite), a variety with up to 12 wt.% $(Y,Ce)_2O_3$. Keilhauite from its type locality (Büo, South Norway) is enriched in the Y group, whereas most other sphenes with lower REE are Ce-rich. Marchenko et al. (1969) found an example of yttrotitanite partly replaced by betafite, both with very similar REE distributions, the Ce-Nd maxima having survived the alteration process. Studying sphenes from different geological environments Fleischer (1978b) showed that in the sequence alkaline pegmatite, alkaline rock, gabbro and pyroxenite, granodiorite, granite and granite pegmatite, the average contents of the lightest REE decrease and the contents of intermediate and heaviest REE and Y increase.

Steenstrupine. $(Ce,La,Na,Mn)_6(Si,P)_6O_{18}(OH)$. Trigonal. A characteristic mineral of highly agpaitic pegmatites of nepheline-sodalite syenites. Occurs in the Julianhaab District, Greenland, and in the Lovozero Massif, U.S.S.R. The total REE contents reported vary between 20 and 30 wt.% and are governed by the isomorphous substitution $Na^+ + Ce^{3+} \rightarrow 2(Ca,Mn)^{2+}$. The LREE predominate (Makovicky and Karup-Møller, 1981).

Stillwellite. $(Ce,La,Ca)BSiO_5$. Trigonal. Occurs in association with allanite and garnet in northwest Queensland as a metasomatic replacement of metamorphosed calcareous sediments (McAndrew and Scott, 1955); also from several other localities. The mineral is Ce-rich with the HREE constituting less than 5% of the total REE.

Tadzhikite — see hellandite.

Thalenite. $Y_3Si_3O_{10}(OH)$. Monoclinic; structurally analogous to thortveitite. A Y silicate containing up to 65 wt.% $(Y + RE)_2O_3$. *Yttrialite* is a variety with Th and U replacing some of the REE; *rowlandite* is a fluoride-bearing variety. The minerals are found in granite pegmatites. In Teller Co., Colorado, crystals of thalenite are often partly enclosed in allanite and both minerals appear to be extensively replaced by fluorite (Adams et al., 1962). This occurrence and that of yttrialite from Iveland, South Norway, reported by Nilssen (1971) show broadly similar HREE enrichment.

Thorite. $ThSiO_4$. Tetragonal but usually metamict. The REE are generally present in thorite to the extent of a few percent (Frondel, 1958) with the Ce earths usually dominant. However, Staatz et al. (1976) have recorded up to 20 wt.% RE_2O_3 in thorite from the Seerie pegmatite, Colorado. In each of the various types of thorite found in this pegmatite the Y group predominates. Thorite is widespread as a primary mineral, chiefly in pegmatites, metasomatized zones in impure limestones, hydrothermal veins, and in detrital deposits.

Thortveitite. $(Sc,Y)_2Si_2O_7$. Monoclinic. A Sc silicate in which part of the Sc is isomorphously replaced by Y and the HREE (up to about 17 wt.% $(Y + RE)_2O_3$). Several of the occurrences show very pronounced Yb maxima leading Adams (1969) to suggest that the mineral probably represents the ultimate in REE fractionation. It occurs in granite pegmatites, often associated with euxenite, ilmenorutile, monazite, beryl, and biotite.

Tombarthite. $Y_4(Si,H_4)_4O_{12-x}(OH)_{4+2x}$. Monoclinic. A hydrous rare earth silicate found associated with perthitic alkali feldspar and thalenite in the Högetveit pegmatite dyke, Evje, Norway. Structurally related to monazite but unusual in that it is enriched in Y and the HREE, whereas other monazite-type minerals are enriched in the LREE and Th. The REE distribution obtained from the data of Neumann and Nilssen (1968) is similar to the distribution obtained for yttrialite (Nilssen, 1971) from a related pegmatite in this area of southern Norway.

Törnebohmite. $(Ce,La)_3Si_2O_8OH$. Hexagonal. A Ce silicate usually found associated with cerite. It is thought to be an intermediate phase in the alteration of cerite to allanite (Geijer, 1921).

Tranquillityite. $Fe_8(Zr,Y)_2Ti_3Si_3O_{24}$. Hexagonal. Found as thin laths in lunar basaltic rocks (Lovering et al., 1971). The grains analysed showed $(Y + RE)_2O_3$ variation up to 5.5 wt.%.

Tritomite-spencite. $(Ce,La,Y,Th)_5(Si,B)_3(O,OH,F)_{13}$-$(Y,Ca,La,Fe)_5(Si,B,Al)_3(O,OH,F)_{13}$. Hexagonal but invariably metamict. These minerals are boro-silicates of Y and the rare earths. Tritomite occurs in nepheline-syenite pegmatites in southern Norway and spencite (now known as tritomite-(Y) following Levinson's (1966) proposals for the nomenclature of rare earth minerals) in granite pegmatites at several North American localities and in a Russian nepheline-syenite pegmatite. Structurally the minerals have been allied with apatite and, in common with that phosphate, will accept the complete range of rare earth ions into their structure. Hogarth et al. (1973) found total rare earth contents of between 28 and 50 wt.% $(Y + RE)_2O_3$ for spencite and 41.44 wt.% for type tritomite. Their analyses of spencite, and those of other authors, show the mineral to have complex REE compositions with major Y, Ce, and La.

Tundrite. $Na_3(Ce,La)_4(Ti,Nb)_2(SiO_4)_2(CO_3)_3O_4(OH)\cdot 2H_2O$. Triclinic. The mineral was formerly known as titanorhabdophane, having been found originally in nepheline-syenite pegmatites in the Lovozero tundra. A later occurrence (Semenov et al., 1967b) is very unusual in containing two varieties of the mineral, tundrite, with a REE maximum at Ce, and tundrite-(Nd) with an Nd maximum at 45% of the REE content.

Zircon. $ZrSiO_4$. Tetragonal. Zircon is isostructural with xenotime YPO_4, and minor amounts of the REE are thought to enter zircon through the isomorphous substitution $Y^{3+} + P^{5+} \rightleftharpoons Zr^{4+} + Si^{4+}$. It follows that the Y group are the dominant REE in zircon, although Khomyakov and Manukhova (1971) have found that the proportion of heavier REE is greater in zircons from nepheline syenites than from granites (see also Chapter 1).

2.8. Silicates with other anions

Agrellite. $NaCa_2Si_4O_{10}F$. Triclinic. Found in pegmatitic pods and lenses in gneisses of regionally metamorphosed agpaitic rocks in Villedieu Township, Quebec (Gittins et al., 1976). 3.8 wt.% RE_2O_3 replaces Ca, with CeO_2 and Y_2O_3 the dominant rare earth oxides.

Caysichite. $(Y,Ca)_4Si_4O_{10}(CO_3)_3 \cdot 4H_2O$. Orthorhombic. Occurs as a whitish coating lining cavities in granite pegmatite at the Evans-Lou feldspar mine, Quebec (Hogarth et al., 1974). 36.16 wt.% $(Y + RE)_2O_3$ is recorded, with REE of the Y group dominant.

Okanoganite. $(Na,Ca)_3(Y,Ce,Nd,La)_{12}Si_{12}B_2O_{27}F_{14}$. Hexagonal. Found in miarolitic cavities in the arfvedsonite granite phase of the Golden Horn Batholith, Okanogan County, Washington (Boggs, 1980). It has a high REE content (RE_2O_3 64.75 wt.%) and is unusual in containing 20.46% Y_2O_3 but with the LREE being dominant (Ce_2O_3 15.42%).

Phosinaite. $H_2Na_3(Ca,Ce)(SiO_4)(PO_4)$. Orthorhombic. Occurs in alkalic pegmatites on Mt. Koashva, Khibina, and in ussingite veinlets cutting alkaline rocks at Mt. Karnasurt, Lovozero, U.S.S.R. (Kapustin et al., 1974). 13 wt.% RE_2O_3 is reported for the mineral, with the Ce group dominant.

Saryarkite. $Ca(Y,Th)Al_5(SiO_4)_2(PO_4,SO_4)_2(OH)_7 \cdot 6H_2O$. Hexagonal. Found in Devonian propylitized acid effusives and in altered granitic rocks in the U.S.S.R. (Krol et al., 1964). 11 wt.% RE_2O_3 is reported for the mineral, with the Y group dominant.

2.9. Phosphates

Apatite group. $A_5(XO_4)_3(F,OH,Cl)_3$ where A = Ca, Be, Ce, Pb, etc., and X = P, As, V, etc. Hexagonal. The commonest member of the group is fluorapatite, $Ca_5(PO_4)_3F$, and this name is often used synonymously with apatite. Minerals of the apatite group are common accessories in many types of rock and are the most abundant P-bearing minerals. Several members of the group (e.g. belovite and britholite) contain REE as major constituents, but in general fluorapatites contain small but important amounts (usually tenths of a percent), rising to as much as 12 wt.% in apatites from alkaline rocks. The REE occupy Ca positions in the apatite structure, in two distinct sites having 7- and 9-fold co-ordination. According to Fleischer (1965) this should lead to complex distributions with both the Ce and Y groups present. Fleischer and Altschuler (1969) have shown that apatites from mafic, ultramafic and alkaline igneous rocks have dominant Ce-group compositions; those from granitic rocks and granitic pegmatites have a wide range of compositions, from Ce-group dominant to those with a high content of the intermediate and heavy REE. Ce is dominant, but some analyses show maxima at Nd, Gd, Dy, or Yb. The main exception found by Fleischer and Altschuler (1969) are apatites of marine sedimentary origin which are so depleted in Ce that either La or Nd is dominant. This is thought to be due to the depletion of Ce in seawater by oxidation of Ce^{3+} to Ce^{4+} and its removal through precipitation in hydrous Mn oxides; Mn nodules are notably enriched in Ce.

Lee et al. (1973) determined the REE compositions of a number of apatites from hybrid granitoid rocks of the southern Snake Range, Nevada, finding that apatites from the more mafic rocks contained rare earth assemblages richer in the lighter REE.

Puchelt and Emmermann (1976) found that almost all the apatites from twelve German localities showed relatively pronounced negative Ce anomalies when the REE data were plotted on a chondrite-normalized basis. A few also showed negative Eu anomalies. The Ce deficiences were in this case thought to be due to the competitive crystallization of various REE-enriched phases, some of which (monazite, allanite, sphene) preferentially accumulate Ce and its neighbours La and Pr. A negative Eu anomaly is regarded by Puchelt and Emmermann (1976) as an indicator of low f_{O_2} provided that the whole rock is not depleted in Eu and apatite crystallizes early. Under such reducing conditions Eu occurs in the divalent state and will not be admitted to the apatite structure (see also Chapter 1).

Åmli (1975) has shown the relative mobility of the REE in apatite by the exsolution of monazite and xenotime inclusions in apatite from the Gloserheia pegmatite, Norway. The exsolution process led to the depletion of the REE in the apatite around the inclusions.

Belovite. $(Sr,Ce,Na,Ca)_5(PO_4)_3(OH)$. Hexagonal. The mineral is structurally related to apatite through the partial substitution of Sr, Na, and the Ce earths for Ca (Borodin and Kazakova, 1954). Analyses give RE_2O_3 near 24 wt.%. It is found in the Lovozero alkali massif in an alkaline pegmatite, enclosed in ussingite.

Brockite. $(Ca,Th,Ce)PO_4 \cdot H_2O$. Hexagonal, related to rhabdophane through the substitution of Ca and Th for the REE. Found as massive reddish aggregates and as earthy yellow coatings in veins and altered granitic rocks in the Wet Mountains, Colorado. Its rare earth content, 7.96 wt.% RE_2O_3 (Fisher and Meyrowitz, 1962), shows an unusual excess of Nd_2O_3 over CeO_2, La_2O_3 and Y_2O_3.

Churchite (syn. weinschenkite). $YPO_4 \cdot 2H_2O$. Monoclinic. Found originally as a coating on siliceous rocks from a Cornish Cu-bearing hydrothermal vein; since recorded as a supergene mineral from several limonite deposits and pegmatites. It has near to 51 wt.% $(Y + RE)_2O_3$ and is enriched in REE of the Y group but always with a moderate concentration of the Ce earths (e.g. Eshkin et al., 1967).

Florencite (syn. koivinite). $CeAl_3(PO_4)_2(OH)_6$. Trigonal. A member of the crandallite group in which the Ca of crandallite is almost entirely replaced by the Ce earths. The mineral is strongly Ce-selective with RE_2O_3 concentrations of up to 30 wt.%. Although Ce is usually the dominant lanthanide in florencite (e.g. Mel'nikova et al., 1975), an occurrence anomalously low in Ce (Nd > Ce) has been recorded in a weathered chert from Sausalito, California (D.J. Milton and Bastron, 1971). Florencite occurs at a number of localities in, or weathered from, granite pegmatites. In the Kangankunde Hill carbonatite, Malawi (McKie, 1962), it occurs with some of the REE content replaced by 9.0 wt.% SrO.

Monazite. $(Ce,La)PO_4$. Monoclinic. REE distributions have been studied more in monazite than any other rare earth mineral. From synthetic studies Carron et al. (1958) found that orthophosphates of La, Ce, Pr, Nd, Sm and Eu each have a monazite-type structure, whereas those of Ho, Er, Tm, Yb, Lu and Y form with the tetragonal xenotime-type structure. Thus monazites have strongly selective Ce-group compositions, as might be expected for a mineral with the REE in 9-fold co-ordination (Ueda, 1967). Jensen (1967) concludes that the monazite structure will accept REE ions with ionic radii between those of La and Eu, based on the existence of a solid-solution series between $(La,Ce)PO_4$ and $CaTh(PO_4)_2$, and implying that the structure will accommodate ions with radii between those of Ce and Ca. She also concluded, from the data then available, that fractionation is minimal among the larger REE, but increases steadily towards the Lu end of the series.

Monazite is isostructural with cheralite $(RE,Th,Ca,U)(P,Si)O_4$ and with huttonite,

$ThSiO_4$, both of which also have highly ceric REE distributions (Bowles et al., 1980; Pabst and Hutton, 1951). Th is usually present in monazite in substitution for the REE, and up to 13 wt.% ThO_2 has been recorded. The charge balance is thought to be maintained through the simultaneous substitution of Si^{4+} for P^{5+}.

Monazite is very widespread, mainly as a detrital mineral in placer deposits and beach sands; in pegmatites associated with granites and syenites; in metamorphic rocks and vein deposits; and as an accessory mineral in granites, alkaline rocks, and carbonatites. Various authors have correlated the REE distributions in monazites with their geological environment. In particular Murata et al. (1953) found that ten monazites from several localities had a near constant (La + Nd) sum of 42 ± 2 at.%, a near constant Pr figure of 5 ± 1 at.% and a near constant (Ce + Sm + Gd + Y) sum of 53 ± 3 at.%. Von Knorring and Clifford (1960) found that the atomic percentages of the more basic REE (La, Ce and Pr) in pegmatitic monazites from South-West Africa varied between 50 and 80% of the total REE whereas from alkaline environments it exceeded 80%. Similarly, Lee and Bastron (1967) showed that the REE in monazites from a series of related granitic rocks in Nevada became progressively enriched in the light elements La, Ce and Pr with increasing Ca content of the host rock.

Vainshtein et al. (1956) compared the REE composition of monazite from granites and their derived pegmatites. In some examples there was no difference in composition, but in others the pegmatitic monazite tended to be higher in Sm, Gd, Tb and other heavier REE. This difference in composition suggests that during the crystallization of a granitic magma lighter REE tend to be precipitated preferentially so that the residual magma becomes relatively richer in heavier REE. Murata et al. (1959) found the same to be the case in the crystallization of monazites in pegmatites from Minas Gerais, Brazil. Fleischer and Altschuler (1969) collected the available analytical data and concluded that monazites from alkaline rocks and carbonatites are significantly richer in (La + Ce + Pr) than those from granitic rocks and granitic pegmatites.

Gramaccioli and Segalstad (1978) noted La enrichment and Nd depletion in a U- and Th-rich monazite from a South Alpine pegmatite at Piona, Italy. Similarly, Petruk and Owens (1977) found that those monazites from the Mount Pleasant deposit, New Brunswick, which were enriched in ThO_2 and U_3O_8 had reduced Nd_2O_3 contents, whereas those enriched in Gd_2O_3 and Sm_2O_3 had reduced La_2O_3.

Rhabdophane. $(Ce,La)PO_4 \cdot H_2O$. Hexagonal. A rather common secondary mineral, formed by the weathering of bastnäsite, belovite, britholite, steenstrupine, or other REE minerals. RE_2O_3 concentrations of over 60 wt.% have been recorded for the mineral, with strong selectivity for the LREE. Two varieties are known, however, one enriched in Ce and the other in La. Adams (1968) and Hildebrand et al. (1957) have recorded Ce-deficient rhabdophane (rhabdophane-La) and Adams attributes this to the partial fractionation of the REE through the oxidation of Ce^{3+} to Ce^{4+} at some stage prior to the formation of rhabdophane. Van Wambeke (1977) has reported both rhabdophane-La and rhabdophane-Ce in the Karonge rare earth deposits of Burundi. He showed that, in these deposits, a process of rare earth differentiation had led to the formation at the end of the hydrothermal deposition stage of, first, rhabdophane-Ce, followed by rhabdophane-La.

Vitusite. $Na_3(Ce,La,Nd)(PO_4)_2$. Orthorhombic. Occurs in the natrolite zone of an alkaline pegmatite at Mt. Karnasurt, Kola Peninsula, associated with steenstrupine, belovite, neptunite, and sazhinite; also at Ilimaussaq, South Greenland, in a melanocratic nepheline syenite with arfvedsonite, albite, microcline, nepheline, sodalite, and steenstrupine. Vitusite from both localities contains close to 40 wt.% RE_2O_3 (Rønsbo et al., 1979) with the Ce earths dominant.

Xenotime. YPO_4. Tetragonal, isostructural with zircon. Xenotime is the most common mineral in the system YPO_4-$YAsO_4$ (chernovite)-YVO_4 (wakefieldite); all crystallize in the tetragonal system and xenotime is known to form at least a partial solid-solution series with chernovite (Graeser et al., 1973). Xenotime is a widespread and important rare earth mineral, occurring in small amounts in acid and alkaline igneous rocks; in pegmatites, metamorphic rocks, Alpine-type vein deposits and as a detrital mineral. It is Y-selective with REE maxima generally at Yb or Dy, and is frequently found associated with monazite, the latter acting as concentrator for the Ce earths. Khomyakov (1970) has considered the potential of coexisting monazite and xenotime as a geothermometer. There are considerable variations in the HREE assemblages in xenotimes from different occurrences. Lyakhovich (1962) has recorded a Dy maximum in granitic xenotimes and a Yb maximum in pegmatitic xenotime; other authors, however, show that this is not universally applicable (Vlasov, 1966, table 144). Vainshtein et al. (1956) recorded a tendency for LREE enrichment in xenotimes from quartz veins. Jefford (1962) reported high Yb enrichment (11.68 wt.%) at the expense of Y and Gd in xenotime concentrates from a Nigerian biotite granite, whereas Sahama et al. (1973) report a Gd maximum in a pegmatitic xenotime from the Morrua mine, Mozambique. Åmli (1975) has described xenotime from the Gloserheia pegmatite, southern Norway, occurring as inclusions in apatite: the xenotime shows marked REE zoning, with LREE enrichment towards the rims of the grains, arising from REE fractionation between xenotime and the apatite from which it has exsolved.

2.10. Arsenates

Agardite. $(Y,Ca)Cu_6(AsO_4)_3(OH)_6 \cdot 3H_2O$. Hexagonal. The rare earth analogue of mixite, in which the Bi of mixite is replaced by REE and Ca. It was found originally at Ouarzazate, Morocco, in the oxidation zone of a copper deposit; now known from several other localities. The mineral has a complex REE distribution; the Moroccan agardite has 8.7 wt.% $(Y + RE)_2O_3$ and, despite major Y, the REE maximum falls at Nd (Dietrich et al., 1969).

Retzian. $Mn_2Y(AsO_4)(OH)_4$. Orthorhombic. A rare mineral containing 10.3 wt.% $(Y + RE)_2O_3$, found associated with jacobsite in dolomite at the Moss mine, Nordmark, Sweden (Moore, 1967).

2.11. Sulphate

Chukhrovite. $Ca_3(Y,Ce)Al_2(SO_4)F_{13} \cdot 10H_2O$. Cubic. Occurs in the secondary oxidation zone of the Kara-Oba molybdenum-tungsten deposit, central Kazakhstan. It has a complex REE composition (18.00 wt.% RE_2O_3) with Ce as the maximum REE (2.70 wt.% Ce_2O_3) but major Y_2O_3 (7.38 wt.% by difference; Ermilova et al., 1960).

References

Adams, J.W., 1968. Rhabdophane from a rare-earth occurrence, Valley County. *U.S. Geol. Surv., Prof. Paper*, 600-B: 48—51.

Adams, J.W., 1969. Distribution of lanthanides in minerals. *U.S. Geol. Surv., Prof. Paper*, 650-C: 38—44.

Adams, J.W., Hildebrand, F.A. and Havens, R.G., 1962. Thalenite from Teller County, Colorado. *U.S. Geol. Surv., Prof. Paper*, 450-D: 6—8.

Akelin, N.A. and Kazakova, M.Ye., 1963. A new find of gagarinite. *Dokl. Acad. Sci. U.S.S.R., Earth Sci. Sect.*, 149: 111—113.

Aleksiev, E. and Ivanov, I.M., 1970. Rare-Earth Elements in the Sredna Gora Granitoids and the Pegmatites Connected with them in the Region of Koprivshtitsa and Strelcha. *Izv. Geol. Inst., Bulg. Akad. Nauk, Ser. Geokhim., Mineral., Petrogr.*, 19: 5—15 (in Bulgarian).

Åmli, R., 1975. Mineralogy and rare earth geochemistry of apatite and xenotime from the Gloserheia granite pegmatite, Froland, southern Norway. *Am. Mineral.*, 60: 607—620.

Bakakin, V.V., Belov, N.V., Borisov, S.V. and Solovyeva, L.P., 1970. The crystal structure of nordite and its relationship to melilite and datolite-gadolinite. *Am. Mineral.*, 55: 1167—1181.

Bender, M., Broecker, W., Gornitz, V., Middel, U., Kay, R., Sun, S.-S. and Biscaye, P., 1971. Geochemistry of three cores from the East Pacific Rise. *Earth Planet. Sci. Lett.*, 12: 425—433.

Bergstøl, S., Jensen, B.B. and Neumann, H., 1977. Tveitite, a new calcium yttrium fluoride. *Lithos*, 10: 81—87.

Bjørlykke, H., 1937. Scheteligite, a new mineral. *Nor. Geol. Tidsskr.*, 17: 47—49.

Boggs, R.C., 1980. Okanoganite, a new rare-earth borofluorosilicate from the Golden Horn batholith, Okanogan County, Washington. *Am. Mineral.*, 65: 1138- 1142.

Borodin, L.S. and Kazakova, M.E., 1954. Belovite — a new mineral from an alkaline pegmatite. *Dokl. Akad. Nauk S.S.S.R.*, 96: 613—616.

Bowles, J.F.W., Jobbins, E.A. and Young, B.R., 1980. A re-examination of cheralite. *Mineral. Mag.*, 43: 885—888.

Brögger, W.C., 1890. Die Mineralien der Syenitpegmatitgänge der Südnorwegischen Augit- und Nephelinsyenite. *Z. Kristallogr.*, 16: 387—396, 462—467.

Brooks, C.K., Henderson, P. and Rønsbo, J.G., 1981. Rare-earth partition between allanite and glass in the obsidian of Sandy Braes, Northern Ireland. *Mineral. Mag.*, 44: 157—160.

Bulakh, A.G., Kondrat'eva, V.V. and Baranova, E.N., 1961. Carbocernaite — a new rare-earth carbonate. *Mem. All-Union Mineral Soc.*, 90: 42—49 (in Russian).

Bussen, I.V., Gannibal, L.F. and Nedorezova, A.P., 1972. Ilmajokite, a new mineral from the Lovozero Tundra. *Zap. Vses. Mineral. O-va*, 101: 75—79 (in Russian).

Butler, J.R., 1958. Rare earths in some niobate-tantalates. *Mineral. Mag.*, 31: 763—780.

Butler, J.R. and Hall, R., 1960. Chemical variations in members of the fergusonite-formanite series. *Mineral. Mag.*, 32: 392—407.

Carron, M.K., Mrose, M.E. and Murata, K.J., 1958. Relation of ionic radius to structures of rare-earth phosphates, arsenates, and vanadates. *Am. Mineral.*, 43: 985—989.

Chao, G.Y., Mainwaring, P.R. and Baker, J., 1978. Donnayite, $NaCaSr_3Y(CO_3)_6 \cdot 3H_2O$, a new mineral from Mont St. Hilaire, Quebec. *Can. Mineral.*, 16: 335—340.

Condie, K.C. and Lo, H.H., 1971. Trace element geochemistry of the Louis Lake batholith of early Precambrian age, Wyoming. *Geochim. Cosmochim. Acta*, 35: 1099—1119.

Dal Negro, A., Rossi, G. and Tazzoli, V., 1975. The crystal structure of ancylite, $(RE)_x(Ca,Sr)_{2-x}(CO_3)_2(OH)_x \cdot (2-x)\,H_2O$. *Am. Mineral.*, 60: 280—284.

Dietrich, J.E., Orliac, M. and Permingeat, F., 1969. L'agardite, une nouvelle espèce minérale et le problème du chlorotile. *Bull. Soc. Fr. Mineral. Cristallogr.*, 92: 420—434.

Dixon, P. and Wylie, A.W., 1951. An unusual distribution of the lanthanons. *Nature*, 167: 526.

Donnay, G., 1953. Roentgenite, $3CeFCO_3 \cdot 2CaCO_3$, a new mineral from Greenland. *Am. Mineral.*, 38: 868—870.

Donnay, G. and Donnay, J.D.H., 1953. The crystallography of bastnaesite, parisite, roentgenite, and synchisite. *Am. Mineral.*, 38: 932—963.

Donnay, G. and Donnay, J.D.H., 1971. Ewaldite, a new barium calcium carbonate. *Tschermaks Mineral. Petrogr. Mitt.*, 15: 185—200.

Edgar, A.D. and Blackburn, C.E., 1972. Eudialyte from the Kipawa Lake area, Temiscamingue Co., Quebec. *Can. Mineral.*, 11: 554—559.

Ermilova, L.P., Moleva, V.A. and Klevtsova, R.F., 1960. Chukhrovite, a new mineral from central Kazakhstan. *Zap. Vses. Mineral. O-va*, 89: 15—25 (in Russian).

Eshkin, V. Yu., Rudenko, S.A. and Baklanova, T.A., 1967. Churchite from pegmatite veins of the south Urals. *Zap. Vses. Mineral. O-va*, 96: 714—720 (in Russian).

Es'kova, E.M. and Nazarenko, I.I., 1960. Pyrochlore from the Vishnevyye Mountains; its paragenesis and chemical composition. *Tr., Inst. Mineral. Geokhim. Kristallokhim. Redk. Elem., Akad. Nauk S.S.S.R.*, 4: 33—50 (in Russian).

Es'kova, E.M., Semenov, E.I., Khomyakov, A.P., Kazakova, M.E. and Siderenko, O.V., 1974a. Laplandite, a new mineral. *Zap. Vses. Mineral. O-va*, 103:571—575 (in Russian).

Es'kova, E.M., Semenov, E.I., Khomyakov, A.P., Kazakova, M.E. and Shumyatskaya, N.G., 1974b. Sazhinite, a new silicate of sodium and rare earths. *Zap. Vses. Mineral. O-va*, 103: 338—341 (in Russian).

Exley, R.A., 1980. Microprobe studies of REE-rich accessory minerals: implications for Skye granite petrogenesis and REE mobility in hydrothermal systems. *Earth Planet. Sci. Lett.*, 48: 97—110.

Ferris, C.S. and Ruud, C.O., 1971. Brannerite: its occurrences and recognition by microprobe. *Q., Colo. Sch. Mines*, 66: 1—35.

Fisher, F.G. and Meyrowitz, R., 1962. Brockite, a new calcium thorium phosphate from the Wet Mountains, Colorado. *Am. Mineral.*, 47: 1346—1355.

Fitzpatrick, J. and Pabst, A., 1977. Burbankite from the Green River Formation, Wyoming. *Am. Mineral.*, 62: 158—163.

Fleischer, M., 1965. Some aspects of the geochemistry of yttrium and the lanthanides. *Geochim. Cosmochim. Acta*, 29: 755—772.

Fleischer, M., 1966. Rare earths in the aeschynite-priorite series. The status of lyndochite. *Mineral Mag.*, 35: 801—809.

Fleischer, M., 1969. The lanthanide elements in fluorite. *Indian Mineral.*, 10: 36—39.

Fleischer, M., 1978a. Relative proportions of the lanthanides in the minerals of the bastnaesite group. *Can. Mineral.*, 16: 361—363.

Fleischer, M., 1978b. Relation of the relative concentrations of lanthanides in titanite to types of host rocks. *Am. Mineral.*, 63: 869—873.

Fleischer, M. and Altschuler, Z.S., 1969. The relationship of the rare-earth composition of minerals to geological environment. *Geochim. Cosmochim. Acta*, 33: 725—732.

Flink, G., 1901. On the Minerals from Narsarsuk on the Firth of Tunugdliarfik in Southern Greenland. *Medd. Grønl.*, 24: 42—56.

Frondel, C., 1958. Systematic mineralogy of uranium and thorium. *U.S. Geol. Surv., Bull.*, 1064.

Gatehouse, B.M., Grey, I.E., Campbell, I.H. and Kelly, P.R., 1978. The crystal structure of loveringite — a new member of the crichtonite group. *Am. Mineral.*, 63: 28—36.

Gatehouse, B.M., Grey, I.E., and Kelly, P.R., 1979. The crystal structure of davidite. *Am. Mineral.*, 64: 1010—1017.

Geijer, P., 1921. The cerium minerals of Bastnäs at Riddarhyttan. *Sver. Geol. Unders., Arb.*, 14(6): 16—20.

Gerasimovskii, V.V., 1964. Bastnäsite and parisite from northern Prebaikalia. *Mineraly S.S.S.R.*, 15: 194—202.

Gibson, S.J. and Ehlmann, A.J., 1970. Annealing characteristics of metamict gadolinite from Rode Ranch, Texas. *Am. Mineral.*, 55: 288—291.

Gittins, J., Brown, M.G. and Sturman, B.D., 1976. Agrellite, a new rock-forming mineral in regionally metamorphosed agpaitic alkalic rocks. *Can. Mineral.*, 14: 120—126.

Glass, J.J., Evans, H.T., Jr., Carron, M.K. and Hildebrand, F.A., 1958. Cerite from the Mountain Pass, San Bernardino County, California. *Am. Mineral.*, 43: 460—475.

Graeser, S., Schwander, H. and Stalder, H.A., 1973. A solid solution series between xenotime ($YtPO_4$) and chernovite ($YtAsO_4$). *Mineral. Mag.*, 39: 145—151.

Graeser, S., Schwander, H., Hänni, H. and Mattioli, V., 1979. Vigezzite, (Ca,Ce) $(Nb,Ta,Ti)_2O_6$, a new aeschynite-type mineral from the Alps. *Mineral. Mag.*, 43: 459—462.

Graham, A.R., 1955. Cerianite, CeO_2: a new rare-earth oxide mineral. *Am. Mineral.*, 40: 560—564.

Gramaccioli, C.M. and Segalstad, T.V., 1978. A uranium- and thorium-rich monazite from a south-alpine pegmatite at Piona, Italy. *Am. Mineral.*, 63: 757—761.

Gurov, E.P., Gurova, E.P., Loginova, L.G. and Lavitskaya, Yu.A., 1975. Fluocerite from pegmatites and metasomatites of the northwestern part of the Ukrainian Shield. *Zap. Vses. Mineral. O-va*, 104: 455—458.

Harris, D.C., 1972. Carbocernaite, a Canadian occurrence. *Can. Mineral.*, 11: 812—818.

Haskin, L.A. and Haskin, M.A., 1968. Rare-earth elements in the Skaergaard intrusion. *Geochim. Cosmochim. Acta*, 32: 433—447.

Hayton, J.D., 1960. The constitution of davidite. *Econ. Geol.*, 55: 1030—1038.

Herrmann, A.G., 1970. In: K.H. Wedepohl (Editor), *Handbook of Geochemistry, II/5.* Springer-Verlag, Berlin, pp. 39,57-71-D-1.

Hess, H.D. and Trumpour, H.J., 1959. Second occurrence of fersmite. *Am. Mineral.*, 44: 1—8.

Hidden, W.E. and Warren, C.H., 1906. On yttrocrasite, a new yttrium-thorium-uranium titanate. *Am. J. Sci., Ser. 4*, 22: 515—519.

Hildebrand, F.A., Carron, M.K. and Rose, H.J., Jr., 1957. Re-examination of rhabdophane (scovillite) from Salisbury, Connecticut. *Geol. Soc. Am. Bull.*, 68: 1744—1745 (abstract).

Hogarth, D.D., 1977. Classification and nomenclature of the pyrochlore group. *Am. Mineral.*, 62: 403—410.

Hogarth, D.D., Chao, G.Y. and Harris, D.C., 1972. New data on hellandite. *Can. Mineral.*, 11: 760—776.

Hogarth, D.D., Steacy, H.R., Semenov, E.I., Proschenko, E.G., Kazakova, M.E. and Kataeva, Z.T., 1973. New occurrences and data for spencite. *Can. Mineral.*, 12: 66—71.

Hogarth, D.D., Chao, G.Y., Plant, A.G. and Steacy, H.R., 1974. Caysichite, a new silico-carbonate of yttrium and calcium. *Can. Mineral.*, 12: 293—298.

Ito, J. and Hafner, S.S., 1974. Synthesis and study of gadolinites. *Am. Mineral.*, 59: 700—708.

Jacob, K.H., 1974. Ph.D. Thesis. Technical University, Berlin.

Jaffe, H.W., Meyrowitz, R. and Evans, H.T., Jr., 1953. Sahamalite, a new rare earth carbonate mineral. *Am. Mineral.*, 38: 741—754.

Jefford, G., 1962. Xenotime from Rayfield, northern Nigeria. *Am. Mineral.*, 47: 1467—1473.

Jensen, B.B., 1967. Distribution patterns of rare earth elements in cerium-rich minerals. *Nor. Geol. Tidsskr.*, 47: 9—19.

Kapustin, Yu. L., 1972. First find of huanghoite in the U.S.S.R. *Dokl. Acad. Sci. U.S.S.R., Earth Sci. Sect.*, 202: 116—119.

Kapustin, Yu. L., Khomyakov, A.P., Semenov, E.I., Es'kova, E.M., Bykova, A.V. and Pudovkina, Z.V., 1974. Phosinaite, a new rare-earth mineral. *Zap. Vses. Mineral. O-va*, 103: 567—570. (in Russian).

Kato, A. and Nagashima, K., 1970. *Introduction to Japanese Minerals.* pp. 85—86.

Keidel, F.A., Montgomery, A., Wolfe, C.W. and Christian, R.P., 1971. Calcian ancylite from Pennsylvania: new data. *Mineral. Rec.*, 2: 18—25.

Khomyakov, A.P., 1970. Rare-earth minerals as potential geothermometers. *Dokl. Acad. Sci. U.S.S.R., Earth Sci. Sect.*, 191: 182—183.

Khomyakov, A.P. and Manukhova, A.A., 1971. Rare earths in zircon from miaskite. *Dokl. Acad. Sci. U.S.S.R., Earth Sci. Sect.*, 196: 221—223.

Komkov, A.I., 1973. Polymorphic relationships in compounds of $TRNbTiO_6$ and $TRTaTiO_6$ types. *Sov. Phys.-Dokl.*, 17: 959—960.

Krol, O.F., Chernov, V.I., Shipovalov, Yu. V. and Khan, G.A., 1964. Saryarkite, a new mineral. *Zap. Vses. Mineral. O-va*, 93: 147—155 (in Russian).

Kuzmenko, M.V. and Kozhanov, S.I., 1959. A new mineral — karnasurtite. *Trans. Inst. Min., Geochem. Crystallochem. Rare Earths, Acad. Sci. U.S.S.R.*, 2: 95—98 (in Russian).

Lee, D.E. and Bastron, H., 1967. Fractionation of rare-earth elements in allanite and monazite as related to geology of the Mt. Wheeler mine area, Nevada. *Geochim. Cosmochim. Acta*, 31: 339—356.

Lee, D.E., Van Loenen, R.E. and Mays, R.E., 1973. Accessory apatite from hybrid granitoid rocks of the southern Snake Range, Nevada. *J. Res., U.S. Geol. Surv.*, 1: 89—98.

Levinson, A.A., 1966. A system of nomenclature for rare-earth minerals. *Am. Mineral.*, 51: 152—158.

Lindström, G., 1910. Om lantaniten. *Geol. Fören. Stockholm Förh.*, 32: 206—214.

Livingstone, A., Atkin, D., Hutchison, D. and Al-Hermezi, H.M., 1976. Iraqite, a new rare-earth mineral of the ekanite group. *Mineral. Mag.*, 40: 441—445.

Lovering, J.F. et al., 1971. Tranquillityite, a new silicate mineral from Apollo XI and Apollo XII basaltic rocks. *Proc. 2nd Lunar Sci. Conf.*, 1: 39—45.

Lyakhovich, V.V., 1962. Rare earth elements in accessory minerals of granitoids. *Geochemistry (U.S.S.R.)*, 1962: 39—55.

Maaskant, P., Coolen, J.J.M.M.M. and Burke, E.A.J., 1980. Hibonite and coexisting zoisite and clinozoisite in a calc-silicate granulite from southern Tanzania. *Mineral. Mag.*, 43: 995—1003.

Makarochkin, B.A., Yes'kova, Ye. M. and Aleksandrov, V.B., 1963. A rare-earth variety of fersmanite. *Dokl. Acad. Sci. U.S.S.R., Earth Sci. Sect.*, 148: 93—97.

Makovicky, E. and Karup-Møller, S., 1981. Crystalline steenstrupine from Tunugdliarfik in the Ilimaussaq alkaline intrusion, South Greenland. *Neues. Jahrb. Mineral., Abh.*, 140: 300—330.

Maksimovic, Z. and Panto, G., 1978. Minerals of the rare-earth elements in karstic bauxites: synchysite-(Nd), a new mineral from the Grebnik deposit. *Proc. 4th Int. Congr. Study of Bauxites, Alumina and Aluminium, Athens*, 13 pp.

Marchenko, Ye.Ya., Khvostova, V.A. and Chashka, A.I., 1969. The replacement of yttrotitanite by betafite. *Dokl. Acad. Sci. U.S.S.R., Earth Sci. Sect.*, 186: 125—128.

Mazzi, F., Ungaretti, L., Dal Negro, A., Petersen, O.V. and Rönsbo, J.G., 1979. The crystal structure of semenovite. *Am. Mineral.*, 64: 202—210.

McAndrew, J. and Scott, T.R., 1955. Stillwellite, a new rare-earth mineral from Queensland. *Nature*, 176: 509—510.

McKie, D., 1962. Goyazite and florencite from two African carbonatites. *Mineral. Mag.*, 33: 281—297.

Mel'nikova, E.M., Kokarev, G.N. and Kyazeva, D.N., 1975. Florencite from hydrothermal metasomatites of the polar Urals region. *Zap. Vses. Mineral. O-va*, 104: 341—343.

Milton, C., Ingram, B., Clark, J.R. and Dwornik, E.J., 1965. Mckelveyite, a new hydrous sodium barium rare-earth uranium carbonate mineral from the Green River Formation, Wyoming. *Am. Mineral.*, 50: 593—612.

Milton, D.J. and Bastron, H., 1971. Churchite and florencite-(Nd) from Sausalito, California. *Mineral. Rec.*, 2: 166—168.

Mineev, D.A., Larrishcheva, T.I. and Bykova, A.V., 1970. Yttrium bastnäsite — a product of gagarinite alteration. *Zap. Vses. Mineral. O-va*, 99: 328—332.

Moore, P.B., 1967. Crystal chemistry of the basic manganese arsenate minerals, 1. The crystal structures of flinkite, $Mn_2^{2+}Mn^{3+}(OH)_4(AsO_4)$ and retzian, $Mn_2^{2+}Y^{3+}(OH)_4(AsO_4)$. *Am. Mineral.*, 52: 1603—1613.

Moore, P.B., Bennett, J.M. and Louisnathan, S.J., 1969. Ashcroftine is not a zeolite. *Mineral. Mag.*, 37: 515—517.

Murata, K.J., Rose, H.J., Jr., and Carron, M.K., 1953. Systematic variation of rare earths in monazite. *Geochim. Cosmochim. Acta*, 4: 292—300.

Murata, K.J., Rose, H.J., Jr., Carron, M.K. and Glass, J.J., 1957. Systematic variation of rare-earths in cerium-earth minerals. *Geochim. Cosmochim. Acta*, 11: 141—161.

Murata, K.J., Dutra, C.V., DaCosta, M.T. and Branco, J.J.R., 1959. Composition of monazite from pegmatites of eastern Minas Gerais, Brazil. *Geochim. Cosmochim. Acta*, 16: 1—14.

Murdoch, J., 1951. Perovskite. *Am. Mineral.*, 36: 573—580.

Nash, W.P., 1972. Apatite chemistry and phosphorus fungacity in a differentiated igneous intrusion. *Am. Mineral.*, 57: 877—886.

Neumann, H. and Nilssen, B., 1968. Tombarthite, a new mineral from Högetveit, Evje, South Norway. *Lithos*, 1: 113—123.

Neumann, H., Jensen, B.B. and Brunfelt, A.O., 1966. Distribution patterns of rare earth elements in mineral. *Nor. Geol. Tidsskr.*, 46: 141—179.

Nilssen, B., 1970. Samarskites. Chemical composition, formula and crystalline phases produced by heating. *Nor. Geol. Tidsskr.*, 50: 357—373.

Nilssen, B., 1971. Yttrialite from Ivedal, Iveland, South Norway. *Nor. Geol. Tidsskr.*, 51: 1—8.

Nilssen, B., 1973. Gadolinite from Hundholmen, Tysfjord, North Norway. *Nor. Geol. Tidsskr.*, 53: 343—348.

Pabst, A. and Hutton, C.O., 1951. Huttonite, a new monoclinic thorium silicate. *Am. Mineral.*, 36: 60—69.

Pashov, I.N. and Kussovski, G.V., 1971. A mineral of the samarskite group in the pegmatites from the Velingrad area. *Rev. Bulg. Geol. Soc.*, 32: 361—365 (in Bulgarian).

Pavlenko, A.S., Orlova, L.P., Akhmanova, M.V. and Tobelko, K.I., 1965. A thorium fluocarbonate, thorbastnäsite. *Mem. All-Union Mineral. Soc.*, 94: 105—113.

Pecora, W.T. and Kerr, J.H., 1953. Burbankite and calkinsite, two new carbonate minerals from Montana. *Am. Mineral.*, 38: 1169—1183.

Peishan, Z. and Kejie, T., 1981. Zhonghuacerite $Ba_2Ce(CO_3)_3F$ — a new mineral. *Sci. Geol. Sin.*, 4: 196.

Petruk, W. and Owens, D.R., 1977. Monazite from the Mount Pleasant deposit, New Brunswick. *Can. Mineral.*, 15: 295—299.

Portnov, A.M., Sidorenko, G.A., Dubinchuck, V.T., Kuznetsova, N.N. and Ziborova, T.A., 1969. Melanocerite from the North Baikal region. *Dokl. Acad. Sci. U.S.S.R., Earth Sci. Sect.*, 185: 107—109.

Pouliot, G., Maxwell, J.A. and Robinson, S.C. 1964. Cenosite from Bancroft, Ontario. *Can. Mineral.*, 8: 1—10.

Puchelt, H. and Emmermann, R., 1976. Bearing of rare earth patterns of apatites from igneous and metamorphic rocks. *Earth Planet. Sci. Lett.*, 31: 279—286.

Raup, O.B., Gude, A.J., 3rd, Dwornik, E.J., Cuttitta, F. and Rose, H.J., Jr, 1968. Braitschite, a new hydrous calcium rare-earth borate mineral from the Paradox Basin, Grand County, Utah. *Am. Mineral.*, 53: 1081—1095.

Roalsdet, E., 1975. Rare earth element distributions in some Precambrian rocks and their phyllosilicates, Numedal, Norway. *Geochim. Cosmochim. Acta*, 39: 455—469.

Rønsbo, J.G., Khomyakov, A.P., Semenov, E.I., Voronkov, A.A. and Garanin, V.K., 1979. Vitusite, a new phosphate of sodium and rare earths from the Lovozero alkaline massif, Kola, and the Ilimaussaq alkaline intrusion, South Greenland. *Neues Jahrb. Mineral., Abh.*, 137: 42—53.

Sahama, Th.G. and Hytönen, K., 1957. Unit cell of mosandrite, johnstrupite and rinkite. *Geol. Fören. Stockholm Förh.*, 79: 791—796.

Sahama, Th.G., Von Knorring, O. and Lehtinen, M., 1970. Cerotungstite, a cerian analogue to yttrotungstite from Uganda. *Bull. Geol. Soc. Finl.*, 42: 223—228.

Sahama, Th.G., Von Knorring, O. and Rehtijärvi, P., 1973. Xenotime from Morrua, Mozambique. *Bull. Geol. Soc. Finl.*, 45: 67—71.

Sawyer, J., Caro, P. and Eyring, L., 1973. Hydroxycarbonates of the lanthanide elements. *Rev. Chim. Miner.*, 10: 93—103.

Schnetzler, C.C. and Philpotts, J.A., 1968. Partition coefficients of rare-earth elements and barium between igneous matrix material and rock-forming mineral phenocrysts, I. In: L.H. Ahrens (Editor), *Origin and Distribution of the Elements*. Pergamon, Oxford, pp. 929—938.

Schnetzler, C.C. and Philpotts, J.A., 1970. Partition coefficients of rare-earth elements between igneous matrix material and rock-forming mineral phenocrysts, II. *Geochim. Cosmochim. Acta*, 34: 331—340.

Segalstad, T.V. and Larsen, A.O., 1978a. Chevkinite and perrierite from the Oslo region, Norway. *Am. Mineral.*, 63: 499—505.

Segalstad, T.V. and Larsen, A.O., 1978b. Gadolinite-(Ce) from Skien, southwestern Oslo region, Norway. *Am. Mineral.*, 63: 188—195.

Semenov, E.I., 1957. Isomorphism and camouflage of rare earths. *Geochemistry (U.S.S.R.)*, 1957: 735—748.

Semenov, E.I., 1958. Relationship between composition of rare earths and composition and structure of minerals. *Geochemistry (U.S.S.R.)*, 1958: 574—586.

Semenov, E.I., 1963. *Mineralogy of the Rare Earths.* Academy of Sciences U.S.S.R., Moscow, 412 pp. (in Russian).

Semenov, E.I. and Chang, P.-S., 1961. Huanghoite — a new rare-earth mineral. *Sci. Sin.*, 10: 1007—1011.

Semenov, E.I., Bukin, V.I., Balashov, Yu.A. and Sørensen, H., 1967a. Rare earths in minerals of the joaquinite group. *Am. Mineral.*, 52: 1762—1769.

Semenov, E.I., Kazakova, M.E. and Aleksandrova, R.A., 1967b. The Lovozero minerals nenadkevichite, gerasimovskite and tundrite, from Ilimaussaq, South Greenland. *Medd. Grønl.*, 181(5): 3—11.

Semenov, E.I., Kazakova, M.E. and Bukin, V.J., 1968. Ilimaussite, a new rare-earth niobium-barium silicate from Ilimaussaq, South Greenland. *Medd. Grønl.*, 181(7): 3—7.

Shilin, L.L. and Yanchenko, M.T., 1964. Knopite from the apatite-nepheline ores of the Khibiny massif. *Dokl. Acad. Sci. U.S.S.R., Earth Sci. Sect.*, 144: 139—142.

Shipovalov, Yu.V. and Stepanov, A.V., 1971. Cappelenite from Kazakhstan. New data on tengerite. *Issled. Obl. Khim. Fiz. Metod. Anal. Mineral. Syr'ya*, pp. 180—185 (in Russian).

Siivola, J., 1975. The lanthanoid content of some minerals from the Pyörönmaa pegmatite in Kangasala, Finland. *Bull. Geol. Surv. Finl.*, 276: 17 pp.

Sitnin, A.A. and Leonova, T.N., 1961. Loparite — a new accessory mineral of albitized and greisenized granites. *Dokl. Acad. Sci. U.S.S.R., Earth Sci. Sect.*, 140: 1090—1093.

Smellie, J.A.T., Cogger, N. and Herrington, J., 1978. Standards for quantitative microprobe determination of uranium and thorium with additional information on the chemical formula of davidite and euxenite-polycrase. *Chem. Geol.*, 22: 1—10.

Smith, W.L., Stone, J., Ross, D.R. and Levine, H., 1960. Doverite, a possible new yttrium fluocarbonate from Dover, Morris County, New Jersey. *Am. Mineral.*, 45: 92—98.

Staatz, M.H., Adams, J.W. and Wahlberg, J.S., 1976. Brown, yellow, orange, and greenish-black thorites from the Seerie pegmatite, Colorado. *J. Res., U.S. Geol. Surv.*, 4: 575—582.

Staatz, M.H., Conklin, N.M. and Brownfield, I.K., 1977. Rare earths, thorium, and other minor elements in sphene from some plutonic rocks in west-central Alaska. *J. Res., U.S. Geol. Surv.*, 5: 623—628.

Stepanov, A.V. and Severov, E.A., 1961. Gagarinite — a new rare-earth mineral. *Dokl. Acad. Sci. U.S.S.R., Earth Sci. Sect.*, 141: 1290—1293.

Ueda, T., 1967. Reexamination of the crystal structure of monazite. *J., Jpn. Assoc. Mineral., Petrol. Econ. Geol.*, 58: 170—179.

Vainshtein, E.E., Tugarinov, A.I. and Turanskaya, N.V., 1956. Regularities in the distribution of rare earths in certain minerals. *Geochemistry (U.S.S.R.)*, 1956: 159—178.

Vainshtein, E.E., Aleksandrova, I.T. and Turanskaya, N.V., 1960. Composition of the rare earths in gadolinite from deposits of different genetic types. *Geochemistry (U.S.S.R.)*, 1960: 596—603.

Van Wambeke, L., 1977. The Karonge rare earth deposits, Republic of Burundi: new mineralogical-geochemical data and origin of the mineralization. *Miner. Deposita*, 12: 373—380.

Vasil'yeva, V.V. and Kupriyanova, I.I., 1972. New data on caryocerite. *Dokl. Acad. Sci. U.S.S.R., Earth Sci. Sect.*, 203: 141—144.

Vlasov, K.A., 1966. *Geochemistry and Mineralogy of Rare Elements and Genetic Types of their Deposits, II. Mineralogy of Rare Elements.* Israel Program for Scientific Translation, Jerusalem.

Von Knorring, O. and Clifford, T.N., 1960. On a skarn monazite occurrence from the Namib desert near Usakos, S.W. Africa. *Mineral. Mag.*, 32: 650—653.

White, A.J.R., Chappell, B.W. and Jakeš, P., 1972. Coexisting clinopyroxene, garnet and amphibole from an "eclogite", Kakanui, New Zealand. *Contrib. Mineral. Petrol.*, 34: 185—191.

Yefimov, A.F., Dusmatov, V.D., Alkhazov, V.Yu., Pudovkina, Z.G. and Kazakova, M.Ye., 1970. Tadzhikite, a new rare-earth borosilicate of the hellandite group. *Dokl. Acad. Sci. U.S.S.R., Earth Sci. Sect.*, 195: 136—139.

Chapter 3

COSMOCHEMISTRY OF THE RARE EARTH ELEMENTS: METEORITE STUDIES

WILLIAM V. BOYNTON

3.1. Introduction

Study of the cosmochemistry of the REE can be said to have started with the early work of Noddack (1935), who determined the abundances of the REE in a composite mixture of meteorites using X-ray spectroscopy. No other work followed, however, until the pioneering work by Schmitt et al. (1960, 1963, 1964), who measured all fourteen REE plus Sc and Y in a wide variety of meteorites by neutron activation analysis. The early studies were aimed at determining the solar or cosmic abundances of the REE, but since that time numerous studies, both theoretical and experimental, have been undertaken to understand other aspects of the cosmochemistry of the REE.

Meteorites provide our best samples of primitive solar system materials. Many of them, particularly the chondrites, appear to have escaped igneous processing, and have preserved evidence of the earliest events responsible for the formation of the solar system. Other meteorites provide examples of igneous processing in systems different from those on the Earth and provide insight into the later processes that occurred on the meteorite parent bodies. This review will first provide a theoretical basis for our understanding of REE patterns related to nebular processes; it will then describe the evidence for such processes in the meteorites, comment on what the evidence may tell us about the formation of the solar system and finally describe the later parent body process.

Before discussing REE cosmochemistry we will very briefly describe current thoughts about the origin of the solar system, in order to provide a framework for considering the cosmochemical processes of interest. Our solar system is thought to have formed out of a large cloud of gas and dust (the pre-solar cloud). The cloud collapsed due to its internal gravity, perhaps initiated by shock waves from some source such as a nearby supernova. As it collapsed, the cloud flattened into a disk and became hot from the release of gravitational energy. Peak temperatures in the disk (the solar nebula) presumably increased with decreasing radial distance while the sun was forming at the center of the disk. The rise in temperature caused a considerable fraction of the dust to vaporize. The actual extent of the temperature rise is a subject of much debate and will be discussed below. Later the gas cooled, forming grains which agglomerated together. These agglomerations gradually grew by mutual accretion to form ever larger bodies, which eventually became planets, satellites, asteroids or comets.

Many of these bodies became warm enough to cause at least partial melting of the silicates, resulting in differentiated bodies. Differentiated bodies from which we have samples include the Earth, the Moon, and parent bodies of the achondrites and iron meteorites. The REE are very useful for studying igneous fractionation processes, such as partial melting and fractional crystallization. These studies can provide information about the petrogenesis of the sample studied, or they can be applied to large-scale planetary processes and properties. The study of REE in lunar samples has been reviewed by Taylor (1975, 1982) and will not be discussed in this review.

Most of the undifferentiated meteorites are referred to as chondrites, which are made up of metal, silicates and sulfides and contain abundances of non-volatile elements in nearly solar proportions. Because REE patterns are commonly normalized to chondrites, their classification will be briefly discussed. The classification into chemical groups, as proposed by Van Schmus and Wood (1967) and modified by Wasson (1974), is presented in Table 3.1. The designation consists of the chemical group, followed by a number referred to as the petrologic type, although the CI and CM meteorites are often cited without the number, since only one petrologic type is known. The CI chondrites are most nearly solar in composition and are generally used for normalization values in applications where one is concerned with the amount of an element in a sample relative to the solar nebula from

TABLE 3.1

Classification of chondrites into chemical groups

Common name	Group	Known petrologic types	Characteristics
Carbonaceous	CI	CI1	no chondrules, high carbon content, most nearly solar in composition
	CM	CM2	well-defined chondrules, high carbon content
	CO	CO3, CO4	small chondrules, Al/Si <0.11
	CV	CV2, CV3	large chondrules, Al/Si >0.11
Ordinary	LL	LL3—LL6	low total Fe, low metal
	L	L3—L6	low total Fe
	H	H3—H6	high total Fe
Enstatite	E	E4—E6	highly reduced silicates with nearly no oxidized Fe

which it formed. In most applications using REE, however, the important aspect of the normalization is the ratio between the REE. Traditionally, abundances in a mixture of chondrites from several classes are used (see section 3.5) and, except for when specifically designated otherwise, the average chondrite values of Wakita et al. (1971) will be used in this chapter.

3.2. Condensation theory

Condensation from a gas of solar composition

Several investigators have considered theoretical models for the condensation of elements from the solar nebula (Lord, 1965; Larimer, 1967; Grossman, 1972). These models have all assumed that the gas was initially at a sufficiently high temperature that all the elements were vaporized, that the gas had solar composition, and that the grains formed in thermodynamic equilibrium with the gas as it cooled. In a nebula of 10^{-3} atm total pressure, the first phase to condense is corundum, Al_2O_3, at 1758 K, followed by perovskite, $CaTiO_3$, at 1647 K; melilite, $Ca_2(Al_2,MgSi)SiO_7$, at 1625 K and spinel, $MgAl_2O_4$, at 1513 K. The most abundant phases, metallic iron and forsterite, Mg_2SiO_4, do not condense until 1473 K and 1444 K, respectively.

Grossman (1973) made similar calculations for a variety of refractory trace elements, assuming that they condensed as pure phases. Because refractory trace elements will condense at higher temperatures by forming solid solutions with major minerals, these calculations yield only a lower limit to the temperature at which the element will begin to condense from the gas. It was found that the trace elements which had condensation temperatures higher than 1473 K, the condensation temperature of iron, were enriched (relative to chondritic abundances) in Ca,Al-rich inclusions in the Allende meteorite. These inclusions will be discussed in more detail later, but for now it is sufficient to note that they have mineral and major element composition consistent with that expected of high-temperature condensates from the solar nebula. Although Grossman (1973) purposely ignored the condensation behavior of the REE, he noted that the REE were also enriched in the inclusions and suggested that they also condensed at high temperatures. Grossman and Ganapathy (1976a), however, calculated the condensation temperature at 10^{-3} atm total pressure for pure REE sesquioxides and found they ranged from 1590 K for Er to 1140 K for Eu. Only Dy, Ho, Er and Lu had condensation temperatures high enough to have started condensation at the point where the Allende Ca,Al-rich inclusions were presumably isolated from the gas. The fact that all of the REE are observed to be present in nearly their solar proportions in the inclusions demonstrates the inadequacy of pure-phase trace element condensation temperatures.

Because the REE will condense in solid solution with other elements, it is difficult to calculate an accurate condensation temperature. In addition, because at least a fraction of each of the REE will condense as soon as any host phase is present, the concept of a trace element condensation temperature that represents the onset of condensation has little significance. Generally, the temperature of 50% condensation, or the relative solid/gas distribution coefficient (discussed below), are better measures of the relative volatility of a trace element.

In order to calculate the amount of a REE condensed at a given temperature, it is necessary to first calculate the thermodynamic activity, *a*, of the element in some reference state, usually the oxide. Given sufficient thermodynamic data, the activity can be calculated from:

$$a_{LaO_{1.5}} = K \cdot P_{LaO} \left(\frac{P_{H_2O}}{P_{H_2}}\right)^{1/2} \tag{3.1}$$

where P is the partial pressure of the gaseous species and K is the equilibrium constant for the condensation reaction:

$$LaO_{(g)} + \tfrac{1}{2}\,H_2O_{(g)} \rightarrow LaO_{1.5\,(s)} + \tfrac{1}{2}\,H_{2(g)} \tag{3.2}$$

Then, to calculate the concentration of the element, it is necessary to know its activity coefficient in every phase present. The activity coefficient, γ, relates the thermodynamic activity of the element in the reference state to the mole fraction of the element, X, as:

$$a = \gamma \cdot X \tag{3.3}$$

The activity coefficient can be considered to be a measure of the ease with which the element can be accommodated by the host crystal lattice. For example, because of the similarity in size and valence, Ho_2O_3 is expected to be easily accommodated in Y_2O_3, and its activity coefficient in this phase should be near unity. However, in most common minerals the REE are considered incompatible elements and their activity coefficients in these minerals will be much greater than unity.

The difficulty in calculating the condensation temperature arises because the activity coefficients of the REE are unknown in any of the condensate phases. Grossman and Ganapathy (1976a) also calculated the fraction of REE condensed in the Ca,Al-rich inclusions, with the assumption that the REE condensed in perovskite with activity coefficients of unity. (This arbitrary assumption is often referred to as "ideal solid solution".) The result predicted that Eu still should not have been present in the Allende inclusions, contrary to observation. In an earlier work, however, Boynton (1975a) noted that much can be learned from studies of relative abundances of REE, even if the absolute abundances cannot be calculated at a given temperature. He noted that because the REE have very similar properties with respect to substitution in minerals, one could consider as a first approximation that the REE activity coefficients are identical; the absolute values are not required to calculate REE ratios. A second approximation was suggested, that the REE activity coefficient ratios vary as a smooth function of size, the variation being related to the mineral host. Boynton calculated relative solid/gas distribution coefficients at 1650 K, the approximate temperature of perovskite condensation at 10^{-3} atm. Perovskite is known to be a favorable host for REE (Borodin and Barinskii, 1960) and has been suggested as the most likely host of the REE in condensate minerals (Grossman, 1973). The activity coefficient ratios required to make a good fit between calculated REE abundances and abundances observed in a Ca,Al-rich inclusion by Tanaka and Masuda (1973) suggested that the host mineral showed a preference for light REE over heavy. This relationship was shown to be consistent with that expected for perovskite and was supported by subsequent measurements of perovskite/liquid distribution coefficients (Nagasawa et al., 1980a).

The relative solid/gas distribution coefficient, D_{M}, relates the element, M, to La ratio in the gas to the ratio in the solid according to:

$$\left(\frac{M}{\mathrm{La}}\right)_{(\mathrm{s})} = D_{\mathrm{M}} \left(\frac{M}{\mathrm{La}}\right)_{(\mathrm{g})} \tag{3.4}$$

The values for D_{M} can be calculated from thermodynamic data and knowledge of the oxygen fugacity. The latter can be readily calculated as a function of pressure and temperature with knowledge of the composition of the gas (which is usually assumed to be solar in composition). Most of the REE exist in the gas as monoxides, and for these elements the relative distribution coefficients are independent of oxygen fugacity. Condensation in gases of non-solar composition, which affect the oxygen fugacity, will be discussed in the next section.

Presentation of the thermodynamic data in the form of distribution coefficients is useful for calculating the REE patterns of condensates. The use of distribution coefficients allows the equations used for calculating REE patterns from partial melting or fractional crystallization to be used for their cosmochemical equivalents of partial vaporization or fractional condensation. By choosing the fractional amount condensed for one element, the amount condensed for each of the other REE can easily be determined. In addition, the distribution coefficients do not change much with temperature and can generally be considered constant over the temperature range of condensation. For example, at 10^{-3} atm total pressure in the nebula, significant condensation begins at about 1650 K, when perovskite appears, and ends at around 1500 K. Over this temperature range, the log of the D values change by only

10%. Thus, one unfamiliar with cosmochemical condensation calculations can easily calculate REE patterns in nebular condensates.

The distribution coefficients calculated by Boynton (1975a) are given in Table 3.2 and Fig. 3.1. The value for Ce is from Boynton (1978a) and reflects the contribution of CeO_2 in the gas which was ignored in the earlier work. Also given in Table 3.2 is the fraction of the gaseous element in the form of the monoxide, F_{MO}. The alternate gaseous form is the monatomic vapor except for Ce, which is CeO_2. As can be seen, the distribution coefficients span a very wide range, over 7 orders of magnitude. This range is much larger than that of conventional solid/liquid distribution coefficients, which seldom exceeds a factor of 100 and is generally less than a factor of 10. In addition, conventional solid/liquid distribution coefficients are always a smooth function of size, except for divalent Eu, but the solid/gas distribution coefficients show many irregularities. For example, one of the most volatile REE, Yb, differs by nearly a factor of 10^5 from the adjacent REE, Lu. As mentioned above, the relative amount of a REE that will condense is determined by both the intrinsic volatility of the element as measured by the solid/gas distribution coefficients, and by the ability of the element to substitute into the host crystal as measured by the activity coefficients. It should be noted that because the activity coefficient ratios are a smooth function of size, except for divalent Eu, the irregularities in the distribution coefficients may be seen in the REE pattern of a nebular condensate. It should also be noted here that Grossman and co-workers (Grossman and Ganapathy, 1976a; Davis and Grossman, 1979) have criticized this use of activity coefficients in the condensation calculations; a discussion of their

TABLE 3.2

Fraction of element present as gaseous monoxides and solid/gas distribution coefficients in the solar nebula at 1550 K

	F_{MO}	D
La	1.000	1.00
Ce	0.294	0.47
Pr	1.000	2.6
Nd	1.000	0.75
Sm	0.80	1.3
Eu	0.34	0.0014
Gd	1.000	65
Tb	1.000	200
Dy	0.954	350
Ho	0.975	380
Er	0.967	4300
Tm	0.51	640
Yb	$<1.5 \times 10^{-6}$	0.54
Lu	1.000	32,000

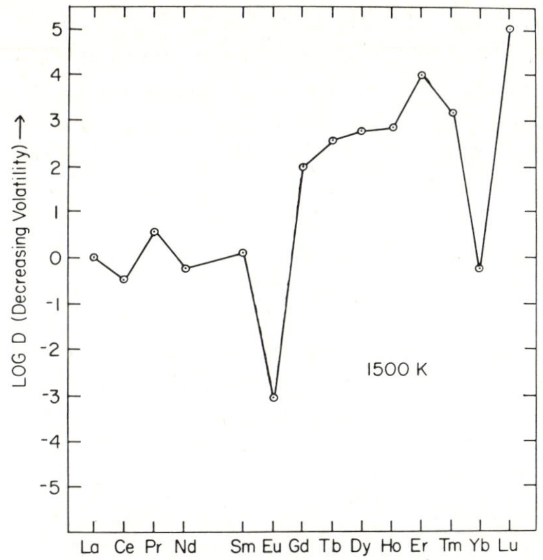

Fig. 3.1. REE solid/gas distribution coefficients; note greatly compressed logarithmic scale. These distribution coefficients determine the order in which the elements condense from the solar nebula and provide a very effective means of fractionating the REE in a way that is not a smooth function of size.

criticism will be deferred until the section on REE data in Ca,Al-rich inclusions.

Condensation in non-solar environments

An interesting aspect of the condensation of REE is that the relative volatility of several of the REE is a function of the oxygen fugacity in the gas (Boynton, 1978a). The reason for this relationship can be seen by considering the following set of equations:

$$LaO_{(g)} + \tfrac{1}{4}\,O_{2(g)} \rightarrow LaO_{1.5(s)} \qquad (3.5)$$

$$CeO_{2(g)} \rightarrow CeO_{1.5(s)} + \tfrac{1}{4}O_{2(g)} \qquad (3.6)$$

$$Yb_{(g)} + \tfrac{3}{4}\,O_{2(g)} \rightarrow YbO_{1.5(s)} \qquad (3.7)$$

Most of the REE exist in the solar nebula as monoxides, and reaction (3.5) describes their condensation behavior; increasing the oxygen fugacity will drive the reaction to the right, making these elements more refractory. However, Ce, which is present in significant amounts in the form of gaseous CeO_2, will actually become more volatile if the oxygen fugacity is increased. For Yb, which is present almost exclusively as the monatomic vapor, an increase in oxygen has a much larger effect than it does for the other REE. Thus, when normalized to La, most of the REE have relative volatilities that are independent of oxygen fugacity, but Ce becomes much more volatile and Yb much less volatile with an increase in oxygen. From inspection of Table 3.2, it can be seen that Sm, Eu and Tm have significant amounts of monatomic vapor present, and they should behave like Yb. This is true for environments which are more reducing than the solar nebula, but in more oxidizing environments, the reaction:

$$Tm_{(g)} + \tfrac{1}{2}\,O_{2(g)} \rightarrow TmO_{(g)} \qquad (3.8)$$

converts the monatomic vapor to monoxide and these elements then behave in the same way as La. Similarly, Ce will become much more volatile than La in an oxidizing environment, but it will be converted from CeO_2 to CeO in a reducing gas and will then behave like La.

The dependence of REE volatility on oxygen fugacity is important because several astrophysical environments with very different oxygen fugacities may have been important in creating grains that were present in our solar system. Several investigators (Hoyle and Wickramasinghe, 1970; Clayton, 1975; Lattimer et al., 1978) have suggested that grains can form in ejecta from a supernova. Such an environment can be highly oxidizing because the hydrogen, which is a strong reducing agent in a gas of solar composition, has been converted to helium, or beyond to carbon and oxygen. The large excess of oxygen over hydrogen yields a highly oxidizing gas. However, under certain conditions, more carbon than oxygen will be produced and the oxygen will be tied up as carbon monoxide, with the excess carbon present as graphite. Such an environment is much more reducing than is a gas of solar composition. Lattimer et al. (1978) give detailed calculations of the temperatures at which various minerals condense in a supernova. They find that the same minerals form in an oxidizing region of the supernova as during condensation in the solar nebula, e.g., perovskite, melilite, and corundum. Thus, these minerals, which are stable in the solar nebula, may have been preserved if peak nebular temperatures were not sufficient to vaporize them, and they may have preserved evidence of their formation in a highly oxidizing environment.

Even within our solar system, regions with variable oxygen fugacity have been proposed. Larimer (1975) has suggested that regions with high C/O ratios will yield the very low oxygen fugacities needed to explain the highly reduced enstatite chondrites. Herndon and Wilkening (1978) noted that oxidizing conditions can obtain if chondritic matter is vaporized after being removed from a large fraction of the gaseous components of the solar nebula. Such an environment has been suggested for the formation of chondrules, small spherical droplets present in chondritic meteorites.

3.3. REE abundances in Ca,Al-rich inclusions

By far the most work in recent years on REE abundances in chondrites has been with the Ca,Al-rich inclusions (CAI) found in carbonaceous chondrites. As mentioned earlier, they are thought to be early condensates from the solar nebula. Many of the inclusions are made up of the minerals expected, based on equilibrium thermodynamics, to be present in early condensates, but the most persuasive arguments for their origin as condensates come from studies of trace elements. In most of the inclusions a wide variety of refractory trace elements are enriched rather uniformly to abundances of approximately 17 times those in CI chondrites, suggesting that these inclusions represent the first 6% (= 100/17) of the elements to condense in the solar nebula (Grossman et al., 1977). The nearly uniform enrichment of the refractory trace elements, which include the REE, suggests that these elements are totally condensed; the enrichment factor is determined by the amount of dilution from major elements, which condense after the refractory elements. Although most investigators consider these inclusions to be early condensates, an equally plausible interpretation is that they are residues from incomplete vaporization of grains from the pre-solar cloud (Chou et al., 1976).

With only a few exceptions all of the CAI studied are from one meteorite, Allende, a member of the CV3 class of carbonaceous chondrites. Unless otherwise stated, all data in this section will refer to Allende CAI. The literature on the CAI is a little confusing because of various nomenclatures that have been used. The Allende inclusions can be grouped into two major types: coarse-grained chondrules and fine-grained aggregates (Clarke et al., 1971). Chondrules are rounded or subrounded, range in size from 3 to 25 mm and have grain sizes of the order of 1 mm. Aggregates are generally very fine-grained, highly irregular in shape and vary in size from 0.1 to 10 mm. Often the chondrules are referred to as coarse-grained inclusions, and the aggregates as fine-grained inclusions. Some of the confusion exists because some of the aggregates are somewhat coarser-grained than the others and some investigators group them with the coarse-grained inclusions, thus mixing aggregates and chondrules together. For example, the low-Rb aggregate of Gray et al. (1973) was re-classified as a coarse-grained inclusion by Grossman and Ganapathy (1975). Both the coarse-grained and the fine-grained inclusions can be sub-divided further. Grossman (1975) grouped the coarse-grained inclusions (including aggregates) into two types: type A have minor amounts of pyroxene, which occurs as diopside, whereas type B have major amounts of pyroxene ($>35\%$), which occurs as a Ti,Al-rich clinopyroxene called fassaite. The characteristics of the coarse-grained inclusions have recently been reviewed by Grossman (1980). The sub-divisions of the fine-grained aggregates are based on their REE patterns and will be discussed below.

Coarse-grained (Group I) inclusions

The abundances of REE in coarse-grained inclusions are generally enriched uniformly around 12 to 15 times chondrites; occasionally, samples have a small (20%) positive anomaly in Eu. Examples of this type have been analyzed for REE by many investigators (Gast et al., 1970; Grossman, 1973; Tanaka and Masuda, 1973; Martin and Mason, 1974; Osborn et al., 1974; Wänke et al., 1974; Grossman and Ganapathy, 1976b; Chou et al., 1976; Conard, 1976; Davis et al., 1978; Taylor and Mason, 1978; Nagasawa et al., 1981). A typical analysis from Tanaka and Masuda (1973) is plotted in Fig. 3.2. This type of REE pattern is referred to as Group I by Mason and co-workers (Martin and Mason, 1974; Mason and Martin, 1977; Taylor and Mason, 1978), who use REE patterns as major classification parameters.

The slight increase in chondrite-normalized abundance from La to Lu observed in the Group I REE patterns in Fig. 3.2 may represent a sampling

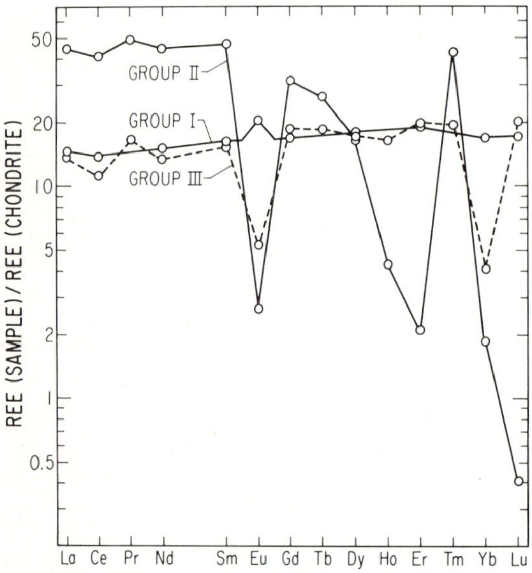

Fig. 3.2. Typical REE patterns found in Allende Ca,Al-rich inclusions. The Group I patterns are generally found in coarse-grained chondrules and represent nearly total condensation of REE; the enrichment to around 15 times chondritic abundances is due to the absence of Si, Mg and Fe, which condense at lower temperature and dilute the REE in chondrites. The Group II and Group III patterns are generally found in fine-grained aggregates. The Group III aggregates are deficient in the two most volatile REE, Eu and Yb, because they were isolated from the gas before Eu and Yb were fully condensed. The Group II aggregates are deficient in the highly refractory REE as well as in Eu and Yb. This material is thought to have condensed from a gas following the condensation of an ultra-refractory condensate which removed most of the refractory REE. Group I pattern from Tanaka and Masuda (1973); Groups II and III from Conard (1976).

problem, since the entire inclusion was almost certainly not analyzed. The enrichment cannot be caused by an incomplete condensation of the more volatile REE because Yb, which is more volatile than La, is enriched over La by the amount expected from interpolating between Er and Lu. Mason and Martin (1974), Onuma et al. (1974) and Nagasawa et al. (1977) have measured REE abundances in separated minerals from coarse-grained Allende inclusions, and both find similar results. Except for Eu, the REE are enriched in the pyroxene relative to the melilite, the enrichment increasing as a smooth function of size towards the HREE. The ratio of REE between the phases agrees very well with calculated pyroxene/melilite distribution coefficients, suggesting that the REE have equilibrated between the phases (Nagasawa et al., 1977). Thus, the enrichment in HREE could represent a preferential sampling of pyroxene over melilite in the material taken for analysis. However, as suggested by Grossman et al. (1977), various ratios of pyroxenes and melilite cannot account for the positive Eu anomaly, as the samples which are the most enriched in Eu are those most deficient in La, just the opposite of what would be expected from the addition of melilite, since melilite is enriched in both Eu and La.

Nagasawa et al. (1977) found that an anorthite separate contained about twice the Eu concentration of the bulk sample but negligible amounts of the other REE. Thus, anorthite can contribute Eu to the bulk sample without affecting the other REE. However, a 20% excess of Eu requires a 20% excess of anorthite, an amount that seems most unlikely considering that the inclusion studied by Nagasawa et al. (1977) had only 5% anorthite. In addition, the average of ten coarse-grained inclusions analyzed by Mason and Martin (1977) also show a positive anomaly, suggesting that sampling of the inclusions for analysis is not a likely explanation for the Eu excess. It should be noted that the average of ten inclusions also shows no significant slope from light to heavy REE, supporting the suggestion that a sampling problem is the explanation for the HREE enrichment shown in Fig. 3.2.

It is likely, therefore, that the positive Eu anomaly represents a real enrichment in the whole inclusion relative to the other REE. This result can be explained by condensation, since Eu is much more volatile than the other REE (Fig. 3.1). If some grains were removed from the nebula after most of the REE had condensed, the remaining gas would be over-abundant in Eu relative to the grains and continued condensation would yield an enrichment of the more volatile elements. It may seem rather ad hoc to invoke such a mechanism, but as we shall see from the following section, there is much evidence suggesting that there were multiple episodes of condensation and isolation of grains from the gas. There are also many examples of inclusions depleted in Eu, presumably because they were isolated from the gas before Eu had fully condensed, suggesting that, indeed, Eu-rich gases must have been present. However, these samples (the Group III inclusions) generally show Yb depleted to the same extent as Eu. If these materials were responsible for enriching the gas in Eu, they should also have enriched the gas in Yb, and Yb should also be enriched in the coarse-grained Group I inclusions, which is contrary to observations.

Another possible explanation of the positive Eu anomaly is that the whole inclusion was composed of various condensate grains which clumped together. The grains in the rounded coarse-grained inclusions probably partially melted during or after the aggregation. Different grains that may have been present in the gas may have contributed different amounts of the various trace elements. Grossman et al. (1977) tried to determine which elements may have been added simultaneously, and concluded that Ca, Sr and Eu were added together in the same phase, which is probably melilite. Although the abundances of refractory trace elements averaged over all the grains present would have uniform enrichments, variable amounts of different component grains could lead to variable enrichments of different elements. The implication of this model is that only a statistically small number of grains must have been sampled to make the inclusions. Because the anomalies are about 20%, no more than 100 grains of each type could be sampled. If many more grains had been aggregated together to form the inclusions, the grains would have been a good statistical sample of the grains present in the gas, and the inclusions would then have had uniform enrichments. Thus, according to this model, the centimeter-sized inclusions must have been made from millimeter-sized grains. It is doubtful that such large grains would form directly from the nebula.

Another difficulty of this model is that a mixture of a small number of condensate grains should contribute other volatility-related irregularities. If several different minerals contain REE, it is unlikely that they started condensing at exactly the same time. The grains which started to condense later would be deficient in the most refractory REE, such as Lu and Er, and enriched in more volatile REE, such as Yb. Non-representative sampling of these grains would be expected to provide volatility-related anomalies between Yb and Lu, which is again contrary to observation.

A solution to this difficulty is to suggest that the coarse-grained chondrules are secondary objects formed from pre-existing coarse-grained material. If equilibration between phases had occurred in the pre-existing material, then the components which mixed together to form the inclusions would have contributed REE and other trace elements according to mineral preference effects rather than volatility effects. The bulk composition of the pre-existing material would, of course, be determined by the volatility of the elements. Several of the models for the formation of ordinary chondrules are based on pre-existing material (Wasson, 1974, p. 198f; Kieffer, 1975; Dodd, 1978). A serious constraint on this model is the presence of ^{26}Al ($t_{1/2}$ = 7.4 × 10^5 years) in the Allende inclusions (Lee et al., 1976), which requires that the inclusions must have formed within a few million years of the last nucleosynthetic event.

Fine-grained (Groups II and III) inclusions

Most of the fine-grained Ca,Al-rich aggregates in Allende can be divided into two groups, called Group II and Group III by Martin and Mason (1974). Typical examples of the REE patterns in these groups are plotted in Fig. 3.2. Analyses of REE in fine-grained aggregates have been reported by many groups (Tanaka and Masuda, 1973; Nagasawa et al., 1977; Conard, 1976; Grossman and Ganapathy, 1976b; Mason and Martin, 1977; Taylor and Mason, 1978). The Group II patterns are characterized by strong enrichment of light REE relative to heavy, accompanied by large anomalies at Eu, Tm and Yb. In addition, smaller anomalies among the LREE are often observed. It was noted by Tanaka and Masuda (1973), who first discovered these peculiar inclusions, that although the Eu and Yb anomalies are in the opposite direction, their chondrite-normalized abundances are similar. Subsequent analyses have shown this observation to be generally true. The Group III REE patterns are quite flat except for depletions at Eu and Yb; again, the chondrite-normalized abundances of Eu and Yb are usually identical in a given sample, but the depletion relative to the other REE can vary widely between various samples.

Models for the formation of these REE patterns have been discussed by several workers. Boynton (1975a) and Davis and Grossman (1979) discussed detailed thermodynamic calculations and applied them mainly to the Group II inclusions, while Boynton (1978b) gave a less rigorous discussion of both Group II and III. The Group III patterns are a bit easier to understand and will be discussed first. The uniform enrichment of most of the REE suggests that these elements are totally condensed, while Eu and Yb are volatile enough to have only partially condensed. Examination of Fig. 3.1 shows that Eu and Yb are indeed the most volatile REE, which is consistent with the above suggestion. However, in detail, there are problems with the model because Yb is only slightly more volatile than the LREE, and Eu is much more volatile than Yb. Yet in the Group III (and Group II) inclusions the Eu/Yb ratio is generally chondritic (Taylor and Mason, 1978; Davis and Grossman, 1979), suggesting that these elements have equal volatility. In addition, La/Yb ratios as large as 50 have been found (Nagasawa et al., 1977), suggesting that Yb is much more volatile than La.

As mentioned above, the activity coefficients indicate that the early condensates show a preference for the light REE relative to the heavy. This preference will tend to increase the volatility of Yb relative to La. Davis and Grossman (1979) made an alternative suggestion to increase the volatility of Yb relative to the LREE by proposing that the condensation took place in a more highly reduced region of the nebula. In section 3.1 it was shown that Yb will indeed become more volatile in a reducing environment, but Sm shows a similar effect, and one would expect a significant Sm depletion in those samples with a large Yb depletion. The Sm depletions are not observed, suggesting that only slightly more reducing conditions can be tolerated. It is possible that a combination of activity coefficient effects and reducing conditions are responsible for the enhanced volatility of Yb.

It is unlikely, however, that these effects would be such that Yb would have a volatility precisely equal to that of Eu. A more reasonable suggestion is that Eu and Yb are sufficiently more volatile than the other REE so that they will not have condensed significantly when the remaining REE have totally condensed. At this time, if the grains are partially separated from the gas (e.g. by settling to the median plane of the nebula) and condensation continues, Eu and Yb can condense totally and maintain their solar proportions. They will be depleted relative to the other REE to the extent that a portion of the original gas was lost. An alternative mechanism in which Eu and Yb were added with a later volatile-rich component was suggested by Davis and Grossman (1979). In either case a rather abrupt separation of grains from gas seems to be required.

The Group II inclusions have an even more complicated history. The REE pattern is unlike any previous pattern found from studies of terrestrial, lunar or meteoritic samples. Patterns generated by igneous processes often yield familiar Eu abundance anomalies due to the stability of Eu in the divalent state. Although one could argue that under sufficiently reducing conditions Yb could also become divalent, it is not possible to reduce Tm to the divalent state without reducing several other REE. Clearly, the Tm anomaly in the Group II inclusions cannot be made by igneous fractionation.

If these inclusions were the first condensates to form from the solar nebula, the REE solid/gas distribution coefficients (Fig. 3.1) would suggest that Lu and Er, the two most refractory REE, would be the first REE to condense and would be the most enriched. This suggestion is contrary to observation; in fact, the opposite appears to be the case. The more refractory elements are the most depleted and the more volatile elements are the most enriched; the only exceptions to this observation are Eu and Yb, which will be discussed later. Boynton (1975a) noted this effect and suggested that the Group II inclusions may have formed from the gas remaining after an earlier, more-refractory condensate formed and depleted the gas in the most refractory REE. A plot of REE in a typical Group II inclusion is shown in Fig. 3.3, with calculated values based on the REE solid/gas distribution coefficients. The calculations assume that the initial condensate has an activity coefficient relationship which shows a preference for light REE relative to heavy. The calculated values shown are not for the initial highly-refractory condensate, but rather for the gas remaining after the initial condensate formed and was isolated from the solar nebula. It can be seen that within the errors in the distribution coefficients (approximately a factor of two to three) the theoretical REE pattern agrees well with the observed pattern; even the peculiar Tm anomaly is explained. Davis and Grossman (1979) present plots of REE data from twenty individual Group II REE

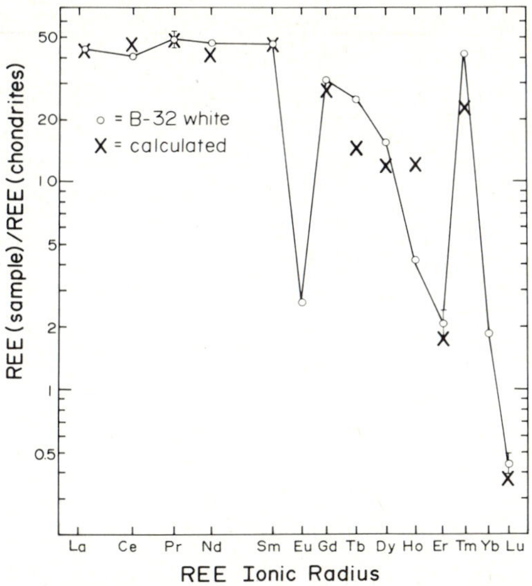

Fig. 3.3. A Group II REE pattern with calculated REE abundances. The values calculated based on REE volatility account very well for the peculiar fractionation pattern. The Tm anomaly is expected in a gas/solid fractionation, but not in any igneous, or parent body, fractionations; it provides strong evidence that this material condensed from the solar nebula.

patterns with calculated values shown as in Fig. 3.3. It is remarkable how well the calculations fit the variety of observed Group II REE patterns.

Calculated values for Eu and Yb have not been plotted in Fig. 3.3 because these elements are so volatile that they did not condense into the Group II inclusions. They are presumably depleted in these inclusions by a mechanism similar to that suggested for the Group III inclusions. Thus, the REE in the Group II inclusions represent the middle of the REE condensation sequence; the more refractory REE were removed by an initial condensate that was isolated from the gas, and the more volatile REE were left behind in the gas from which the Group II inclusion was isolated.

Boynton (1978b) discussed the implications of these REE patterns. He noted that they gave strong evidence for multiple episodes of condensation and isolation from the gas. The isolation mechanism must have been very efficient; it is quite common for Lu to be depleted by a factor of 100 relative to La. This depletion requires that 99% of the Lu was removed and isolated from the solar nebula before the second condensation episode, the one which produced the grains that made the Group II inclusions. The Lu must have remained isolated throughout the entire formation period of the Group II inclusions.

The Group II inclusions are also significant in that they are the only samples which are unequivocally condensates. As mentioned earlier, many of the Ca,Al-rich inclusions have a trace element pattern that is consistent with having an origin from either condensation or partial vaporization. If the formation process occurs at equilibrium, the results will be identical. The Group II inclusions, with their strange abundance anomalies, require a gas/solid fractionation process to generate their REE patterns. Furthermore, because of the inverse relationship between abundance and volatility, the Group II inclusions must have been made from material that was originally the gas component of the fractionation. The fact that these REE were in the gas phase places strong constraints on the temperature reached in the region of the solar nebula where these inclusions formed.

Without knowing the amount of the various host phases and the REE activity coefficient in these phases, one cannot calculate the temperature precisely. However, a lower limit to the temperature can be calculated by assuming that the REE condense as pure phases; condensation in solution with other phases can only raise the temperature. The calculations for the Group II patterns require that essentially all the LREE are in the gas phase. Grossman and Ganapathy (1976a) calculated condensation temperatures of pure REE oxides in a solar nebula of 10^{-3} atm and found that the LREE began to condense at temperatures from 1312 to 1353 K. Thus, 1300 K can be taken as a lower temperature limit if the nebula was at 10^{-3} atm. At 10^{-6} atm, about the lowest pressure possible to account for the mass of the planets in the solar system, this temperature limit will drop to about 1150 K.

This temperature of 1150 K is a very rigorous limit. The REE will almost certainly condense out at higher temperatures by forming solutions with other phases. Wasson (1978) suggested that if the REE condensed in perovskite, the limit at 10^{-6} atm is 1440 K. This temperature is a reasonable, although less rigorous, estimate to the lower limit of the peak temperatures achieved in this part of the solar nebula.

The temperature of 1440 K is difficult to understand in terms of proposed models for the formation of the solar nebula (e.g., Cameron, 1978). Although there is sufficient gravitational energy to heat the solar nebula during the collapse, the process is generally thought to be slow enough that the heat is quickly radiated away. Currently, the evidence for high temperatures in parts of the solar nebula is very strong; finding a physical environment in which these high temperatures can occur still remains a problem.

Other models for the solar nebula, in which the gas of the nebula is very hot but the grains are very cool, have been suggested (e.g., Arrhenius and Alfven, 1971; Alfven and Arrhenius, 1976). Under these conditions the gas is hot enough that most elements are ionized; the elements are thought to condense in the order in which they became neutral, which is determined largely by their ionization potentials. Boynton (1975b) noted that the ionization potentials of the REE vary as a smooth function of size except for large anomalies at Gd and Lu. Condensation under such non-thermal conditions as proposed by Arrhenius and co-workers should result in samples with large abundance anomalies at Gd and Lu. Such samples have not been found. The Group II inclusions argue strongly for condensation under conditions near or at thermodynamic equilibrium.

Grossman and Ganapathy (1976b) have proposed another model for the Group II and Group III inclusions. They suggested that Group II and Group III inclusions all belonged to the same class of aggregates with similar condensation temperatures. They suggested that the different inclusions could be formed by incorporating different proportions of several phases, each of which preferentially accepted different REE into its structure during condensation. The gas from which these grains condensed was thought to have a composition intermediate between the Group II and Group III compositions. This gas was that which remained after the coarse-grained inclusions condensed and were isolated from the solar nebula. The amount of a given element left behind in the gas was determined not by thermodynamic properties of the element but by a property called the accretion efficiency of the element. No explanation for different accretion efficiencies of individual REE was suggested. The model has difficulty explaining how an assemblage of many fine grains, with a composition determined by non-equilibrium processes, can be sampled to produce an assemblage with solar abundances of REE (the Group III inclusions), and an assemblage that mimics an equilibrium fractionation (the Group II inclusions).

Davis and Grossman (1979) agreed with the model of Boynton (1975a) for the formation of the Group II REE patterns, but suggested two changes in the method of calculations. They did not allow for non-ideal solid solution, but preferred to assume that the activity coefficients were equal to unity. (Such values are not excluded in the original Boynton (1975a) model, of course; it was simply shown that, with the inclusion studied, the fit was better with non-ideal solutions.) Davis and Grossman also allowed for a second REE component with chondritic abundance ratios in the Group II inclusions, as suggested by Tanaka and Masuda (1973). The chondritic component can often account for a large fraction of Lu, the most depleted element, but generally contributes a negligible amount of the LREE. The two-component, ideal-solution model was found to have a slightly better fit than the one-component, non-ideal solution model (a mean deviation of 8.0% vs. 11.4%) when only the elements Gd, Tb, Dy, Ho, Er and Lu were considered in twenty Group II inclusions. The ideal model, however, did not provide a good fit for Tm.

Davis and Grossman (1979) noted that the non-ideal, one-component model required large differences in the ratio of the activity coefficients between Lu and Sm. The ratio was found to vary from unity to 2093. Such a variation is not expected if only one mineral phase, such as perovskite, is the host for the REE. Even if, as now seems likely, other phases such as hibonite, with a different activity coefficient versus size relationship, were an additional host for REE, such a change in the activity ratio seems large.

Davis and Grossman (1979) concluded that such a variable ratio argued strongly against non-ideal solid solutions and they preferred their two-component, ideal-solution model. However, because they changed two parameters between the models (ideal vs. non-ideal and one vs. two components), they had no way to determine which parameter accounted for the slight improvement in the fit for the subset of REE which they considered. They failed to realize that all of the variability in the activity coefficient ratio can be attributed to the absence of the chondritic component in the test of the non-ideal calculations. Clearly, a two-component, non-ideal model would give a better fit to the data than would the two-component, ideal model, even if the Lu/Sm activity ratio were fixed at some value. The ideal model is simply a special case of the non-ideal model in which the ratio is fixed at unity.

The presence of the chondritic component was first suggested by Tanaka and Masuda (1973), who showed that the REE abundances in one of their inclusions could be duplicated almost exactly by a mixture of 55% of another inclusion and 45% chondritic material. How this chondritic component, which Davis and Grossman (1978) show to be present in most Group II aggregates, came to be mixed into the inclusions is still a mystery. It provides one more example of how complicated was the formation of the solar system.

Mason and Martin (1977) have suggested a few other groups of Ca,Al-rich inclusions in Allende. They designate Group V inclusions as those inclusions which have flat REE patterns with no significant anomalies. The Group VI inclusions are those with flat REE patterns, except for positive Eu and Yb anomalies. Apparently the Eu and Yb fractionation that occurs relative to the other REE in the Group III inclusions can be either enrichments or depletions. There does not appear to be enough data in the literature to determine whether these groups are statistically significant, or whether they represent a continuum from Group III though Group V to Group VI.

The Group IV inclusions of Mason and Martin (1977) are not Ca,Al-rich inclusions. They generally have fairly flat chondrite-normalized REE patterns at values of about 3—4. The patterns are similar to the amoeboid olivine aggregates of Grossman et al. (1979), who also measured REE abundances. The Group IV inclusions may represent later stages in the condensation sequence after olivine condensed and diluted the refractory trace elements.

None of the inclusions discussed so far have the REE pattern which is complementary to that of the Group II inclusions. In fact, the author is aware of 96 analyses of Allende Ca,Al-rich inclusions (including 11 unpublished analyses by H. Palme), in which at least a partial REE pattern was determined, and none of them contains the REE pattern expected of the hypothetical initial condensate that is complementary to the Group II pattern.

Inclusions from other carbonaceous chondrites: ultra-refractory condensates

Relatively few Ca,Al-rich inclusions from carbonaceous chondrites other than Allende have been studied for REE abundances. Examples of the ultra-refractory component apparently missing from Allende have been observed, however. The only fairly complete REE pattern available is of an inclusion, MH-115, from the Murchison meteorite, a member of the CM class of carbonaceous chondrites (Boynton et al., 1980). The REE pattern is given in Fig. 3.4. The sample is strongly enriched in the heavy REE, but the enrichment is clearly not a smooth function of size. The most volatile elements, Yb and Eu, are strongly depleted, while Ce and Tm are moderately depleted relative to their neighbors. As can be seen from Fig. 3.4, the abundances agree well with those calculated using the solid/gas distribution coefficients. The abundances of Eu and Yb are higher than expected based on the solid/gas distribution coefficients and, as with the Allende fine-grained aggregates, the elements are present in their solar ratios. As was noted for Allende aggregates, this observation suggests that either the grains may have settled or otherwise become partially isolated from the gas, or Eu and Yb may have been added later by an additional component.

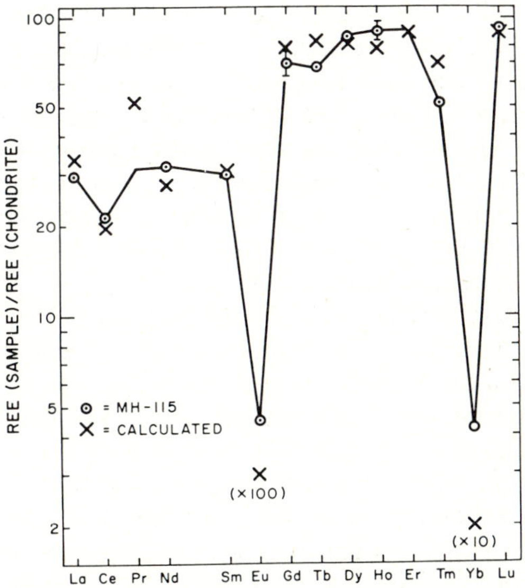

Fig. 3.4. The missing ultra-refractory component has not been found in Allende, but an example has been found in Murchison (Boynton et al., 1980). It is similar to that which may have depleted the gas from which the Group II Allende inclusions formed. This inclusion may represent the first 1% of the solid matter to condense from the solar nebula. Again, the anomalies are expected from a nebula fractionation but not from a planetary fractionation.

Although inclusion MH-115 is not exactly complementary to the Group II inclusions, it has all of the right properties. It is enriched in the highly refractory REE, those depleted in the Group II inclusions, and is depleted in the LREE and Tm. Similar material, isolated at a slightly earlier stage in the condensation sequence, would be a very good example of the Group II complementary material.

The Ca,Al-rich inclusions in Murchison have been the subject of several mineralogical investigations and consist mainly of spinel and hibonite, with small amounts of perovskite (Fuchs et al., 1973; Macdougall, 1979; MacPherson et al., 1980). These minerals are more refractory than those found in the Allende inclusions; this fact supports the conclusion that MH-115 is made of ultra-refractory materials. The enrichment of about a factor of 100 in the most refractory REE, Dy, Ho and Lu, suggests that MH-115 may represent the first 1% of the total condensation sequence.

Tanaka et al. (1980) have analyzed similar inclusions from Murchison but found different REE patterns. One inclusion has a typical Group III REE pattern with the REE enriched to 17 times chondritic abundances except for Eu and Yb, which are enriched to only 11 and 14 times chondritic abundances, respectively. Another sample, actually two inclusions that were mixed together, had a typical Group II pattern. The only exception is that Gd was nearly as depleted as Eu. This result was found by instrumental neutron activation analysis (Grossman et al., 1977), and confirmed in the above work by radiochemical analysis. The possibility that the result was due to a variation in the isotopic ratio of Gd (Gd was the only element determined via a *p*-process nuclide, ^{153}Gd) was dismissed. They concluded that the Gd anomaly was due to chemical processes. Another possibility not considered by Tanaka et al. (1980) is that a reaction interference in the high flux reactor used for the analysis tends to make large amounts of ^{153}Gd from ^{151}Eu (Kramar, 1980; Boynton, 1979); this interference would make ^{153}Gd a measure of Eu in the sample and would account for the apparent Gd depletion.

Kurat (1975) determined Y, Dy and Gd with an electron microprobe in perovskite from a Ca,Al-rich inclusion from Lancé, a carbonaceous chondrite from the class CO3. The chondrite-normalized enrichments range from 3000 to about 6000. These data are too incomplete to tell much about the origin of the inclusion, but they do indicate that perovskite is a favorable host for REE. Noonan et al. (1977) measured the REE with an ion microprobe in a Zr-Y oxide found in an inclusion from Ornans, also a CO3 chondrite. The LREE were undetectable (<100 times chondritic), but the HREE had enrichment factors from 5000 times chondritic abundances for Gd to 35,000 for Er. The more volatile elements Tm and Yb are enriched by factors of 9000 and <2000, respectively. Except for the most refractory REE, Lu, which had an enrichment factor of 9000, all these abundances are in direct proportion to their solid/gas distribution coefficients. This sample may also be an example of an early ultra-refractory condensate. Why Lu is not the most enriched REE is not clear; it may be due to an even more refractory condensate scavenging a significant fraction of the Lu or it may be simply a mineral preference effect tending to exclude Lu from the Zr-Y oxide.

Recently, Palme and Wlotzka (1981) found a Ca,Al-rich phase, also from Ornans, in which Dy, Ho, Er and Lu are progressively enriched from about 1500 to 10,000 times chondritic abundances. The more volatile REE, La, Sm, Eu and Yb, could not be detected; their upper limits are less than 100 times chondritic abundances except for Yb, for which the limit is about 1000. The abundances of the four REE, as well as the four upper limits, are consistent with an origin of the inclusion as an ultra-refractory condensate from the nebula. The observed abundances agree quite well with those predicted based on the REE solid/gas distribution coefficients. However, Palme and Wlotzka (1981) prefer an origin as a refractory residue from a partial vaporization event. Although an equilibrium configuration should be identical regardless of whether equilibrium is approached from above or below, a refractory residue can form in a hydrogen-poor environment. Under such conditions, the vaporized gas becomes oxidizing, and relative volatilities of the trace elements will change (see section 3.2). Although none of the REE determined by Palme and Wlotzka (1981) are sensitive to this effect, they found that W, which is sensitive to oxygen, was more depleted than expected from a reduced, hydrogen-rich gas. It is for this reason that they consider the inclusion to be an evaporation residue rather than a condensate. A determination of the LREE abundances, including Ce, would give further support to the hypothesis that, if the inclusion is a residue, Ce would be highly depleted relative to the other LREE. The Murchison inclusion studied by Boynton et al. (1980) has no large negative Ce anomaly and must therefore have formed under reducing conditions.

The presence of these inclusions indicates that materials can be isolated from the nebula at very early stages in the condensation sequence. The chondrite-normalized Lu value of 100 in MH-115 suggests that this inclusion represents the first 1% of the solid matter to condense; the Ornans inclusion, with a chondrite-normalized Lu value of 10,000, may be the first 0.01% to condense. Alternatively, of course, the inclusions could be the last 1% and last 0.01% that remained after partial vaporization.

In any event, it is clear that very efficient, very high-temperature fractionation events occurred early in the history of the solar system. Finding a plausible mechanism for such events remains a problem.

FUN inclusions

There is another group of Allende inclusions, consisting of three members, that have been very actively studied. The three samples, designated C1, EK 1-4-1 and HAL, are referred to as FUN inclusions because the isotopic abundances of several elements are Fractionated relative to terrestrial ratios and also have a component of Unknown Nuclear origin (Wasserburg et al., 1977). Although all Ca,Al-rich inclusions have oxygen isotopic abundances that are distinct from terrestrial values in a way that cannot be due to chemical fractionation effects (Clayton et al., 1977), the FUN inclusions have isotopic abundance anomalies in nearly all of the elements studied. These abundance anomalies require incomplete mixing of material from a variety of nucleosynthetic sources (e.g., different supernovae). The whole topic of isotopic anomalies has received much attention and a detailed discussion is beyond the scope of this chapter. The interested reader is referred to a review by Lee (1979).

Analyses of Sm and Nd isotopic composition in C1 and EK 1-4-1 (McCulloch and Wasserburg, 1978a,b; Lugmair et al., 1978) have figured prominently in the interpretation of the nucleosynthetic components present. Inclusion EK 1-4-1 has an excess of *r*-process and *p*-process components relative to *s*-process. (The processes designate rapid neutron capture, formation of proton-rich nuclides, and slow neutron capture, respectively.) Inclusion C1 has only an excess of the *p*-process component. The excesses are small, on the order of 0.1% or less. These components are thought to come from an injection of material from a supernova (Cameron and Truran, 1977; McCulloch and Wasserburg, 1978a), but incomplete mixing of material from many past events has also been suggested (Lugmair et al., 1978; Clayton, 1979).

In addition to their peculiar isotopic composition, at least two of the FUN inclusions have another unique aspect; they apparently formed under highly oxidizing conditions. This suggestion was first made by Boynton (1978a), who noted that inclusion C1 had a large negative Ce anomaly, unaccompanied by a negative Eu anomaly, indicating this oxidizing condition. Inclusion HAL was later shown to also have a large negative Ce anomaly (Tanaka et al., 1979). It was also shown that these inclusions have low abundances of other trace elements that, like Ce, become more volatile in an oxidized environment (Boynton and Cunningham, 1981). Most recently, however, the REE pattern of the third FUN inclusion, EK 1-4-1, was found to have normal Ce abundances (Nagassawa et al., 1981).

The REE patterns of these inclusions are shown in Fig. 3.5. Inclusion C1 appears to have a normal Group I REE pattern except for the Ce depletion. Because Ce is slightly more volatile than the other LREE, even in a reducing environment, the small Ce anomaly by itself does not require highly oxidizing conditions. However, the fact that Ce is depleted less than Eu, which is normally more volatile than Ce, does require a highly oxidizing environment. The REE pattern of HAL, both the core and the bulk inclusions, show a very large Ce anomaly requiring the highly oxidized conditions independent of consideration of the Eu abundance. The bulk HAL has a fairly flat pattern except for Ce and smaller depletions of Eu and possibly Gd. The core, on the other hand, except for Ce, has a REE pattern like the Group II inclusions

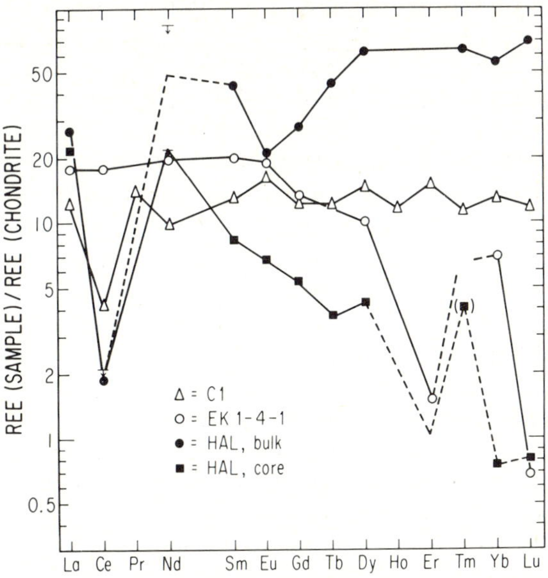

Fig. 3.5. REE patterns from three FUN inclusions in Allende, C1 (Conard, 1976), HAL (Tanaka et al., 1979) and EK 1-4-1 (Nagasawa et al., 1981). The large Ce anomalies in HAL and C1 indicate that each inclusion was made from grains which were found in a highly oxidizing environment. The lack of a Ce depletion in EK 1-4-1 does not rule out an oxidizing environment because Eu is also not depleted and the material may have equilibrated with an oxidized gas to lower temperatures. The isotopic anomalies and the Ce anomalies found in the FUN inclusions suggest that the grains may have formed in an oxidized zone of a supernova, but there are difficulties with this suggestion. The Tm anomaly in the HAL core may be an analytical error (A.M. Davis, private communication).

with the strong depletions of the HREE and the positive Tm anomaly. It was noticed that Tm was not so enriched relative to the LREE as it is in the Group II inclusions; this result also was thought to be indicative of a change in Tm volatility due to the oxidized conditions (Tanaka et al., 1979). (Tm becomes more refractory in oxidizing conditions; hence it will be more strongly depleted in Group II type material.) More recent data, however, indicate that the Tm anomaly may have been due to analytical error, and that the HREE are smoothly depleted, probably due to mineral preference effects (A.M. Davis, private communication, 1981).

The lack of a Ce anomaly in EK 1-4-1, which could be expected from the data for the other FUN inclusions, does not necessarily argue against an oxidized environment. Because Eu is also not depleted, it is possible that EK 1-4-1 could have formed in an oxidizing environment but was isolated from the gas at low enough temperatures that all the Ce condensed, even though Ce may have been more volatile than Eu. The pattern of EK 1-4-1 is like that of the Group II inclusions, but Tm was not analyzed. An analysis of Tm, showing it to be depleted relative to the LREE, would be indicative of the oxidizing conditions, for reasons discussed above with respect HAL (Tanaka et al., 1979).

Boynton (1978a) suggested that the grains which aggregated to form inclusion C1 condensed in an oxidizing environment that may have been in the ejecta of a supernova. Such an environment was shown to be highly oxidizing (because H has been fused to form He), and could be conducive to grain formation (Lattimer et al., 1978). However, a problem with this model is that to preserve such a large Ce anomaly, the inclusions must be made of a large fraction of supernova grains, and one would expect much larger isotopic anomalies in Sm and Nd than are observed. Although it is possible to generate oxidizing conditions in a supernova with only a minimum of nucleosynthesis, very special conditions appear to be required (Dearborn and Boynton, 1981). Finding other astrophysical environments with oxidizing conditions is difficult because of the ubiquity of hydrogen. A partial vaporization event generating a gas that displaces the hydrogen, as suggested by Palme and Wlotzka (1981) for the Ornans inclusion, is certainly possible.

3.4. REE in other components of chondrites

Chondrules and other inclusions

Graham et al. (1976) analyzed REE in a Ca,Al-rich glass inclusion from the Bovedy (L3) chondrite. Although this glass is Ca,Al-rich, it has little in common with the inclusions discussed above. It has major element abundances similar to a plagioclase of composition An_{85}, and is thought to have formed by shock from a pre-existing plagioclase. The chondrite-normalized REE abundances in the glass inclusion decrease smoothly from 0.73 at La to 0.20 at Lu, with a large positive Eu anomaly of a factor of 30. This Eu anomaly was compared to plagioclases from basaltic achondrites (eucrites) and lunar mare basalts and was found to be much larger. The size of the Eu anomaly could be matched only by lunar anorthosites. Graham et al. (1976) suggested that the Bovedy glass was actually a shocked fragment of an anorthosite. If this suggestion is true, igneous rocks needed to be present before the material in the Bovedy chondrite accreted to form the stone.

Tanaka et al. (1975) determined the REE abundances in a large (7 mm) olivine chondrule from Allende (Fig. 3.6). These abundances were much lower than those previously found in a smaller Allende olivine chondrule analyzed by Tanaka and Masuda (1973). The smaller chondrule has a flat REE pattern at about 1.8 times chondritic abundances with no significant anomalies. The large chondrule is about a factor of 40 lower in REE abundances, with a chondrite-normalized pattern which decreases from 0.06 at La to 0.045 at Lu. In addition, this chondrule has a positive Eu anomaly of a

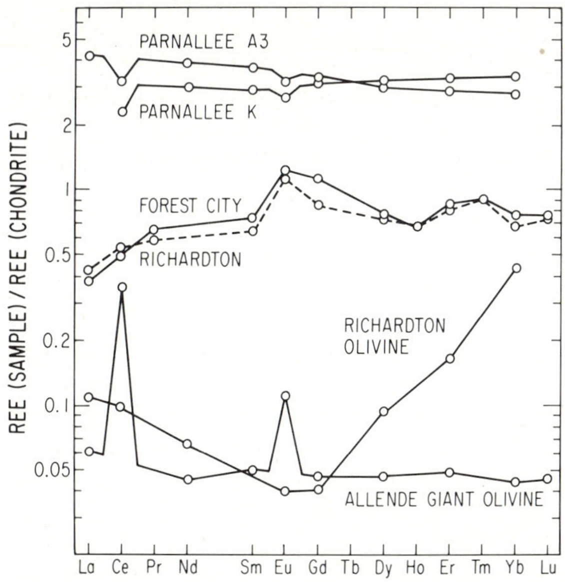

Fig. 3.6. REE patterns in chondrules. The olivine chondrule from Allende (Tanaka et al., 1975) has large positive Ce and Eu anomalies, perhaps due to addition of late condensing REE in an oxidizing environment. An olivine chondrule from Richardton (Evensen et al., 1979a) has an unusual shape similar to olivine in pallasites. A composite of many chondrules from the equilibrated chondrites Richardton and Forest City (Schmitt et al., 1968) show a linear fractionation and positive Eu anomaly thought to be due to migration of trivalent REE from the chondrules to the matrix. The chondrules from the unequilibrated chondrite Parnallee (Hamilton et al., 1979) show a linear fractionation with Ce and Eu anomalies perhaps indicative of partial vaporization under oxidizing conditions.

factor of 2.3 and a large positive Ce anomaly of a factor of 6.2. The chondrule is also remarkable in that its chondrite-normalized Ba abundance is about 4.5. The authors suggested that this large chondrule may have sampled some of the late condensing refractory elements from the solar nebula. They noted that even though Eu is expected to be more volatile than Ce in the solar nebula, the volatility of the pure oxides decreases in the order Ba > Ce > Eu. Although this order of volatilities is not expected in the solar nebula, it is exactly the order expected under oxidizing conditions (Boynton and Cunningham, 1981). It is possible, therefore, that this chondrule received some extra component, containing Ba, Ce and Eu, which condensed from a gas generated by a partial vaporization event. As discussed above, a partial vaporization event, either after the hydrogen-rich nebula had dissipated, or if rapid enough to displace the hydrogen, can generate the necessary oxidizing conditions.

Schmitt et al. (1968) determined the REE in 30 combined chondrules from Mokoia (CV2), 30 from Chainpur (LL3), 35 from Forest City (H5) and 50 from Richardton (H5). They found that the Mokoia chondrules were enriched to about twice the whole-rock values and showed a slight increase from La to Lu of about 20% with no anomalies. The Chainpur chondrules had unfractionated abundances at chondritic levels. Chondrules from the two equilibrated chondrites, Forest City and Richardton, had significant linear fractionations showing about a factor of 1.9 increase from light to heavy REE (Fig. 3.6). The absolute amounts of REE were lower in these chondrules (La = 0.40 times chondritic abundances), but positive anomalies were observed with Eu being present at nearly chondritic levels. The authors noted that the fractionated pattern is found only in the chondrules from the two equilibrated, H5, chondrites, and suggested that during the period of metamorphism, the LREE preferentially migrated into the matrix. The positive Eu anomalies resulted from a lack of mobilization of divalent Eu.

Evensen et al. (1979a) determined the REE in nine individual chondrules from Richardton. Their average result agreed quite well with the values found by Schmitt et al. (1968), but the individual analyses differed widely. Five of the chondrules had positive Eu anomalies, two showed no anomalies, and one had a negative Eu anomaly; the abundances of LREE ranged over a factor of 7.5. The remaining chondrule (Fig. 3.6) had low REE abundances and an unusual U-shape, similar to that found for olivine in the Brenham pallasite (Masuda, 1968). Hamilton et al. (1979) analyzed four individual chondrules from the unequilibrated chondrite, Parnallee (LL-3). Two of the chondrules, J and E, have patterns similar to those from Richardton, although the absolute abundances are higher. One chondrite-normalized pattern is flat at about 2.5, and the other is fractionated with Ce at 1.0 and Yb at 1.5. Both chondrules have small positive Eu anomalies. The other two chondrules, A3 and K, each have negative Ce and Eu anomalies (Fig. 3.6) and chondrule K also shows a significant linear fractionation with the LREE enriched.

Because Parnallee is an unequilibrated chondrite, the non-chondritic REE patterns cannot easily be explained by metamorphism, as suggested for the Richardton and Forest City chondrules. Hamilton et al. (1979) suggested that the Ce and Eu anomalies in chondrules A3 and K are related and reflect a partition of REE between a gas phase and a solid or liquid. As discussed above, such Ce anomalies could be expected if the gas were oxidizing. Loss of Ce and Eu by partial vaporization could generate the necessary oxidizing conditions, leaving the chondrules as the residue. Such a process, however, cannot explain the presence of the linear REE fractionations observed in the chondrules. If the vaporization event were rapid enough to be controlled by diffusion, a linear fractionation could result, but such a mechanism does not explain why both positive and negative slopes are observed. Because each of the individual chondrules were split and only a portion analyzed, some of the fractionation and anomalies may represent non-random sampling of the phases in the chondrule.

The origin of chondrules is still a mystery, but the ability of the REE to distinguish, in principle, nebular fractionations (gas/solid or gas/liquid) from planetary fractionations (solid/solid or solid/liquid) may eventually shed some light on the problem.

Mineral separates

Several investigators have studied REE abundances in mineral separates of chondrites. The techniques used for generating the separates were selective dissolution (Honda and Shima, 1967; Shima and Honda, 1967) and heavy liquid separation (Schnetzler and Bottino, 1971; Allen and Mason, 1973; Curtis and Schmitt, 1979). With the exception of Shima and Honda, who studied Abee (E4), all investigators studied equilibrated, type 6 chondrites. The minerals in the equilibrated meteorites presumably have lost any record of primitive REE distributions, so such studies will mainly address questions related to metamorphism.

The most detailed work on REE abundances in mineral separates is that of Curtis and Schmitt (1979), who studied three L6 chondrites and determined the REE in olivine, orthopyroxene, clinopyroxene, feldspar and phosphates; metal and troilite separates were prepared, but the REE were so low in abundance that the analyzed quantity was due largely to incomplete separation from other minerals. They found that the REE patterns are as expected for closed system equilibration between minerals. Based on mineral/liquid distributions, the absolute abundances of REE were found to be about a factor of ten higher than those expected if the minerals had equilibrated with a liquid having chondritic REE abundances. Curtis and Schmitt also found that the REE abundance ratios between phosphates and pyroxenes differed by a factor of four in two of the meteorites. Because they found that the mineral compositions were virtually identical between the two meteorites, they attributed this difference in abundance ratio to metamorphic temperature or pressure effects.

3.5. Whole-rock chondrite analyses

Solar abundances

The early work on REE analyses was concerned exclusively with whole rock measurements (Schmitt et al., 1960, 1963, 1964). Most of the samples studied were chondrites. It was known at that time that the chondrites were relatively unfractionated samples of solar system material. In order to test models of nucleosynthesis, one needs to know the solar abundances, in at least relative quantities, over a wide range of mass and atomic number. The REE, which span a range of atomic number from 57 to 71 and a range of mass from 138 to 176, are ideally suited for such studies. Since the REE were known to migrate as a group (Goldschmidt, 1954), it was thought that in chondrites they would very precisely reproduce solar ratios. As is well known, the abundances of REE in chondrites are indeed quite similar, but it will be shown in this section that, at levels on the order of 10%, the REE abundances in chondrites are in fact quite variable, both in absolute abundances and in ratios of elements. The early REE studies provided considerable input to theories of nucleosynthesis (Suess and Urey, 1956). The solar abundances of REE are still providing input for such models. Recently, for example, Steinberg and Wilkens (1978) suggested that the smooth peak in abundances in the REE region near mass 165 was most likely due to the fission of a superheavy element at the termination of *r*-process nucleosynthesis, and that this termination must have occurred beyond mass 300.

The early measurements of the REE abundances in chondrites cited above were made using neutron activation analysis (NAA). Most of the more recent measurements (Masuda et al., 1973; Nakamura and Masuda, 1973; Nakamura, 1974; Evensen et al., 1978) have been made using isotope dilution mass spectrometry (IDMS). The latter technique produces higher precision, about 2—3%, than the older NAA data, which has a precision of 10—15%, although newer NAA techniques are capable of precision comparable to that of IDMS. In an excellent review of whole-rock chondrite REE analyses, Evensen et al. (1978) found that of 77 analyses of 50 chondrites, 55 were found to have one or more abundance anomalies. (An anomaly was defined as an abundance which differed by more than 3% (IDMS) or 20% (NAA) from a smooth curve through the remaining REE.) Even when Eu anomalies are excluded, half of the samples still have abundance anomalies. It is clear that chondrites with "chondritic" REE abundances are the exception rather than the rule.

Evensen et al. (1978) compiled a table of average REE abundances in CI chondrites for normalization purposes. The values, reproduced in Table 3.3, are based on IDMS data for the multi-isotopic elements and NAA data for the mono-isotopic elements; the NAA data are normalized based on elements analyzed in common by the two techniques. As mentioned in section 3.1, CI normalization values are the best estimates of solar abundances and are recommended for cosmochemists concerned with condensation of the elements. For others interested in igneous fractionation processes, whether in terrestrial or extraterrestrial materials, the absolute values of the REE used for normalization purposes are less important than the ratios. In the past most investigators have chosen to normalize to "average chondrites" rather than to CI chondrites. For this reason the normalizations in this book have been made to the chondrite composite NAA values of Wakita et al. (1971) or Haskin et al. (1968), which are also given in Table 3.3. These normalization values, as well as a few others that have been commonly used, are plotted in Fig. 3.7. It can be seen that the older NAA data do scatter quite

TABLE 3.3

Rare earth element normalization values

	CI average[a]	Wakita et al. composite[b]	Haskin et al. composite[c]	Recommended chondrite
La	0.2446	0.34	0.330	0.310
Ce	0.6379	0.91	0.88	0.808
Pr	0.09637	0.121	0.112	0.122
Nd	0.4738	0.64	0.60	0.600
Sm	0.1540	0.195	0.181	0.195
Eu	0.05802	0.073	0.069	0.0735
Gd	0.2043	0.26	0.249	0.259
Tb	0.03745	0.047	0.047	0.0474
Dy	0.2541	0.30	—	0.322
Ho	0.05670	0.078	0.070	0.0718
Er	0.1660	0.20	0.200	0.210
Tm	0.02561	0.032	0.030	0.0324
Yb	0.1651	0.22	0.20	0.209
Lu	0.02539	0.034	0.034	0.0322

[a]Evensen et al. (1978).
[b]Wakita et al. (1971).
[c]Haskin et al. (1968).

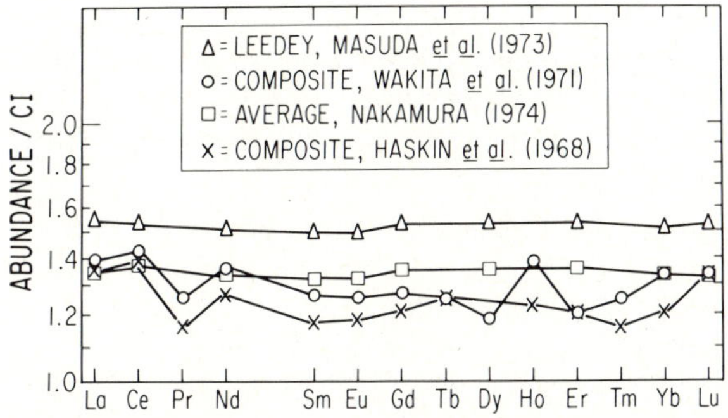

Fig. 3.7. Commonly used chondrite normalization values relative to CI chondrite values from Evensen et al. (1978). The older data scatter more than the more recent data but include data on the mono-isotopic REE. Recommended chondrite normalization values (Table 2) are based on the CI values of Evensen et al. adjusted to the difference between CI chondrites and average chondrites.

badly compared to the IDMS data or the more recent NAA data on which the CI normalization values are in part based. For the future, this author has calculated a "recommended" set of chondrite normalization values derived from the CI normalization values of Evensen et al. (1978) by multiplying by the factor of 1.267, the average of the Wakita et al. (1971) and Haskin et al. (1968) normalization values.

The absolute abundances of REE clearly vary between the different groups of chondrites. Evensen et al. (1978) plotted median abundances in the various chondrite groups and showed that the abundances ranged from about CI abundance levels (in CI and E chondrites) to about 2.3 times CI levels for the CV chondrites. These abundance differences have two origins. One simply relates a dilution by different amounts of volatile components, and the other represents a nebula-wide fractionation of refractory elements. The conventional way of expressing chondritic abundances to eliminate the effect of variable amounts of volatiles is to normalize to Si. Nakamura (1974) normalized his data for the various chondritic classes in this manner and found that abundances decreased in the order CV $>$ CM $\approx$ CI $>$ L $\approx$ H $>$ E. This order is the same as that found for a variety of refractory elements and is generally thought to be caused by a nebula-wide fractionation of refractory elements (Larimer and Anders, 1970; Wasson 1978). The REE appear to be typical refractory elements in this regard.

In spite of all of the high-temperature fractionations observed in the Ca,Al-rich inclusions of the carbonaceous chondrites, the average REE patterns of the various classes are quite flat, suggesting that this large-scale nebular fractionation of refractory elements took place after the REE were totally condensed. Exceptions to this lack of flatness are evident only in the CV group (represented by Allende and Mokoia), which appears to have an excess of Group II type material, as discussed in section 3.3, and possibly the CO group, which may have an excess of the ultra-refractory condensates. Sampling uncertainties, however, prohibit a firm conclusion regarding the CO group fractionation. Enough analyses of Allende have been performed to ensure adequate sampling of that meteorite, but it is not clear that Allende is an adequate sample of its parent body. Allende is known to be heterogeneous on a scale of tens of centimeters and it is certainly possible that its parent body is heterogeneous on a scale of meters or more. If these analyses are indicative of the meteorite parent body composition, then apparently the higher-temperature refractory element fractionations, as discussed in section 3.3, operated over a region of the solar nebula comparable in size to the zone which accreted to form the CV parent body.

Chondrites with anomalous REE patterns

It was noted above that the majority of chondrite analyses have one or more abundance anomalies at a level of 4% or more. The fact that average or median abundances are quite regular suggests that much of the problem may be due to inadequate sampling. Considering the wide range of REE patterns in individual chondrules (section 3.4), it should not be surprising that small samples would yield variable results. In fact, Keil (1962) suggested that typically 10—70 g would be required for most chondrites, but Dodd and Jarosewich (1980) suggest that 10 g is an adequate amount in most cases.

Typically, analyses are made on less than a gram; six of the eighteen analyses made by Evensen et al. (1978) were on samples under 0.1 g. Thus many of the anomalies suggested by Evensen et al. may just be indicative of the heterogeneity of the meteorites.

Some meteorites have very large anomalies that appear to be far beyond those reported from sampling of chondrules. Nakamura and Masuda (1973) published several such analyses. For example, two pieces of Kohar (L3) weighing 0.9 and 0.2 g, each had a positive Ce anomaly of around a factor of four and a positive Gd anomaly of about 10% in an otherwise flat pattern. They also analyzed two chips of Abee (E4); a 0.2-g chip had a flat REE pattern, but a 0.75-g chip had a "zig-zag" pattern with alternating enrichments and depletions of about 20% (only multi-isotopic REE were determined) and a large positive Yb anomaly of nearly a factor of three. Two samples of Atlanta (E5) were analyzed and found to have Eu and Ce anomalies on a curved pattern which appeared to be smoothly partitioned as a function of size, suggesting an igneous fractionation.

Some of these anomalies are very difficult to understand. The large Ce anomalies in Kohar can be explained by consideration of solid/gas fractionations in an oxidizing atmosphere, but the Gd anomaly would not be expected under such conditions. Neither nebular nor planetary fractionations appear to offer an easy way to fractionate Gd from the other REE. The apparent igneous fractionation of Atlanta is difficult to understand except that Atlanta is a find. Frazier and Boynton (1981) found a substantial amount of REE in Abee to be water leachable, presumably from oldhamite, CaS. The smooth pattern in Atlanta could be a weathering phenomenon.

The Abee pattern, with its large Yb anomaly and alternating anomalies in the other REE, is probably the most difficult to understand. In preliminary work Frazier and Boynton (1980, 1981) found no large Yb anomalies in four clasts or mineral separates of matrix material from Abee. The Abee enstatite chondrite is composed of clasts which are internally equilibrated (Rubin and Keil, 1980), but based on magnetic measurements, the clasts have not equilibrated with each other (Sugiura and Strangway, 1981). Perhaps the larger sample analyzed by Nakamura and Masuda came from a rare clast with peculiar REE abundances. Again, these abundances are not easily explained. Perhaps because the highly reduced nature of the enstatite chondrite allows the REE to go into phases such as sulfides (or phosphides?), the REE partitioning between phases would permit such irregularities. (Our experience of terrestrial rocks is mainly with fractionation between oxides. When sulfides or phosphides are involved, the difference in affinity of the individual REE for sulfur or phosphorous may be important.) It is also possible that the REE condensation from the nebula into phosphides or sulfides may generate such peculiar patterns. Unfortunately sufficient thermodynamic data to permit calculation of solid/gas distribution coefficients for phosphides or sulfides are lacking.

It is clear in any event that the REE abundances in chondrites are not as featureless as once thought. With recent developments in analytical capability to measure REE with high precision on small samples, it is hoped that future work will shed some light on the detailed processes that were operating when our solar system formed.

3.6. Analyses of achondrites

Eucrites, howardites and diogenites

The eucrites, howardites, and diogenites are discussed as a group because they are considered by many investigators to be genetically related. Analyses of REE in this group have been published by Schmitt et al. (1963, 1964), Haskin et al. (1966), Schnetzler and Philpotts (1969), Gast et al. (1970), Gast and Hubbard (1970), Jérome (1970), Fukuoka et al. (1977), Palme et al. (1978), Hamet et al. (1978), Ma and Schmitt (1979), Mittlefehldt (1979), Mittlefehldt et al. (1979) and Grossman et al. (1981). These meteorites, like all achondrites, appear to have formed by igneous processes. The eucrites and howardites have been called the pyroxene-plagioclase achondrites after their dominant minerals. The eucrites have been distinguished from the howardites by several criteria. Prior (1920) classified them on the type of pyroxene present; eucrites were those meteorites in which pigeonite was dominant over hypersthene, whereas howardites had hypersthene dominant over clinopyroxene. Mason (1962) showed that the howardites and eucrites plotted in different fields on a plot of CaO vs. FeO/(FeO + MgO); eucrites had higher values of both parameters. Duke and Silver (1967) defined eucrites as monomict breccias (made from material with a common origin) and howardites as polymict breccias. Grossman et al. (1981) showed that the Mason (1962) chemical criterion and the Prior (1920) mineralogical criterion were consistent. From fig. 1 in Grossman et al. (1981) it can be seen that the FeO/(FeO + MgO) parameter is not required, since the CaO content of all eucrites is greater than 9% and in all howardites it is less than 9%. There are no unbrecciated or monomict howardites but there are five polymict eucrites.

Most of the eucrites have REE patterns which are flat and enriched from 6 to 10 times chondritic abundances, within the range shown for Pasamonte and Sioux County in Fig. 3.8. A few eucrites have greater enrichments of REE and have negative Eu anomalies, while a few have lower REE abundances and positive Eu anomalies (Fig. 3.8). Stolper (1977) suggested, on the basis of experimental petrology, that most of the eucrites could be formed as partial melts of primitive chondritic material. A few of the eucrites, of which Moore County and Serra de Magé are examples, could have formed as cumulates from the same partial melt liquid. Consolmagno and Drake

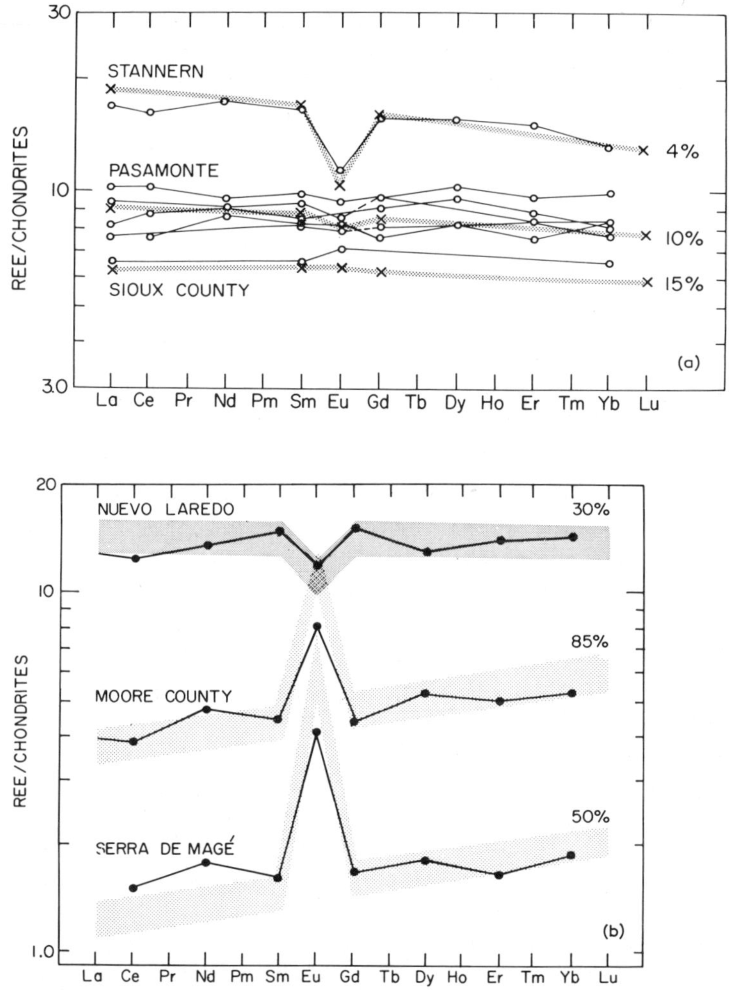

Fig. 3.8. REE data on eucrites with calculated values from Consolmagno and Drake (1977). (a) Most of the eucrites fall within the range defined by Pasamonte and Sioux County and represent 6—10% partial melting. (b) The cumulate eucrites Moore County and Serra de Magé could have formed after 85% or 50% fractional crystallization of a Pasamonte-like partial melt, although other constraints argue against this hypothesis (see text). Nuevo Laredo could have formed as the residual liquid after 30% crystallization. Note that these degrees of melting or crystallization are a function of the assumed initial REE concentration, here taken to be 1.0 times chondritic abundances. Ma and Schmitt (1979) have noted some difficulty with the Consolmagno and Drake model (see text).

(1977) calculated REE patterns for this model using REE mineral/liquid distribution coefficients. They noted that the REE pattern in partial melts is a sensitive indicator of the nature of the source region. Various minerals leave a "signature" in the melt REE pattern; for example, pyroxenes in the source yield an enrichment in LREE, and plagioclase yields a negative Eu anomaly. Consolmagno and Drake found that the model proposed by Stolper (1977) does indeed account for the observed REE patterns. Most of the eucrites could form as a partial melt of material with chondritic REE abundances. Most of the eucrites require 10% melting of the source, but others require less (e.g., Stannern — 4%) or more (e.g., Sioux County — 15%). Calculated REE patterns from Consolmagno and Drake (1977) are also shown in Fig. 3.8. The negative Eu anomaly with 4% partial melting results from the presence of plagioclase which has been consumed in the source after 4% but before 10% partial melting.

The REE in the cumulate eucrites, Moore County and Serra de Magé, are also consistent with the Stolper (1977) model. The calculated REE patterns agree quite well with the observations (Fig. 3.8). These patterns are generated in a pyroxene-plagioclase cumulate which formed from a starting liquid similar in composition to that of typical eucrites, 8—10 times chondritic REE abundances. The pattern of Serra de Magé formed after 50% fractional crystallization of this liquid and Moore County after 85%. The pattern of Nueveo Laredo represents the residual liquid after 30% fractional crystallization has occurred.

Consolmagno and Drake (1977) noted that the age of the eucrites is similar to the age of the solar system and thus the eucrites may be interpreted as having formed in a single-stage melting event. Because they were able to determine the composition of the source of this melting event, they concluded that they had actually determined the bulk composition of the eucrite parent body, assuming the body was homogeneous. They compared their inferred composition of the eucrite parent body with others determined by largely independent means and found good agreement. They noted that the disappearance of plagioclase in the source put a tight constraint on the amount of this mineral present in the parent body. They could not distinguish olivine from metal in the source (both have very low REE distribution coefficients), but the sum of these two components was reasonably well-known. The amount of pyroxene was constrained by the sum of the other components.

Fukuoka et al. (1977), however, pointed out the difficulty in modeling the composition of the source materials by such a technique. They noted that the composition is dependent on the assumed absolute concentration of REE in the starting materials. Depending on assumed initial chondrite-normalized concentrations from 1 to 3, they found a wide range of partial melting and mineral compositions that would be consistent with the observed REE abundances in the non-cumulate eucrites. Apparently of the three parameters — initial abundance, degree of fractionation, and composition of the source — the REE patterns of the melt allow any two to be determined.

Ma and Schmitt (1979) performed a detailed study of REE and other elemental abundances in two cumulate eucrites, Serra de Magé and Moore County. They used REE data on plagioclase separates to calculate the REE abundances in the equilibrium magmas, the liquids from which each of the eucrites formed. They noted that it is difficult to calculate the REE pattern of the parental magma, the liquid present before a fractional crystallization had occurred, unless an independent estimate of the degree of fractional crystallization is available. The estimates made by Consolmagno and Drake (1977) are not useful because they *assumed* that the parental magma was 8—10 times chondritic abundances in order to calculate the degree of crystallization. The problem is similar to that noted by Fukuoka et al. (1977) for the non-cumulate eucrites. In particular, Fukuoka et al. noted that although the Consolmagno and Drake (1977) calculations for the origin of the cumulate eucrites were internally consistent, the model could not explain the observed FeO/(FeO + MgO) ratios. Following fractional crystallization the FeO/(FeO + MgO) ratio would reach higher values. They concluded that there was no means to generate Serra de Magé and Moore County by a fractional crystallization process from a common parent magma. A possible relationship between the two meteorites was suggested in which their formation involved a difference, by a factor of about six, in the degree of partial melting of the same source. (The process of partial melting has only a small effect on FeO/(FeO + MgO) ratios and makes this parameter easier to model.) Fukuoka et al. concluded that the source material must contain >30% high-Ca pyroxene and have a chondrite-normalized La/Yb ratio about 3; this material is quite different from that proposed for the non-cumulate eucrites. The two meteorites could then form as cumulates from these two different partial melt liquids. If such a model is correct, the relationship between the cumulate eucrites and non-cumulate eucrites must be distant at best.

It can be seen from this discussion of eucrite genesis that the REE can provide strong constraints on models of petrogenesis. Like all studies on samples, the REE data can often easily be used to eliminate a model but cannot be used to choose between a variety of models that may all be consistent with the observations. Ma and Schmitt (1979) argued that the Consolmagno and Drake (1977) model for the cumulate eucrites is not consistent with all the available data, but they left us with a complicated model that requires many steps in order to account for the data. It appears that the work of the model makers will be to use their ingenuity to find the simplest model that is consistent with the data, but to recognize that Mother Nature may not have felt compelled to be simple.

The diogenites are nearly monomineralic meteorites composed almost entirely of hypersthene. With the exception of Tatahouine, all are brecciated. They have been studied in detail by Fukuoka et al. (1977). The REE in the diogenites, except for Roda, are highly fractionated; the patterns show strong enrichments from LREE to HREE and negative Eu anomalies ranging from

2 to 4. The chondrite-normalized Lu/Sm ratio ranges from over 100 in Tatahouine to 6 in Johnstown; Lu is at nearly chondritic abundance. Roda, on the other hand, has a much flatter chondrite-normalized REE pattern with Sm = 1.4 and Lu = 2.3. Fukuoka et al. suggested that the diogenites are cumulates formed from a melt having nearly chondritic-relative REE abundances, but with chondrite-normalized abundances between 2 and 10; the mineral/liquid distribution coefficients were not known well enough to determine the absolute amount with high precision. The high LREE abundances in Roda, compared to the other diogenites, were thought to be due to mixing of eucrite material or the presence of trapped intercumulus liquid. The latter hypothesis was preferred, since no eucrites were known with a large enough negative Eu anomaly to account for that observed in Roda. A separation of Johnstown, into coarse pyroxene crystals and a fine fraction which would presumably contain any mesostasis, indicated that the LREE were present in significant amounts in accessory phases. The highly fractionated REE pattern in Tatahouine led Fukuoka et al. (1977) to suggest that this meteorite must have crystallized from a melt which was already highly fractionated. A melt derived from other diogenites was suggested.

Consolmagno (1979) attempted a more detailed modelling of the diogenites with the hope that they could be related to the eucrites by a simple means. He was unable to devise a simple genetic mechanism and had particular difficulty in trying to account for the large Eu anomalies if the diogenites formed from material related to the eucrites.

Although the REE data do not suggest a simple relationship between the eucrites and diogenites, there is one group of meteorites that indirectly suggests a relationship. This group is the howardites. The howardites are generally considered to represent mechanical mixtures between diogenites and eucrites, presumably in a parent body regolith (Jérome, 1970; Jérome and Goles, 1971; McCarthy et al., 1972; Dreibus et al., 1977; Fukuoka et al., 1977). Fukuoka et al. (1977) showed that only two components, diogenites and eucrites, could account for the major, minor and trace elements of the howardites, except for siderophile elements; the siderophile elements required a few percent of a chondritic component. Recent REE analyses of howardites are reported by Fukuoka et al. (1977) and Mittlefehldt et al. (1979). The chondrite-normalized REE patterns are typically flat at between 2 and 8. Mittlefehldt et al. (1979) found one sample of Malvern with a flat pattern and another sample with a negative Eu anomaly of a factor of two. They emphasized the problem of sampling these polymict meteorites.

Jérome (1970) and Jérome and Goles (1971) first suggested that a compositional and textural continuum exists from diogenites through howardites to eucrites. Mittlefehldt (1979) found a large ($\sim$4 mm) plagioclase clast in the diogenite Johnstown and using REE analyses and mineral/liquid partition coefficients, concluded that the clast could not have formed from the same liquid as the Johnstown pyroxene. The presence of a foreign clast indicates that Johnstown is polymict and thus shares some characteristics with the howardites. Fukuoka et al. (1977) noted that siderophile elements in Johnstown were also higher than in other diogenites, which further supports the idea of a continuum. With this evidence linking eucrites, diogenites, and howardites, it is perhaps surprising that a simple igneous model cannot account for the REE patterns in the diogenites and eucrites. It may be that because our sample of the eucrite parent body is probably non-representative, other unsampled meteorite types may exist which contain other pieces to the puzzle.

Angra dos Reis

Angra dos Reis is a unique achondrite which has been the subject of a detailed consortium study by a number of investigators (Keil, 1977). The REE were studied by Ma et al. (1977), who measured whole-rock samples and olivine, pyroxene, and whitlockite mineral separates. This meteorite consists mainly of pyroxene (fassaite) and is thought to have formed as a cumulate (Prinz et al., 1977). Ma et al. (1977) found that the whole-rock REE pattern and the clinopyroxene REE pattern were virtually identical. The patterns are enriched to 20—30 times chondritic abundances, are concave downwards, and have a small negative Eu anomaly. The whitlockite separate is highly enriched in REE, with chondrite-normalized values decreasing from La = 1100 to Lu = 70. Two olivine separates were made; in each the chondrite-normalized value for Lu is about 9, but for La the values are 0.12 and 0.67 in the different separates. This difference was attributed to the presence of variable amounts of kirschsteinite ($CaFeSiO_4$) in the olivine.

Using the clinopyroxene REE data and mineral/liquid distribution coefficients, Ma et al. (1977) calculated the composition of the magma from which Angra dos Reis could have crystallized. They found it to be enriched in REE and fractionated linearly with chondrite-normalized values from La = 200 to Lu = 35. Using these data, they considered various models to relate Angra dos Reis to other known meteorite types, but all their models were unsuccessful. Angra dos Reis could not easily be related to the non-cumulate eucrites because its highly fractionated magma could not be generated from the eucrites without extensive fractional crystallization. The fractional crystallization would raise the FeO/(FeO + MgO) ratio above that observed in the eucrites, but Angra dos Reis requires a lower ratio. Ma et al. also considered the formation of Angra dos Reis from nakhlites but, for a similar reason, found it to be unlikely. The formation of Angra dos Reis from the cumulate eucrites or by a process similar to that which formed the cumulate eucrites was also dismissed mainly on the basis of major element data. They concluded that Angra dos Reis was unrelated to any other achondrites.

Ma et al. proposed a model for the origin of Angra dos Reis but cautioned that, as discussed above with respect to the eucrites, the model is a function of the assumed initial REE abundances. Their preferred model is as follows: Angra dos Reis crystallized following 70% fractional crystallization of a melt generated from 7 to 10% partial melting of source materials with REE abundances between 3.5 and 4.7 times chondrites. They note that estimates of the initial chondrite-normalized REE abundances for the Moon are between 3 and 5, and thus a high initial REE abundance is not unreasonable.

Ureilites

The ureilites are a group of achondrites with very low REE abundances. They have been analyzed for REE by Wänke et al. (1972b) and Boynton et al. (1976). They are composed mainly of olivine and pigeonite and are thought by Berkley et al. (1976) to have formed as cumulates. The REE data, however, argue against an origin as a cumulate and suggest an origin as a residue of partial melting. The REE patterns of the ureilites are plotted in Fig. 3.9. They are all V-shaped, or, in the case of Haverö, U-shaped. From Kenna, both an acid-treated and an untreated sample were analyzed. It was

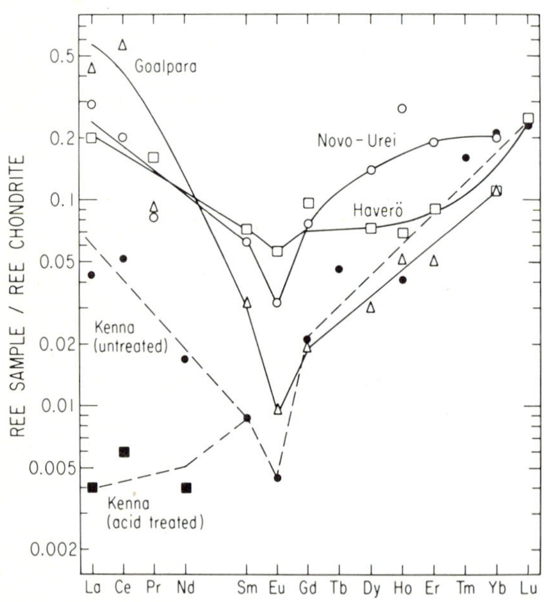

Fig. 3.9. REE patterns of ureilites from Boynton et al. (1976) and Wänke et al. (1972b). The peculiar V-shaped patterns require addition of two components. The HREE component is thought to be present in the olivine and pyroxene, which was a residue of a fractional partial melting event. The LREE are thought to have been added later, but the source of the material and the mechanism for this addition are unclear.

found that the LREE were removed in the acid treatment (15 minutes in concentrated HNO_3), suggesting that they are located in a phase different from the HREE. As noted by Boynton et al. (1976), however, one would not expect such a sharp V-shaped REE pattern to be made from REE in one phase, since minerals are expected to have REE distribution coefficients which are a smooth function of ionic size, without sharp discontinuities.

Boynton et al. (1976) considered both fractional crystallization and partial melting to model the observed REE patterns. They expressed the calculated REE pattern in terms of the chondrite-normalized Lu/Sm ratio. The quantity is equal to 27 in Kenna, which has the most fractionated pattern. Boynton et al. found that because the REE distribution coefficients are much less than unity for olivine and pigeonite, the dominant minerals in the ureilites, the liquid did not change much in Lu/Sm ratio during fractional crystallization. Over a wide range of crystallization, from 10% to 90%, the chondrite-normalized Lu/Sm ratio in the solid ranged between 3.6 and 6.3, considering both fractional crystallization, with continuous removal of crystals, or equilibrium crystallization, with removal at the end of the process. The process of partial melting provided the desired chondrite-normalized Lu/Sm ratio of 27 with only 7% melting, provided the liquid was removed from the solid as soon as it formed. They noted that complete removal of liquid as soon as it formed was unreasonable and cited 7% as a strict lower limit. They suggested, therefore, that a more extensive partial melting with a less efficient loss of liquid was responsible for the formation of the ureilites. The partial melting event could only explain the HREE. They suggested that the LREE were added by some later event although the nature of the event is obscure.

Because the REE pattern is so highly fractionated in the ureilites, it provides a strong constraint on the origin of these stones. The conclusion of an origin by partial melting seems very sound, but a more elaborate mechanism cannot be excluded. The petrographic studies, however, show crystals oriented as expected in a cumulate which has experienced flow (Berkley et al., 1976). This discrepancy between cumulate or partial melt residue can be resolved if the partial melting residue is allowed to flow, e.g., by extrusion or convection, and then solidify as proposed for the cumulate. Such a mechanism can account for the petrographic observations as well as the trace element data.

Chassignites and nakhlites

The Chassigny meteorite is composed mainly of an iron-rich (92 vol.%, Fa_{31}) olivine with minor pyroxene feldspar, chromite and melt inclusions, and is thought to be a cumulate (Prinz et al., 1974). Its REE pattern has been determined by Mason et al. (1976) and Boynton et al. (1976). The chondrite-normalized pattern found by Mason et al. is fractionated from 1.3

at La to 0.5 at Yb, with a minimum of 0.39 at Dy. Only five REE were determined by Boynton et al., but they showed a progressively greater enrichment toward the LREE, suggesting sampling heterogeneity. Boynton et al. noted that if the trend were real, it could not be due to differences in amount of trapped liquid in the two samples, since the ratio of LREE to HREE is unaffected by the ratio of olivine to trapped liquid, as olivine is expected to contain negligible amounts of REE. They suggested that a significant fraction of the clinopyroxene may be a cumulus phase.

The nakhlites are composed mainly of augite (79 wt.%) and olivine (15.5 wt.%), with minor plagioclase, magnetite, and K-feldspar; they are also thought to be cumulates, with the olivine being an intercumulus phase (Bunch and Reid, 1975). Schmitt and Smith (1963) determined the REE patterns in Nakhla and Lafayette. The patterns are fractionated with chondrite-normalized values of La = 5, Ce = 6, followed by a linear decrease to Lu = 1.

Boynton et al. (1976) calculated the chondrite-normalized REE abundances in the liquid from which the nakhlites could have formed and found a decrease from La = 16 to Lu = 2.2. They found that even with 15% trapped liquid in the nakhlites, the cumulus phase, augite, contains a significant fraction of the HREE. They noted the similar enrichment in LREE relative to HREE in the liquids from which Chassigny and the nakhlites formed and suggested that similar processes may have been responsible for generation of the liquid. They did not suggest a mechanism for this enrichment, but did indicate that Chassigny and the nakhlites could not have formed from the same source, because of differences in their oxygen isotope ratios.

Shergottites

The shergottites are volatile-rich achondrites composed mainly of pyroxene (pigeonite plus augite = 70%) and maskelynite (23%); the pyroxenes are thought to be cumulus phases (Stolper and McSween, 1979). They have been studied for REE abundances by Schnetzler and Philpotts (1969), A.J. Irving (unpublished data; in Stolper and McSween, 1979), Ma et al. (1981a, b), and Shih et al. (1981). The two shergottites, Shergotty and Zagami, and the related meteorite, Allan Hills 77005, have similar REE patterns. They are concave downwards with the maximum near Dy. In Shergotty and Zagami, the chondrite-normalized REE are at about 5—10, but in Allan Hills 77005 they range from 1 to 4.

Ma et al. (1981b) noted that it would be difficult to produce this type of pattern if the meteorite is either a cumulate from a melt with chondritic or LREE-enriched liquids, or a residue from partial melting. They suggested that the LREE part of the pattern was not caused by igneous differentiation but may represent an episode of metasomatism with a volatile-rich fluid.

Ma et al. were unable to model this process quantitatively, however. Shih et al. noted that Nd isotope ratios preclude Allan Hills 77005 and Shergotty from being co-magnetic.

3.7. REE in other differentiated meteorites

As expected, most studies of REE in meteorites have been concerned with the stony meteorites; chondrites provide information about early solar system formation processes and achondrites provide information about later planetary fractionation events. A few studies have been made of stony-irons, irons and a few unique meteorite types. These studies have generally been aimed at trying to understand the origin of the silicate component in the meteorite, although a few investigations have been concerned with silicate-free meteorites.

Mesosiderites

Mesosiderites are meteorites composed of approximately equal parts of silicates and metal. The silicates are considered to be igneous in origin and are similar to the silicates of eucrites and howardites (Prior, 1920). The silicate portion of the mesosiderite has been analyzed for REE by Schmitt et al. (1963, 1964), Wänke et al. (1972a), Mittlefehldt (1979), and Mittlefehldt et al. (1979). In particular, Mittlefehldt and co-workers were comparing the silicates of mesosiderites to those of howardites.

Mittlefehldt et al. (1979) analyzed six mesosiderites and found that in all but one of them, the LREE were enriched relative to the HREE by a factor of about 2, and only small or non-existent Eu anomalies were observed. In Morristown, however, La was depleted relative to Lu by a factor of 3, and a large positive Eu anomaly of a factor of 3 was observed. They noted that it was difficult to get the LREE enrichment with a fractional crystallization model and proposed several models to form the mesosiderite silicates via partial melting. They chose a small degree of partial melting, 2.5—4%, of a source depleted to about 0.3—0.4 times chondritic abundances. The source needed to have a flat REE pattern, and they claimed that this was expected of a eucrite residue if it contained some trapped liquid. Thus, according to their model, the eucrite source material, after losing its liquid to form the eucrites, melted a few percent more to generate a liquid from which the mesosiderite silicates formed.

The Mittlefehldt et al. (1979) data on Morristown is similar to the Wänke et al. (1972a) data on three other mesosiderites and suggests a cumulative origin. They calculated that 80% equilibrium crystallization from a melt, which was enriched to 3.3 times chondritic abundances, gave a good fit to the data. Mittlefehldt et al. (1979) concluded that the REE data from the mesosiderites indicate that these meteorites are not closely related to the howardites or eucrites. Most of the howardites and eucrites have flat REE patterns, whereas the mesosiderites have patterns indicative of a more extensive igneous history. Although the models for the formation of the REE patterns in the mesosiderites may not necessarily be correct, the conclusion of, at best, a distant relationship between these mesosiderites and the eucrites or howardites seems quite reasonable.

Pallasites

The pallasites are stony-iron meteorites with about equal amounts of metal and olivine. They are generally thought to have formed at the core-mantle interface in the interior of the parent body. Only two pallasites have been studied for REE, Brenham and Thiel Mountains by Schmitt et al. (1964), and Brenham by Masuda (1968). Both patterns show a U-shape, though it is much more pronounced in Brenham. The REE in Thiel Mountains are more abundant that in Brenham by factors of 2—10.

Consolmagno (1979) attempted to model the REE patterns in the pallasites with little success. He concluded that these meteorites probably did not come from the eucrite parent body, but this conclusion was based more on petrographic evidence, and the idea that the eucrite parent body is still intact so that its core-mantle interface could not have been sampled.

Irons

Several classes of iron meteorites have silicate inclusions. Bild (1977) determined REE abundances in four IAB iron meteorites. In two of the meteorites, Copiapo and Landes, the HREE were flat at chondritic levels with no Eu anomaly, but La was depleted by about a factor of 2. Woodbine had a pattern that was slightly concave down, was enriched from about 1.5 to 2.5 times chondritic abundances and had a negative Eu anomaly of about a factor of 2. Masuda (1969) also analyzed silicates from Woodbine and found a similar pattern. The shape was better determined by Masuda, who analyzed ten REE compared to five by Bild. Based on a variety of other trace elements, Bild concluded that the silicates were chondritic; the cause of the LREE depletion was not clear, but a loss of a small amount of partial melt was suggested as a possibility.

Fukuoka et al. (1978) determined the REE in silicates from Lodran, a unique meteorite containing nearly equal amounts of metal, olivine and pyroxene. They found a chondrite-normalized pattern with the LREE being flat at about 0.5, a negative Eu anomaly of a factor of 2.7 and a monotonic increase in HREE to 1.2. They also interpreted the pattern as being due to the loss of a small amount of partial melt from chondritic parent material.

Two inclusions from Campo del Cielo, an IA iron meteorite with inclusions of silicate, graphite and troilite, have REE patterns which are concave upwards with large positive Eu anomalies (Bild, 1977). The average chondrite-normalized pattern has La = 0.1, Sm = 0.07, Eu = 1, and Lu = 0.7. The pattern was attributed to a mixture of plagioclase and diopside (The sample contained 57% orthopyroxene, 27% olivine, 16% plagioclase and 0.3% diopside.)

Evensen et al. (1979b) measured REE abundances in five inclusions from Weekeroo Station, a IIE iron meteorite. The five inclusions have very different chondrite-normalized REE patterns. Two are fairly flat at values of 0.2 and 5, with positive Eu anomalies of a factor of 10 and 2, respectively. Three others show the HREE enriched relative to the LREE by a factor of 2—5, with both positive and negative Eu anomalies. Evensen et al. claimed that to a first approximation four of the five patterns could be related to a single parent liquid but that this liquid was highly evolved and fractionated relative to chondritic material. They did not give any further details on the nature of the fractionations. It is clear, however, that Weekeroo Station silicates are not at all chondritic and are quite different from those studied by Bild (1977) in the IAB irons.

Abundances of REE in iron meteorites without silicate inclusions have also been studied and, as expected, are found to be very low. Schmitt et al. (1963) measured REE in Odessa and Yardymly (= Aroos) and found chondrite-normalized abundances at about 10^{-4}. Because this value, although quite low, was about a factor of 10^3 higher than that predicted from thermodynamic data, they attributed the REE concentrations to contamination or to a small silicate inclusion. The thermodynamic calculations, however, did not include effects of non-ideal solution, either in the silicates or the metal, so the calculations could easily be in error by a factor of 10^3.

Kaiser et al. (1981) and Nozette and Boynton (1981) measured REE in iron meteorites for another reason. They were looking for a possible decay product of an extinct siderophile superheavy element. The superheavy element is expected to decay by spontaneous fission and would contribute neutron-rich isotopes of the REE to the meteorite. Kaiser et al. measured Sm and Nd in Santa Clara, Cañon Diablo, and Piñon by mass spectrometry and found abundances of the various isotopes on the order of $5—100 \times 10^8$ atoms per gram. This value translates to about 2×10^{-12} grams of Sm per gram, or 10^{-5} times chondritic abundance. Nozette and Boynton (1981) found $(10.8 \pm 1.8) \times 10^8$ atoms of ^{152}Sm per gram (10^{-12} g/g, or 5×10^{-6} times chondritic abundance) in Santa Clara. These meteorites were selected because studies of Ag isotopic ratios showed an excess of ^{107}Ag, presumably due to the decay of ^{107}Pd ($t_{1/2} = 6.5 \times 10^6$ years), suggesting that the meteorites are very old and may preserve a record of other extinct radionuclides (Kelly and Wasserburg, 1978). Values or upper limits for heavier REE were determined in Santa Clara by Nozette and Boynton (1981) and ranged from 42 for Eu to 0.39 for Lu in units of 10^8 atoms per gram.

Neither group found any evidence for a significant amount of a superheavy element.

3.8. Summary of REE cosmochemistry and future directions

As in studies of terrestrial samples, the analysis of REE in extraterrestrial materials provides a powerful tool in understanding the processes that were active in forming the sample. In studies of material formed at the very beginning of our solar system the REE provide a unique set of elements which are highly sensitive to the condensation process, but only moderately sensitive to mineral preference effects. Any later planetary fractionation cannot erase abundance anomalies that are imprinted during condensation. The REE have provided certain evidence of very efficient gas/solid fractionation mechanisms which occurred in the early history of solar system formation. Studies of individual chondrules have shown that their REE patterns are fractionated in a way unlike that expected for gas/solid fractionation and strongly suggest that the chondrules received their REE pattern as signatures of pre-existing minerals.

In the achondrites, which have undergone processes familiar to geochemists, the REE patterns provide strong constraints on the processes that formed the rocks. For example, the non-cumulate eucrites (Fig. 3.8a) must certainly represent partial melt liquids, even if the exact extent of partial melting or the initial concentrations in the source are not tightly constrained. Likewise, the cumulate eucrites (Fig. 3.8b) must, indeed, be cumulates, even if the exact origin of the magma from which they formed is unclear. In studies of these planetary fractionation processes, the constraints provided by REE studies can be particularly important because we necessarily do not have the field data that can provide constraints in studies of terrestrial systems. Interpreting REE data on clasts and inclusions from rare meteorites is more difficult, of course, because only a few samples are available for comparison, but nevertheless, constraints on the origin of the material can usually be provided.

Future directions in REE studies of meteorites are difficult to predict. In the area of Ca,Al-rich inclusions, it seems clear that much of the future work will be done on inclusions from meteorites other than Allende. It seems certain that a variety of different REE patterns will be found which will provide much information on early nebula-fractionation processes. The origin of chondrules has always intrigued cosmochemists and it is likely that more work in this area will provide constraints on chondrule formation processes.

Advances in techniques are opening up new areas of study. The study of milligram-sized chondrules by isotope dilution mass spectrometry would not have been possible a few years ago. New radiochemical neutron activation analysis techniques permit studies of microgram-sized inclusions with precisions for most elements of 1—2%, comparable to those achieved by mass spectrometry. These improvements, and ones which will certainly follow, will permit more detailed examination of smaller samples. It is now possible, for example, to extract a 30-μm-sized grain from a thin section and determine its REE pattern.

For studies of achondrites, programs sponsored by both the United States and Japan to find meteorites in Antarctica are providing much-needed new material. At least two meteorites similar to the shergottites have been found (doubling the world's inventory) and promise to aid our understanding of this important group of meteorites. Several polymict eucrites have been found. The clasts in these eucrites may provide new types of material with different relationships to the igneous fractionation processes which occurred on the eucrite parent body. These new materials may help to compensate for what currently seems to be the limited sampling of the parent body.

The prospect for REE cosmochemistry in the coming decade looks good. We can all look forward to significant advances in our understanding of the origin and evolution of the solar system. These advances will certainly come in part from REE studies.

Acknowledgements

This work was supported in part by NASA grant NSG-7436. The manuscript was typed by J. Henderson.

References

Alfven, H. and Arrhenius, G., 1976. *Evolution of the Solar System*. NASA SP 345, 503 pp.

Allen, R.O., Jr. and Mason, B., 1973. Minor and trace elements in some meteoritic minerals. *Geochim. Cosmochim. Acta*, 37: 1435—1456.

Arrhenius, G. and Alfven, H., 1971. Fractionation and condensation in space. *Earth Planet. Sci. Lett.*, 10: 253—267.

Berkley, J.L., Brown, G.H., IV and Keil, K. 1976. The Kenna ureilite: an ultramafic rock with evidence for igneous, metamorphic, and shock origin. *Geochim. Cosmochim. Acta*, 40: 1429—1437.

Bild, R.W., 1977. Silicate inclusions in group IAB irons and a relation to the anomalous stones Winona and Mt. Morris (Wis.). *Geochim. Cosmochim. Acta*, 41: 1439—1456.

Borodin, L.S. and Barinskii, R.L., 1960. Rare earths in perovskites (knopites) from massifs of ultrabasic alkalic rocks. *Geochemistry (U.S.S.R.)*, 4: 343—351.

Boynton, W.V., 1975a. Fractionation in the solar nebula: condensation of yttrium and the rare earth elements. *Geochim. Cosmochim. Acta*, 39: 569—584.

Boynton, W.V., 1975b. The application of rare-earth elements to infer gas-grain temperature differentials in the solar nebula. *Meteoritics*, 10: 369—371.

Boynton, W.V., 1978a. Rare-earth elements as indicators of supernova condensation. In: *Lunar Science IX*. Lunar Science Institute, Houston, Texas, pp. 120—122.

Boynton, W.V., 1978b. The chaotic solar nebula: evidence for episodic condensation in several distinct zones. In: T. Gehrels (Editor), *Protostars and Planets*. University of Arizona Press, Tucson, Ariz., pp. 427—438.

Boynton, W.V., 1979. Neutron activation analysis. In: K.A. Gschneidner, Jr. and L. Eyring (Editors), *Handbook on the Physics and Chemistry of Rare Earths, 4*. North-Holland, Amsterdam, pp. 457—470 (Chapter 37F).

Boynton, W.V. and Cunningham, C.C., 1981. Condensation of refractory lithophile trace elements in the solar nebula and in supernovae. In: *Lunar Science XII*. Lunar Science Institute, Houston, Texas, pp. 106—108.

Boynton, W.V., Starzyk, P.M. and Schmitt, R.A., 1976. Chemical evidence for the genesis of the ureilites, the achondrite Chassigny and the nakhlites. *Geochim. Cosmochim. Acta*, 40: 1439—1447.

Boynton, W.V., Frazier, R.M. and Macdougall, J.D., 1980. Identification of an ultra-refractory component in the Murchison meteorite. In: *Lunar Science XI*. Lunar Science Institute, Houston, Texas, pp. 103—105.

Bunch, T.E. and Reid, A.M., 1975. The nakhlites, I. Petrography and mineral chemistry. *Meteoritics*, 10: 303—315.

Cameron, A.G.W., 1978. Physics of the primitive solar accretion disk. *Moon Planets*, 18: 5—40.

Cameron, A.G.W. and Truran, J.W., 1977. The supernova trigger for the formation of the solar system. *Icarus*, 30: 447—461.

Chou, C.-L., Baedecker, P.A. and Wasson, J.T., 1976. Allende inclusions: volatile-element distribution and evidence for incomplete volatilization of presolar solids. *Geochim. Cosmochim. Acta*, 40: 85—94.

Clarke, R.S., Jr., Jarosewich, E., Mason, B., Nelen, J., Gomez, M. and Hyde, J.R., 1971. The Allende Mexico meteorite shower. *Smithsonian Contrib. Earth Sci.* 5: 1—53.

Clayton, D.D., 1975. Extinct radioactivities: trapped residuals of presolar grains. *Astrophys. J.*, 199: 765—769.

Clayton, D.D., 1979. On isotopic anomalies in samarium. *Earth Planet. Sci. Lett.* 42: 7—12.

Clayton, R.N., Onuma, N., Grossman, L. and Mayeda, T.K., 1977. Distribution of the pre-solar component in Allende and other carbonaceous chondrites. *Earth Planet. Sci. Lett.*, 34: 209—244.

Conard, R., 1976. *A study of the chemical composition of Ca-Al rich inclusions from the Allende meteorite*. M.S. Thesis, Oregon State University, Corvallis, Oreg., 129 pp.

Consolmagno, G.J., 1979. REE patterns versus the origin of the basaltic achondrites. *Icarus*, 40: 522—530.

Consolmagno, G.J. and Drake, M., 1977. Composition and evolution of the eucrite parent body: evidence from rare earth elements. *Geochim. Cosmochim Acta*, 41: 1271—1282.

Curtis, D.B. and Schmitt, R.A., 1979. The petrogenesis of L-6 chondrites: insights from the chemistry of minerals. *Geochim. Cosmochim. Acta*, 43: 1091—1103.

Davis, A.M. and Grossman, L., 1979. Condensation and fractionation of rare earths in the solar nebula. *Geochim. Cosmochim. Acta*, 43: 1611—1632.

Davis, A.M., Grossman, L. and Allen, J.M., 1978. Major and trace element chemistry of separated fragments from a hibonite-bearing Allende inclusion. *Proc. 9th Lunar Sci. Conf.*, pp. 1235—1247.

Dearborn, D.S. and Boynton, W.V., 1981. Oxygen isotopes, rare-earth elements, supernovae and Wolf-Rayet stars. *Abstr., 44th Annu. Meteorit. Soc. Meet., Bern*, p. 40.

Dodd, R.T., 1978. Compositions of droplet chondrules in the Manych (L-3) chondrite and the origin of chondrules. *Earth Planet. Sci. Lett.*, 40: 71—82.

Dodd, R.T. and Jarosewich, E., 1980. Chemical variations among L-group chondrites, I. The Aïr, Apt and Tourinnes-La-Grosse (L6) chondrites. *Meteoritics*, 15: 69—83.

Dreibus, G., Kruse, H., Rammensee, W., Spettel, B. and Wänke, H., 1977. Cosmochemical constraints on planetary compositions: Earth, moon, eucrite parent body. In: *Lunar Science VIII*. Lunar Science Institute, Houston, Texas, pp. 251—253.

Duke, M.B. and Silver, L.T., 1967. Petrology of eucrites, howardites and mesosiderites. *Geochim. Cosmochim. Acta*, 31: 1637—1665.

Evensen, N.M., Hamilton, P.J. and O'Nions, R.K., 1978. Rare-earth abundances in chondritic meteorites. *Geochim. Cosmochim. Acta*, 42: 1199—1212.

Evensen, N.M., Carter, S.R., Hamilton, P.J., O'Nions, R.K. and Ridley, W.I., 1979a. A combined chemical-petrological study of separated chondrules from the Richardton meteorite. *Earth Planet. Sci. Lett.*, 42: 223—236.

Evensen, N.M., Hamilton, P.J., Harlow, G.E., Klimentidis, R., O'Nions, R.K. and Prinz, M., 1979b. Silicate inclusions in Weekeroo Station: planetary differentiates in an iron meteorite. In: *Lunar Science X*. Lunar Science Institute, Houston, Texas, pp. 376—378.

Frazier, R.M. and Boynton, W.V., 1980. Rare-earth element abundances in separates from the enstatite chondrite Abee. *Meteoritics*, 15: 291.

Frazier, R.M. and Boynton, W.V., 1981. Trace element concentration in Abee density separates and minerals. *Meteoritics*, 16: 315.

Fuchs, L.H., Olsen, E. and Jensen, K.J., 1973. Mineralogy, mineral-chemistry and composition of the Murchison (C2) meteorite. *Smithsonian Contrib. Earth Sci.*, 10: 39 pp.

Fukuoka, T., Boynton, W.V., Ma, M.-S and Schmitt, R.A., 1977. Genesis of howardites, diogenites, and eucrites. *Proc. 8th Lunar Sci. Conf.*, pp. 187—210.

Fukuoka, T., Ma, M.-S., Wakita, H. and Schmitt, R.A., 1978. Lodran: the residue of limited partial melting of matter like a hybrid between H and E chondrites. In: *Lunar Science IX*. Lunar Science Institute, Houston, Texas, pp. 356—361.

Gast, P.W. and Hubbard, N.J., 1970. Rare earth abundances in soil and rocks from the Ocean of Storms. *Earth Planet. Sci. Lett.*, 10: 94—101.

Gast, P.W., Hubbard, N.J. and Wiesmann, H., 1970. Chemical composition and petrogenesis of basalts from Tranquillity Base. *Proc. Apollo 11 Lunar Sci. Conf.*, pp. 1143—1163.

Goldschmidt, V.M., 1954. *Geochemistry*. Clarendon, Oxford, pp. 306—318.

Graham, A.L., Easton, A.J. and Hutchison, R., 1976. The Bovedy meteorite; mineral chemistry and origin of its Ca-rich glass inclusions. *Geochim. Cosmochim. Acta*, 40: 529—535.

Gray, C.M., Papanastassiou, D.A. and Wasserburg, G.J., 1973. The identification of early condensates from the solar nebula. *Icarus*, 20: 213—239.

Grossman, L., 1972. Condensation in the primitive solar nebula. *Geochim. Cosmochim. Acta*, 36: 597—619.

Grossman, L., 1973. Refractory trace elements in Ca-Al-rich inclusions in the Allende meteorite. *Geochim. Cosmochim. Acta*, 37: 1119—1140.

Grossman, L., 1975. Petrography and mineral chemistry of Ca-rich inclusions in the Allende meteorite. *Geochim. Cosmochim. Acta*, 39: 433—454.

Grossman, L., 1980. Refractory inclusions in the Allende meteorite. *Annu. Rev. Earth Planet. Sci.*, 8: 559—608.

Grossman, L. and Ganapathy, R., 1975. Volatile elements in Allende inclusions. *Proc. 6th Lunar Sci. Conf.*, pp. 1729—1736.

Grossman, L. and Ganapathy, R., 1976a. Trace elements in the Allende meteorite, I. Coarse-grained, Ca-rich inclusions. *Geochim. Cosmochim. Acta*, 40: 331—334.

Grossman, L. and Ganapathy, R., 1976b. Trace elements in the Allende meteorite, II. Fine-grained, Ca-rich inclusions. *Geochim. Cosmochim. Acta*, 40: 967—977.

Grossman, L., Davis, A.M., Olsen, E. and Santoliquido, P.M., 1972. Chemical studies of condensates in the Murchison type 2 carbonaceous chondrite. In: *Lunar Science VIII*. Lunar Science Institute, Houston, Texas, pp. 377—379.

Grossman, L., Ganapathy, R. and Davis, A.M., 1977. Trace elements in the Allende meteorite, III. Coarse-grained inclusions revisited. *Geochim. Cosmochim. Acta*, 41: 1647—1664.

Grossman, L., Ganapathy, R., Methot, R.L. and Davis, A.M., 1979. Trace elements in the Allende meteorite, IV. Amoeboid olivine aggregates. *Geochim. Cosmochim. Acta*, 43: 817—829.

Grossman, L., Olsen, E., Davis, A.M., Tanaka, T. and MacPherson, G.J., 1981. The Antarctic achondrite ALHA 76005: a polymict eucrite. *Geochim. Cosmochim. Acta*, 45: 1267—1279.

Hamet, J., Noboru, N., Unruh, D.M. and Tatsumoto, M., 1978. Origin and history of the adcumulate eucrite Moama as inferred from REE abundances, Sm-Nd and U-Pb systematics. *Proc. 9th Lunar Sci. Conf.*, pp. 1115—1136.

Hamilton, P.J., Evensen, N.M. and O'Nions, R.K., 1979. Chronology and chemistry of Parnallee (LL-3) chondrules. In: *Lunar Science X*. Lunar Science Institute, Houston, Texas, pp. 494—496.

Haskin, L.A. and Paster, T.P., 1979. Geochemistry and mineralogy of the rare earths. In: K.A. Gschneidner, Jr. and L. Eyring (Editors), *Handbook on the Physics and Chemistry of Rare Earths, 3*. North-Holland, Amsterdam, pp. 1—80. (Chapter 21).

Haskin, L.A., Frey, F.A., Schmitt, R.A. and Smith, R.H., 1966. Meteoritic, solar and terrestrial rare-earth distributions. In: L.H. Ahrens, F. Press, S.K. Runcorn and H.C. Urey (Editors), *Physics and Chemistry of the Earth, 7*. Pergamon, Oxford, pp. 167—321.

Haskin, L.A., Haskin, M.A., Frey, F.A. and Wildeman, T.R., 1968. Relative and absolute terrestrial abundances of the rare earths. In: L.H. Ahrens (Editor), *Origin and Distribution of the Elements*. Pergamon, Oxford, pp. 889—912.

Herndon, J.M. and Wilkening, L.L., 1978. Conclusions derived from the evidence on accretion in meteorites. In: T. Gehrels (Editor), *Protostars and Planets*. University of Arizona Press, Tucson, Ariz., pp. 502—515.

Honda, M. and Shima, M., 1967. Distribution of rare earths among component minerals of Bruderheim chondrite. *Earth Planet. Sci. Lett.*, 2: 344—348.

Hoyle, F. and Wickramasinghe, N.C., 1970. Dust in supernova explosions. *Nature*, 226: 62—63.

Jérome, D.Y., 1970. *Composition and origin of some achondritic meteorites*. Ph.D. Dissertation, University of Oregon, Eugene, Oreg., 156 pp.

Jérome, D.Y. and Goles, G.G. 1971. A reexamination of relationships among pyroxene-plagioclase achondrites. In: A.O. Brunfelt and E. Steinnes (Editors), *Activation Analysis in Geochemistry and Cosmochemistry*, Universitetsforlaget, Oslo, pp. 261—266.

Kaiser, T., Piepgras, D. and Wasserburg, G.J., 1981. A search for evidence of a fissionable nuclide in iron meteorites and implications on heat sources in planetary cores. *Earth Planet. Sci. Lett.*, 52: 239—250.

Keil, K., 1962. Quantitativ-erzmikroskopische Integrationsanalyse der Chondrite (Zur Frage des mittleren Verhaltnisses von Nickeleisen-:Troilit-:Chromit-:Silikat-Anteil in den Chondriten). *Chem. Erde*, 22: 281—348.

Keil, K., 1977. The Angra Dos Reis consortium. *Earth Planet. Sci. Lett.*, 35: 271.

Kelly, W.R. and Wasserburg, G.J., 1978. Evidence for the existence of ^{107}Pd in the early solar system. *Geophys. Res. Lett.*, 5: 1079—1082.

Kieffer, S.W., 1975. Droplet chondrules. *Science*, 189: 333—340.

Kramar, U., 1980. The importance of second order activation in the determination of trace elements in geological samples by instrumental neutron activation analysis. *Geochim. Cosmochim. Acta*, 44: 379—382.

Kurat, G., 1975. Der kohlige Chondrit Lance: Eine petrologische Analyse der komplexen

Genese eines Chondriten. *Tschermaks Mineral. Petrogr. Mitt.*, 22: 38—78.

Larimer, J.W., 1967. Chemical fractionations in meteorites, I. Condensation of the elements. *Geochim. Cosmochim. Acta*, 31: 1215—1238.

Larimer, J.W., 1975. The effect of C/O ratio on the condensation of planetary material. *Geochim. Cosmochim. Acta*, 39: 389—392.

Larimer, J.W. and Anders, E., 1970. Chemical fractionations in meteorites, III. Major element fractionations in chondrites. *Geochim. Cosmochim. Acta*, 34: 367—387.

Lattimer, J.M., Schramm, D.N. and Grossman, L., 1978. Condensation in supernova ejecta and isotopic anomalies in meteorites. *Astrophys. J.*, 219: 230—249.

Lee, T., 1979. New isotopic clues to solar system formation. *Rev. Geophys. Space Phys.*, 17: 1591—1611.

Lee, T., Papanastassiou, D.A. and Wasserburg, G.J., 1976. Demonstration of ^{26}Mg excess in Allende and evidence for ^{26}Al. *Geophys. Res. Lett.*, 3: 109—112.

Lord, H.C., III, 1965. Molecular equilibria and condensation in a solar nebula and cool stellar atmospheres. *Icarus*, 4: 279—288.

Lugmair, G.W., Marti, K. and Scheinin, N.B., 1978. Incomplete mixing of products from *r*-, *p*- and *s*-process nucleosynthesis: Sm-Nd systematics in Allende inclusions EK 1-4-1. In: *Lunar Science IX.* Lunar Science Institute, Houston, Texas, pp. 672—674.

Ma, M.-S. and Schmitt, R.A., 1979. Genesis of the cumulate eucrites Serra de Mage and Moore County: a geochemical study. *Meteoritics*, 14: 81—89.

Ma, M.-S., Murali, A.V. and Schmitt, R.A., 1977. Genesis of the Angra Dos Reis and other achondritic meteorites. *Earth Planet. Sci. Lett.*, 35: 331—346.

Ma, M.-S., Laul, J.C. and Schmitt, R.A., 1981a. Analogous and complementray rare earth element patterns on meteorite parent bodies and the Earth inferred from a study of the achondrite ALHA 77005, other unique achondrites and the shergottites. In: *Lunar Science XII.* Lunar Science Institute, Houston, Texas, pp. 634—636.

Ma, M.-S., Laul, J.C. and Schmitt, R.A. 1981b. Complementary rare earth element patterns in unique achondrites, such as ALHA 77005 and shergottites, and in the earth. *Proc. 12th Lunar Planet. Sci. Conf.*, pp. 1349—1358.

Macdougall, J.D., 1979. Refractory-element-rich inclusions in CM meteorites. *Earth Planet. Sci. Lett.*, 42: 1—6.

MacPherson, G.J., Bar Matthews, M., Tanaka, T., Olsen, E. and Grossman, L., 1980. Refractory inclusions in Murchison: recovery and mineralogical description. In: *Lunar Science XI.* Lunar Science Institute, Houston, Texas, pp. 660—662.

Martin, P.M. and Mason, B., 1974. Major and trace elements in the Allende meteorite. *Nature*, 249: 333—334.

Mason, B., 1962. *Meteorites.* Wiley, New York, N.Y., 274 pp.

Mason, B. and Martin, P.M., 1974. Minor trace element distribution in melilite and pyroxene from the Allende meteorite. *Earth Planet. Sci. Lett.*, 22: 141—144.

Mason, B. and Martin, P.M., 1977. Geochemical differences among components of the Allende Meteorite. *Smithsonian Contrib. Earth. Sci.*, 19: 84—95.

Mason, B., Nelen, J.A., Muir, P. and Taylor, S.R., 1976. The composition of the Chassigny meteorite. *Meteoritics*, 11: 21—27.

Masuda, A., 1968. Lanthanide concentrations in the olivine phase of the Brenham pallasite. *Earth Planet. Sci. Lett.*, 5: 59—62.

Masuda, A., 1969. Lanthanides in the silicate inclusion of the Woodbine meteorite. *Nature*, 224: 164—165.

Masuda, A., Nakamura, N. and Tanaka, T., 1973. Fine structures of mutually normalized rare-earth patterns of chondrites. *Geochim. Cosmochim. Acta*, 37: 239—248.

McCarthy, T.S., Ahrens, L.H. and Erlank, A.J., 1972. Further evidence in support of the mixing model for howardite origin. *Earth Planet. Sci. Lett.*, 15: 86—93.

McCulloch, M.T. and Wasserburg, G.J., 1978a. Barium and neodymium isotopic anomalies in the Allende meteorite. *Astrophys. J.*, 220: L15—L19.

McCulloch, M.T. and Wasserburg, G.J., 1978b. More anomalies from the Allende meteorite: samarium. *Geophys. Res. Lett.*, 5: 599—602.

Mittlefehldt, D.W., 1979. Petrographic and chemical characterization of igneous lithic clasts from mesosiderites and howardites and comparison with eucrites and diogenites. *Geochim. Cosmochim. Acta*, 43: 1917—1935.

Mittlefehldt, D.W., Chou, C.-L. and Wasson, J., 1979. Mesosiderites and howardites: igneous formation and possible genetic relationships. *Geochim. Cosmochim. Acta*. 43: 673—688.

Nagasawa, H. and Onuma, N., 1979. High temperature heating of the Allende meteorite, II. Fractionation of the rare earth elements. In: *Lunar Science X*. Lunar Science Institute, Houston, Texas, pp. 884—886.

Nagasawa, H., Schreiber, H.D. and Blanchard, D.P., 1976. Partition coefficients of REE and Sc in perovskite, melilite, and spinel and their implications for Allende inclusions. In: *Lunar Science VII*. Lunar Science Institute, Houston, Texas, pp. 588—590.

Nagasawa, H., Blanchard, D.P., Jacobs, J., Brannon, J.C. and Philpotts, J.A., 1977. Trace element distribution in mineral separates of the Allende inclusions and their genetic implications. *Geochim. Cosmochim. Acta*, 41: 1587—1600.

Nagasawa, H., Yamakoshi, K. and Shimamura, T., 1979. Trace element concentrations in silicate spherules from oceanic sediments. *Geochim. Cosmochim. Acta*, 43: 267—272.

Nagasawa, H., Schreiber, H.D. and Morris, R.V., 1980a. Experimental mineral/liquid partition coefficients of the rare earth elements (REE), Sc and Sr for perovskite, spinel and melilite. *Earth Planet. Sci. Lett.*, 46: 431—437.

Nagasawa, H., Yamakoshi, K., Korotev, R.L., Haskin, L.A. and Glass, B.P., 1980b. Trace element concentrations in glassy silicate spherules and microtektites from deep-sea sediments. In: *Lunar Science XI*. Lunar Science Institute, Houston, Texas, pp. 780—781.

Nagasawa, H., Blanchard, D.P., Shimizu, H. and Masuda, A., 1981. Rare-earths in the Allende inclusion, EK 1-4-1 and the origin of Allende inclusions. In: *Lunar Science XII*. Lunar Science Institute, Houston, Texas, pp. 744—746.

Nakamura, N., 1974. Determination of REE, Ba, Fe, Mg, Na and K in carbonaceous and ordinary chondrites. *Geochim. Cosmochim. Acta*, 38: 757—775.

Nakamura, N. and Masuda, A., 1973. Chondrites with peculiar rare-earth patterns. *Earth Planet. Sci. Lett.*, 19: 429—437.

Noddack, I., 1935. Die Haufigkeiten der seltenen Erden in Meteoriten. *Z. Anorg. Allg. Chem.*, 225: 337—364.

Noonan, A.F., Nelen, J., Fredriksson, K. and Newbury, D., 1977. Zr-Y oxides and high-alkali glass in an ameboid inclusion from Ornans. *Meteoritics*, 12: 332—334.

Nozette, S. and Boynton, W.V., 1981. Superheavy elements: an early solar system upper limit for elements 107 to 110. *Science*, 214: 331—333.

O'Nions, R.K., Carter, S.R., Evensen, N.M., Hamilton, P.J. and Ridley, W.I., 1978. Chemical-petrological studies of individual chondrules from the Richardton meteorite. In: *Lunar Science IX*. Lunar Science Institute, Houston, Texas, pp. 826—828.

Onuma, N., Tanaka, T. and Masuda, A., 1974. Rare-earth abundances in two mineral separates with distinct oxygen isotopic composition from an Allende inclusion. *Meteoritics*, 9: 387—388.

Osborn, T.W., Warren, R.G., Smith, R.H., Wakita, H., Zellmer, D.L., and Schmitt, R.A., 1974. Elemental composition of individual chondrules from carbonaceous chondrites, including Allende. *Geochim. Cosmochim. Acta*, 38: 1359—1378.

Palme, H. and Wlotzka, F., 1981. Iridium-rich phases in Ornans. *Abstr., 44th Annu. Meteorit. Soc. Meet., Bern*, p. 54.

Palme, H., Baddenhausen, H., Blum, K., Cendales, M., Dreibus, G., Hofmeister, H., Kruse, H., Palme, C., Spettel, B., Vilcsek, E. and Wänke, H., 1978. New data on lunar samples and achondrites and a comparison of the least fractionated samples from the earth, the moon and the eucrite parent body. *Proc. 9th Lunar Planet Sci. Conf.*, pp. 25—27.

Prinz, M., Hlava, P.H. and Keil, K., 1974. The Chassigny meteorite: a relatively iron-rich cumulate dunite. *Meteoritics*, 9: 393—394.

Prinz, M., Keil, K., Hlava, P.F., Berkley, J.L., Gomes, C.B. and Curvello, W.S., 1977. Studies of Brazilian meteorites, III. Origin and history of the Angra Dos Reis achondrite. *Earth Planet. Sci. Lett.*, 35: 317—330.

Prior, G.T., 1918. On the mesosiderite-grahamite group of meteorites: with analyses of Vacsa Muerta, Hainholz, Simondium, and Powder Mill Creek. *Mineral. Mag.*, 18: 151—172.

Prior, G.T., 1920. The classification of meteorites. *Mineral. Mag.*, 19: 21—52.

Rubin, A.E. and Keil, K., 1980. Mineralogy and petrology of the Abee enstatite chondrite. *Meteoritics*, 15: 358—359.

Schmitt, R.A. and Smith, R.H. 1963. Implications of similarity in rare-earth fractionation of nakhlitic meteorites and terrestrial basalts. *Nature*, 199: 550—551.

Schmitt, R.A., Mosen, A.W., Suffrendini, C.S., Lasch, J.E., Sharp, R.A. and Olehy, D.A., 1960. Abundances of the rare-earth elements, lanthanum to lutetium, in chondritic meteorites. *Nature (London)*, 186: 863.

Schmitt, R.A., Smith, R.H., Lasch, J.E., Mosen, A.W., Olehy, D.A. and Vasilevskis, J., 1963. Abundances of the fourteen rare-earth elements, scandium, and yttrium in meteoritic and terrestrial matter. *Geochim. Cosmochim. Acta*, 27: 577—622.

Schmitt, R.A., Smith, R.H. and Olehy, D.A., 1964. Rare-earth, yttrium and scandium abundances in meteoritic and terrestrial matter, II. *Geochim. Cosmochim. Acta*, 28: 67—86.

Schmitt, R.A., Smith, R.H. and Olehy, D.A., 1968. Rare earth abundances in meteorite chondrules. In: L.H. Ahrens (Editor), *Origin and Distribution of the Elements.* Pergamon, Oxford, pp. 273—282.

Schnetzler, C.C. and Bottino, M.L., 1971. Some alkali, alkaline earth, and rare earth element concentrations and the Rb-Sr age of the Lost City meteorite and separated phases. *J. Geophys. Res.*, 76: 4061—4066.

Schnetzler, C.C. and Philpotts, J.A., 1969. Genesis of the calcium-rich achondrites in light of rare-earth and barium concentrations. In: P. Millman (Editor), *Meteorite Research.* D. Reidel, Dordrecht, pp. 206—216.

Shih, C.-Y., Nyquist, L.E., Bansal, B.M., Wiesmann, H., Wooden, J.L. and McKay, G., 1981. REE, Sr and Nd isotopic studies on shocked achondrites — Shergotty, Zagami and ALHA 77005. In: *Lunar Science XII.* Lunar Science Institute, Houston, Texas, pp. 973—975.

Shima, M. and Honda, M., 1967. Distributions of alkali, alkaline earth and rare earth elements in component minerals of chondrites. *Geochim. Cosmochim. Acta*, 31: 1995—2006.

Steinberg, E.P. and Wilkins, B.D., 1978. Implications of fission mass distributions for the astrophysical *r*-process. *Astrophys. J.*, 223: 1000—1014.

Stolper, E., 1977. Experimental petrology of eucritic meteorites. *Geochim. Cosmochim. Acta*, 41: 587—611.

Stolper, E. and McSween, H.Y. Jr., 1979. Petrology and origin of the shergottite meteorites. *Geochim. Cosmochim. Acta*, 43: 1475—1498.

Suess, H.E. and Urey, H.C., 1956. Abundances of the elements. *Rev. Mod. Phys.*, 28: 53—74.

Sugiura, N. and Strangway, D.W., 1981. Magnetic properties of the Abee meteorite. *Abstr., 44th Annu. Meteorit. Soc. Meet., Bern*, p. 34.

Tanaka, T. and Masuda, A., 1973. Rare-earth elements in matrix, inclusions and chondrules of the Allende meteorite. *Icarus*, 4: 523—530.

Tanaka, T., Nakamura, N., Masuda, A. and Onuma, N., 1975. Giant olivine chondrule as a possible later-stage product in the nebula. *Nature*, 256: 27—28.

Tanaka, T., Davis, A.M., Grossman, L., Lattimer, J.M., Allen, J.M., Lee, T. and Wasserburg, G.J., 1979. Chemical study of an isotopically-unusual Allende inclusion. In: *Lunar Science X.* Lunar Science Institute, Houston, Texas, pp. 1203—1205.

Tanaka, T., Davis, A.M., Hutcheon, I.D., Bar Matthews, M., Olsen, E., MacPherson, G.J. and Grossman, L.A., 1980. Refractory inclusions in Murchison: chemistry and Mg isotopic composition. In: *Lunar Science XI.* Lunar Science Institute, Houston, Texas, pp. 1122—1124.

Taylor, S.R., 1975. *Lunar Science: A Post-Apollo View.* Pergamon, Oxford, 372 pp.

Taylor, S.R., 1982. *Planetary Science: A Lunar Perspective.* Lunar and Planetary Institute, Houston, Texas, 481 pp.

Taylor, S.R. and Mason, B.H., 1978. Chemical characteristics of Ca-Al inclusions in the Allende meteorite. In: *Lunar Science IX.* Lunar Science Institute, Houston, Texas, pp. 1158—1160.

Van Schmus, W.R. and Wood, J.A. 1967. A chemical-petrologic classification for the chondritic meteorites. *Geochim. Cosmochim. Acta*, 31:747—765.

Wakita, H., Rey, P. and Schmitt, R.A., 1971. Abundances of the 14 rare-earth elements and 12 other trace elements in Apollo 12 samples: five igneous and one breccia rocks and four soils. *Proc. 2nd Lunar Science Conf.*, pp. 1319—1329.

Wänke, H., Baddenhausen, H., Balascescu, A., Teschke, F., Spettel, B., Dreibus, G., Palme, H., Quijano-Rico, M., Kruse, H., Wlotzka, F. and Begemann, F., 1972a. Multielement analyses of lunar samples and some implications of the result. *Proc. 3rd Lunar Sci. Conf.*, pp. 1251—1268.

Wänke, H., Baddenhausen, H., Spettel, B., Teschke, F., Quijano-Rico, M., Dreibus, G. and Palme, H. 1972b. The chemistry of Havero ureilite. *Meteoritics*, 7: 579—589.

Wänke, H., Baddenhausen, H., Palme, H., Spettel, B., 1974. On the chemistry of the Allende inclusions and their origin as high temperature condensates. *Earth Planet. Sci. Lett.* 23: 1—7.

Wark, D.A., Hughes, T.C., Lucas, M. and McKenzie, C.D., 1980. Proton microprobe analysis of Perovskite in Allende Ca-Al-rich inclusions. In: *Lunar Science XI.* Lunar Science Institute, Houston, Texas, pp. 1205—1207.

Wasserburg, G.J., Lee, T. and Papanastassiou, D.A., 1977. Correlated and Mg isotopic anomalies in Allende inclusions, II. Magnesium. *Geophys. Res. Lett.*, 4: 299—302.

Wasson, J.T., 1974. *Meteorites.* Springer-Verlag, Berlin, 316 pp.

Wasson, J.T. 1978. Maximum temperatures during the formation of the solar nebula. In: T. Gehrels (Editor), *Protostars and Planets.* University of Arizona Press, Tucson, Ariz., pp. 488—501.

Chapter 4

PETROGENETIC MODELLING — USE OF RARE EARTH ELEMENTS

LARRY A. HASKIN

4.1. Introduction

The geochemical behavior of REE is intrinsically interesting, but use of REE abundances in determining evolutionary histories of suites of rocks is even more interesting and important. Consider, for example, a suite of lavas. Did they stem from a common magma or a common solid source? What was the composition of the initial solid source and how does it relate to that of average mantle or crust? By what process was the magma produced, and how did the REE behave during its formation? Was the composition of the magma modified before the lavas erupted? If so, by what process, and how did the REE behave? Was there further compositional modification during solidification once the lava was emplaced, by what processes, and with what REE behavior? Was there subsequent compositional alteration, by what processes, and with what REE behavior? Similar sets of questions can be written for intrusive igneous rocks, for metamorphic rocks, and for sedimentary rocks.

We are seldom favored to make direct observations of the processes governing rock compositions, because of where those processes occur, the times at which they occur, and the time scale over which they take place. Instead, we rely on circumstantial evidence in accessible materials and attempt to construct self-consistent pictures of events that took place and the nature of the starting materials. One important type of evidence is REE abundances; interpretation of observed REE abundances depends on modelling. By "modelling" is meant a well-defined description of REE behavior during the process of magma or rock formation. Predictions of a model may be compared with observed REE abundances in a rock or magma, if the model includes a description of the starting material. Alternatively, the modelling may be inverted and the REE abundances in the rock or mineral used to predict the nature of the starting material. In either case, the conclusions of the exercise are "model dependent", i.e., valid only to the extent that the composition-producing process was correctly identified and the chemical behavior of the REE during that process was adequately described.

A model need not provide a detailed description of a process to be useful. Archeologists use trace element "fingerprints" to determine from which geological sources obsidian artifacts found at ancient camp sites might have come. Their underlying model merely states that each geological source of obsidian has a unique composition. Insofar as this has been tested by analyses of obsidians from different sources it has proved to be correct. To a geochemist, this observation might imply that obsidians are an extreme end product of some process that exaggerates small differences in starting material or conditions of separation. Alternatively, obsidians might be less extreme products of a process acting on starting materials that happen to be compositionally unique. The geochemist should wish to know just what processes are responsible for the compositions of obsidians. Until he has determined this and has attained a quantitative description of those processes, he cannot predict for the archeologist the circumstances under which different sources of obsidian might have analytically indistinguishable compositions. Such an occurrence could lead to erroneous archeological conclusions about the distances over which migration or trading might have taken place.

In fact, the geochemist will probably not know what processes are responsible for obsidian compositions until he has developed and tested a number of models for those processes. Qualitative observations may be adequate to eliminate certain processes from consideration and semi-quantitative observations might eliminate others. The process of formation of a given obsidian body will probably not be convincingly identified until a quantitative model for that body has been produced, one that includes the type and composition of the starting material. Then comes the question of how well the model for that obsidian can be extrapolated to others.

A "quantitative" model here means one that predicts REE (and other elemental) concentrations to approximately the accuracy with which they can be measured. The materials and processes envisioned in such a model must be consistent with other information, e.g., that for major elements, isotopes, petrography, field relationships, geophysical constraints, and results of experimental petrology. Many investigators believe that such accurate predictions will never be possible or, at least, not be worthwhile because real rocks are just too complicated to describe that well. Certainly, such accurate descriptions have been developed for few systems so far. This writer believes, however, that the potential value of REE and other trace element abundances for developing and constraining our understanding of how compositions of rocks are produced and modified has not nearly been realized. It is common practice to rationalize away discrepancies between predictions of models and observed REE abundances as insignificant on the basis that close agreement is too much to expect. This is tantamount to discarding key information about the processes by which rocks really attain their compositions. This seems to occur because investigators are confident that they know a priori what chemical-petrologic processes are principally responsible for the characteristics of the rocks under study. This is despite the failure of the models for those processes to predict the observed compositions.

The point is not that the selection of chemical-petrologic processes is wrong if the results of the modelling disagree with observed compositions. In fact, the models used have usually been too simple to describe accurately the processes envisioned. More sophisticated versions, tailored to the rock suites under consideration, need to be developed. The point is that *as long as a discrepancy exists between predicted and observed REE concentrations and no specific reason for the discrepancy can be defended, the study is incomplete* and substantial uncertainty remains. Recent careful studies are indicating that the fault of the discrepancy does not always lie entirely with the modelling but sometimes with the choice of process as well. This penetrates to the core of our notions of how rocks form. Perhaps the occurrence of processes almost dogmatically accepted as predominant during rock formation is not as widespread as believed.

4.2. Modelling of separation processes

Most processes of rock or mineral formation involve either physical separation or physical addition of two or more materials. Modelling begins with selecting the chemical process that distributes the elements among the resulting materials. Then a mathematical description that accounts for both the chemical and physical processes is developed. Consider the crystallization of a mineral grain from a melt. Depending on the time scale of the separation into mineral grain plus residual melt, the elemental distribution might be fully thermodynamically controlled such that the crystal was not zoned with respect to any constituent; all elements would have equilibrium concentrations relative to the residual parent liquid. If the time scale were shorter, the elemental distribution might be thermodynamically controlled only at the growing crystal surface, so that changes in the concentrations of elements in the liquid would be reflected only in the layer of crystal currently at the surface. This would lead to a zoned crystal whose interior was effectively isolated from further reaction with the parent liquid. If the time scale were faster still, concentration gradients might develop at the surface of the growing crystal because diffusion and convection rates were too slow to maintain homogeneity of the residual liquid. The final rock as analyzed might be a mixture of mineral grains plus frozen residual liquid, but in proportions different from the average for the entire system. As the rock solidified, late-stage growth of the mineral grain might take place with access to a disproportionate amount of residual liquid. Some element redistribution might occur because of changes in temperature or recrystallization. This would lead to a fairly complicated model for such a seemingly simple event, but a model composed of rather simple parts. If the task of producing such a model seems overwhelming, it usually is not, and must in any case be done if quantitative agreement is sought. The range of useful models will extend from the purely qualitative to the fully quantitative. The degree of certainty in conclusions about the history of formation of a rock suite will be in proportion to the degree to which a quantitative description has been attained.

In principle, all trace elements should be useful in this sort of modelling. In practice, REE are more useful than most. Mainly, this is because their geochemical behavior changes gradually from the lightest lanthanide, La, to the heaviest, Lu. Thus, separation parameters vary gradually so that when a separation occurs relative REE abundances in both products furnish an intuitive as well as quantitative indication of the nature and extent of the separation process. This situation is intermediate between that for, say, Rb and Sr which are subject to far more drastic separations, and that for Zr and Hf for which hardly any separation is usually observed. In addition, the REE can indicate the extent to which oxidizing or reducing conditions have prevailed; anomalies for Ce and Eu can be measured against the "background" of the other REE. Better than those for most groups of elements, relative REE abundances in a material can be usefully related to a common baseline, for example, chondritic meteorites. Good analytical procedures have been developed (especially neutron activation and mass spectrometric isotope dilution) so precise data are available. Errors of analysis are easier to discern than for other elements; for most materials they appear as relative abundance patterns that are not smooth functions of atomic number relative to chondritic abundances. Relative REE abundances are also useful in discerning samples that are aberrant compared with most members of a sample suite. Finally, REE are relatively refractory, relatively resistant to weathering and alteration, and are believed to be fractionated by mainstream rock-forming processes. Thus, their relative abundances appear to record reliably the effects of primary differentiation processes in igneous geochemistry.

Because most modelling has been done on igneous systems, most of the following discussion is about igneous rocks. The purpose of this chapter is to set forth the principles of modelling and to demonstrate the kinds of results attainable. It is left to the reader to evaluate the results of modelling for all kinds of rocks as described in the subsequent chapters, or to develop the specific models needed to describe the systems of interest to him.

4.3. Simple chemical partitioning — the Nernst D

Suppose the rock- or mineral-forming process involves separation of a homogeneous material into two separate phases. For example, consider a mafic liquid losing heat at constant temperature to form a crystal of diopside. The elements Ca, Mg, Si, and O are essential constituents of the crystal. Other elements such as Fe, Al, and Cr may comprise a few percent of the crystal. Other elements such as the REE are trace constituents which become distributed between the crystal and the parent liquid. Given sufficient time, the crystal and liquid will come to equilibrium and all elements will be distributed between them in accordance with the energy condition for each element in each phase.

The energy condition can be described in terms of the Gibbs free energy (at constant temperature, pressure, and composition, i.e., T, P, n). This can be done by noting that for each element the chemical potential in the crystal must equal that in the liquid, or simply by noting that for each element the activity in the solid and the activity in the liquid will satisfy the expression for the equilibrium constant. Using the element Sm as an example:

$$\underset{\text{liquid}}{Sm^{3+}} + \underset{\text{liquid}}{CC^{1+}} + \underset{\text{liquid}}{2\,SiO_3^{2-}} = \underset{\text{diopside, solid solution}}{SmCC(SiO_3)_2} \qquad (4.1)$$

That is, Sm^{3+} ions in the liquid phase plus charge compensating ions (written arbitrarily here as 1+ ion, but other reasonable combinations could be considered) plus two metasilicate ions react together to form a silicate species that in reality is part of a solid solution in the crystal of diopside. The equilibrium condition at those T, P, and n is given by the following expression in which K_{eq} is the equilibrium constant and the symbol a stands for the activity of the various species in equation (4.1):

$$K_{eq} = \frac{a_{SmCC(SiO_3)_2}}{a_{Sm^{3+}}\, a_{CC^{1+}}\, a^2_{SiO_3^{2-}}} \qquad (4.2)$$

If a chemical analysis is made of the crystal and its liquid, the result is usually given in terms of concentration, not activity, and usually in terms of weight fraction (e.g., parts per million). Concentrations on a molar basis are related to activities through the multiplicative activity coefficient, γ. The change from molar to weight basis involves simple stoichiometry and amounts to a constant, here symbolized by W, which includes representing $SmCC(SiO_3)_2$ as a constant times the weight fraction of Sm in the diopside. Replacing activities by products of activity coefficients and concentrations and solving for the ratio of concentrations for the trace element, Sm, gives the following expression:

$$\frac{[Sm^{3+}, \text{solid}]}{[Sm^{3+}, \text{liquid}]} = \frac{\gamma_{Sm^{3+}} K_{eq} W\, a_{CC^{1+}} a^2_{SiO_3^{2-}}}{\gamma_{SmCC(SiO_3)_2}} = D_{Sm} \qquad (4.3)$$

If the free energies of formation of the entities in this reaction (or the free energy for the reaction) and the activity coefficients were known, the equilibrium ratio of Sm concentration in the crystal to that in the liquid could be calculated. Neither set of quantities is known. It is then common practice to make two simplifying assumptions. First, the ratio of activity coefficients of Sm in the two phases is assumed to be constant. Second, the activities of the charge compensators and the metasilicate ions are assumed constant. This suggests that a trace ion such as Sm^{3+} is present in such small quantities that its behavior never noticeably affects that of major elements, charge compensating ions, or ions competing with Sm^{3+} for sites in the crystal, and that these are present in such overwhelming quantities compared with Sm^{3+} that their effect on the behavior of Sm^{3+} is constant. The right-hand side of equation (4.3) is then constant, is usually given the symbol D, and is called the Nernst distribution coefficient. Derived here for the specific case of Sm^{3+} entering diopside, it can be generalized that, subject to equivalent assumptions about activity coefficients and effects of other elements, the concentration ratio of a trace element between two chemical phases in equilibrium is constant. Over some range of conditions, the assumptions will be valid and the distribution coefficient will be constant. In practice, the concept of the Nernst distribution coefficient will be useful in modelling only if D is not a sensitive function of T, P, and n or if variations with T, P, and n are known. This is because there is substantial variability in the trace and major element compositions of liquids capable of crystallizing a given mineral in equilibrium and the T and P under which they do so. McIntire (1963) and Banno and Matsui (1973) further discuss the nature of distribution coefficients.

Masuda (1965, 1967) called attention to the systematic variation of REE D's with atomic number. Values of D for REE for minerals relative to their parent liquids have been determined on natural systems and on synthetic systems, as discussed in Chapter 1. The first useful measurements by this technique were done by Schnetzler and Philpotts (1968, 1970) who showed that rock-forming minerals were consistently selective in their uptake of REE. This demonstrated that partitioning of REE between common rock-forming minerals and liquids followed by physical separation of the minerals from the liquids could be an effective means of changing both REE concentrations and REE relative abundances in natural systems.

Note that for an element such as Eu which can exist in two different oxidation states, the ratio of Eu^{2+} to Eu^{3+} and thus the oxygen fugacity affects the distribution coefficient (e.g., Drake and Weill, 1975; Weill and McKay, 1975). However, the ratio of Eu^{2+} to Eu^{3+} also is strongly affected by the composition of the liquid (Morris and Haskin, 1974). This suggests that values of D can be sensitive to bulk liquid composition as well, since the energy condition of the individual Eu species clearly changes with composition.

The state of the art is that D values have been accurately determined experimentally only for a few systems. Values are known to vary significantly with temperature. Use of D values in REE modelling is straightforward under circumstances where conclusions are not invalidated by uncertainties in level of magnitude of as much as a factor of ten or relative values of as much as a factor of two (for most minerals). Conclusions dependent on better knowledge of D values require that careful attention be paid to what is known about values appropriate to the system in question. The reader is referred to Chapter 1 and the review by Irving (1978) and to other articles in that issue and the recent literature.

Before showing how D is used as a parameter in REE modelling it is worth reminding ourselves about the assumptions that must hold for the right-hand side of equation (4.3) to be constant. A variety of liquid compositions can yield diopside or any given named mineral. A D value determined accurately for a liquid of one composition and its equilibrium diopside cannot be presumed valid for some other liquid and its equilibrium diopside. This is partly because the temperature of crystallization is likely to be different, which changes the values of K_{eq} in equations (4.2) and (4.3) and which may change the values of activity coefficients. For liquids with the same temperatures and major element compositions and producing diopsides with identical major element compositions, the D value could be affected by different concentrations of compensating and competing ions. Finally, the value of the activity coefficient of the REE ion cannot remain constant in the liquid or the solid phase for all conceivable values of REE concentration. At some sufficiently high concentration, the availability of sites in the crystal will be seriously diminished, or the quantity of charge compensating ions will be diminished, or the REE will be so close to each other that they interact, or some other effect will occur. The question is, are such effects important in natural systems?

When the value of D for a given system changes as a function of concentration for the trace element, Henry's law, which states that the concentration in one phase is proportional to that in another, no longer holds. Possible failure of REE to obey Henry's law under conditions of concentration found in rock-forming systems has been a subject of recent controversy in the literature (e.g., Mysen, 1976, 1978; Drake and Holloway, 1978). Common rock-forming minerals seem capable of taking up substantial quantities of REE (at least 1000 ppm in olivine, orthopyroxene, and clinopyroxene). This was demonstrated experimentally in studies of partitioning between aqueous fluids and such minerals (Cullers et al., 1970, 1973; Zielinski and Frey, 1974). The nature of the sites occupied in Ca-free minerals is unknown, but the electron paramagnetic resonance spectra for Gd^{3+} in minerals synthesized in the presence of an aqueous fluid are the same as those for minerals grown from an anhydrous silicate melt (Morris, 1975) so lack of available sites does not appear to be a problem. Apparent violations of Henry's law for REE (see Mysen, 1976, 1978) in olivine, pyroxene, and amphibole have not been confirmed by other studies (e.g., Drake and Holloway, 1978).

An example of failure of REE to obey Henry's law is the experimental study of melt/crystal partitioning of Sm^{3+} and Tm^{3+} in garnet and diopside (Harrison and Wood, 1980) and in diopside (Harrison, 1977). In these cases, D values found at very low REE concentrations decrease to constant values at concentrations encountered in natural systems (see Chapter 1). That the values remain constant over several orders of magnitude (at a given T, P, n + REE) suggests that charge-balancing compensation rather than availability of defect sites probably controls incorporation of REE into these minerals at the higher REE concentrations. Lack of sensitivity of D to REE concentration in the range found for most natural systems simplifies the task of accurate modelling. Harrison (1981) has recently discussed effects of non-Henry's law behavior of REE distribution coefficients in garnet. This behavior occurs in the concentration range expected to be important in mantle partial melting.

4.4. Equilibrium melting and crystallization

In this and following sections mathematical formulations are presented to describe REE behavior during various idealized processes thought to be important in determining REE concentrations in rocks. These formulations have their origins in the chemical and geochemical literature (e.g., McIntire, 1963) and have been further developed for application to REE (and other trace elements) by several investigators. Of particular use are the works by Schilling and Winchester (1967), Gast (1968), Greenland (1970), Shaw (1970), Haskin et al. (1970), Helmke et al. (1972), Paster et al. (1974), Hertogen and Gijbels (1976) and Allègre and Minster (1978). In some cases, presentations of these models are accompanied by calculations of REE patterns, and the reader is referred to these works for a full understanding of derivations, nuances, etc. Given below are illustrative formulations and their derivations using a common set of parameters for all the models. The first is that for equilibrium melting or crystallization, i.e., the simple partitioning of a REE between fixed quantities of two equilibrium phases.

Consider a solid melting slowly enough that the resulting liquid is in equilibrium with the residue. Consider a magma crystallizing slowly enough that the solid is in equilibrium with the residual liquid. In either case, the ratio of concentration of a REE in the solid phase to that in the liquid phase would be given by the value for D (for the prevailing conditions of T, P, n). The concentration of the element in the solid ($C_{E,\mathrm{s}}$) or the liquid ($C_{E,\mathrm{l}}$) depends not only on D but on the average concentration in the entire solid/liquid system ($C_{E,\mathrm{t}}$) and the fraction by weight of that system that is liquid (X_l). To relate these quantities requires a mass balance, as follows, in which $M_{E,\mathrm{t}}$ is the total mass of the element in the system, M_t is the total mass of the system, and $M_{E,\mathrm{s}}$, M_s, $M_{E,\mathrm{l}}$, and M_l are analogous quantities for the solid and liquid. From the expression for D:

$$C_{E,1} = \frac{C_{E,s}}{D} = \frac{1}{D}\,\frac{M_{E,t} - M_{E,1}}{M_t - M_1} = \frac{1}{D}\,\frac{C_{E,t}M_t - C_{E,1}M_1}{M_t - M_1} \tag{4.4}$$

Since $X_1 = M_1/M_t$,

$$C_{E,1} = \frac{1}{D}\,\frac{C_{E,t} - C_{E,1}X_1}{1 - X_1} \tag{4.5}$$

Gathering of coefficients for $C_{E,1}$ and $C_{E,t}$ gives:

$$C_{E,1}\left(1 - X_1 + \frac{X_1}{D}\right) = C_{E,t}\frac{1}{D} \tag{4.6}$$

and:

$$C_{E,1} = \frac{C_{E,t}}{D + X_1\,(1-D)} \text{ or } \frac{C_{E,t}}{D\,(1 - X_1) + X_1} \tag{4.7}$$

Of course,

$$C_{E,s} = DC_{E,1} \tag{4.8}$$

In this formulation, D is that for whatever solid is present. If the solid is monomineralic, the value of D is just that for element E and mineral m ($D_{E,m}$). In more common cases, the solid will be polymineralic, but with all phases still in mutual equilibrium. Then the value of D for the overall solid (D_{WR}, for "whole rock") is the summed effect of all minerals, namely:

$$D_{\mathrm{WR}} = \sum_m f_m D_{E,m} \tag{4.9}$$

in which f_m is the fraction by weight of the solid that is comprised of mineral m. The equation that gives the concentration for a particular mineral is given in a later section (equation (4.23)). This treatment assumes constant mineral proportions in the residue. Hertogen and Gijbels (1976) discuss non-modal melting.

To illustrate how REE would presumably behave during a simple process of equilibrium melting, the following calculations are presented. Assume a peridotite, idealized for present convenience, that consists of 80% (wt.) olivine, 10% plagioclase, and 10% clinopyroxene and contains REE at a concentration level of three times that of chondrites (see Table 4.1). Furthermore, assume for convenience that the melting removes all minerals in proportion to their mass fractions in the peridotite, i.e., that melting is mass-modal. The starting concentrations are represented by the horizontal line at 3.0 in Fig. 4.1. Because the REE are incompatible ($D < 1$) for the

TABLE 4.1

Concentrations of REE in hypothetical rocks

	Peridotite (ppm)	Basalt (ppm)
La	0.99	13.2
Ce	2.64	32
Pr	0.34	3.7
Nd	1.80	18.0
Sm	0.54	4.4
Eu	0.21	1.52
Gd	0.75	5.0
Tb	0.141	0.85
Dy	0.95	5.2
Ho	0.21	1.05
Er	0.60	2.7
Tm	0.090	0.37
Yb	0.60	2.2
Lu	0.102	0.31

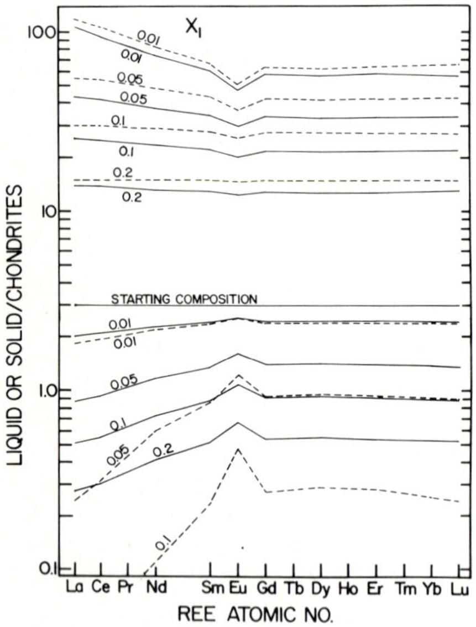

Fig. 4.1. Concentrations in liquid and solid phases after modal partial melting of a source with 80% olivine, 10% plagioclase, and 10% clinopyroxene as a function of the fraction melted, X_l. D values used were those shown in Haskin and Paster (1979, fig. 21.19). Solid lines are for equilibrium melting (or crystallization) and dashed lines are for incremental partial melting.

minerals of the solid, they will be transferred efficiently to the liquid fraction as it forms. The whole-rock values for the distribution coefficients (equation (4.9); data from Haskin and Paster, 1979) range from 0.020 for La to 0.043 for Dy to 0.041 for Lu, with the value for Eu anomalous at 0.054.

When only 1% of the solid has melted, the liquid has REE concentrations between 47 and 118 times those of chondrites, or 15 and 40 times those of the starting composition. Note that the LREE are the most enriched in the liquid and Eu is selectively retained by the solid. That is a result mainly of the higher values of D for HREE in clinopyroxene than for LREE in any mineral, plus the high value for Eu in plagioclase. Note that the conversion of only 1% of the system to liquid has depleted the REE concentrations in the solid substantially by factors of 0.84 (Eu) to 0.61 (La). Note also that initially the change in relative REE abundances away from that of the starting material is more striking for the liquid, which contains a relatively small fraction of the REE in the system, than for the solid. By the time the extent of melting has reached 20%, most of each REE is in the liquid and the relative REE abundances in that phase have become very similar to those of the starting material. The concentrations have become about four times higher than those in the original peridotite. As melting proceeds, relative REE abundances in the residual solid deviate significantly from those of the initial solid and concentrations become very low ($<$ 0.1 times that of the starting material for La, 0.4 times for Eu, and 0.3 times for the HREE). Note that the size of the Eu anomaly is greatest for that phase with the smaller portion of the REE in the system.

To apply such calculations to natural materials, many other considerations must enter in. For instance, if melting occurs deep within the mantle, plagioclase will not be present. Melting is unlikely to be approximately modal, and the proportion of olivine in the residue should increase with increasing X_l. This would diminish the size of the Eu anomaly as X_l increased and widen the gap between $C_{E,l}$ and $C_{E,s}$ for a given extent of partial melting. Equations for non-modal melting have been given by Hertogen and Gijbels (1976). As long as the mineral composition of the solid is known for a given value of X_l, equation (4.9) can be used to obtain the value of D needed in equation (4.7).

Some REE concentrating mineral might actually be present in the starting solid, e.g., apatite (which seems to favor MREE) or zircon (which favors HREE) (e.g., Nagasawa, 1970). Should either of these minerals be retained in the residual solid, the concentrations of REE would be substantially reduced in the liquid and correspondingly increased in the solid, as those minerals have values of D for REE much greater than one. It is important to note, however, that their presence in the initial solid loses its importance entirely in this model as soon as those minerals have melted. The equilibrium is between the liquid and the residual solid. Thus, the constituents of completely melted minerals only affect the partitioning of a trace element through the dependence of the value of D on liquid composition. How large these effects are has not been systematically demonstrated.

How closely does this model describe what actually occurs in nature? That some form of melting occurs within planetary mantles seems beyond dispute, in view of the widespread phenomenon of volcanism. That melting processes are at least approximately equilibrium processes seems likely when equilibration times for experimental charges and diffusion times (e.g., Hofmann and Hart, 1978) are taken into account. Whether a melting solid can actually stay in equilibrium with more than one percent or so of liquid before that liquid migrates away is not at all certain (e.g., Turcotte and Ahern, 1978).

That Earth as a whole has approximately chondritic relative REE abundances is reasonable, based on general observations of REE in many materials. Few peridotites have chondritic relative REE abundances, however. Frey and coworkers have shown that most peridotites whose mineralogy and bulk compositions are such that the major constituents of basalts are available in partial melting have low REE concentrations and relative abundances enriched in LREE compared with chondrites (e.g., Frey and Prinz, 1978). Approximately chondritic relative REE abundances were found in peridotites whose mineralogy was that of possible refractory residue rather than of fertile source material. Thus, there is no strong basis for assuming chondritic relative abundances in source regions of basalts.

Based on the mineralogy of the source used in the example above (Fig. 4.1), REE relative abundances cannot be sufficiently fractionated from chondritic ones to produce the kinds of patterns observed for most basalts. In particular, a negative slope for the LREE (relative to chondrites) appears only at very low extents of partial melting and no negative slope for HREE is produced at all. For the equilibrium melting process to be the dominant determiner of REE abundances in basalts, either the source must be non-chondritic in REE concentrations and relative abundances, or have different mineralogy, or both. (The values of three times chondritic concentrations for REE in the above example were chosen because most previous modelling exercises of this type demonstrated that chondritic concentrations were just too low.) This problem is usually "solved" by assuming the presence of garnet in the source region. This serves to retain HREE selectively and introduce a negative slope. Certainly, the source regions for at least some basalts should contain garnet. Whether equilibrium melting with a solid residue containing garnet is the principal reason for the negative slopes of REE patterns found in most basalts (but not those typical of ocean floors) remains conjectural. Such a hypothesis is strained in trying to account for the very steep patterns of strongly alkalic basalts. Some of those have quite high LREE concentrations (e.g., Kay and Gast, 1973) and their REE patterns grade smoothly into those for many carbonatites and kimberlites, which are surely not produced by simple melting of garnet-bearing solids. In short, a convincing demonstration that the process just modelled is the principal determiner of REE abundances of basalts, even prior to modification by subsequent processes, and to the exclusion of other possible processes, has not been made. A particularly thorough study demonstrating that it might be responsible for a series of volcanic lavas is the work of Frey et al. (1978). As pointed out in the introduction, many considerations other than REE abundances must be taken into account before such a demonstration can be made.

The principal value so far of this type of model has been to indicate what range of REE abundances could arise from operation of an equilibrium partial melting process on source regions of various compositions. This was done in the pioneering paper of Gast (1968), which set in motion a plethora of similar calculations. Gast showed, for example, that ocean floor tholeiite could not arise from partial melting of primitive mantle material with chondrite-like REE abundances, but might come from mantle material that had been selectively depleted in LREE during its prior history. That conclusion found excellent support in Sm-Nd isotopic studies (e.g., DePaolo and Wasserburg, 1976; Richard et al., 1976). Gast also showed that small extents of partial melting could readily produce substantial changes in relative REE abundances that only unreasonably large extents of crystallization could effect. The reader is referred to Gast's paper as an excellent demonstration of how modelling can constrain the capability of a process to produce compositional change. Demonstrating the constraints of capability, however, does not demonstrate that partial melting actually occurred in the manner implied by such a simple formulation of the model as was used. Some subsequent works, unfortunately, blurred this distinction and attributed characteristics of various lavas directly to a partial melting process because of rough agreement with model predictions. Some other works, equally unfortunately, denied a significant role to partial melting on the basis that REE abundances did not conform to predictions of such a simple model.

(Gast, in his 1968 work, did not actually use this model, but one more akin to that for incremental partial melting, discussed later. The results of his work are discussed here because most subsequent investigators have used the above model for equilibrium partial melting and because that pioneering effort has so strongly influenced our ideas about generation of REE patterns in igneous melts.)

Equations (4.7) and (4.8) also describe equilibrium crystallization. This process is usually not regarded as important because it does not lead to the more extensive changes observed in fractional crystallization, the process that has enjoyed popular acceptance as the principal process by which compositions of magmas, once formed, are modified. Fractional crystallization is the subject of the next two sections. It is worth noting first, however, that in Fig. 4.1 the curves for X_1 = 0.01, 0.05, and 0.20 correspond to 99%, 95% and 80% crystallization of an initial liquid with REE concentrations equal to three times those of chondrites. Such a liquid would be a curiosity, but the figure illustrates that small amounts of crystallization produce strongly fractionated REE patterns only in the solid, large amounts only in the liquid.

4.5. Fractional crystallization: continuous removal of crystals

Suppose a magma cools to produce successive sets of crystals, each set amounting in mass to no more than an increment of the mass of the parent magma. Also, let each set, on forming, be removed and isolated from any interaction with the evolving magma. The concentration of the REE in the currently forming increment of solid will be equal to the increment of the mass of the element divided by the increment of mass of the solid, i.e.:

$$\frac{dM_{E,s}}{dM_s} = DC_{E,l} = D\frac{M_{E,l}}{M_l} \tag{4.10}$$

Replacing terms for masses in the liquid by their equivalents in terms of masses in the solid and in the entire system gives:

$$\frac{dM_{E,s}}{M_{E,t} - M_{E,s}} = D\frac{dM_s}{M_t - M_s} \tag{4.11}$$

Integration yields the following expression:

$$\ln\,(M_{E,t} - M_{E,s}) = D\,\ln\,(M_t - M_s) + \text{const.} \tag{4.12}$$

Since when $M_s = 0$, $M_{E,s} = 0$, the constant equals $\ln\,(M_{E,t}/M_t^D)$. Substituting that into equation (4.12), taking the antilog, and rearranging gives:

$$\frac{M_{E,s}}{M_s} = \bar{C}_{E,s} = \frac{M_{E,t}}{M_s}\left[1 - \left(1 - \frac{M_s}{M_t}\right)^D\right] \tag{4.13}$$

Letting X_s be the fraction of the system that is solid (i.e., the fraction that has crystallized) and converting mass ratios to concentrations gives:

$$\bar{C}_{E,s} = C_{E,t}\,\frac{1-(1-X_s)^D}{X_s} \tag{4.14}$$

In equation (4.14), $\bar{C}_{E,s}$ is the *average* concentration of the element over all the incremental sets of crystals produced up to the point at which X_s gives the fraction of the system that is solid. The concentration of the element in the residual liquid ($C_{E,l}$) at that point is found as follows:

$$M_{E,l} = M_{E,t} - M_{E,s} = C_{E,t}M_t - \bar{C}_{E,s}M_s \tag{4.15}$$

Substituting for the average concentration in the solid gives:

$$M_{E,1} = C_{E,\mathrm{t}}M_\mathrm{t} - M_\mathrm{s}C_{E,\mathrm{t}}\,\frac{1-(1-X_\mathrm{s})^D}{X_\mathrm{s}} = C_{E,\mathrm{t}}M_\mathrm{t}\,(1-X_\mathrm{s})^D \tag{4.16}$$

Division by M_1 to obtain concentrations leads to the expression:

$$C_{E,1} = C_{E,\mathrm{t}}\,\frac{(1-X_\mathrm{s})^D}{1-X_\mathrm{s}} = C_{E,\mathrm{t}}\,(1-X_\mathrm{s})^{D-1} \tag{4.17}$$

The concentration of the element in the increment of solid immediately forming, of course, is just that in equilibrium with the liquid as in equation (4.8).

What will be the concentrations of REE in each different kind of mineral in those cases where the D's in equations (4.14) and (4.17) are whole-rock values? Each mineral crystal forming at any time is in equilibrium with the liquid at that time. Also, since each type of mineral is in equilibrium with the liquid, each type is in equilibrium with every other type in the same crystal set. The concentration of a REE in a given mineral in a given set of crystals (i.e., for a given value of X_s) cannot, however, be calculated from that of the liquid obtained by using equation (4.7) with the D just for that mineral, because the compositional evolution of the liquid is guided by the combined effects of all the precipitating minerals. Once the composition of the liquid has been obtained for the appropriate values of D_WR and X_s, the concentration of a REE in the crystals of a particular mineral in the currently forming set is just the product of $D_{E,m}$ for that element in that mineral and the current concentration of that element in the liquid as in equation (4.8). The average of all of the incremental sets of crystals of one mineral for the concentration of a REE is seldom of interest, but can be calculated by use of equation (4.22) in the next section.

Crystal fractionation as described by equations (4.14) and (4.17) has usually been considered responsible for changes in REE concentrations in magmas prior to eruption as lavas. If primary mantle melts are considered as being in equilibrium with olivine of Fo ~ 90, having Ni concentrations of 300—400 ppm, and having relatively low concentrations of incompatible elements such as REE, perhaps with relative abundances not too different from chondritic ones, then few lavas represent primary melts. If the precursors of lavas are believed to have been primary melts as considered above, then some process subsequent to producing these melts must have modified their compositions substantially prior to their eruption as lavas. Since melts can be expected to crystallize as they cool, since many lavas contain phenocrysts, and in view of the nature of "simple" layered intrusions such as the Skaergaard, most petrologists and geochemists regard crystal fractionation as the most likely process responsible for modification of magma composition.

Examples of modelling of this process include the works by Zielinski and Frey (1970), Zielinski (1975), Ferrara and Treuil (1974), Barberi et al. (1975), Allègre et al. (1977), and many others.

An example for a hypothetical liquid (Table 4.1) crystallizing to form 20% (wt.) olivine, 40% plagioclase, and 40% clinopyroxene is shown in Fig. 4.2. Values of D used are from Haskin and Paster (1979). Values of D_{WR} range from 0.056 for La through 0.128 for Gd to 0.106 for Yb, with 0.186 for Eu. The dashed curves represent $C_{E,1}$ for *complete equilibrium crystallization* according to equations (4.7) and (4.8). Note that the REE, as incompatible elements, become concentrated in the liquid as crystallization progresses. Note also that the slope of the REE pattern by the time 80% of the original liquid has crystallized ($X_s = 0.80$) has scarcely changed from its original value. Eu, because of its anomalously high value of D for plagioclase, is significantly less enriched than are neighboring Sm and Gd; the size of this Eu anomaly grows as X_1 decreases. No easily noticeable change in slope of the REE pattern occurs until the liquid is nearly frozen ($X_s \gtrsim 0.98$). (Complementary changes in composition are observed in the solid. This is equivalent to the interpretation given for Fig. 4.3 for crystallization, but with a more realistic starting composition.)

The solid lines are for crystal fractionation. The trends of change in REE

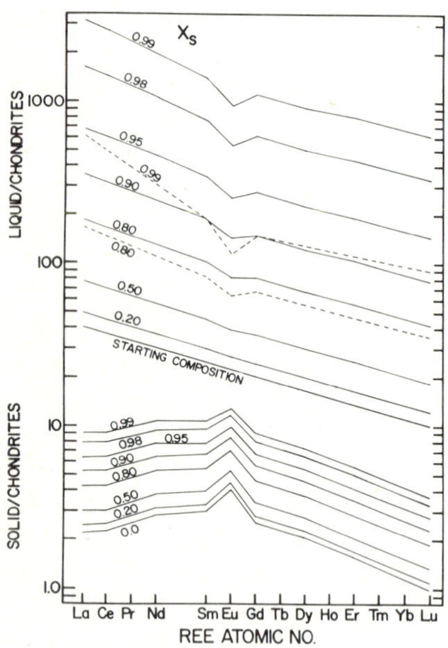

Fig. 4.2. Concentrations in liquid and solid phases after crystallization of a liquid of the starting concentrations shown. Mineralogy of the solid is 20% olivine, 40% plagioclase, and 40% clinopyroxene. D values used are from Haskin and Paster (1979, fig. 21.19). Solid lines are for fractional crystallization, dashed lines for equilibrium crystallization.

concentrations and patterns are qualitatively similar to those for complete equilibrium but quantitatively more drastic. REE concentrations at 99% solidified (X_s = 0.99, X_l = 0.01) for equilibrium crystallization lie between those for 80% and 95% solidified for crystal fractionation. The average concentration in the solid for a given value of X_s is lower for crystal fractionation than for equilibrium crystallization.

Change in REE concentrations in the liquid as modelled for increasing extents of crystal fractionation qualitatively and even semi-quantitatively resemble those seen for many suites of basalt or suites of volcanic rocks ranging in composition from basalt to more silicic rocks. This no doubt accounts for the popularity and uncritical acceptance of such REE patterns as supporting evidence for the occurrence of substantial fractional crystallization during production of lavas. Many types of data can be brought to bear on this question, however, including major element composition, other trace element concentrations, experimental petrological studies, isotopic measurements, and field relationships. Few studies have so far been comprehensive enough to provide quantitative REE modelling and to demonstrate the consistency of the results with sufficient other data to make a strong and unequivocal case for fractional crystallization as a major determiner of compositional variation among volcanic rocks whose compositions span a wide range. Several recent, carefully done studies suggest that fractional crystallization may govern REE (and other elemental) composition much less generally than expected.

An example of a careful and somewhat inconclusive study of related volcanic rocks is that of 39 lavas of Craters of the Moon National Monument (Leeman et al., 1976). Many parts of the study suggest that fractional crystallization was principally responsible for the compositions, which ranged from ferrobasalt to ferrolatite. Quantitative modelling of major and trace elements was consistent with loss of minerals with the same compositions as observed microphenocrysts from more primitive lavas to produce more evolved lavas. However, results of experimental petrology did not reproduce the inferred compositional trends. Also, Sr isotope ratios were high and somewhat variable, suggesting significant crustal interaction. The more primitive ferrobasalts could not be easily related to typical olivine tholeiite of the Snake River Plain or to primary melts (in the sense indicated above). There was no correlation between stratigraphic position and composition that might imply a simple setting for fractional crystallization such as a single magma chamber (even a periodically refilled one of the type suggested by O'Hara, 1977).

In a recent study of lavas of the Reykjanes Peninsula, Jacobsson et al. (1978) concluded that low-pressure fractional crystallization could not account for the compositional differences between suites of shield lavas and those of fissure lavas or for the volume-composition relationships between the suites. Brannon et al. (1979a, b) in an extensive study of 150 well-preserved, stratigraphically successive Keweenawan lavas of the North Shore Volcanic Group found substantial correlations between stratigraphic position and major and trace element composition over a wide compositional range (factor of ~ 50 for La). Although relationships among these lavas have commonly been attributed to fractional crystallization, that mechanism fails to account for compositional relationships among samples of similar composition and stratigraphic location as well as across a wider range of composition. Fractional crystallization does not account for major and trace element relationships among volcanic rocks of the Taos Plateau of the Rio Grande Rift (L.A. Haskin et al., unpublished data).

The idea that fractional crystallization should be a major process for modifying magma compositions is plausible and deeply entrenched in petrologic thinking. Thus, it is difficult to gain acceptance for the notion that the process might not be as dominant as believed and that extensive evidence needs to be provided to substantiate that it did dominate in any given situation. There are strong tendencies to react to failures of model predictions to match observed compositions as resulting because fractional crystallization is a more complicated process than that envisioned in a simple model, or that maybe the process occurred along conduit walls instead of in magma chambers, etc. In part, these difficulties are semantic, and the complicated forms of fractional crystallization are really mixed processes, e.g., fractional crystallization plus magma mixing or plus assimilation. These combined processes can be modelled quantitatively also, so it is not sufficient to pass off the disagreement between results of modelling and results of analysis in such a shallow manner. Also, fractional crystallization is a rather simple and well-defined chemical process and, if it controls compositions of lavas, the results will be independent of physical setting. Stopping the explanation of REE abundances at the stage of qualitative or semi-quantitative agreement of modelling with analytical results and ancillary information does not necessarily support (or refute) that the modelled process was important. It merely leaves the explanation in limbo.

Another type of system for which crystal fractionation is believed to be an important process is the development of layered intrusions. Petrologic, compositional, and field evidence all support some such mechanism for producing the compositional trends observed in such relatively simple bodies as the Skaergaard and Kiglapait intrusions. REE data, insofar as determined, support this in a general way (e.g., Paster et al., 1974). Equation (4.17) is adequate to describe the evolution of REE abundances in the residual liquids, but neither equation (4.8) nor equation (4.14) can explain the compositions of the individual layers of solid or of the average solid. That is because the crystals in a narrow layer, which might correspond to a single incremental set of crystals as defined in equation (4.10), entrap part of the parent liquid, which becomes part of the rock as sampled. This is a combined process of different events of fractional crystallization, and the brief description of it is deferred to the next section, which concerns fractional crystallization without removal of crystals.

4.6. Fractional crystallization: crystal zoning-surface equilibrium

Consider a homogeneous liquid in which mineral crystals form but are not physically removed. Suppose, however, that the rate of crystallization is rapid enough that the cores of these crystals are effectively isolated from the evolving liquid in which they are suspended by the layers of the crystals that have built around the cores. In the somewhat extreme case in which the interior of the crystal is effectively isolated as quickly as it has formed, the evolving liquid will be in equilibrium only with the surface layer of each crystal. This, as in the previous case of fractional crystallization, is described by equation (4.17), but the increment of newly formed solid in this case corresponds to the most recent layers formed on the participating, zoned crystals rather than to a set of crystals of constant composition that are removed from the liquid. The average composition of the entire solid consisting of these zoned crystals is given by equation (4.14), and the average composition of the currently forming layer is just that in equilibrium with the liquid (equation (4.8)).

If the zoned crystals of a particular mineral were removed at some point, what would be the average, or overall, concentration of a REE? The concentration of a REE, E, in the solid will equal the contributions from all the minerals, each according to its fraction by weight in the solid; i.e.,

$$C_{E,s} = \sum_{m} C_{E,m} f_m \tag{4.18}$$

At any point during crystallization:

$$C_{E,m} = D_{E,m} C_{E,1} \text{ and } C_{E,n} = D_{E,n} C_{E,1} \qquad (4.8)\ (4.19)$$

so:

$$C_{E,m} = C_{E,n} \frac{D_{E,m}}{D_{E,n}} \tag{4.20}$$

If we select mineral n as that of interest, single it out from equation (4.18), and substitute for $C_{E,n}$ from equation (4.20):

$$C_{E,s} = C_{E,n} \left(f_n + \frac{\sum_{m \neq n} D_{E,m} f_m}{D_{E,n}} \right) \tag{4.21}$$

But the average concentration $C_{E,\mathrm{s}}$ must also be given by equation (4.14); eliminating $C_{E,\mathrm{s}}$ between these equations gives the following expression for $C_{E,n}$:

$$C_{E,n} = \frac{D_{E,n} C_{E,\mathrm{t}} [1-(1-X_{\mathrm{s}})^{D_{E,\mathrm{WR}}}]}{X_{\mathrm{s}} D_{E,\mathrm{WR}}} \tag{4.22}$$

The right-hand side of equation (4.22) can be regarded as two factors. The first is $D_{E,n}/D_{E,\mathrm{WR}}$ which describes the fraction incorporated into mineral n of the element entering the solid phase. The second is all the rest, which gives the overall average concentration of the solid as a function of the amount of solid that has formed, X_{s} (as in equation (4.14)). In other words, the average concentration of a REE in some mineral n is in the same proportion to the whole-rock concentration of the REE as the value of D for that REE in that mineral is to the overall D for all the minerals, each mineral being properly weighted for its proportion in the overall solid. An equation analogous to (4.22), but for complete equilibrium, can be derived by solving equations (4.7), (4.8), and (4.21) to eliminate $C_{E,\mathrm{s}}$ and $C_{E,1}$. The result is as follows:

$$C_{E,n} = \frac{C_{E,\mathrm{t}} D_{E,n}}{D_{\mathrm{WR}} + X_1 (1 - D_{\mathrm{WR}})} \tag{4.23}$$

Equation (4.23) can be used to represent any type of phase, i.e., gas, solid, or liquid, in equilibrium with one phase, here taken as the residual parent liquid. Shaw (1978) has discussed partitioning among liquid, solid, and immiscible fluid phases in mutual equilibrium using an approach related to this one.

If crystallization proceeds to completion, equation (4.22) becomes (for $X_{\mathrm{s}} = 1$):

$$C_{E,n} = C_{E,\mathrm{A}} \frac{D_{E,n}}{D_{E,\mathrm{WR}}} \tag{4.24}$$

The maximum possible change in *overall average* concentration of a REE in a mineral is just the ratio of the average concentration at the end of crystallization ($X_{\mathrm{s}} = 1$) to that when crystallization began ($X_{\mathrm{s}} = 0$). I.e.:

$$C_{E,n,1}/C_{E,n,0} = \frac{1}{D_{E,\mathrm{WR}}} \tag{4.25}$$

Note that the extent of this change is the same for every mineral in the solid. From this equation it is apparent that the maximum change that can occur in the ratio of one REE to another is just the inverse of the ratios of the D_{WR}

values for the two elements. A more extensive discussion of this can be found in Helmke et al. (1972).

In principle, REE abundances in the minerals in a quietly freezing lava might develop according to equation (4.22) if no separation of crystals from residual parent liquid occurred. Figs. 4.2 and 4.3 show the changes in average REE concentrations for the residual liquid, average solid, and individual minerals in a hypothetical basalt that begins as pure liquid and crystallizes to completion as a closed system in the constant proportions of 20% (wt.) olivine, 40% plagioclase, and 40% clinopyroxene. The starting REE concentrations for the liquid are given in Table 4.1. It is assumed that all three minerals grow by surface equilibrium fractional crystallization. The changing composition of the residual liquid (equation (4.17)) and the crystallized portion (equation (4.14)) are shown in Fig. 4.2. The changes in concentrations for the minerals are shown in Fig. 4.3. At $X_s = 0$, the newly forming minerals are in equilibrium with the starting liquid, whose composition will, of course, be that of the finally solidified basalt. As soon as significant solidification has occurred, the portion of each mineral currently crystallizing will be in equilibrium with evolved parent liquid. The concentrations of REE in the newly forming layers of solid can be obtained by calculating the composition of the residual liquid (equation (4.17)) then multiplying by the appropriate values of $D_{E,m}$ (equation (4.8)). REE abundances shown in Fig. 4.3 are the averages for the entire quantity of each mineral crystallized so far. Note that for olivine, plagioclase, and clinopyroxene there are no first-order changes in relative REE abundances throughout the first 99% of crystallization. The size of the Eu anomaly increases with X_s for olivine and clinopyroxene because that element is anomalously removed by plagioclase. On the other hand, the size of the Eu anomaly in plagioclase decreases with increasing X_s because the partitioning is with a liquid whose relative Eu abundance continuously decreases.

By $X_s = 0.99$, the average REE concentrations in each mineral have increased by factors of 4.1 for La to 3.7 for Lu, and 3.1 for Eu. There have been attempts to estimate values for distribution coefficients by separating minerals from basalts, analyzing them for REE, and dividing the concentrations by the values for the whole rock on the assumption that the whole-rock composition represented that of the liquid from which the minerals formed. From Figs. 4.2 and 4.3 it should be obvious why such a practice can lead to serious error; the average mineral was never in equilibrium with liquid of the composition of the starting liquid, as represented by the whole rock.

The most drastic changes in composition occur when the liquid phase disappears. At that point the REE are contained entirely within the major minerals, which are in equilibrium with each other. Note the substantial changes in REE pattern that result. The *relative* abundances of the REE in any mineral are no longer approximately those expected for that mineral in

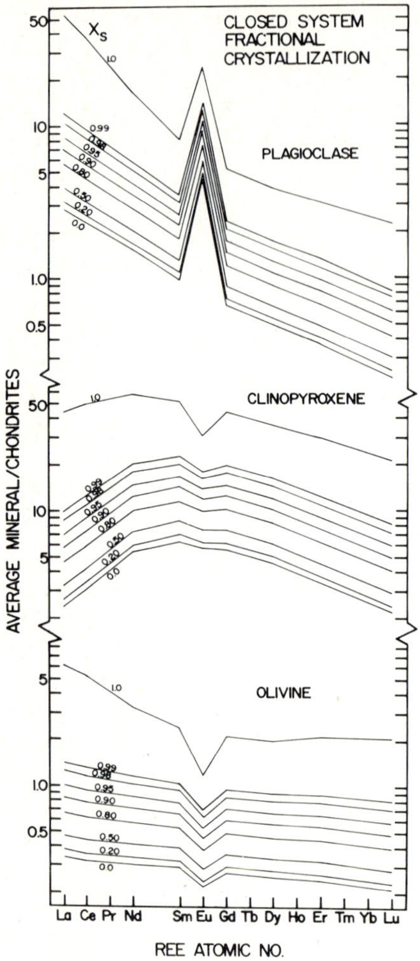

Fig. 4.3. Mineral concentrations as a function of the fraction solidified, X_s, during closed system fractional crystallization. Conditions are the same as for Fig. 4.2 and the solid lines in that figure show the changes for liquid and overall solid that correspond to the mineral REE concentrations represented here.

equilibrium with a liquid of the overall composition of the whole rock, as they still were prior to $X_s = 0.99$. This is because the D_{WR} between the overall solid and the liquid favored selective retention of LREE by the liquid. As the liquid phase disappears, these LREE enter entirely into the minerals, substantially changing their average REE abundances.

Of course, this model is far too simple to describe exactly the behavior of a solidifying lava. Among other things, the mineralogy will not stay constant throughout solidification, nor will the D values remain constant throughout the changes in T and n. The last portion of the liquid will probably form mesostasis which may contain REE concentrating minerals, etc. (For a discussion of the effects of these phenomena, see Helmke et al., 1972 and Haskin and Korotev, 1977.)

Disappearance of a liquid phase as described above can lead to relative REE abundances in minerals that are substantially different from those expected intuitively on the basis of D values. A possible example of this phenomenon is an eclogite analyzed by Haskin and Frey (1966). To a first approximation, all the REE in the eclogite are contained in the two major minerals, garnet and clinopyroxene. A graph of REE atomic number versus ratios of REE concentrations in early crystallized garnet and clinopyroxene to those of their parent liquids would necessarily resemble a plot of D values, in accordance with equation (4.8). If the entire liquid solidified so that all the REE ended up in garnet and clinopyroxene, in accordance with equation (4.24), however, the plot of ratios of REE concentrations in the overall garnet and clinopyroxene to those of the whole rock (hypothetical parent liquid) would look like those shown in Fig. 4.4. Since on that graph the whole-rock distribution corresponds to values of unity for every REE, one might be tempted to conclude that clinopyroxene preferred LREE, garnet HREE. In fact, as shown in Chapter 1, both minerals prefer HREE. Because garnet has higher D values for HREE than clinopyroxene does, when the two minerals are in equilibrium with each other, the garnet will have the greater HREE concentrations. Also, since garnet discriminates against LREE more strongly than clinopyroxene does, the LREE will be concentrated preferentially into clinopyroxene when the last portions of liquid solidifies.

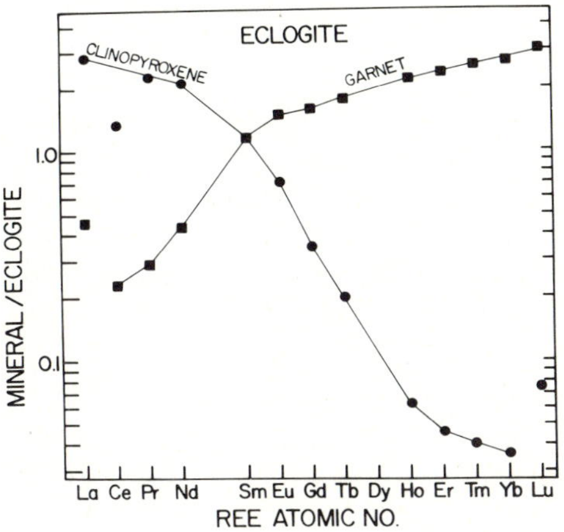

Fig. 4.4. REE relative abundances in garnet and clinopyroxene, normalized to the parent eclogite. This shows how REE abundances of clinopyroxene can become very different from those corresponding to equilibrium with a hypothetical liquid with the REE abundances of the eclogite, if such a liquid solidified entirely to produce only garnet and clinopyroxene as mineral hosts for the REE.

Of course, the eclogite of Fig. 4.4 may not have formed in this manner. Conceivably, both minerals may have precipitated from some liquid that was somehow entirely separated from the whole-rock solid as sampled. However, the La concentration in the hypothetical liquid in equilibrium with the garnet and clinopyroxene would have been ~85 ppm and that of Yb ~0.18 ppm. This corresponds to an enrichment of La over Yb compared to chondritic relative abundances of nearly 300 times! That is higher than observed even for carbonatites and kimberlites, which are among the most fractionated rocks known. Thus it seems more plausible that the eclogite, if it formed from a liquid, did so by a mechanism more similar to that described by equations (4.22) and (4.25) than one involving equilibrium with a liquid that was then removed.

The model for surface equilibrium crystallization in a closed system has been applied to a single sample of basalt by Haskin and Korotev (1977), who found that it gave reasonably self-consistent results for whole rock, mineral separates, and $D_{E,m}$ values. It provided the first step toward understanding the reasons for heterogeneity among basalt fragments from a single hand specimen or within single, undifferentiated flows.

It is hoped that at this point the essential qualities of the chemical portions of REE modelling have been adequately described. From here on, discussions will be more brief, giving only the essential features of other chemical mechanisms and their coupling to the appropriate physical settings for application to geochemical problems. Readers are referred to the literature for details.

4.7. Incremental partial melting

Consider a source rock being slowly heated so that melting begins. If the rock material is porous, or by the time a small fraction of it has melted, melt may begin migrating away from the source. If each increment of melt forms in equilibrium with the whole of the residual source material, the mathematical description is analogous to that for fractional crystallization, but with the roles of the liquid and solid reversed. The evolution of the composition of the residual solid is of the same form as equation (4.17), but with the inverse of the distribution coefficient D, etc., as follows:

$$C_{E,s} = C_{E,t}\,(1 - X_l)^{1/D - 1} \qquad (4.26)$$

Note that the melting may deplete certain minerals more rapidly than others, so the mode of the residual solid may change as a function of the fraction of the original system that has been melted and removed, X_l. The distribution coefficient for the solid thus changes as a function of X_l. Equations taking this into account have been presented by Hertogen and Gijbels

(1976) for equilibrium melting and could in principle be adapted for this case as well. This change may be fairly smooth or may be abrupt as a significant mineral is depleted. Partitioning of a REE is controlled by the combination of minerals still present in the solid phase, and to some poorly known extent, by the composition of the liquid. Sialic liquids appear to have less affinity for REE than mafic liquids do. The extent to which this is an effect of liquid composition as opposed to temperature or equilibrium mineral composition is not well established.

The concentration of a REE in the increment of liquid in equilibrium with the solid phase at any particular stage of melting and removal of the liquid is equal to the concentration in the solid divided by the value of D. In the case for which successive increments of liquids might collect and mix in some sort of holding tank (e.g., a magma chamber), the average concentration of a REE after the fraction X_1 of the system has melted is given by an equation analogous to equation (4.14), with the changes shown below:

$$\bar{C}_{E,1} = C_{E,t} \frac{1-(1-X_1)^{1/D}}{X_1} \tag{4.27}$$

Results of calculations for this process are included in Fig. 4.1 (dashed lines) for comparison with those for equilibrium melting. For the liquid phase, the results are not strongly different and, for this reason, most melting calculations are equilibrium calculations. The maximum differences occur at intermediate extents of melting. For small extents ($X_1 < 0.01$) the equilibrium melt is not appreciably different from being a single increment, and for $X_1 > 0.10$, essentially the entire quantity of REE is already in the liquid portion of the system. For the residual solid, however, the results are strongly different. Extraction of incompatible REE by a series of liquids in small batches is far more efficient than extraction once with a large batch. Incremental partial melting has been used for example, by Haskin et al. (1970) for lunar basalts and by Smewing and Potts (1976) in studies of the Troodos Massif.

Incremental partial melting may be more akin to natural melting processes if it is true that not more than 1% or so of liquid will remain trapped within its source region (e.g., Turcotte and Ahern, 1978). In the absence of a magma chamber to collect successive increments, might these increments not appear as individual flows of basalt? The sequence of REE abundances generated by successive melts amounting to 1% of the solid source used in previous melting calculations (Table 4.1, Fig. 4.1) and calculated in a manner that adjusts D_{WR} values and residual mineralogy and uses equations (4.7) and (4.8) for equilibrium melting are shown in Fig. 4.5. It is worth noting that very few series of related lavas show the trends of LREE depletion seen in Fig. 4.5. This suggests that this mechanism is unimportant in producing series of lavas. A possible case for this mechanism for very extensive melting

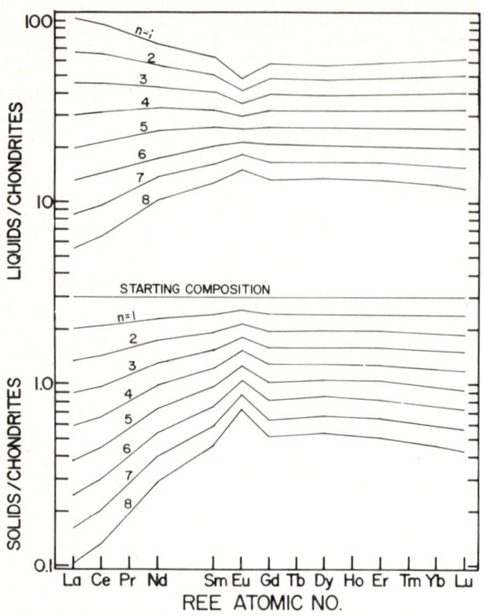

Fig. 4.5. REE abundances for successive modal melts of a source like that shown in Fig. 4.1. Each liquid corresponds to 1% of the mass of the original source.

has been suggested by Arndt (1977). Another possible case is a suite of lavas from the South Sandwich islands (Hawkesworth et al., 1977), interpreted, however, in terms of fractional crystallization.

4.8. Zone melting

Suppose a source region is heated from underneath in a manner that produces near the bottom a melt that works its way to the top. An extreme case would be a wholly molten zone that rises from the bottom through the solid source and reacts continuously. The initial melt $C_{E,1}$ will have the same REE concentration as the source average ($C_{E,t}$) or the solid phase ($C_{E,s}$). Let the total thickness of source through which the zone of liquid must travel be s, with movement of the zone in increments ds. Assume for convenience a constant thickness h for the zone and a constant horizontal cross section a^2 for the source column; thus, the zone has a constant volume ha^2. The zone begins with a mass of a given REE (E) of $C_{E,t}ha^2\rho$, in which ρ is the density. As it rises through a distance ds, it dissolves a volume a^2ds of the solid above it, gaining a mass of E equal to $C_{E,t}a^2$ds. At the same time an equal volume of solid is deposited beneath the zone. This solid removes a mass of E equal to $D_{E,\mathrm{WR}}C_{E,1}a^2\,\mathrm{d}s$, assuming the densities of the solids above and below the zone to be equal. The net change in concentration is:

$$\frac{dM_{E,1}}{\rho a^2 h} = dC_{E,1} = \frac{\rho a^2 (C_{E,t} - D_{E,WR}\, C_{E,1})\, ds}{\rho a^2 h} = \frac{C_{E,t} - D_{E,WR}\, C_{E,1}}{h}\, ds \tag{4.28}$$

This equation can be integrated between the limits of $C_{E,1} = C_{E,t}$ at $s = 0$ and $C_{E,1}$ at $s = s$ to give:

$$C_{E,1} = \frac{C_{E,t}}{D_{E,WR}} + \frac{C_{E,t}(D_{E,WR} - 1)}{D_{E,WR}} \exp(-D_{E,WR}\, s/h)$$

$$= \frac{C_{E,t}}{D_{E,WR}} [1 - (1 - D_{E,WR}) \exp(-D_{E,WR}\, s/h)] \tag{4.29}$$

A condition of saturation is reached when the exponent in the above equation is small, leading to a steady-state concentration independent of h (the height of the source column) and equal to $C_{E,t}/D_{E,WR}$. For incompatible elements ($D_{WR} \sim 0.01$), the exponent will have a value of ~0.1 by the time the ratio s/h is ~250. That is, the zone will not approach steady-state saturation for a strongly incompatible element until it has traversed some 250 times its volume of the source solid; after about 70 passes it will reach the half-saturation concentration. For a moderately incompatible element ($D_{WR} \sim 0.1$), saturation will be approached after the zone has traversed some 25 times its volume. For a compatible element ($D_{WR} \sim 1.5$) (not likely to be a REE under suspected mantle conditions) steady state occurs quickly because it is just the equilibrium between the liquid formed by melting the source material and the solid that solidifies behind the moving zone.

Fig. 4.6 shows the REE abundances produced in a moving zone as a

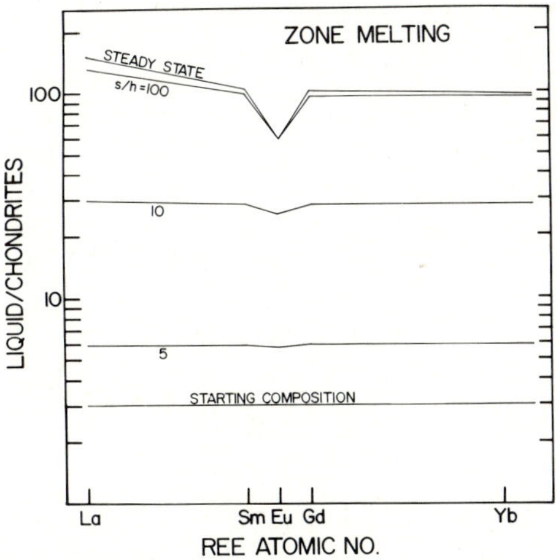

Fig. 4.6. REE abundances in liquids as a function of the thickness s of solid source region that has been traversed by a molten zone of thickness h. Conditions are those of Fig. 4.1.

function of a distance (in terms of zone thickness) which the zone has traversed. Relatively little use of this mechanism has been made in modelling studies (e.g., Harris, 1957; Schilling and Winchester, 1967). Balashov et al. (1970) have discussed the importance of zone melting and used an experimental technique to obtain estimates of D values.

4.9. Extraction

Suppose that a solid is porous and that a solution percolates through it, continuously coming into equilibrium with the solid phase and either extracting REE or depositing them, depending on the D values and the REE concentrations of the fluid phase. Such a model for REE has possible application to igneous, metamorphic, and sedimentary geochemistry.

Let the concentration of a REE in the incoming liquid be $C^{\circ}_{E,1}$. The concentration of the liquid contained within the solid will be that in equilibrium with the solid at that point and will be given by equation (4.8). The loss of E from the solid for each increment of liquid passing out will be as follows:

$$-\frac{dM_{E,s}}{dM_1} = \frac{C_{E,s}}{D_{E,WR}} - C^{\circ}_{E,1} \tag{4.30}$$

The change in concentration in the solid will equal $-dM_{E,s}/M_s$ (with M_s, the mass of the solid, presumed to remain constant). Also, rather than consider the mass of the liquid (M_L) that has passed into or through the solid, it is convenient to consider the number of changes of liquid that have passed through the solid, i.e., $M_L/M_1 = n$, where M_1 is the mass of liquid in equilibrium with the solid at any time. Thus:

$$-dC_{E,s} = \frac{M_1}{M_s}\left(\frac{C_{E,s}}{D_{E,WR}} - C^{\circ}_{E,1}\right) dn \tag{4.31}$$

Converting M_s and M_1 to the fraction of liquid in the equilibrating system gives:

$$-dC_{E,s} = \frac{X_1}{1-X_1}\left(\frac{C_{E,s}}{D_{E,WR}} - C^{\circ}_{E,1}\right) dn \tag{4.32}$$

in which X_1 is constant.

This equation can be integrated, using as the initial condition that of the first filling of the solid with liquid, for which the concentrations correspond to those for equilibrium, namely, for $n = 0$:

$$C_{E,s} = D_{E,\mathrm{WR}} \frac{X_1 C^\circ_{E,1} + (1 - X_1)C^\circ_{E,s}}{D_{E,\mathrm{WR}} + X_1(1 - D_{E,\mathrm{WR}})} \tag{4.33}$$

Then:

$$C_{E,s} = D_{E,\mathrm{WR}}\, C^\circ_{E,1}\left[1 - \exp\left(\frac{-X_1 n}{(1-X_1)D_{E,\mathrm{WR}}}\right)\right]$$

$$+ D_{E,\mathrm{WR}} \frac{X_1 C_{E,1} + (1 - X_1)C^\circ_{E,s}}{D_{E,\mathrm{WR}} + X_1(1 - D_{E,\mathrm{WR}})} \exp\left(\frac{-X_1 n}{(1-X_1)D_{E,\mathrm{WR}}}\right) \tag{4.34}$$

This equation served as the basis for considering what interaction might occur for a liquid magma seeping upward from its source in a manner that kept it in equilibrium with the material through which it flowed, as applied to studies of Keweenawan lavas ("overburden equilibration" model; Haskin and Brannon, 1980). An illustrative example is that of a basalt (Table 4.1) travelling through a solid of its same composition. No change in basalt mineralogy or mode is assumed: X_1 = 0.02. REE abundances in successive batches of liquid, each amounting to 2% of the mass of the solid/liquid system, are shown in Fig. 4.7. The basaltic liquid strives to convert its frozen equivalent to its equilibrium solid. In the process, it extracts incompatible

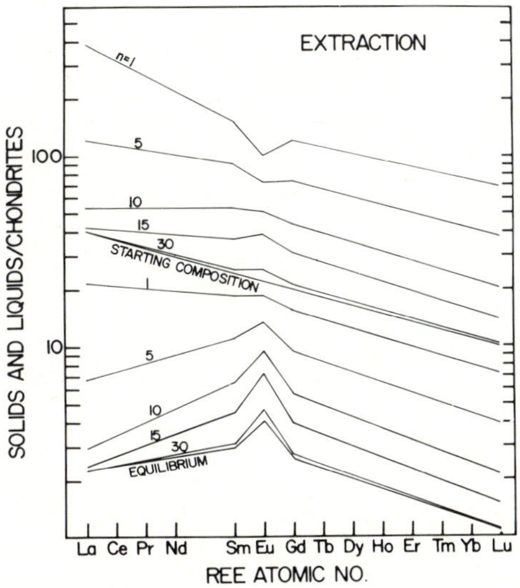

Fig. 4.7. REE abundances in successive liquids and residual solid after extraction of n portions of the solid by 2% of its mass of the liquid. This is purely illustrative and not yet proved to apply to any natural system. The starting composition of the liquid and solid was taken as that of the basalt used in Fig. 4.2. D values used are those shown by Haskin and Paster (1979).

elements from the solid and leaves behind compatible ones. The changes in REE pattern for the solid are also shown in Fig. 4.7.

To provide a rough idea of how such a model might be applied to other geological settings, Fig. 4.8 is a hypothetical instance of weathering or hydrothermal alteration of a rhyolite by a solution with relative REE abundances similar to those in seawater, and with distribution coefficients for rock/aqueous phase of the order of those found by Cullers et al. (1973). An anomalous value of D for Ce was chosen arbitrarily to account for possible oxidizing conditions. Changes of the general type shown near the top of Fig. 4.8 have been observed recently in weathering profiles for rhyolite and diabase (B. Curtis and Haskin, unpublished).

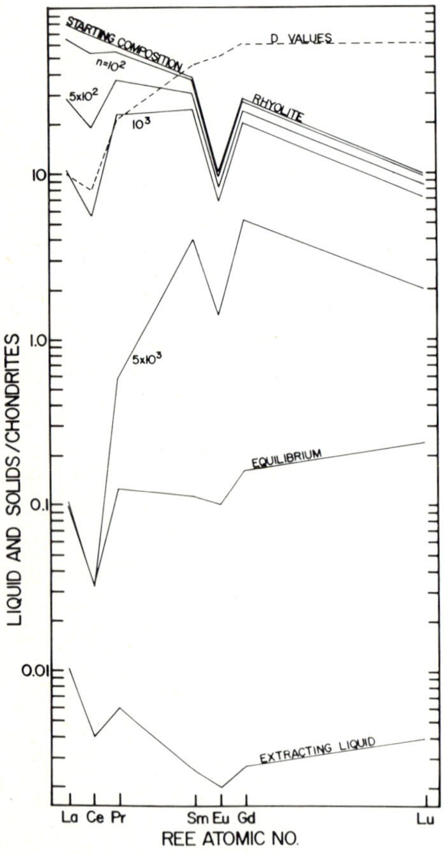

Fig. 4.8. REE abundances in a hypothetical rhyolite subjected to extraction during weathering or alteration by an aqueous fluid. Starting compositions and D values are shown. The abundance changes are given in terms of n equilibrations of the liquid with the solid, the amount of liquid in each equilibration corresponding to 2% of that of the solid.

4.10. Mixing

Whenever two materials A and B are merely mixed together, the resulting REE concentrations are those resulting from a mass balance. The total mass of a REE in a mixed system, $M_{E,\mathrm{t}}$, is given as follows:

$$M_{E,\mathrm{t}} = C_{E,A} M_A + C_{E,B} M_B \quad (4.35)$$

Converting to concentration involves dividing by the mass of the system, $M_A + M_B$, and using mass fractions for the relative proportions of materials A and B; i.e.:

$$C_{E,\mathrm{t}} = C_{E,A} f_A + C_{E,B} (1 - f_A) \quad (4.36)$$

This self-evident result is useful in magma mixing or simple assimilation. Langmuir et al. (1978) present a discussion of equations for mixing.

4.11. References to combined models, and other comments

An important aspect of REE modelling is to blend the chemical part of the process in a suitable manner with the physical process that brings about the compositional change, i.e., to adapt the chemical parts of the model to use on real rocks. The simplest forms of the models might at first seem adequate. If fractional crystallization is the process that has caused the compositional differences between two samples of related basalt, should any equation other than (4.17) be needed to describe the process? After all, it was mentioned that the equation holds whether the fractional crystallization occurred within a magma chamber or along conduits during ascent of a liquid to the surface. To make a strong case for fractional crystallization, several observations must be taken into account. First, what minerals have been crystallized and removed, and in what proportions? This is usually determined through petrographic examination of phenocrysts, followed by electron microprobe determination of phenocryst major element compositions, followed by use of a chemical mixing model (e.g., Bryan et al., 1969) which uses those compositions plus whole-rock compositions to determine the amount of each mineral crystallized. This also leads to a value for X_s, the fraction of the parent liquid that has apparently crystallized. Then, if the D values for crystallizing minerals are known, the change in REE concentrations can be calculated for comparison with observed values. The observed values for REE concentrations will not, however, always be the same as those obtained by whole-rock analysis. The equation for fractional crystallization is for one liquid evolving to another liquid. If either of the two volcanic rocks contains phenocrysts, the composition of that rock does not represent that of the liquid. It must be corrected according to the modal abundance (by weight) and the contributions of the phenocrysts, both for major elements and for REE. This is a simple case of combining the equations for equilibrium crystallization (to get phenocryst REE concentrations), mixing (to combine, or to separate, contributions from phenocrysts and groundmass) and fractional crystallization (to predict the REE concentrations of the derived liquid from that of the parent liquid). The initial two steps must be done iteratively, first assuming that the entire amount of REE is in the groundmass portion, next calculating the concentrations in phenocrysts from such a liquid, then subtracting off the estimated phenocryst contribution to get an improved estimate for the groundmass portion, etc., until no further compositional change arises. For this to work accurately, values of $D_{E,m}$ must be known with reasonable accuracy, more than is sometimes available. If, finally, the modelled results for major elements, REE, and other trace elements all agree to within approximately the expected uncertainty, the case for fractional crystallization is self-consistent and can only be compromised by some other form of evidence. The level of uncertainty expected depends on the particular situation. In all cases, some minimum level of uncertainty is introduced by the analytical uncertainties to which compositions are known. Knowledge of the D values can make a significant contribution to the uncertainty. Finally, sampling may affect the uncertainty, since volcanic rocks are somewhat heterogeneous; also, they are subject to post-fractionation alteration. All of these factors must be taken into account if the quality of the case for or against fractional crystallization (or any other mechanism) is to be properly evaluated.

A combined model has been used to determine the nature of the heterogeneity in single basalt flows (Lindstrom and Haskin, 1981; "short-range segregation" model). It was shown that the small-scale heterogeneities in REE concentrations in a sample of Iceland tholeiitic basalt were consistent with short-range migration of phenocrysts, groundmass, and residual liquid (at $X_s \sim 0.75$) as the basalt cooled. The equations for fractional crystallization were used to estimate mineral and residual liquid concentrations. A mixing model for trace and major element concentrations in phenocrysts, groundmass minerals, and residual liquid was used to estimate the relative proportions of each phase in the individual analyzed samples.

That the process acted randomly seemed apparent from the lack of correlation of sample composition with location in the basalt flow. This was affirmed through a model for random processes, a Monte Carlo model, which was used to select random portions of the different phases to produce a group of hypothetical sample compositions. The trends of correlation between concentrations of certain elements (e.g., Sm and Eu) and the lack of such trends among others estimated by the Monte Carlo model matched those observed among the actual basalt samples.

Studies of the Skaergaard intrusion (Paster et al., 1974) also made use of combined modelling. Rocks from the Skaergaard layered series consist of minerals with cores of purely cumulus origin, overgrowths of adcumulus origin, and portions arising from closed system crystallization of trapped residual parent liquid. Any simple model for fractional crystallization gives a poor fit to the data for the rocks. By taking proper account of the contribution of trapped liquid, from which most of the REE in the rocks actually derives, Paster et al. were able to model approximately the compositions of the solids. The number of samples used was small, and the uncertainties in the results were relatively large. Large layered intrusions are more complicated and the REE concentrations in their rocks are poorly understood (e.g., the Stillwater intrusion; Kosiewicz, 1973).

O'Hara (1977) and O'Hara and Mathews (1981) have produced models for magma chambers that are periodically recharged with fresh magma. This involves both mixing and crystallization. Also, they consider assimilation of roof materials. One suite of lavas for which there is stratigraphic control plus data that may support a mechanism of periodic recharging is discussed by Yanagi and Ishizaka (1978). No REE analyses were done in that work, however, nor was there any discussion of the effects of the phenocrysts and xenocrysts, which are abundant in many lavas from that region (Hayatsu, 1977).

Langmuir et al. (1977) produced a model (dynamic partial melting) to account for (among other things) the variation in REE patterns and the lack of correlation between the extent of LREE enrichment and HREE concentrations among ocean floor basalts from the FAMOUS area. The model includes partial melting, continuous removal of part of the liquid and retention of the rest in the residue, solidification and remelting of part of the melt, and zone melting, in different regions of upwelling mantle material. The model is capable of producing the variety of REE abundances observed and providing insight into the apparent decoupling of REE and major elements. The model may not describe the only important processes that affect compositions of ocean floor basalts. Magma mixing is important in the generation of such basalts (e.g., Rhodes et al., 1978). The Soret effect, which involves diffusion through a thermal gradient, may also play a strong role (Walker et al., 1981); a model for REE behavior during Soret separation has not yet been worked out. Hildreth (1979) has invoked Soret separation to explain REE patterns in the Bishop Tuff.

DePaolo (1981) has recently presented a model combining assimilation (mixing) with fractional crystallization and has applied it mainly to isotopic data from Andean volcanic rocks. Results were consistent with observed Nd isotopic data when wall-rock contamination of mantle-derived magma was assumed. Depending on the relative rates of fractional crystallization and assimilation, the model is approximated by simple binary mixing or by a combination of fractional crystallization and zone refining. The model for more extreme reaction with wall rock (or overburden) considered by Haskin and Brannon (1980) has many similar features. It differs from the usual zone refining models in that successive batches of fresh magma pass upward through the same region, so that later batches are less modified by the process than are initial ones. Each batch brings in fresh material of constant composition, in contrast to repeated-pass zone melting. This process offers some possibility of accounting for trends of REE concentrations in successive lavas with stratigraphic position, a parameter inadequately considered so far in modelling studies.

All of the models here are based at least in part on thermodynamic considerations. Surface equilibrium is kinetically controlled, and the treatment that suggests perfect zoning of crystals does not take into account diffusion rates, crystal growth rates, or post-crystallization equilibration. Albarède and Bottinga (1972) have considered the effects of various non-equilibrium processes on trace element distribution. Experimental studies by D. Lindstrom (personal communication) on rapidly grown diopside do not indicate a strong deviation in REE concentrations from the equilibrium values. Evidence that disequilibrium processes exert important control over REE distributions has not been developed.

Much work remains to be done on REE modelling. The extent of development of quantitative procedures is greatest for igneous systems and is meagre in low-temperature and hydrothermal systems. In some cases, the level of mathematical treatment can even be regarded as sophisticated. For example, Minster et al. (1977) present an inversion procedure to obtain best fits of data to a fractional crystallization model. More realistic, combined models lend themselves less readily to careful fitting of parameters and to analysis of uncertainties, but that does not diminish the importance of doing such analyses. Real rock systems, if complicated, are still comprised of materials that meticulously obey the laws of chemistry and physics. For those cases in which a particular process can be certainly identified as that predominantly responsible for compositional trends, insight gained from quantitative measurement and modelling of REE, in conjunction with other types of evidence, can be expected to make significant changes in our notions of how rocks get their compositions.

References

Albarède, F. and Bottinga, Y., 1972. Kinetic disequilibrium in trace element partitioning between phenocrysts and host lava. *Geochim. Cosmochim. Acta*, 36: 141.

Allègre, C.J. and Minster, J.-F., 1978. Quantitative models of trace element behavior in magmatic processes. *Earth Planet. Sci. Lett.*, 38: 1.

Allègre, C.J., Treuil, M., Minster, J.-F., Minster, B. and Albarède, F., 1977. Systematic use of trace element in igneous process, I. Fractional crystallization processes in volcanic suites. *Contrib. Mineral. Petrol.*, 60: 57.

Arndt, N.T., 1977. Ultrabasic magmas and high-degree melting of the mantle. *Contrib. Mineral. Petrol.*, 64: 205.

Balashov, Yu.A., Frenkel', M.Ya and Yaroshevskiy, A.A., 1970. The influence of the crystal chemical factor on the separation of rare-earth elements in crystallization differentiation of silicates. *Geochem. Int.*, 7: 611.

Banno, S. and Matsui, Y., 1973. On the formulation of partition coefficients for trace elements distribution between minerals and magma. *Chem. Geol.*, 11: 1.

Barberi, F., Ferrera, G., Santacroce, R., Treuil, M. and Varet, J., 1975. A transitional basalt-pantellerite sequence of fractional crystallization, the Boina Centre (Afar Rift, Ethiopia). *J. Petrol.*, 16: 22.

Brannon, J.C., Haskin, L.A. and Green, J.C., 1979a. A stratigraphic regularity in compositions of olivine tholeiites of the North Shore Volcanic Group. *Geol. Soc. Am., Abstr.*, 11: 226.

Brannon, J.C., Haskin, L.A. and Green, J.C., 1979b. A geochemical study of successive Keweenawan olivine tholeiite flows. *EOS, Trans. Am. Geophys. Union*, 60: 409.

Bryan, W.B., Finger, L.W. and Chayes, F., 1969. Estimating proportions in petrographic mixing equations by least-squares approximation. *Science*, 163: 926.

Cullers, R.L., Medaris, L.G. and Haskin, L.A., 1970. Gadolinium: distribution between aqueous and silicate phases. *Science*, 169: 580.

Cullers, R.L., Medaris, L.G. and Haskin, L.A., 1973. Experimental studies of the distribution of rare earths as trace elements among silicate minerals and liquids and water. *Geochim. Cosmochim. Acta*, 37: 1499.

DePaolo, D.J., 1981. Trace element and isotopic effects of combined wallrock assimilation and fractional crystallization. *Earth Planet. Sci. Lett.*, 53: 189.

DePaolo, D.J. and Wasserburg, G.J., 1976. Inferences about magma sources and mantle structure from variations of $^{143}Nd/^{144}Nd$. *Geophys. Res. Lett.*, 3: 743.

Drake, M.J. and Holloway, J.R., 1978. "Henry's law" behaviour of Sm in a natural plagioclase/melt system: importance of experimental procedure. *Geochim. Cosmochim. Acta*, 42: 679.

Drake, M.J. and Weill, D.J., 1975. The partition of Sr, Ba, Ca, Y, Eu^{2+}, Eu^{3+} and other REE between plagioclase feldspar and magmatic silicate liquid: an experimental study. *Geochim. Cosmochim. Acta*, 39: 689.

Ferrera, G. and Treuil, M., 1974. Petrological implications of trace elements and Sr isotope distributions in basalt-pantellerite series. *Bull. Volcanol.*, 38: 548.

Frey, F.A. and Prinz, M., 1978. Ultramafic inclusions from San Carlos, Arizona: petrologic and geochemical data bearing on their petrogenesis. *Earth Planet. Sci. Lett.*, 38: 129.

Frey, F.A., Green, D.H. and Roy, D.S., 1978. Integrated models of basalt petrogenesis: a study of quartz tholeiites to olivine melilitites from south eastern Australia utilizing geochemical and petrological data. *J. Petrol.*, 19: 463.

Gast, P.W., 1968. Trace element fractionation and the origin of tholeiitic and alkaline magma types. *Geochim. Cosmochim. Acta*, 32: 1057.

Greenland, L.P., 1970. An equation for trace-element distribution during magmatic crystallization. *Am. Mineral.*, 55: 455.

Harris, P.G., 1957. Zone refining and the origin of potassic basalts. *Geochim. Cosmochim. Acta*, 12: 195.

Harrison, W.J., 1977. Rare earth element partitioning between garnets, pyroxenes, and melts at low trace element concentration. *Carnegie Inst. Washington Yearb.*, 77: 682.

Harrison, W.J., 1981. Partition coefficients for REE between garnets and liquids: implications of non-Henry's law behaviour for models of basalt origin and evolution. *Geochim. Cosmochim. Acta*, 45: 1529.

Harrison, W.J. and Wood, B.J., 1980. An experimental investigation of the partitioning of REE between garnet and liquid with reference to the role of defect equilibria. *Contrib. Mineral. Petrol.*, 72: 145.

Haskin, L.A. and Brannon, J.C., 1980. OBEQ, an alternative model for modifying magma compositions. In: *Lunar Science XI.* Lunar and Planetary Institute, Houston, Texas, p. 407.

Haskin, L.A. and Frey, F.A., 1966. Dispersed and not-so-rare earths. *Science*, 152: 299.

Haskin, L.A. and Korotev, R.L., 1977. Test of a model for trace element partitioning during closed-system crystallization of a silicate liquid. *Geochim. Cosmochim. Acta*, 41: 921.

Haskin, L.A. and Paster, T.P., 1979. Geochemistry and mineralogy of the rare earths. In: K.A. Gschneider, Jr. and L. Eyring (Editors), *Handbook on the Physics and Chemistry of Rare Earths, 3.* North-Holland, Amsterdam, p. 1.

Haskin, L.A., Frey, F.A., Schmitt, R.A. and Smith, R.H., 1966. Meteoritic, solar, and terrestrial rare-earth distributions. *Phys. Chem. Earth*, 7: 167.

Haskin, L.A., Allen, R.O., Helmke, P.A., Paster, T.P., Anderson, M.R., Korotev, R.L. and Zweifel, K.A., 1970. Rare earths and other trace elements in Apollo 11 lunar samples. *Proc. Apollo 11 Lunar Sci. Conf.*, p. 1213.

Hawkesworth, C.J., O'Nions, R.K., Pankhurst, R.J., Hamilton, P.J. and Evensen, N.M., 1977. A geochemical study of island-arc and back-arc tholeiites from the Scotia Sea. *Earth Planet. Sci. Lett.*, 36: 253.

Hayatsu, K., 1977. Geologic study of the Myoko volcanoes, central Japan, 2. Petrography. *Mem. Fac. Sci. Kyoto Univ., Ser. Geol. Mineral.*, XLIII: 1.

Helmke, P.A., Haskin, L.A., Korotev, R.L. and Ziege, K.E., 1972. Rare earths and other trace elements in Apollo 14 samples. *Proc. 3rd Lunar Sci. Conf.*, p. 1275.

Hertogen, J. and Gijbels, R., 1976. Calculation of trace element fractionation during partial melting. *Geochim. Cosmochim. Acta*, 40: 313.

Hildreth, W., 1979. The Bishop Tuff: evidence for the origin of compositional zonation in silicic magma chambers, *Geol. Soc. Am. Spec. Paper*, 180: 43—75.

Hofmann, A.W. and Hart, S.R., 1978. An assessment of local and regional isotopic equilibrium in the mantle. *Earth Planet. Sci. Lett.*, 38: 44.

Irving, A.J., 1978. A review of experimental studies of crystal/liquid trace element partitioning. *Geochim. Cosmochim. Acta*, 42: 743.

Jacobssen, S.P., Jansson, J. and Shido, F., 1978. Petrology of the western Reykjanes Peninsula, Iceland. *J. Petrol.*, 19: 669.

Kay, R.W. and Gast, P.W., 1973. The rare-earth content and origin of alkali-rich basalts. *J. Geol.*, 81: 653.

Kosiewicz, S.T., 1973. *Rare-earth elements in U.S.G.S. rocks SCo-1 and STM-1, basalts from the Servilleta and Hinsdale Formations, and rocks from the Stillwater and Muskox Intrusions.* Ph.D. Thesis, University of Wisconsin, Madison, 135 pp.

Langmuir, C.H., Bender, J.F., Bence, A.E., Hanson, G.N. and Taylor, S.R., 1977. Petrogenesis of basalts from the FAMOUS area: Mid-Atlantic Ridge. *Earth Planet. Sci. Lett.*, 36: 133.

Langmuir, C.H., Vocke, R.D., Jr., Hanson, G.N. and Hart, S.R., 1978. A general mixing equation with applications to Icelandic basalts. *Earth Planet. Sci. Lett.*, 37: 380.

Leeman, W.P., Vitaliano, C.J. and Prinz, M., 1976. Evolved lavas from the Snake River Plain: craters of the Moon National Monument, Idaho. *Contrib. Mineral. Petrol.*, 56: 35.

Lindstrom, M.M. and Haskin, L.A., 1981. Compositional inhomogeneities in a single Icelandic tholeiite flow. *Geochim. Cosmochim. Acta*, 45: 15.

Masuda, A., 1965. Variation of partition coefficient. *Nature*, 205: 1098.

Masuda, A., 1967. Lanthanide concentration ratios between pyroxene and garnet. *Earth Planet. Sci. Lett.*, 3: 25.

McIntire, W.L., 1963. Trace element partition coefficients — a review of the theory and application to geology. *Geochim. Cosmochim. Acta*, 27: 1209.

Minster, J.-F., Minster, J.B., Treuil, M. and Allègre, C.J., 1977. Systematic use of trace elements in igneous processes, II. Inverse problem of the fractional crystallization process in volcanic suites. *Contrib. Mineral. Petrol.*, 61: 49.

Morris, R.V., 1975. Electron paramagnetic resonance study of the site preferences of Gd^{3+} and Eu^{2+} in polycrystalline silicate and aluminate minerals. *Geochim. Cosmochim. Acta*, 39: 621.

Morris, R.V. and Haskin, L.A., 1974. EPR measurement of the effect of glass composition on the oxidation states of europium. *Geochim. Cosmochim. Acta*, 38: 1435.

Mysen, B.O., 1976. Partitioning of samarium and nickel between olivine, orthopyroxene and liquid: preliminary data at 120 kbar and 1025°C. *Earth Planet. Sci. Lett.*, 31: 1.

Mysen, B.O., 1978. Limits of solution of trace elements in minerals according to Henry's law: review of experimental data. *Geochim. Cosmochim. Acta*, 42: 871.

Nagasawa, H., 1970. Rare earth concentrations in zircons and apatites and their host dacites and granites. *Earth Planet. Sci. Lett.*, 9: 359.

Nagasawa, H., 1973. Rare-earth distribution in alkali rocks from Oki-Dogo Island, Japan. *Contrib. Mineral. Petrol.*, 39: 301.

Nagasawa, H. and Schnetzler, C.C., 1971. Partitioning of rare earth, alkali and alkaline earth elements between phenocrysts and acidic igneous magma. *Geochim. Cosmochim. Acta*, 35: 953.

O'Hara, M.J., 1977. Geochemical evolution during fractional crystallization of a periodically refilled magma chamber. *Nature*, 266: 503.

O'Hara, M.J. and Mathews, R.E., 1981. Geochemical evolution in an advancing, periodically replenished, periodically tapped, continuously fractionated magma chamber. *J. Geol. Soc., London*, 138: 237.

Paster, T.P., Schauwecker, D.S. and Haskin, L.A., 1974. The behavior of some trace elements during solidification of the Skaergaard layered series. *Geochim. Cosmochim. Acta*, 38: 1549.

Rhodes, J.M., Dungan, M.A., Blanchard, D.P. and Long, P.E., 1978. Magma mixing at mid-ocean ridges: evidence from basalts drilled near 22°N on the Mid-Atlantic Ridge. *Tectonophysics*, 55: 35.

Richard, P., Shimizu, N. and Allègre, C.J., 1976. $^{143}Nd/^{144}Nd$, a natural tracer: an application to oceanic basalts. *Earth Planet. Sci. Lett.*, 31: 269.

Schilling, J.-G. and Winchester, J.W., 1967. Rare-earth fractionation and magmatic processes. In: S.K. Runcorn (Editor), *Mantles of the Earth and Terrestrial Planets.* Interscience, New York, N.Y., p. 267.

Schnetzler, C.C. and Philpotts, J.A., 1968. Partition coefficients of rare-earth elements and barium between igneous matrix material and rock forming phenocrysts, I. In: L.H. Ahrens (Editor), *Origin and Distribution of the Elements.* Pergamon, Oxford, p. 929.

Schnetzler, C.C. and Philpotts, J.A., 1970. Partition coefficients of rare-earth elements between igneous matrix material and rock-forming mineral phenocrysts, II. *Geochim. Cosmochim. Acta*, 34: 331.

Shaw, D.M., 1970. Trace element fractionation during anatexis. *Geochim. Cosmochim. Acta*, 34: 237.

Shaw, D.M., 1978. Trace element behaviour during anatexis in the presence of a fluid phase. *Geochim. Cosmochim. Acta*, 42: 933.

Smewing, J.D. and Potts, P.J., 1976. Rare earth abundances in basalts and metabasalts from the Troodos Massif, Cyprus. *Contrib. Mineral. Petrol.*, 57: 245.

Turcotte, D.L. and Ahern, J.L., 1978. A porous flow model for magma migration in the asthenosphere. *J. Geophys. Res.*, 83: 767.

Walker, D., Shibata, T. and Delong, S., 1979. Abyssal tholeiites from the Oceanographer Fracture Zone. *Contrib. Mineral. Petrol.*, 70: 111.

Walker, D., Lesher, C.E. and Hays, J.F., 1981. Soret separation of lunar liquid. In: *Lunar and Planetary Science XII*, Lunar and Planetary Institute, Houston, Texas, p. 1130.

Weill, D.F. and McKay, G.A., 1975. The partitioning of Mg, Fe, Sr, Ce, Sm, Eu, and Yb in lunar igneous systems and a possible origin of KREEP by equilibrium partial melting. *Proc. 6th Lunar Sci. Conf.*, p. 1143.

Yanagi, T. and Ishizaka, K., 1978. Batch fractionation model for the evolution of volcanic rocks in an island arc: an example from central Japan. *Geochim. Cosmochim. Acta*, 39: 713.

Zielinski, R.A., 1975. Trace element evaluation of a suite of rocks from Reunion Island, Indian Ocean. *Geochim. Cosmochim. Acta*, 39: 713.

Zielinski, R.A. and Frey, F.A., 1970. Gough Island: evaluation of a fractional crystallization model. *Contrib. Mineral. Petrol.*, 29: 242.

Zielinski, R.A. and Frey, F.A., 1974. An experimental study of the partitioning of a rare earth element (Gd) in the system diopside—aqueous vapour. *Geochim. Cosmochim. Acta*, 38: 545.

Chapter 5

RARE EARTH ELEMENT ABUNDANCES IN UPPER MANTLE ROCKS

FREDERICK A. FREY

5.1. Introduction

The Earth's mantle comprises approximately two-thirds the mass of the earth; therefore, in evaluating bulk Earth composition and compositional evolution of the crust-mantle-core system, it is imperative to know the composition of the mantle. Much of what is deduced about mantle composition is based on indirect approaches such as: (1) analogy with chondritic meteorites; (2) geophysical data, principally seismic and heat flow data; and (3) study of liquids, principally basalts, derived from the mantle by partial melting. Each of these indirect approaches has significant problems. For example, chondritic meteorites are not uniform in composition and the Earth apparently has non-chondritic abundances of volatile elements; geophysical data are not sensitive indicators of composition; and our knowledge of melting degree and liquid/solid partitioning at mantle pressures and temperatures is presently too limited to infer unambiguously the nature of the mantle source composition for basalts or the residual material created in the mantle by the partial melting process.

The only direct approach for determining mantle compositions is to study samples of mantle rock that are exposed within the crust by geologic processes. Fortunately, within the crust there are many occurrences of rocks with a mineralogy and mineral compositions that reflect equilibration within the upper mantle. The majority of these rocks are peridotites of several distinct types (e.g., Wyllie, 1967, chapter 1), but in this review I discuss only the types that appear to be solids derived directly from the upper mantle by a variety of tectonic processes.

Four major types of peridotites are discussed:

(1) Ultramafic inclusions occurring in alkalic basalts and kimberlites.

(2) Alpine peridotites occurring as isolated masses.

(3) Peridotites forming the basal tectonite unit of ophiolite sequences.

(4) Peridotites occurring on the ocean floor.

While there are some textural and compositional similarities among these types of ultramafic rock (e.g., Wilshire and Pike, 1975), in this review each type listed above will be discussed separately.

Prior to detailed discussion of REE abundances in mantle rocks it is useful to outline what we might predict for REE abundances in the mantle. As discussed in Chapters 3 and 11 there is convincing evidence that the bulk Earth has a *relative* REE distribution similar to that of chondritic meteorites. Thus, primordial mantle is expected to have the same relative REE abundances as chondritic meteorites; i.e., horizontal lines on chondrite-normalized REE diagrams. Furthermore, because the crust is relatively enriched in LREE, i.e., $(\mathrm{LREE/HREE})_{\mathrm{cn}}^{\mathrm{crust}} > 1$ (see Chapter 6, 7 and 8), and in a gross manner the crust has been derived from the mantle, we can predict that residual mantle remaining from generation of crustal material should have $(\mathrm{LREE/HREE})_{\mathrm{cn}} < 1$.

We are less certain about the absolute REE content expected in primordial mantle because different chondrite types vary considerably in REE content (see Chapter 3). Ringwood (1975 p. 195) noted that, as in this chapter, most chondrite-normalized diagrams for REE are based on data for ordinary chondrites. He emphasized that relative to Si atoms, REE abundances in carbonaceous chondrites are enriched relative to those in ordinary chondrites (factors of ~1.25—1.7 for L, H and enstatite chondrites; see Chapter 3). In addition, Ringwood observed that REE are likely to be deficient in the earth's core; thus, the bulk mantle should be enriched by approximately a factor of 1.5 relative to the bulk Earth composition. Therefore, if the REE content of the bulk Earth is similar to that of carbonaceous chondrites on a volatile-free basis, then the REE content of primordial mantle is expected to be enriched in REE by a factor of 1.9—2.6 compared to ordinary chondrites.

A major feature of the early papers (e.g., Nagasawa et al., 1969; Frey et al., 1971; Philpotts et al., 1972) on REE abundances in ultramafic rocks was the finding that REE abundances vary widely in these rocks including samples with LREE/HREE abundance ratios greater *and* less than in chondrites. It was immediately apparent that REE abundances in mantle rocks exposed in the earth's crust are much more variable than anticipated. Initial interpretations of this variability were hampered by the variety of peridotite types and locations studied in these early papers. Subsequent more systematic detailed studies of peridotites from a single locality have led to much greater understanding. Where possible this review focuses on REE abundance studies of samples that are well-characterized in terms of field relationships, mineralogy and major element composition. After review of the major peridotite types listed, the implications for upper mantle heterogeneity and mantle processes are discussed.

5.2. Alpine peridotites

In general, alpine peridotites are dominantly composed of lherzolite and harzburgite with lesser amounts of dunite and pyroxenite (nomenclature is defined in Fig. 5.1). In most alpine peridotites the aluminous phase is

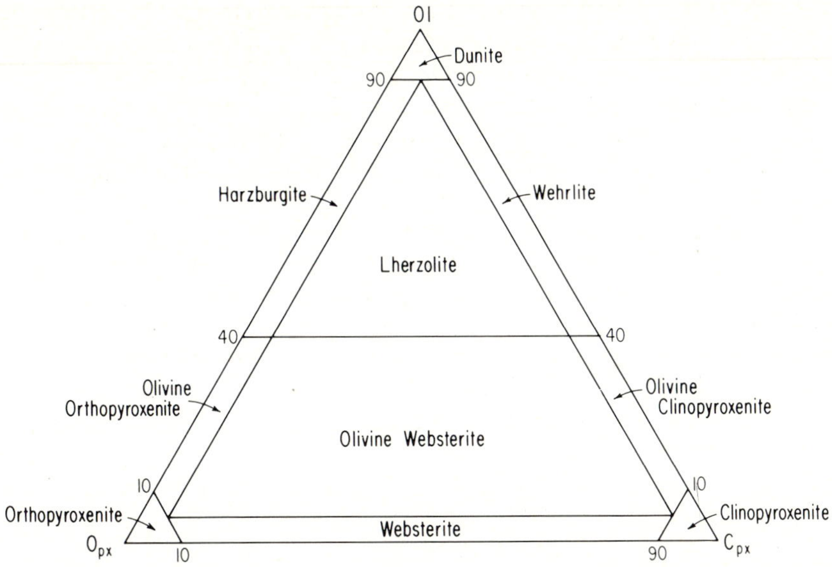

Fig. 5.1. I.U.G.S. nomenclature (based on modal mineral proportions) for ultramafic rocks with <5% spinel and lacking plagioclase, hornblende and garnet (Streckeisen, 1976). Peridotites contain more than 40% olivine and pyroxenites contain less than 40% olivine.

spinel rather than plagioclase or garnet, but in some areas such as Norway, garnet lherzolites dominate. REE abundance studies of alpine peridotites have focussed on spinel-bearing lherzolites and these ultramafics will be discussed first.

Lherzolites

Haskin et al. (1966, fig. 10) first showed that the spinel peridotites forming the bulk of the ultramafic masses at Lizard in Cornwall, England, Mount Albert in Quebec, Canada and Tinaquillo, Venezuela are strongly depleted in LREE relative to chondrites (Fig. 5.2). These peridotites were grouped as high-temperature peridotites by Green (1967) because of their well-developed metamorphic aureoles, although this interpretation has been debated, especially for the Lizard occurrence.

Subsequently, Frey (1969, 1970a) studied in detail the REE abundances in Lizard rocks and constituent minerals. Although there was considerable variability in REE content from the primary to recrystallized peridotites, he found that all the Lizard peridotites have LREE/HREE abundance ratios less than that of chondrites. This result led to the interpretation that these peridotites formed as residues from partial melting. Specifically Frey calculated that for a bulk Earth with chondritic REE abundances, 25% of the mantle must have REE abundances similar to the Lizard primary spinel-

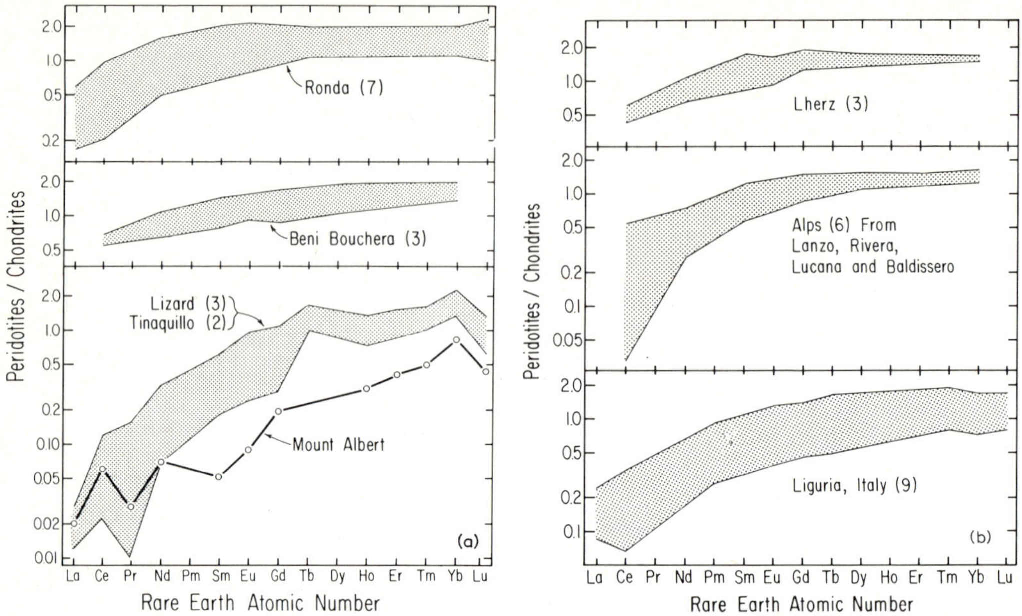

Fig. 5.2. REE abundances in alpine peridotites with number of samples defining range indicated within parentheses. In this chapter REE abundances are always normalized to the chondritic average of Haskin et al. (1968). (a) Lower diagram — Lizard, Cornwall, England (Frey, 1969), Tinaquillo, Venezuela (Haskin et al., 1966; Philpotts et al., 1972) and Mount Albert, Quebec, Canada (Haskin et al., 1966). Middle diagram — Beni Bouchera, Morocco (Loubet et al., 1975). Upper diagram — Ronda, Spain (Dickey et al., 1977, 1979, Frey and Suen, 1983). (b) Lower diagram — Liguria, Italy (Ottonello et al., 1979). Middle diagram — Alps (Loubet et al., 1975). Upper diagram — Lherz, France (Loubet et al., 1975). Additional REE abundance data for thirteen alpine peridotites from the western Alps are in fig. 3 of Ernst (1981). These samples also have $(LREE/HREE)_{cn} < 1$.

peridotite in order to balance the strong relative LREE enrichment in the crust. Frey (1969) also analyzed the constituent minerals of the Lizard peridotites and concluded that the strong relative depletion of LREE in clinopyroxene required that the coexisting melt also had a low LREE/HREE ratio relative to chondrites. Such a feature is characteristic of most mid-ocean ridge basalts (MORB).

Within the last five years REE data have been presented for many alpine peridotites (mostly spinel-bearing) from Ronda, Spain (Suen and Frey, 1977, Dickey et al., 1977, 1979; Menzies et al., 1977; Frey and Suen, 1983), Beni Bouchera, Morocco (Loubet et al., 1975; Menzies et al., 1977; Loubet and Allègre, 1979), Lherz, France (Loubet et al., 1975, 1980), several locations in the Alps (Loubet et al., 1975, 1980; Menzies et al., 1977) and from Liguria, Italy (Ottonello et al., 1979). Ranges of chondrite-normalized REE abundances are shown in Fig. 5.2. It is evident that all of these alpine peridotites are characterized by LREE depletion compared to chondrites, and that HREE abundances typically range from 1 to 2 times chondrites. Only the single poorly-documented sample from Mount Albert has HREE abundances considerably less than in chondrites and it is likely that these data are for a harzburgite (MacGregor and Basu, 1979). In alpine peridotites, the abundances of LREE range (~0.01—0.6 times chondrites with the lowest abundances in the Lizard, Tinaquillo and Mount Albert peridotites) much more widely than the HREE (Fig. 5.2). This wide variation in LREE abundances with relatively uniform HREE contents is expected for residues created by various degrees of melting of a peridotite with uniform REE abundances, ~2 times ordinary chondrites.

From these REE data it can be inferred that wide regions of the mantle were uniform, within a factor of two, in REE content prior to the melting event creating alpine peridotites. Moreover, the inferred REE, particularly HREE, content of the mantle prior to melting was within a factor of two of REE abundances in ordinary chondrites. HREE abundances in basalts are consistent with these inferences based on alpine peridotites, although the peridotite sources of basalts apparently range widely in LREE content (see Chapters 4, 6 and 7). Thus, as concluded by Ringwood (1975), there is no evidence requiring that the upper mantle has been enriched in REE, especially HREE, by mantle processes or that the bulk Earth has an atomic REE/Si ratio distinct from that of carbonaceous chondrites.

The degree of LREE/HREE fractionation in alpine peridotites is most readily seen in Fig. 5.3 where the uniformity of Yb abundance contrasts with the variability of Ce abundance. During melting of mantle materials containing olivine, pyroxenes, garnet, spinel and plagioclase, Ce is a strongly incompatible element whose abundance in residual peridotites depends upon the degree of melting and the nature of the melting process (Fig. 5.4). In particular, if each infinitesimal melt increment is immediately segregated from the residue, i.e., fractional melting, the residues become very depleted in incompatible elements, such as Ce, even at low degrees of melting.

In contrast to LREE, during partial melting of the upper mantle the HREE may be only moderately incompatible, i.e., $D^{\text{solid/liquid}}$ of ~0.1 as in spinel or plagioclase-peridotites, or slightly compatible, i.e., $D^{\text{solid/liquid}}$ ~1 as in garnet-peridotites. Therefore variations in HREE contents, such as Yb abundances (Fig. 5.3), provide information about the mineralogy of the residue during the melting process. For example, based on a small number of samples from several areas Loubet et al. (1975) used Ce-Yb abundance variations to infer that peridotites from the Alps and Lizard represent a series of residues developed from a garnet-peridotite source. Because no garnet is found in these peridotites, they suggested that the residual garnet had been eliminated during low-pressure recrystallization. Because of the greater variation in HREE content for a given range of LREE content in Beni Bouchera and Lherz peridotites, Loubet et al. inferred that these

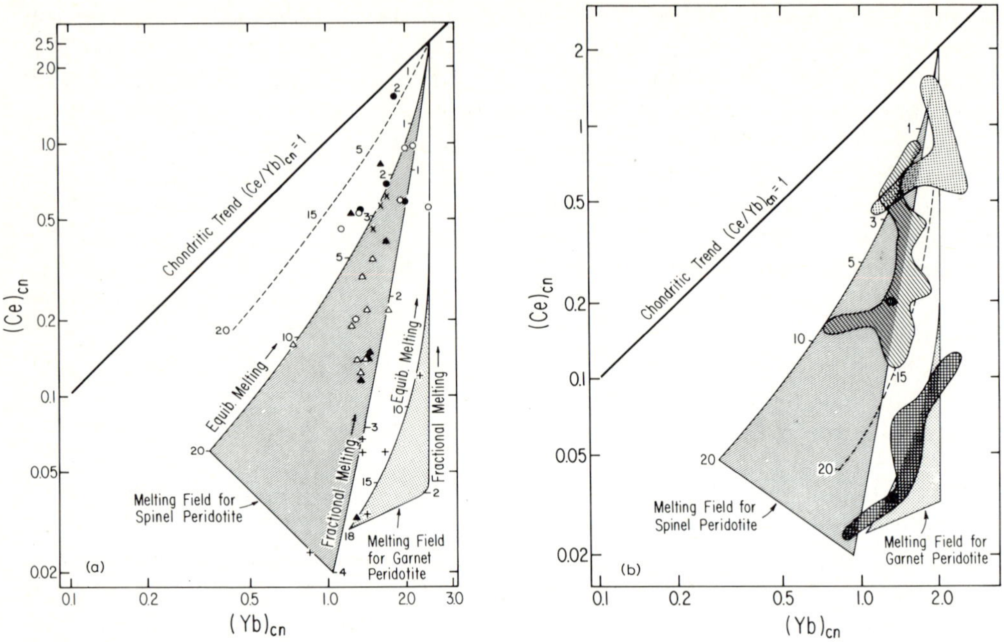

Fig. 5.3. (a) Chondrite-normalized Ce vs. Yb for alpine peridotites. Ce is used as a LREE, rather than La, because La is not usually determined by isotope dilution. • = Beni Bouchera (Loubet et al. 1975; Menzies et al., 1977), ○ = Ronda (Frey and Suen, 1983), + = Tinaquillo, Lizard and Mount Albert (Haskin et al., 1966; Frey, 1969; Philpotts et al., 1972), × = Lherz (Loubet et al., 1975), ▲ = Alps (Loubet et al., 1975; Menzies et al., 1977), △ = Liguria (Ottonello et al., 1979). Model fields are calculated with the following parameters:

Partition coefficients: set 2 of Frey et al. (1978, p. 511)

Initial mode: spinel peridotite — 55% ol, 25% opx, 20% cpx
garnet peridotite — 55% ol, 25% opx, 10% cpx, 10% garnet

Melting proportions: spinel peridotite — cpx:opx:ol = 0.6:0.2:0.2
garnet peridotite — cpx:gt:opx:ol = 0.4:0.4:0.1:0.1

Initial REE content assumed at 2.5 × ordinary chondrites. Spinel peridotite stippled-field boundaries formed by trajectories of 0—20% equilibrium melting and 0—4% fractional melting. Garnet peridotite stippled-field boundaries formed by trajectories of 0—18% equilibrium melting and 0—2% fractional melting. Melting models described by Shaw (1970). Note that fractional melting results in very rapid depletion of Ce even at low degrees of melting. Dashed line indicates equilibrium melting trajectory using REE partition coefficients of Loubet et al. (1975) for spinel peridotite but using other parameters as indicated above.

(b) Chondrite-normalized Ce vs. Yb for alpine peridotites. Upper field (dotted) for Beni Bouchera and Ronda. NW-SE slashed field for Alps and Liguria. Cross-hatched field for Lizard, Tinaquillo and Mount Albert. Model fields same as left except that source peridotite has REE = 2 × ordinary chondrites and dashed line is for 1—20% equilibrium melting of garnet peridotite using REE partition coefficients of Loubet et al. (1975) Note that by changing the source to 2 × ordinary chondrites the high-temperature peridotites lie within the melting field for garnet peridotite, especially if the clinopyroxene/melt REE partition coefficients of Loubet et al. (1975) are used.

peridotites formed as residues during melting of spinel peridotite. However, distinguishing between residual spinel- and garnet-bearing peridotites on the basis of Ce-Yb abundances is difficult, because clinopyroxene and garnet REE distribution coefficients are poorly known. For example, some clinopyroxene/melt distribution coefficients for HREE approach unity (e.g., distribution coefficient set 2 of Frey et al., 1978), and Harrison and Wood (1980) and Harrison (1981) have argued for complex variations in HREE garnet/melt distribution coefficients during the melting of peridotite. If clinopyroxene/melt HREE distribution coefficients are <0.5 and those of garnet/melt are $\geqslant 4$ (as in Fig. 5.3) then a distinction can be made between spinel- and garnet-bearing residual peridotites.

As illustrated in Fig. 5.3 the data for each peridotite locality tend to form elongated groups of variable Ce and relatively uniform Yb. Most of these data lie within the boundaries of the theoretical melting models in Fig. 5.3 (i.e., fractional and equilibrium melting of spinel- and garnet-bearing peridotites with initial Ce and Yb contents of 2.5 times ordinary chondrites). Boundaries in this figure are for a specific set of model parameters, and data lying outside the solid-line envelopes could be interpreted as indicating different (lower) source Ce-Yb contents; alternatively, these data can be easily modelled by choosing different clinopyroxene/melt distribution coefficients (see dashed line in left part of Fig. 5.3). Thus, the REE data for these peridotites cannot unambiguously define the REE concentrations prior to the melting event. However, the data are most readily modelled by abundances of 1.5—2.5 times ordinary chondrites in the peridotites prior to melting.

In theory the data in Fig. 5.3 could be used to infer the melting process that created these alpine peridotites. For example, if the mantle source is assumed to have 2.5 times ordinary chondritic abundances for Ce and Yb, the trends for $<4\%$ fractional melting or 1—15% equilibrium melting are consistent with most of the data. Thus, in order to identify the appropriate melting model, it is necessary not only to estimate the original LREE abundance but also to know the degree of melting required to create a particular alpine peridotite. Additional complexities arise because the trajectories of equilibrium melting and fractional melting models are very dependent on the choice of REE distribution coefficients, the initial mode of the source rock and the proportions of phases entering the melt. The effects of changing REE distribution coefficients are indicated by dashed lines in Fig. 5.3.

Nevertheless, a qualitative conclusion is that the majority of alpine peridotites, all except the high-temperature peridotites, from Lizard, Tinaquillo and Mount Albert, may have formed as residues from less than 15% melting of a peridotite with REE abundances $\leqslant 2.5$ times ordinary chondrites (Fig. 5.3). Consequently, if tholeiitic basalts such as MORB are derived by $>20\%$ melting as commonly believed (e.g., Green et al., 1979) then many alpine peridotites are not residues from generation of MORB, but they may represent source compositions for MORB.

In genetically related basalts the abundance range of incompatible elements is the best indicator of relative degrees of melting and crystallization. However, as shown for Ce in Figs. 5.3 and 5.4, incompatible element abundances in residues depend also on the source abundance and the type of melting process. In peridotites formed as residues from varying degrees of partial melting of a homogeneous source, an estimate of the relative degrees of melting can be inferred from the abundances of major elements (e.g., Ca and Al) that are preferentially enriched in basaltic magmas relative to a peridotite residue. For example, the CaO and Al_2O_3 abundance ranges are 2.34—3.37% and 2.46—3.72%, respectively in six Ronda peridotites and 1.14—2.53% and 1.90—3.69%, respectively, in nine Ligurian peridotites. Note that the samples with ~3.5% CaO and Al_2O_3 are similar to the commonly proposed pyrolite model for mantle composition (Ringwood, 1975, table 5-2). It is presumed that at each locality peridotites with lower CaO and Al_2O_3 contents are residues from higher degrees of melting than those with higher CaO and Al_2O_3 contents. Indeed there is covariance of CaO, Al_2O_3 and

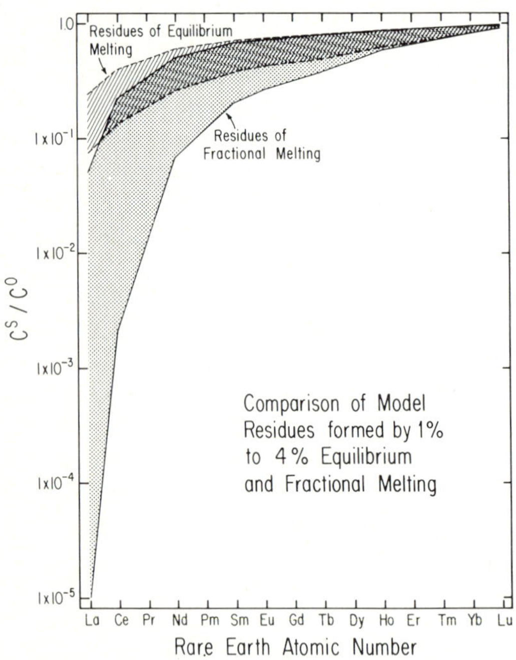

Fig. 5.4. Comparison of model residues generated by 1—4% non-modal equilibrium melting (slashed field) and by 1—4% non-modal fractional melting (stippled field). C^s/C^0 on vertical axis indicates REE abundance in the residue (C^s) relative to the abundance in the source before melting (C^0). Note that LREE abundances are *very* dependent on melting model. (Models calculated for partition coefficient set 1 of Frey et al., 1978; initial mode of 58% ol, 22% opx, 15% cpx and 5% gt; melting mode of 5% ol, 5% opx, 45% cpx and 45% gt; melting models were presented by Shaw, 1970).

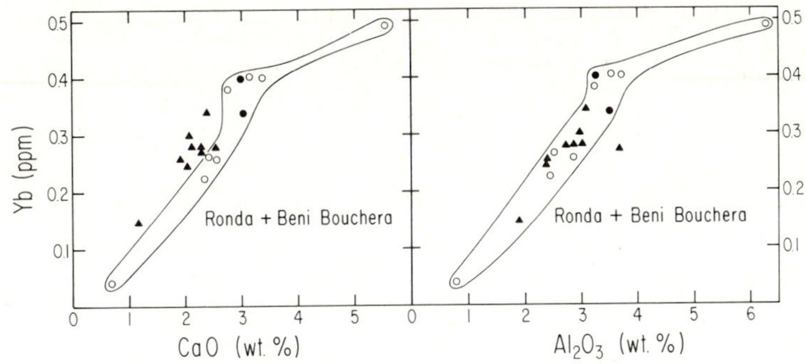

Fig. 5.5. Yb vs. CaO and Al_2O_3 for alpine peridotites from Ronda (○), Beni Bouchera (●) and Liguria (▲). Enclosed field encompasses data for Ronda and Beni Bouchera samples. Low-CaO-Al_2O_3 Ronda sample is a harzburgite. High-CaO-Al_2O_3 Ronda sample is an olivine websterite. Data from Dickey et al. (1977, 1979); Suen and Frey (1978); Frey and Suen (1983); Loubet and Allègre (1979); Ottonello et al. (1979); and Ernst and Piccardo (1979).

HREE contents (Fig. 5.5) in these alpine peridotites which is consistent with an origin as partial melting residues. In terms of a partial melting model, the near factor of two range for CaO and Al_2O_3 abundance in each suite requires considerable variation in melting degree to generate each residue suite. For example, if the Ronda peridotite with the highest CaO and Al_2O_3 content is assumed to be representative of the peridotite before melting, mass balance with the typical CaO (10%) and Al_2O_3 (14%) contents of basalts requires 0 to ~12% melting to account for the CaO and Al_2O_3 abundance range in Ronda peridotites. This melting range is similar to the variation in melting degree estimated from REE data for equilibrium melting of spinel peridotite (Fig. 5.3). Of course, much higher degrees of melting would create herzburgites rather than lherzolites. Thus, the REE and major element abundances of the Ronda peridotites are consistent with an origin as residues generated by ~0 to 10—15% melting.

In the Ronda—Beni Bouchera suite there is only a rough positive correlation of La and Ce abundances with CaO and Al_2O_3 content, and in the Ligurian suite there is poor correlation (Fig. 5.6). A strong positive correlation is expected if these peridotite suites have been derived as residues from varying degrees of partial melting of a compositionally homogeneous source. The poor correlations for LREE (Fig. 5.6) compared to HREE (Fig. 5.5) may result from several possibilities including: complexities in the melt segregation process such as variable amounts of trapped melt or melting processes intermediate between equilibrium and fractional melting; heterogeneity in La and Ce abundances in the original peridotite which has already been inferred from studies of basalts (see Chapters 6 and 7); and alteration processes since most alpine peridotites are serpentinized to varying degrees.

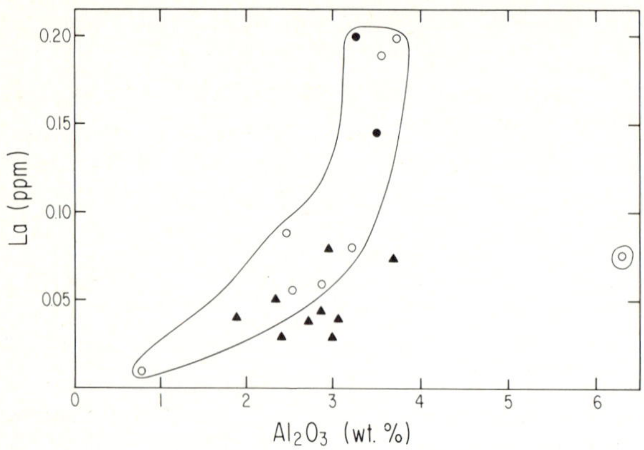

Fig. 5.6. La vs. Al_2O_3 for alpine peridotites. Symbols and references same as in Fig. 5.5.

Garnet lherzolites

Garmann et al. (1975) reported data for three garnet peridotites from West Norway. The data are scattered (Fig. 5.7) but are similar in range to those found in spinel lherzolites and Garmann et al. concluded an origin as partial melting residues was possible. Also, one of the Ronda peridotites in Fig. 5.2 is a garnet lherzolite and it is plotted in Fig. 5.7. At present, there are too few analyses of garnet lherzolites from alpine peridotites to evaluate their trajectories in Fig. 5.3, but the limited data are consistent with an origin as partial melting residues.

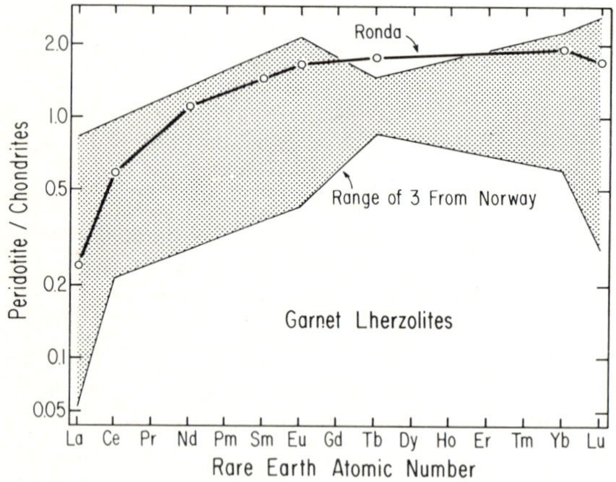

Fig. 5.7. Chondrite-normalized REE abundances in garnet lherzolites from alpine peridotites. Range indicated for three samples from Norway (Garmann et al., 1975) and ×'s indicate Ronda, sample 501 (Frey and Suen, 1983).

Herzburgites and dunites

Harzburgites and dunites are dominant in many alpine peridotites, but there have been no thorough REE studies of alpine peridotites, dominated by harzburgites. REE data for two harzburgites from the Josephine peridotite, Oregon, are in Fig. 5.8. The best studied alpine dunite and harzburgite are the U.S.G.S. standard ultramafic rocks, DTS-1, a dunite from Twin Sisters, Washington, and PCC-1, a serpentinized harzburgite from Cazadero, California. These samples have some of the lowest REE contents found in peridotites. Thus, they are excellent samples to test the analyst's mettle and the data in Fig. 5.8 provide an estimation of the data quality for these rocks. Since clinopyroxene is the major REE host phase in most peridotites, the low REE abundances (0.01—0.2 times chondrites) in

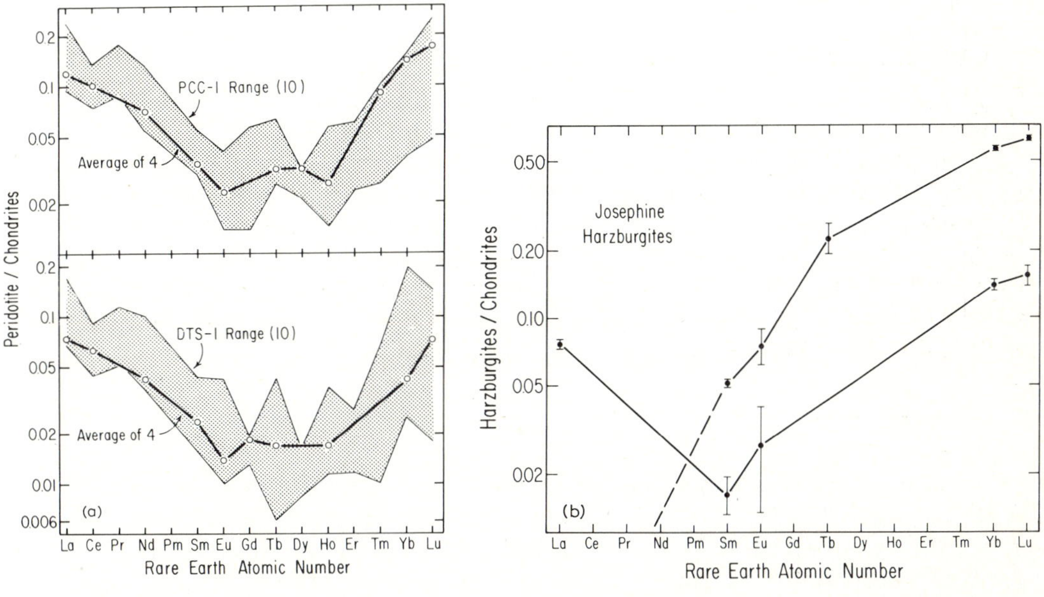

Fig. 5.8. (a) Range of chondrite-normalized REE abundances in U.S.G.S. standard dunite, DTS-1 and standard peridotite, PCC-1. Data range is from ten laboratories using three different analytical techniques (neutron activation — Higuchi et al., 1969; Rey et al., 1970; Frey et al; 1971; Brunfelt et al., 1974; Garmann et al., 1975; Kolesov, 1976; Smet and Roelandts, 1978; G. Ottenello, personal communication, 1978; Ottenello et al., 1979; Laul, 1979 and Stosch and Seck, 1980; isotope dilution mass spectrometry — Hooker et al., 1975; spark source mass spectrometry — Strelow and Jackson, 1974). Even using the four most consistent data sets, whose average is shown in figure, it is apparent that REE abundances in these peridotites are not known to better than ±10%. Nevertheless, all data sets clearly show the anomalous convex-downwards pattern. (b) Chondrite-normalized REE abundances in two harzburgites from the Josephine alpine peridotite, Oregon. La in LREE-depleted sample is less than 0.002 × chondrites. (Data from R. Hickey, personal communication, 1981.)

these clinopyroxene-poor peridotites are not surprising, but their U- or V-shaped chondrite-normalized REE patterns *are unexpected* (Fig. 5.8). Despite the considerable uncertainty in the REE contents of DTS-1 and PCC-1, there is no doubt that these unusual REE patterns are characteristic of these peridotites.

Several other alpine dunites and harzburgites have been analyzed for REE. Most have lower REE content than lherzolites but a variety of chondrite-normalized patterns are found. Menzies et al. (1977) found that dunites from Lanzo and Beni Bouchera have LREE/HREE abundance ratios less than chondrites (Lanzo, LREE = 0.03 and HREE = 0.3 times chondrites; Beni Bouchera, LREE = 0.1 and HREE = 0.6 times chondrites) and a dunite from Ronda has all REE at ~0.1 times chondrites. Frey and Suen (1983) studied a Ronda harzburgite that ranged from 0.03 to 0.2 times chondrites for LREE and HREE, respectively. Obviously, these peridotites with low REE contents and relative LREE depletion may represent residues from very high degrees of melting. In contrast, a dunite from Shikoku, Japan has REE abundances similar to the Twin Sisters dunite (Frey et al., 1971), and three dunites from Norway have REE abundances ranging from 0.2 to 5 × chondrites with $(\mathrm{LREE/HREE})_{cn} > 1$ (Garmann et al., 1975).

It is likely that alpine dunites and harzburgites can be generated by a variety of processes. Certainly, their REE abundance features are more variable than those of alpine lherzolites. Most perplexing are the U- or V-shaped chondrite-normalized REE patterns such as in DTS-1 and PCC-1 (Fig. 5.8). Similar REE patterns occur in harzburgites and dunites from ophiolite sequences and possible interpretations for such REE patterns are discussed in section 5.3 on ultramafics from ophiolites.

Pyroxenites

As noted by Kornprobst (1969), Dickey (1970), Wilshire and Pike (1975) and Conquéré (1977), mafic layers are common in many alpine peridotites These layers are mineralogically diverse, ranging in the Ronda peridotite from olivine gabbro to garnet clinopyroxenite. A surprisingly wide variety of REE abundances have been found in the pyroxenites but almost all samples have LREE/HREE abundance ratios less than chondrites (Ronda — Dickey et al., 1977, 1979; Suen and Frey, 1978; Obata et al., 1980; Beni Bouchera — Loubet and Allègre, 1979; Lanzo and Freychinede bodies — Loubet et al., 1980). Garnet pyroxenites in particular, have LREE/HREE ratios less than the host peridotites (Fig. 5.9). Consequently, REE data preclude the earlier interpretation that these mafic layers represent completely crystallized proto-basalt magmas. Loubet et al. (1976, 1980) and Loubet and Allègre (1979) interpreted some of the layers as cumulates precipitated from basic magma and Suen and Frey (1978) and Obata et al. (1980) suggested that the cumulate compositional features result from

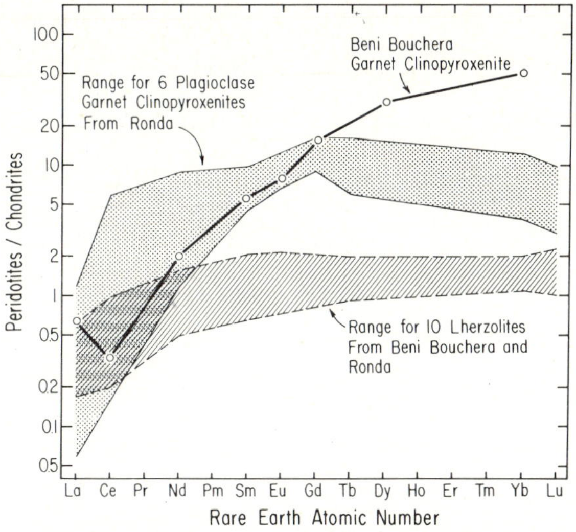

Fig. 5.9. Chondrite-normalized abundances in garnet pyroxenites from Ronda and Beni Bouchera (Suen, 1978; Loubet and Allègre, 1979; Obata et al., 1980). Note that the garnet pyroxenites have lower La/Yb than host lherzolites and some have a lower LREE content. Additional REE data for pyroxenites in alpine peridotites were presented by Suen (1978), Loubet and Allègre (1979), Loubet et al. (1980), and Obata et al. (1980).

fractional crystallization and wall-rock reactions of magmas along conduit walls. The garnet clinopyroxenites with their very low LREE abundances and LREE/HREE ratios less than the host peridotites (Fig. 5.9) apparently have a more complex history. For example, Loubet et al. (1976) and Loubet and Allègre (1979) concluded that the Beni Bouchera garnet pyroxenites are residual from a second melting event that partially melted the previously formed mafic layers. Although their petrogenesis is not understood in detail, it is evident that mafic layers in peridotites have resulted from *multiple* magmatic events in the mantle (e.g., Loubet et al., 1980). With more thorough study they should provide important information about the evolution of the upper mantle.

The complex history of these pyroxenites indicated by their REE abundances is succinctly shown in Fig. 5.10. Key features in this figure are that several generations of melt can be created; e.g., melts from the initial peridotite, later melts derived from the residual peridotite and later melts derived from melting of a partially or totally crystallized melt (i.e., the mafic layers). Also, the pyroxenites may represent completely crystallized melts, solid segregations formed from melts or residues formed by partial fusion of mafic layers.

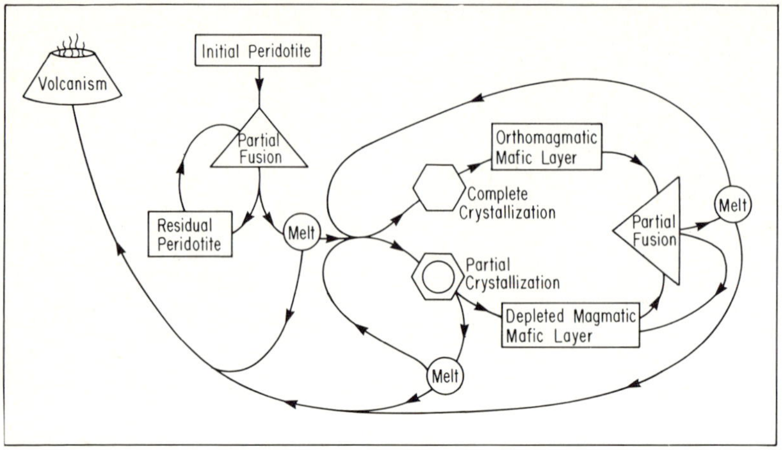

Fig. 5.10. Flow diagram for evolution of an alpine peridotite containing mafic layers (adapted from Dickey et al., 1977, 1979). Key features are that residues from the initial fusion event may be remelted; liquids formed by partial fusion may be modified by crystal segregation; and layers formed by partial or complete crystallization of liquids may be partially melted at a later time. Consequently, multiple magmatic events are represented in such an alpine peridotite.

5.3. Ultramafics associated with ophiolites

Metamorphic peridotites form the basal member of ophiolite sequences, and their textures clearly reflect a mantle history of subsolidus recrystallization. I discuss these peridotites as a separate group because in ophiolites that have been studied in detail, harzburgite is the major rock type (e.g., Nicolas and Jackson, 1972; Coleman, 1977, p. 21), whereas lherzolites dominate in most of the alpine peridotites discussed earlier.

REE abundances in peridotites from the Othris, Vourinos, Pindos, Troodos, Samail and Bay of Islands ophiolites are indicated in Fig. 5.11. Like harzburgites and dunites from isolated peridotite masses (e.g., Fig. 5.8), the harzburgites and dunites from ophiolites have very low REE contents, typically ~0.001—0.5 times chondrites. A few of the volumetrically subordinate lherzolites in ophiolites have also been studied and they have REE abundances which overlap with lherzolites from alpine peridotites, such as Ronda (cf. Fig. 5.2 and 5.11). Consequently, similar petrogenetic inferences can be made for alpine lherzolites and the rare lherzolites occurring in the basal peridotite section of ophiolites.

Petrogenetic interpretations for the harzburgites and dunites from ophiolites and alpine peridotites are more difficult because of their unexpected convex-downwards, chondrite-normalized REE patterns. Firstly, it must be recognized that several of these peridotites have not been analyzed

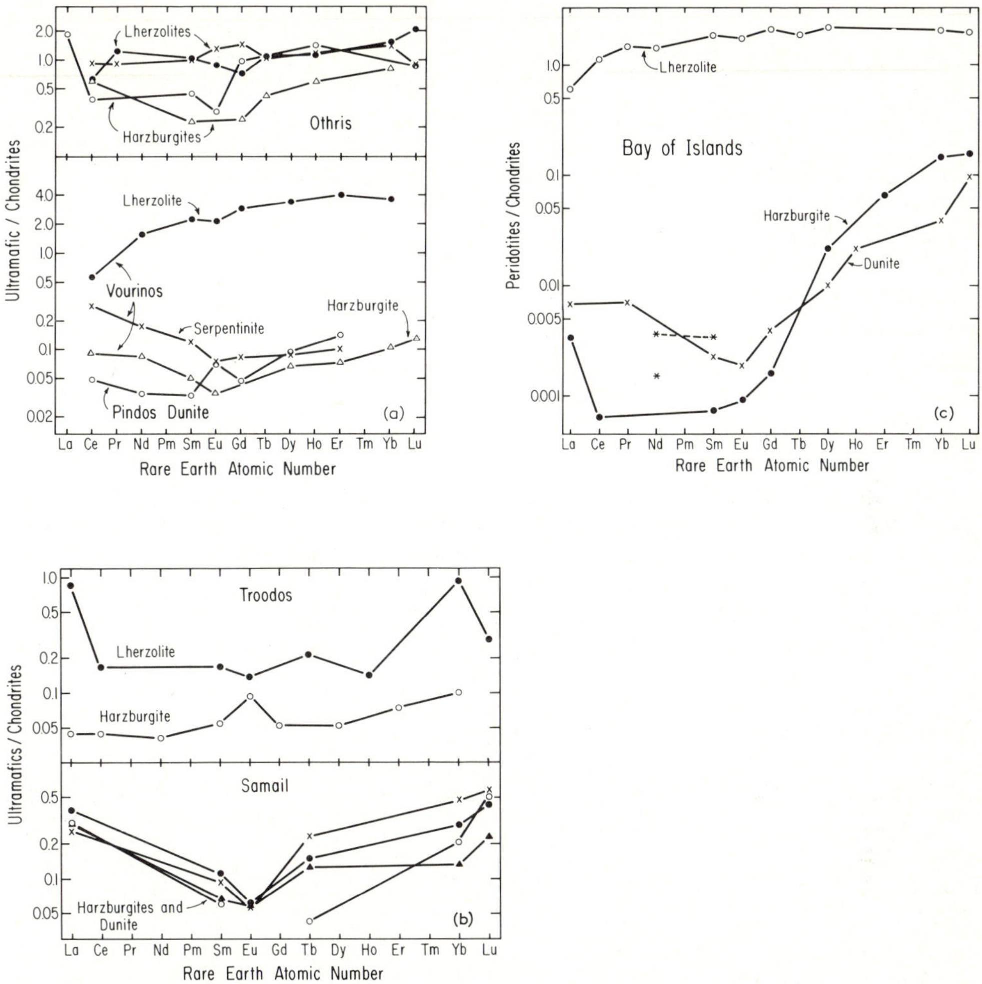

Fig. 5.11. Chondrite-normalized REE abundances in harzburgites and dunites from various ophiolites. (a) Othris, Greece (Menzies, 1976); Vourinos, Greece (Allègre et al., 1973; Noiret et al., 1981); Pindos, Greece (Montigny et al., 1973). (b) Troodos, Cyprus (Kay and Senechal, 1976; Menzies, 1976): Samail, Oman (Pallister and Knight, 1981). (c) Bay of Islands (Suen et al., 1979). Another lherzolite from Suen et al. was not plotted because its REE abundances have been affected by metasomatism. Asterisks indicate Sm-Nd data for two other Bay of Islands harzburgites (Jacobsen and Wasserburg, 1979).

by the optimum techniques for determination of low REE abundances; e.g., the Troodos lherzolite and Othris samples were analyzed by instrumental neutron activation analysis and the Samail samples were analyzed by neutron activation but only after pre-irradiation chemical processing. Almost certainly, some of the scatter in the REE data for the Troodos lherzolite and Othris rocks results from analytical errors. However, the preferential depletion in intermediate atomic number REE (Figs. 5.8 and 5.11), with La/Sm greater than in chondrites, is found in almost every ophiolite and is independent of analytical technique.

Several alternative explanations are possible for convex-downwards, chondrite-normalized REE patterns in harzburgites and dunites:

(1) A genetic relationship between the peridotites and basic rocks, such as the overlying lavas, dikes and gabbros in ophiolites —

(a) Allègre et al. (1973) and quoted in fig. 34 of Coleman (1977) suggested that the V-shaped chondrite-normalized REE patterns found in harzburgites and dunites are consistent with these rocks originating as residues from the melting process creating magmas similar to those represented by the overlying ophiolitic pillow lavas. The relative LREE enrichment in the model harzburgites resulted because Allègre et al. (1973) used LREE spinel/liquid distribution coefficients greater than unity. However, Suen et al. (1979) and Allègre et al. (1980) concluded that these spinel distribution coefficients are inappropriate. Thus, Suen et al. (1979) and Pallister and Knight (1981) concluded that the harzburgites and dunites forming the basal ophiolite section are *not* simply related to the overlying pillow lavas.

However, McCulloch et al. (1981) found that a basal peridotite from the Samail ophiolite has a $^{143}Nd/^{144}Nd$ ratio within error of the ratios in the overlying basic rocks. Based on the REE contents of olivine from other rocks, they argued that olivine preferentially enriches LREE, thereby causing $(La/Sm)_{cn} > 1$ in harzburgites and dunites. Because minerals such as olivine do *not* have characteristic REE contents,* the basis of their argument is incorrect. However, if $D_{La}^{ol/melt} > D_{Sm}^{ol/melt}$ their explanation may be valid, but in an experimental study of olivine/melt REE distribution coefficients at 1 atm McKay (1981) found that $D_{LREE}^{ol/melt} \ll D_{HREE}^{ol/melt}$. Olivine/melt distribution coefficients for REE are the subject of considerable controversy and they will be discussed further in section 5.5.

(b) As outlined by Frey et al. (1971, p. 2063) such REE patterns in dunites may be expected in olivine-rich segregations formed from a magma with a high LREE/HREE ratio relative to chondrites but with uniform enrichment in HREE. This interpretation requires a small increase in olivine/melt distribution coefficients from La to Lu so that the relative LREE enrichment is inherited from the coexisting melt and the relative HREE enrichment results from the preference of olivine for HREE. While this hypothesis might be valid for alpine dunites, such as the Twin Sisters, it is an unlikely explanation for dunites and harzburgites in ophiolites where the associated basic melts have $(LREE/HREE)_{cn} < 1$.

(2) LREE enrichment resulting from alteration and metamorphism —

*In an igneous system at equilibrium the REE content of olivine is controlled by the REE content of the melt and $D_{REE}^{ol/melt}$.

Late-stage REE mobility in crustal rocks is discussed in Chapter 9, and it may be an important process in determining the REE content of serpentinized and metamorphosed ultramafics from alpine peridotites and ophiolites. If LREE are mobilized during alteration of ultramafic rocks, the convex-downwards chondrite-normalized REE patterns may reflect LREE introduction into a residual rock with a low LREE/HREE abundance ratio. Because of the low REE content of dunites and harzburgites they are particularly susceptible to alteration effects. Such an explanation is plausible for PCC-1 (32% serpentine, 58% olivine, 9% orthopyroxene), and serpentinized harzburgite and dunite from ophiolites; e.g., the Samail samples in Fig. 5.11 are 60—80% serpentinized, the Pindos dunite contains 12.7% H_2O, and the ignition loss for the Bay of Islands samples ranges from 9 to 15%. Thus, these rocks have had a very complex history of metamorphism and alteration superimposed on their igneous history.

The possibility of REE mobility during serpentinization and metamorphism of peridotites has been recognized by most researchers but no one has dealt with the problem in a rigorous and meaningful manner. The problem was recognized by Frey (1969) who attempted REE mass balances for the Lizard peridotites by using bulk rock and mineral separate data, and concluded that LREE were removed during serpentinization. The Lizard rocks are extensively serpentinized (50—60% serpentine, ~10% H_2O^+). In retrospect, this mass balance test for the effects of serpentinization on REE abundance is ambiguous because the minerals were not acid-washed (see section 5.5 for effects of acid-washing on REE abundances in pyroxenes, and olivine) and the serpentine was not analyzed. In particular, the anomalously high LREE contents in the Lizard olivine and spinel and variable olivine/spinel distribution coefficients (see table 3 of Frey, 1970a) led to the conclusion that LREE were lost during serpentinization. It is likely that different conclusions would be reached if the Lizard mineral separates were acid-washed.

More recently, Frey and Suen (1983) studied Ronda peridotites ranging from <5% to ~30% serpentine and found no correlation of REE abundances with degree of serpentinization which supports the assumption by Loubet et al. (1975) that serpentinization has no significant effects on REE abundances. However, preferential mobility of Eu^{2+} during serpentinization (Sun and Nesbitt, 1978) may explain the positive Eu anomaly occurring in the Pindos dunite and Troodos harzburgite (Fig. 5.11) which both lack plagioclase. Moreover, there is circumstantial bulk-rock evidence that REE, especially LREE, can be mobilized during hydrous alteration and metamorphism of peridotites. Frey (1969) inferred that the LREE enrichment observed in the recrystallized Lizard peridotites relative to the primary peridotite was a result of LREE transport by aqueous fluids. Potts and Condie (1971) argued for a similar phenomenon in the altered zone of the Preacher Creek proto-stratiform ultramafic intrusion, and Suen et al. (1979)

found relative LREE enrichment in an altered lherzolite from the Bay of Islands ophiolite. In addition, Shih (1972) found that 60% of the Ce and 38% of the Nd were leached from a serpentinized ocean floor peridotite by 1*N* nitric acid. Finally, Menzies (1976, p. 647) states that serpentinization results in loss of LREE but no supporting data are given.

The most thorough consideration of REE mobility during hydrous alteration has been provided by Ottonello et al. (1979), who used a thermodynamic approach to conclude that LREE were mobile during serpentinization of the Ligurian peridotites. However, it is doubtful that appropriate thermodynamic data are available for rigorous evaluation of REE mobility during serpentinization. Moreover, G. Ottonello (personal communication, 1981) has recently found that the Ce anomalies reported by Ottonello et al. (1979) may be incorrect.

Until proven differently, it is best to consider that peridotites with $(La/Sm)_{cn} > 1$ may result from metamorphic, metasomatic and alteration processes, such as serpentinization, but obviously, much more research is required to define the effects of various types of serpentinization, alteration and metamorphism on REE abundances in peridotites. Of course, attributing the V-shaped REE patterns to late-stage processes does not explain the similar pattern in the nearly anhydrous Twin Sisters dunite (Fig. 5.8) or the serpentinized alpine lherzolites which uniformly have $(La/Sm)_{cn} < 1$.

(3) No genetic relationship between the tectonite harzburgites and dunites and overlying basic rocks in ophiolites — Complex REE patterns can be developed in mantle peridotites by mantle metasomatism (see section 5.5 and Hickey and Frey, 1982). Consequently, significant portions of the mantle might be characterized by $(La/Sm)_{cn} > 1$. This interpretation is bolstered by igneous rocks that require such a source (e.g., Betts Cove lower pillow lavas, Coish et al., 1982 and boninites, Hickey and Frey, 1982).

At present, it is not evident if one or more of the above hypotheses are the correct interpretation of the convex-downwards, chondrite-normalized REE patterns in dunites and harzburgites from alpine bodies and the basal sections of ophiolites. Obviously, petrogenetic interpretation remains an open question to be resolved in part by detailed analyses of carefully selected mineral separates.

5.4. Ocean floor peridotites

The best studied peridotites from an oceanic setting are those from the subaerial islets of St. Paul's Rocks in the equatorial Atlantic. They are composed of a heterogeneous, upper mantle-derived, peridotite intrusion emplaced in the solid state. The rocks are mylonitized and at least three principal rock types occur, spinel ± pargasitic amphibole-peridotite which is dominant, brown hornblende-rich alkaline plutonic rocks which are interbanded with the spinel peridotite mylonites and rare gabbro (Melson et al., 1972). In a REE study of St. Paul's rocks Frey (1970b) found that the spinel and pargasitic amphibole-bearing peridotites are relatively enriched in LREE with La/Yb greater than the chondritic ratio. I concluded that these peridotites are *not* samples of source material for MORB or partial melting residues related to MORB. St. Paul's peridotites vary widely in K content (factor of 10) and recent analyses of samples with <150 ppm K show that even these peridotites have $(La/Sm)_{cn}$ and $(La/Yb)_{cn} > 1$ (Fig. 5.12). These new data are consistent with the previous conclusion that these oceanic peridotites are not simply related to MORB.

Key information bearing on the petrogenesis of St. Paul's Rocks is provided by REE abundances in the brown-hornblende mylonites which are interbanded with the spinel peridotite mylonites. REE abundances of some brown hornblende mylonites are nearly identical to a basanitoid dredged from near St. Paul's Rocks (Frey, 1970b, fig. 5). The geochemical features of these hornblende-rich rocks are most consistent with a previous history as an alkaline magma or as hornblende-rich segregations from such a magma. The latter interpretation is most likely for the hornblende-rich (>80%) samples with high K/Rb (~900) and convex upwards chondrite-normalized REE patterns.

Thus, St. Paul's peridotite mylonites are intimately associated with alkaline magmatism. Interestingly, inclusions of peridotite mylonite similar

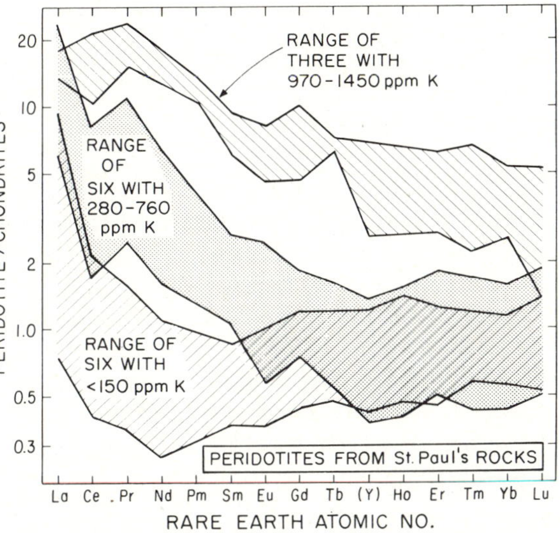

Fig. 5.12. Chondrite-normalized REE abundances in 15 spinel peridotite mylonites from St. Paul's Rocks in the equatorial Atlantic Ocean. Samples grouped according to K content; those with relatively high K contain abundant pargasitic amphibole. REE data from Frey (1970b) and F.A. Frey (unpublished).

to St. Paul's peridotites occur in a basanitoid dredged from near St. Paul's Rocks (Sinton, 1979a). This close association suggests that these mantle peridotites may be related to or affected by alkaline volcanism. While there is apparently no simple relationship such as mixing or partial melting between the St. Paul's peridotites and the brown-hornblende mylonites (Frey, 1970b; Melson et al., 1972), these peridotites could yield alkaline basalts by partial melting. Alternatively, their REE, K, Rb and Sr contents are consistent with an origin as solid segregations from alkalic magma.

Another subaerial exposure of oceanic lithosphere is Zabargad Island in the Red Sea. As at St. Paul's Rocks, the rock types include spinel lherzolite, amphibole-bearing spinel peridotites and gabbroic-troctolitic rocks. Bonatti et al. (1981) found that a spinel-lherzolite has Nd to Lu abundances at ~2 times chondrites with LREE depletion to ~0.7 times chondrites. In contrast, an amphibole-bearing peridotite is strongly enriched in LREE (~20 times chondrites) similar to amphibole-bearing spinel peridotites from St. Paul's Rocks.

Dick (1978) and Dick and Bullen (1983) emphasized that spinel- or plagioclase-bearing harzburgites and lherzolites are the major peridotite type obtained from the Atlantic, Indian and Caribbean Sea floors by dredging. Also, Hamlyn and Bonatti (1980) noted that peridotites from the Owen fracture zone on the Carlsberg Ridge, Indian Ocean, are dominantly spinel lherzolites. These authors emphasize that harzburgite dominates the upper mantle section of most well-studied ophiolites; therefore, the dominant mantle rock type in ophiolites is deficient in clinopyroxene compared to most of the studied ocean floor peridotites. Moreover, the relict mineral compositions in partially serpentinized ocean floor ultramafics are similar to mineral compositions in lherzolites occurring as alpine peridotites and as inclusions in alkalic basalts (Prinz et al., 1976; Sinton, 1979b; Dick and Bullen, 1983).

Few bulk-rock geochemical studies have been published for dredged ocean floor peridotites, primarily because most are highly altered and metamorphosed to low-pressure/low-temperature minerals such as serpentine; thus, it is difficult to infer their pre-serpentinization composition. Thus far, most REE studies of dredged oceanic peridotites have been described only in abstracts (Shih and Gast, 1971; Coish et al., 1979). The range for ten serpentinized peridotites (6—12% H_2O) from the Islas Orcadas fracture zone in the South Atlantic (Fig. 5.13) is typical of ultramafics dredged from Atlantic fracture zones; i.e., HREE at ~1 times chondrites and LREE contents less than chondritic abundances. The individual analyses from serpentinized peridotites at 24°N and the Vema fracture zone (Fig. 5.13) are also typical in that low-CaO rocks have lower REE contents with scattered LREE abundances. Based on these whole-rock data it is not possible to unambiguously infer the pre-serpentinization REE contents. However, by analogy with less-serpentinized alpine peridotites (Fig. 5.2), the increasing

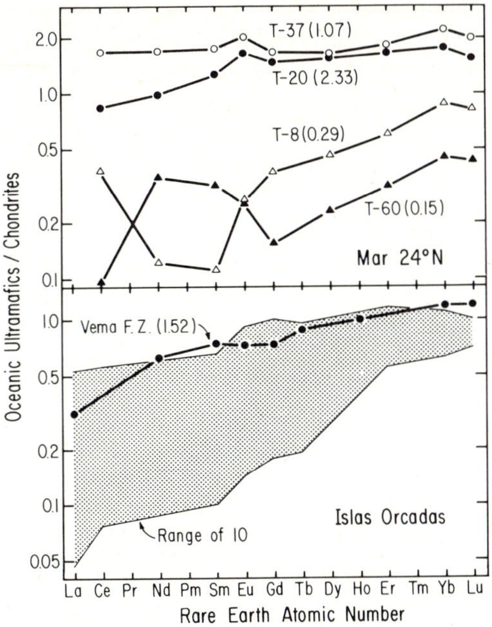

Fig. 5.13. Lower diagram — Range of chondrite-normalized REE abundances in ten serpentinized peridotites (6—12% H_2O) from the Islas Orcadas fracture zone in the South Atlantic (Coish et al., 1979; F.A. Frey, unpublished). Sample AII-20-9-1 from Vema fracture zone in Atlantic (CaO = 1.52%, H_2O = 12.88%, 15% primary minerals including 1% plagioclase; Melson and Thompson, 1971). REE data from F.A. Frey (unpublished). Upper diagram — chondrite-normalized REE abundances in four serpentinites from the Mid-Atlantic Ridge at 24°N. CaO content indicated within parentheses. H_2O content is from 11.95 to 13.31% (Miyashiro et al., 1969). REE data from Shih (1972). Four samples from the Romanche fracture zone and a fracture zone at 30°N on the Mid-Atlantic Ridge have a similar range in REE content (Shih, 1972; F.A. Frey, unpublished).

depletion of LREE from a HREE abundance level at 1—2 times chondrites is probably a primary feature.

The similarity of REE abundances in highly serpentinized ultramafics to those in less serpentinized alpine peridotites is exemplified by the REE abundance variations in Puerto Rico serpentinites from the AMSOC drill hole (cf. Fig. 5.14 with 5.2 and 5.8). In particular, the serpentinized lherzolites have REE contents like alpine lherzolites, whereas the serpentinized dunites are like DTS-1. Serpentinites from the north wall of the Puerto Rico trench (Bowin et al., 1966) are associated with MORB and these ultramafics may be exposures of oceanic upper mantle shortly before subduction. Except for positive Eu anomalies the REE content of Puerto Rico trench serpentinites are qualitatively similar to Islas Orcadas peridotites (Fig. 5.15).

Seven peridotites dredged from the Indian Ocean floor have the same range in REE abundance features seen in peridotites from the Atlantic; that

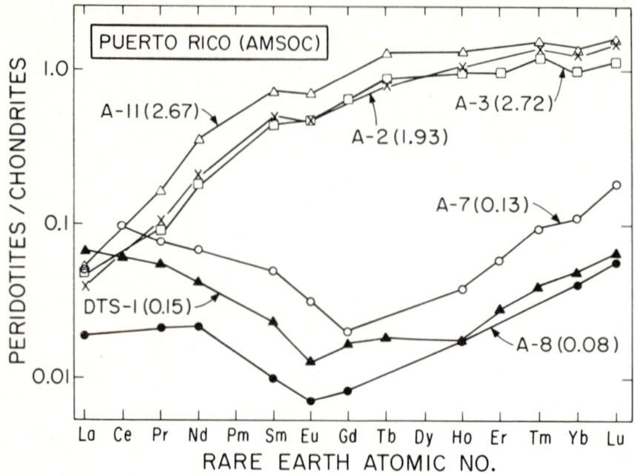

Fig. 5.14. Chondrite-normalized REE abundances in five serpentinites (10.8—17.2% H_2O) from the AMSOC drill hole in Puerto Rico (REE data of F.A. Frey, unpublished). CaO contents shown within parentheses. Samples A-2, A-3 and A-11 interpreted as serpentinized lherzolites and samples A-7 and A-8 as serpentinized dunites (Mattson, 1964). DTS-1, the U.S.G.S. standard dunite shown for comparison. Note the similarity to data in Figs. 5.2, 5.8 and 5.11.

is three samples are strongly depleted in LREE (<0.1 times chondrites) with HREE contents at 0.5—1 times chondrites, whereas four other samples have convex-downwards chondrite-normalized REE patterns with all REE abundances <0.8 times chondrites (Shih, 1972; Shimizu and Hart, 1974).

In summary, despite the highly serpentinized nature of most dredged ocean floor peridotites, their REE abundances are similar to the range of REE contents in alpine lherzolites, harzburgites and dunites. Broadly the chondrite-normalized REE patterns divide into two groups: (a) samples with HREE at ~0.5—2 times chondrites with relative LREE depletion ranging from 0.015—1 times chondrites, and (b) samples with all REE contents less than chondrites but with convex-downwards chondrite-normalized REE patterns.

The interpretations presented for alpine peridotites are applicable to these oceanic peridotites, i.e., peridotites in group (a) are partial melting residues which in general have lower HREE abundances than those in alpine lherzolites. This difference may reflect mantle heterogeneity prior to melting or possibly that the oceanic peridotites are residues from higher degrees of melting. However, petrogenetic inferences based only on REE contents of a few samples must be viewed with caution. For example, peridotites drilled in the Atlantic during the Deep Sea Drilling Project also have REE contents within the range established by other oceanic peridotites and alpine lherzolites (cf. Figs. 5.2, 5.13 and 5.15), but these peridotites have been interpreted as ultramafic cumulates (Hodges and Papike, 1976; Clarke and Loubat, 1977). Therefore, if the textural interpretations are correct that these

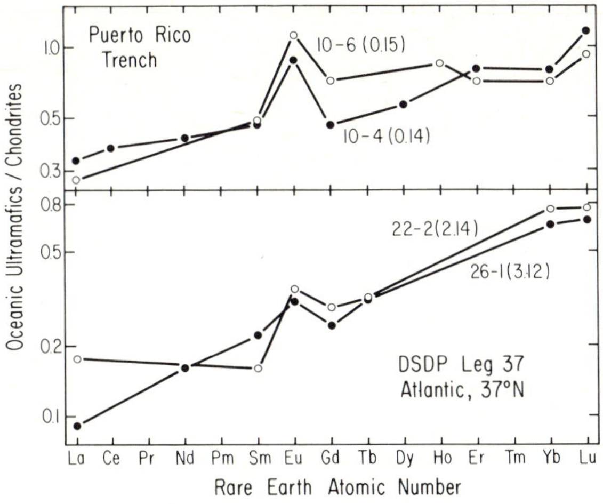

Fig. 5.15. Lower diagram — Chondrite-normalized REE abundances in two serpentinized peridotites (6.94 and 7.80% H_2O) drilled at site 334 of DSDP Leg 37. CaO contents indicated within parentheses. These peridotites have been interpreted as having cumulate texture (see text). REE data from Dostal and Muecke (1978). Upper diagram — Chondrite-normalized REE abundances in two serpentinites (13.40 and 13.74% H_2O) dredged from the Puerto Rico Trench (Bowin et al., 1966). REE data from Jibiki and Masuda (1974) and F.A. Frey (unpublished).

rocks formed as cumulates, the similarities in REE contents suggests that REE data are not sufficient to distinguish between a magnesium-cumulate rock and a partial melting residue.

REE abundances in group (b) peridotites are not easily explained, but the alternative explanations presented in section 5.3 for ophiolite harzburgites and dunites are also applicable for these oceanic peridotites which are generally poor in CaO. An additional interpretation presented by Shih and Gast (1971) and Shih (1972) is that the positive Eu anomaly and $(La/Sm)_{cn} > 1$ in several oceanic peridotites results from residual plagioclase. Plagioclase is the only common mineral which has higher LREE/HREE abundance ratios than a coexisting melt (see Chapter 1), but plagioclase/melt REE distribution coefficients are small (<0.1) so that this model requires 5—15% residual plagioclase and only lesser amounts of residual pyroxene. It is difficult to infer the primary mineralogy of these partially serpentinized oceanic peridotites, but there is no evidence that samples with $(La/Sm)_{cn} > 1$ contained 5—15% plagioclase prior to serpentinization. Moreover, positive Eu anomalies may result from serpentinization (Sun and Nesbitt, 1978).

Finally, none of the dredged peridotites have LREE contents as high as some subaerial oceanic peridotites from St. Paul's Rocks and Zabargad Island. These high LREE contents and the convex downwards nature of the chondrite-normalized REE patterns are similar to that in lherzolite inclusions in alkaline basalts, which are discussed next.

5.5. Ultramafic inclusions in basalts and kimberlites

A wide variety of ultramafic rock types are found as inclusions in alkalic basalts and kimberlites and REE abundances have been determined in more than 150 inclusions. Relative to alpine and ocean floor peridotites two important features of these inclusions are: (1) most inclusions are *not* serpentinized; thus, they have experienced much less late-stage alteration; and (2) inclusions occur over a wide geographic region; thus, they provide upper mantle samples from many diverse areas. Of course, their main disadvantages are the small size, usually <0.5 m, and the lack of stratigraphic relations.

The following discussion is organized according to rock type and the sequence will be: spinel lherzolites and harzburgites belonging to the Cr-diopside group of Wilshire and Shervais (1975) and Group I of Frey and Prinz (1978); garnet lherzolites and garnet harzburgites; pyroxenites and garnet pyroxenites; and relatively Fe and Ti-rich peridotites belonging to the Al-augite group of Wilshire and Shervais (1975) and Group II of Frey and Prinz (1978).

Spinel lherzolites and harzburgites in alkalic basalts

The most abundant ultramafic rock type in many inclusion suites within alkalic basalts are spinel lherzolites and harzburgites which are characterized by Cr-rich and Ti-poor pyroxenes and bulk rock $Mg/(Mg + \Sigma Fe) > 0.85$ (see descriptions of Cr-diopside group in table 1 of Frey and Prinz, 1978). These peridotites have well-defined, systematic major element abundance trends (Maaløe and Aoki, 1977), much like the Ronda alpine peridotites, and at each inclusion locality such trends have been interpreted as reflecting various degrees of melting of a compositionally uniform source with the Ti-, Al- and Ca-poor harzburgites as residues from relatively high degrees of melting and the relatively Ti-, Al- and Ca-rich lherzolites resulting from lower degrees of melting.

The early REE analyses of such inclusions from diverse areas showed a wide variety of REE abundances (Nagasawa et al., 1969; Frey et al., 1971; Philpotts et al., 1972) which were difficult to interpret. Most puzzling were the high relative enrichments in LREE found in the majority of these inclusions (e.g., Fig. 5.16). Similar chondrite-normalized REE patterns are *not* characteristic of alpine peridotites which typically have relative depletion in LREE (Fig. 5.2). Within the last eight years inclusion suites at several different localities have been studied in detail with REE analyses being

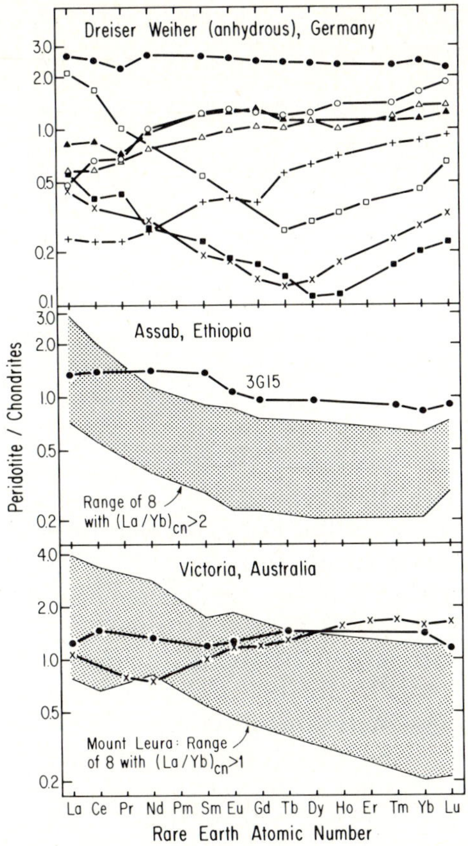

Fig. 5.16. Lower diagram — Chondrite-normalized REE abundances in ten spinel lherzolite inclusions from Mount Leura in Victoria, Australia. Data for other Victorian localities are in Frey and Green (1974). Nearly horizontal patterns are for sample 2642 (×; Frey and Green, 1974) and SL7 (•; Chen and Frey, 1980). Range for eight others determined by sample 2640 (Frey and Green, 1974), GSFC 20 (Philpotts et al., 1972) and six from Chen and Frey (1980). Middle diagram — Chondrite-normalized REE abundances in nine spinel harzburgite inclusions from Assab, Ethiopia (Ottonello, 1980). Upper diagram — Chondrite-normalized REE abundances in eight anhydrous spinel lherzolite inclusions from Dreiser Weiher, Germany (Stosch and Seck, 1980). Note the wide variety of REE patterns; particularly, the V-shaped patterns in the three samples with the lowest HREE content which also have the lowest CaO contents (0.39—1.22% CaO). V-shaped patterns for Dreiser Weiher inclusions were also found by Frey et al. (1971), Philpotts et al. (1972) and a LREE-depleted pattern was reported by Jagoutz et al. (1979b).

accompanied by major element, mineral composition and petrographic studies. These detailed studies have greatly increased our understanding of upper mantle compositional variations and demonstrated the complexity of upper mantle processes. Following, I specifically discuss the inclusion suites from Victoria, Australia (Frey and Green, 1974; Chen and Frey, 1980), Western United States (Frey and Prinz, 1978; BVSP, 1981), Kapfenstein, Austria (Kurat et al., 1980), Dreiser Weiher, Germany (Stosch and Seck, 1980), Assab, Ethiopia (Ottonello et al., 1978; Ottonello, 1980), Salt Lake Crater, Hawaii (Nagasawa et al., 1969; Reid and Frey, 1971; Philpotts et al., 1972; Frey, 1980; Irving, 1982) and Itinome-gata, Japan (Tanaka and Aoki, 1981). REE determinations have also been made on small numbers of inclusions from other regions by Varne and Graham (1971), Flower (1971), Philpotts et al. (1972), Menzies (1976), Morten, (1978), Jagoutz et al. (1979a, b) and BVSP (1981).

Ultramafic inclusions from Victoria, Australia were the first group studied in detail for major and trace element contents of the bulk rocks and constituent minerals. Frey and Green (1974) studied six samples from five scoria cones in Victoria and Chen and Frey (1980) are studying an additional seven samples from the Mount Leura locality. The REE abundances in Victorian lherzolites and harzburgites are typical for inclusion suites from diverse geographic areas (Fig. 5.16). Obviously, the surprising result is the diversity of chondrite-normalized REE patterns in ultramafic rocks of similar mineralogy and with only small differences in major element composition.

There are, however, some systematic correlations between abundances of REE and major elements. Specifically, HREE abundances are positively correlated with CaO and Al_2O_3 contents in inclusion suites from Victoria, Australia; Western United States; Dreiser Weiher, Germany; Kapfenstein, Austria; Salt Lake Crater, Hawaii; and Itinome-gata, Japan (Fig. 5.17). Inclusions with the highest CaO and HREE contents are richest in clinopyroxene, and are interpreted to be residues from the smallest degree of melting, thus, they are the closest approximation to the upper mantle composition in each region prior to the partial melting event. However, note that all of these lherzolite inclusions have $(LREE/HREE)_{cn} < 1$; moreover, their REE contents are similar to those of alpine lherzolites (cf., Figs. 5.2 and 5.18).

Another correlation between REE abundances and CaO content is that LREE contents are higher in the inclusions with the lowest CaO contents. Although the data do not closely define a line, there is no doubt that in most inclusion suites $(LREE/HREE)_{cn}$ generally increases as CaO content decreases (Fig. 5.19). Moreover, at most localities the majority of inclusions have $(La/Yb)_{cn} > 1$. As discussed in section 5.1 these results are inconsistent with the expectation that residual peridotites have $(La/Yb)_{cn} < 1$. Also based on mineral/melt REE distribution coefficients for olivine, pyroxenes, spinel and garnet (Chapter 1), it is expected that during partial melting $(La/Yb)_{cn}$ in residual peridotites should *decrease* with decreasing CaO content; i.e., opposite to the trends in Fig. 5.19.

It is significant that inclusion suites from diverse tectonic regions such as Japan and Hawaii have REE abundances which exhibit the same general

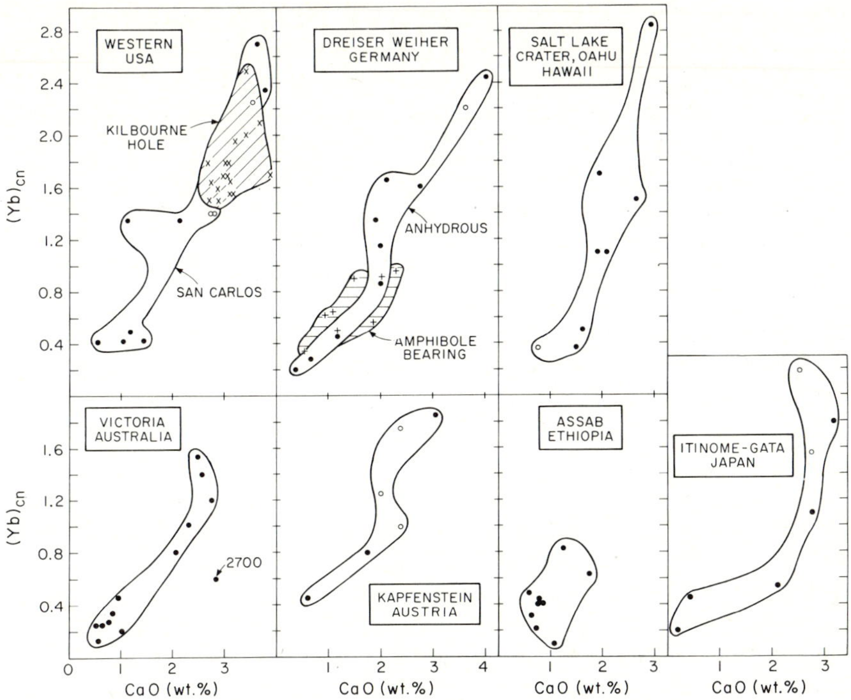

Fig. 5.17. CaO versus $(Yb)_{cn}$ for harzburgite-lherzolite suites from different localities. *Victoria, Australia* — Data from Frey and Green (1974), and Chen and Frey (1980). Point labelled "2700" is for an unusual lherzolite with volume % clinopyroxene exceeding that of orthopyroxene. *Western U.S.A.* — Solid circles with enclosed field for San Carlos, Arizona (Frey and Prinz, 1978; Jagoutz et al., 1979b; BVSP, 1981). Crosses with enclosed field for Kilbourne Hole, New Mexico (Jagoutz et al., 1979b; A.J. Irving, personal communication, 1982). Open circles are for single samples from Potrillo, New Mexico, Cochise Crater, Arizona and Dish Hill, California (Jagoutz et al., 1979b; BVSP, 1981). *Kapfenstein, Austria* — Open circles are for hornblende-bearing lherzolites (Kurat et al., 1980). *Dreiser Weiher, Germany* — Solid circles and enclosed field for amphibole-free peridotites, plus symbols and enclosed field for amphibole-bearing peridotites (Stosch and Seck, 1980). Open circle for inclusion from Massif Central, France (Jagoutz et al., 1979b). *Assab, Ethiopia* — Data from Ottonello (1980). *Salt Lake Crater, Oahu, Hawaii* — Solid symbols from Nagasawa et al. (1969), Philpotts et al. (1972), Frey (1980), and BVSP (1981). Also shown is a clinopyroxene-poor sample (open circle) from Hualalai, Hawaii (Nagasawa et al., 1969). *Itinome-gata, Japan* — Open circles for garnet lherzolites, closed circles for spinel peridotites (Tanaka and Aoki, 1981). Not shown are data for two samples anomalously rich in CaO (6.62 and 7.24%) with Al_2O_3/CaO ratios <0.75.

features as inclusion suites from continental regions. A particularly important result is that samples of oceanic mantle from beneath Hawaii have major element and REE abundances similar to mantle samples from beneath continents (Figs. 5.17, 5.19 and 5.20; Reid and Woods, 1978).

Two inclusion suites in Figs. 5.17 and 5.19 exhibit much less variation

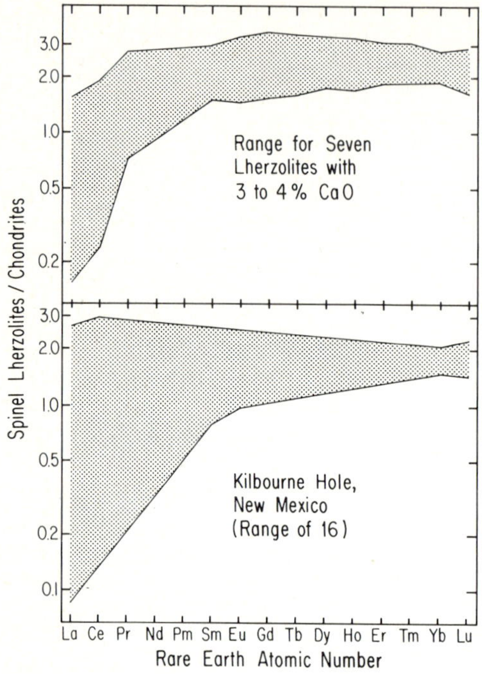

Fig. 5.18. Lower diagram - Range of chondrite-normalized REE abundances in sixteen spinel lherzolite inclusions from Kilbourne Hole, New Mexico (Irving, 1980). CaO contents range from 2.70 to 3.96%. Upper diagram - Range of chondrite-normalized REE abundances in seven spinel lherzolite inclusions each containing 3—4% CaO. Samples SC-1 (3.81% CaO) and 15A (3.66% CaO) from San Carlos, Arizona (Frey and Prinz, 1978; Jagoutz et al., 1979b), KH-1 (3.43% CaO) from Kilbourne Hole, New Mexico, Fr 1 (3.64% CaO) from Massif Central, France, PO 1 (3.55% CaO) from Potrillo, New Mexico, Ka 168 (3.04% CaO) from Kapfenstein, Austria (all from Jagoutz et al., 1979b) and Ib/8 (4.05% CaO) from Dreiser Weiher (Stosch and Seck, 1980).

than other suites. All peridotite inclusions from Assab, Ethiopia, have $(La/Yb)_{cn} > 1$, and there are no correlations of REE and major element abundances (Fig. 5.17 and 5.19). Relative to the other suites, all the Assab peridotites are CaO-poor (<1.8%) and are mostly harzburgites. In contrast, the lherzolite inclusion suite from Kilbourne Hole, New Mexico, is unique because all samples are CaO-rich (16 samples with CaO contents of 2.70–3.96% range from $(La/Yb)_{cn} = 0.05$ to 1.5; A.J. Irving, personal communication, 1982) (Figs. 5.18 and 5.19). These bulk rock lherzolite data show no correlations between CaO and REE contents (Figs. 5.17 and 5.19). However, clinopyroxenes from the Kilbourne Hole harzburgites studied by Reid and Woods (1978) have higher LREE contents than clinopyroxenes from Kilbourne Hole lherzolites (F.A. Frey and J.B. Reid, unpublished). Thus, when a wider range of major element contents is considered, there also appears to be a generalized inverse correlation between bulk rock CaO content and $(La/Yb)_{cn}$ in the Kilbourne Hole suite.

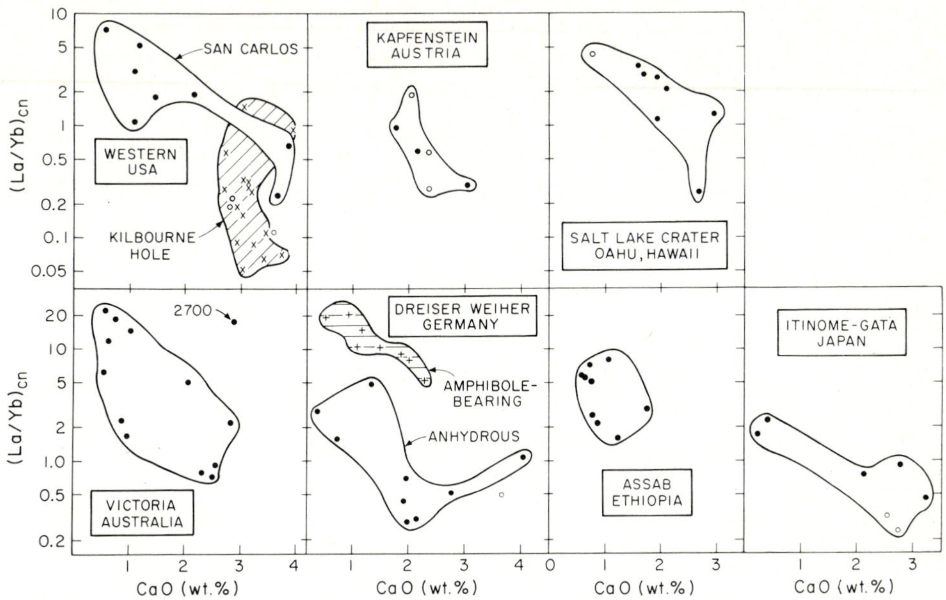

Fig. 5.19. CaO versus $(La/Yb)_{cn}$ for harzburgite-lherzolite suites from different localities. For explanation of fields and symbols see caption for Fig. 5.17.

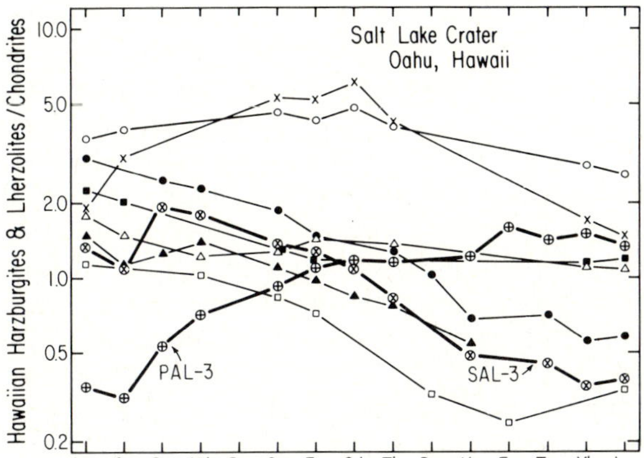

Fig. 5.20. Chondrite-normalized REE abundances in nine spinel harzburgite and lherzolite inclusions from Salt Lake Crater, Hawaii. Data sources: Schilling (1966), 1 sample; Nagasawa et al. (1969), 3 samples; Reid and Frey (1971), 1 sample; Philpotts et al. (1972), 1 sample; Frey (1980), 2 samples; BVSP (1981), 1 sample. Data for the samples PAL-3 and SAL-3 are identified (Frey, 1980).

An important aspect of the studies by Nagasawa et al. (1969) and Frey and Green (1974) was the observation that the LREE enrichment in inclusions with high La/Yb results largely from the constituent clinopyroxenes. This conclusion has been substantiated for other suites by Jagoutz et al. (1979a, b), Stosch and Seck (1980), Ottonello (1980) and Kurat et al. (1980). Kleeman et al. (1969) had earlier shown that clinopyroxenes from some Victoria inclusions have high U contents (~0.3 ppm) with U uniformly distributed throughout the clinopyroxene. Frey and Green found a strong positive correlation between U and LREE contents in such clinopyroxenes, and inferred that the LREE were structurally incorporated within the clinopyroxenes. Further evidence that all the REE are incorporated within the clinopyroxene structure is that HCl-HF acid leaching of clinopyroxene separates, both LREE-enriched and -depleted types, from Victoria and San Carlos have very little effect on the REE content of the clinopyroxenes (Zindler and Jagoutz, 1980; Chen and Frey, 1980). Consequently, there is *no doubt* that the LREE and U enrichment in these inclusions reflects a mantle process.

Frey and Green (1974) interpreted the data shown in Fig. 5.16 as resulting from a mixture of two geochemically distinct components; that is, the lherzolites and harzburgites are composed of component A which determines the major mineralogy, major element composition and abundances of compatible trace elements such as Ni, Co and also the HREE, and component B which determines the abundance of minor and trace incompatible elements such as K, P, LREE, Th and U. They further suggested that component A is a partial melting residue which is modified by addition of a migrating fluid, component B; however, components A and B are not genetically related, i.e., the liquid originally in equilibrium with and lost from component A differs from component B. Since component B has $(La/Yb)_{cn} > 1$ and is responsible for the LREE enrichment, Frey and Green proposed that it formed in the low-velocity zone of the mantle as a small percent melt of garnet peridotite which migrated upwards into residual peridotite where it crystallized in part to amphibole, clinopyroxene and accessory phases such as apatite. However, recent experiments of REE partitioning between minerals and H_2O and CO_2 fluids at high pressure show that these fluids preferentially enrich LREE (Mysen, 1979; Wendlandt and Harrison, 1979); thus, component B may have been a H_2O- or CO_2-rich fluid which permeated component A. In recent literature this proposed alteration of pre-existing mantle has been termed "mantle metasomatism".

In addition to clinopyroxene and accessory phases, amphibole may be a major crystalline host for component B because many elements may be accommodated in its crystal structure (see also Menzies and Murthy, 1980a, table 3; Menzies and Murthy, 1980b). In fact, Varne and Graham (1971) proposed that the marked relative LREE enrichment (La > 10 times chondrites) in a hornblende-bearing lherzolite from Ataq, South Yemen, resulted from upward migration of a hydrous liquid which precipitated as hornblende, thereby forming an incompatible-element-enriched region of the upper mantle. In an effort to further characterize component B, several research groups have studied amphibole-bearing inclusions which are abundant in some suites. For example, at Dreiser Weiher, two spinel peridotite suites are recognized, an anhydrous group and a group that is characterized by the presence of amphibole or its breakdown products. This amphibole-bearing group is strongly enriched in LREE and there is a well-defined inverse correlation of La/Yb with CaO (Fig. 5.19). Stosch and Seck (1980) interpreted these data as a result of metasomatism in an open system; i.e., only part of component B is presumed to have crystallized because several amphibole-bearing peridotites have normalized REE abundance patterns which flatten in the LREE region (Stosch and Seck, 1980, fig. 3a).

An anhydrous and an amphibole-bearing peridotite inclusion suite also occurs at Nunivak Island, Alaska. Roden et al. (1980) found that the amphibole-bearing peridotites have the highest $(La/Yb)_{cn}$, ~3.5—25, whereas the relatively CaO-rich, amphibole-free, granuloblastic-equant inclusions have $(La/Yb)_{cn}$ ~0.4. The high LREE enrichment in clinopyroxenes from these amphibole-bearing inclusions results in part from apatite inclusions within the clinopyroxene.

Additional evidence that mantle metasomatism can be caused by fluids that precipitate amphibole is provided by the occurrence of discrete amphibole- and apatite-rich inclusions that are very enriched in LREE (50—2000 times chondrites, Wass and Rogers, 1980). At Kiama, Australia, these amphibole- and apatite-rich inclusions are accompanied by spinel peridotites that are veined by and altered to minerals identical to those in the amphibole and apatite-rich inclusions (Wass et al., 1980).

Although some of the relative LREE enrichment in inclusions results from the major minerals, acid-leaching studies and mass balances using bulk-rock and constituent mineral compositions consistently show that not all the incompatible elements, such as LREE, are accommodated in the major minerals. Especially, inclusions rich in component B appear to have incompatible trace elements concentrated on mineral surfaces, in fluid inclusions, in unidentified rare phases and as internally-derived glass resulting from decompression melting (e.g., see Frey and Green, 1974; Kurat et al., 1980; Ottonello, 1980; Stosch and Seck, 1980; Tanaka and Aoki, 1981; Stosch, 1982). These more complex hosts for LREE may have formed during the same event creating LREE enrichment in major minerals, such as clinopyroxene, but it is also possible that some inclusions have been affected by several mixing or metasomatic events. Also, some of the surficial contaminants may have been introduced in a crustal environment.

The time of mixing can be estimated only by combining Sm and Nd abundance data with determination of $^{143}Nd/^{144}Nd$ isotopic ratios (see Chapter 11). An important result is that clinopyroxenes from several lherzolite inclusions have Nd isotopic ratios similar to MORB and that clinopyroxenes with $(Sm/Nd)_{cn} < 1$ have Nd isotopic ratios requiring a time-integrated source with $(Sm/Nd)_{cn} > 1$ (Jagoutz et al., 1980; Stosch et al., 1980; Zindler and Jagoutz, 1980). These data require that component B has been added within the last 200 Ma. This conclusion is consistent with studies of alkalic basalts which have Nd isotopic ratios requiring a time-average history in an environment with Sm/Nd greater than chondrites, but whose high degree of LREE enrichment is most easily interpreted as resulting from melting of a source with Sm/Nd less than chondrites (e.g., Menzies and Murthy, 1980a; Roden et al., 1981).

A decoupling of incompatible trace element (K, P, Th, U, LREE) abundances and CaO-Al_2O_3 abundances has been found in all peridotite inclusion suites that have been studied in detail. Most investigators have accepted the general idea of a mixing model such as outlined by Frey and Green (1974). Exceptions are Tanaka and Aoki (1981) and Ottenello (1980). Tanaka and Aoki considered that all Itinome-gata peridotite inclusions were related as residues from melting a compositionally homogeneous source to various degrees. Consequently, they were unable to explain the trend for La/Sm to increase in residues formed by the highest degrees of melting, and they concluded that an equilibrium partitioning model may not be applicable for the melting process creating residual peridotite inclusions.

Ottonello (1980) found evidence that the Assab harzburgite inclusions contain an interstitial, incompatible element-rich component, but he pursued an alternative explanation for the relative LREE enrichment in the constituent minerals. Specifically, he assumed that an equilibrium model is appropriate for explaining REE abundances in residual peridotite inclusions, but that, especially for olivine and orthopyroxene, the conventional REE mineral/melt distribution coefficients used in modelling are *not* appropriate. His model outlined in figs. 7 and 8 of Ottonello (1980) has the important but controversial result that residues created by high degrees of equilibrium melting have *higher* La/Sm ratios than residues created by smaller degrees of equilibrium melting. The key point in the model is that olivine and orthopyroxene can preferentially incorporate La relative to smaller REE, such as Sm. Therefore, La/Sm in the residue begins to increase when clinopyroxene is nearly eliminated by melting; at this point, the residue REE content is controlled primarily by the abundant residual olivine and orthopyroxene.

It is obviously important to critically evaluate Ottonello's (1980) alternative model because, if valid, it provides a simple explanation for the relatively high La/Sm abundance ratios commonly found in harzburgites and dunites. On the basis of crystal chemical considerations the small cation sites in olivine and orthopyroxene should lead to low LREE/HREE ratios in these minerals relative to a coexisting melt. However, if REE in these minerals do *not* substitute at structural sites, the larger radius of La relative to Sm may not be a hindrance to preferential incorporation of La into olivine and orthopyroxene. Nevertheless, experimental studies of olivine/ melt (McKay, 1981), orthopyroxene/melt (e.g., Mysen, 1978) and orthopyroxene/matrix distribution coefficients (see Chapter 1) show that LREE/ HREE abundance ratios in these minerals are very low relative to a coexisting melt, although the relevance of the experimental studies to melting of a natural peridotite can be questioned.

Ottonello's (1980) inferred high La/Sm ratios in olivine and orthopyroxene relative to a coexisting melt are based on mineral/mineral REE partitioning data determined from water-washed mineral separates from Assab inclusions. However, it is apparent that determination of the intrinsic REE content within olivine and orthopyroxene is not straightforward. Difficulties arise not only from obtaining a pure mineral separate devoid of discrete foreign grains, but also in determining the quantity of REE that are loosely bound to grain surfaces and held as foreign inclusions within olivine or orthopyroxene. Indications of such problems are: Stosch and Seck (1980) found that HCl washing of a previously hand-picked and water-washed orthopyroxene reduced the La content from 0.55 to 0.069 ppm, but only changed the Lu content by $<10\%$; Jagoutz et al. (1979a) found that 1-hour digestion with cold 3*N* HCl removed significant amounts of La and Eu from a carefully hand-picked olivine, and they concluded that most REE "in" olivine are on the grain surfaces; Kurat et al. (1980) found that olivines with abundant "black dots" on freshly broken surfaces are strongly enriched in La and Sm relative to the bulk olivine fraction from the same inclusion; Zindler and Jagoutz (1980) found that Sm and Nd abundances in olivine and orthopyroxene depend sensitively on the criteria used for generating pure mineral separates; finally, Stosch (1982) in a study of acid-washed olivine and pyroxene from a single peridotite inclusion found that olivine and orthopyroxene separates with abundant fluid inclusions are significantly enriched in LREE relative to separates with fewer fluid inclusions. In contrast, several of these investigators found that acid-leaching of clinopyroxene had little effect on REE contents.

These results are critical to the proper interpretation of REE abundances in clinopyroxene-poor ultramafic rocks, such as harzburgite and dunites!

Firstly, it is evident that at the current "state of the art" we do *not* accurately know REE partition coefficients for pure olivine and orthopyroxene; i.e., involving only the intrinsic REE content of these minerals. Moreover, there is abundant evidence that REE may be easily leached from these minerals, probably indicating high surface concentrations of REE, and that inclusions within olivine and orthopyroxene may have much higher REE, especially LREE, contents than the intrinsic levels in olivine and orthopyroxene.

Secondly, we must know more about the origins of surface concentrations of REE and REE-rich inclusions within mafic minerals. For example, do these phenomena post-date the melting event as envisioned in the mixing model, or are they formed during melting event, or do they pre-date the melting event. If either of the last two possibilities are valid, accurate petrogenetic models based on REE cannot be developed until we understand how properly to incorporate high surface concentrations of REE and REE-rich inclusions into melting models.

Garnet lherzolite inclusions

Localities with garnet lherzolite inclusions are relatively rare, but REE studies have been made on samples from South Africa (Philpotts et al., 1972; Shimizu, 1975; Mitchell and Carswell, 1976; Morgan et al., 1980; Nixon et al., 1981; BVSP, 1981), Lashaine volcano, Tanzania (Ridley and Dawson, 1975; BVSP, 1981) and the Navajo volcanic field in the southwest United States (Ehrenberg, 1982). Garnet lherzolites are an abundant inclusion type in many South African kimberlites. They are important because these inclusions are compositionally distinct from spinel lherzolites (Maaløe and Aoki, 1977), and this has led to different models for the major element composition of the upper mantle. In addition, garnet instead of spinel as the aluminous phase suggests that garnet peridotites provide deeper samples of the mantle than spinel peridotites.

Because inclusions in kimberlites commonly contain minerals formed by late-stage alteration, REE analyses of mineral separates are most reliable for determining upper mantle REE abundances. Shimizu (1975) analyzed acid-washed clinopyroxene and garnet separates from three sheared and three granular-textured inclusions. These granular inclusions have lower CaO and Al_2O_3 contents than the sheared inclusions, and Shimizu found that the two groups are also distinguished by REE and Sr abundances. Specifically, garnet and clinopyroxene from the granular inclusions have higher Sr contents and LREE/HREE abundance ratios than in the sheared inclusions. As a result, estimated bulk-rock REE contents of granular inclusions have $(Ce/Yb)_{cn} > 1$ whereas the sheared inclusions have $(Ce/Yb)_{cn} \leqslant 1$. Similar bulk-rock $(Ce/Yb)_{cn}$ differences between Ca- and Al-poor granular garnet lherzolite inclusions and Ca- and Al-rich sheared garnet lherzolite inclusions were noted in BVSP (1981). This discordance between major element trends and REE abundance trends is similar to that found in spinel lherzolite inclusions (Fig. 5.19). Shimizu (1975) suggested that these granular garnet lherzolites formed as garnet-free lherzolite cumulates from a kimberlite-like liquid, but as discussed earlier these data can also be interpreted as reflecting a mixing model.

Mitchell and Carswell (1976) found that four granular garnet lherzolites are quite variable in La/Yb with three samples having $(La/Yb)_{cn} > 1$. Based on discrepancies between whole-rock REE analyses and mass balances based on garnet and clinopyroxene, Mitchell and Carswell concluded that whole-rock REE abundances for inclusions within kimberlite *cannot* be used to estimate upper mantle REE abundances.

Nevertheless, Nixon et al. (1981) determined whole-rock REE abundances in twenty garnet-lherzolite inclusions from South African kimberlites. On the basis of major element content they divided this suite into three groups: fertile with Al_2O_3 >2% and CaO >1%, depleted with Al_2O_3 <2% and CaO <1% and metasomatic with K_2O >1%. It is important to note that the previously mentioned correlation of composition with texture, granular or sheared, is not valid for many inclusion localities. However, as found in spinel lherzolite inclusions in alkalic basalts, Nixon et al. (1981) found that the depleted inclusions have high $(La/Yb)_{cn}$ > 7 and the fertile inclusions have lower $(La/Yb)_{cn}$ ~1—3. Thus, despite their complex mineralogy and altered state (1—7% H_2O), garnet lherzolites exhibit major element/REE abundance correlations similar to those in spinel lherzolites (cf. Figs. 5.17, 5.19 with 5.21). Consequently, Nixon et al. (1981) adopted a model similar to that proposed by Frey and Green (1974); i.e., a two-component model whereby a garnet lherzolite lithosphere is depleted in basaltic constituents by igneous events, but is subsequently enriched in elements such as LREE by a poorly characterized component derived from greater depth.

Garnet lherzolite inclusions are also abundant in some minettes of the Navajo volcanic field especially near Shiprock, New Mexico. Ehrenberg (1979) found that the textural varieties of these inclusions are similar to those found in South African kimberlites. Moreover, Ehrenberg (1982) found that REE abundances in thirteen garnet lherzolites vary widely in $(La/Yb)_{cn}$ from ~0.5 to 7. These samples vary considerably in CaO and Al_2O_3, and as in garnet lherzolites from South Africa, samples with high CaO and Al_2O_3 contents have high HREE abundances, but samples with the lowest CaO and Al_2O_3 abundances have higher La/Yb ratios (Fig. 5.21).

Inclusions of garnet peridotite, principally harzburgites, also occur in ankaramitic scoria and carbonatite tuff at Lashaine volcano, Tanzania. Three garnet peridotites and a spinel peridotite have been analyzed for REE (Ridley and Dawson, 1975; BVSP, 1981); three are relatively enriched in LREE, $(La/Yb)_{cn}$ ~ 15, but the sample richest in CaO and Al_2O_3 has $(La/Yb)_{cn}$ = 0.58.

Pyroxenite inclusions

Pyroxenite inclusions are usually less abundant than olivine-rich inclusions in alkalic basalts and kimberlites, but a wide variety of pyroxenites have been found. These include garnet pyroxenites (often referred to as eclogites), pyroxenites belonging to the Cr-diopside group of Wilshire and Shervais (1975) and pyroxenites belonging to the Al-augite group of Wilshire and Shervais (1975). Although olivine-rich peridotites are believed to form the major part of the upper mantle, several researchers have suggested that

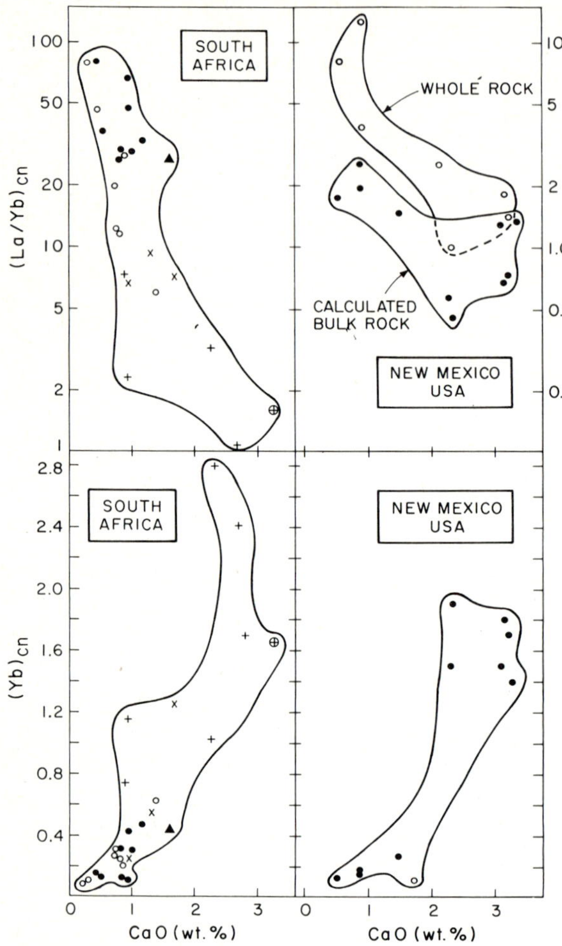

Fig. 5.21. Lower diagrams — CaO versus $(Yb)_{cn}$ for garnet-bearing peridotite inclusions. *South Africa.* Circles are depleted group of Nixon et al. (1981) with open circles indicating six samples from Matsoku, Lesotho (1 from Nixon et al., 1981; 5 from BVSP, 1981). Solid triangle indicates metasomatized inclusion (1.99% K_2O) from Nixon et al. (1981). Plus symbols indicated undepleted group of Nixon et al. (1981). Note overlap in CaO with depleted group. Encircled plus sign indicates fertile inclusion PHN 1611 from Thaba Putsoa (Morgan et al., 1980). X's indicate samples from Mitchell and Carswell (1976). *New Mexico, U.S.A.* Data for garnet peridotites included in a minette near Shiprock, New Mexico (Ehrenberg, 1982). Points are for whole-rock data calculated from mineral analyses. Open circle for sample 105 whose CaO content has been increased by introduction of secondary calcite. Upper diagram — CaO versus $(La/Yb)_{cn}$ for garnet-bearing peridotite inclusions. Note semi-log scale and different scales on right and left. Symbols as in lower plot, except both whole-rock and calculated whole-rock data are shown for garnet peridotites from New Mexico. Both data sets show a general inverse correlation. Calcite-contaminated sample, 105, is not plotted.

pyroxene-rich rocks constitute important parts of the upper mantle (e.g., Anderson, 1981).

REE abundances in garnet pyroxenites and eclogites found as inclusions in mafic rocks and as lenses associated with ultramafic rocks have been reported by Haskin et al. (1966), Reid and Frey (1971), Philpotts et al. (1972), White et al. (1972), Montigny and Allègre (1974), Garmann et al. (1975), Morten et al. (1979), Kurat et al. (1980, Frey (1980), Leeman et al. (1980), Evans et al. (1980), and Ernst (1981). Most true eclogites, i.e., containing omphacitic pyroxene, have REE contents similar to basalts, especially MORB (e.g., Ernst, 1981). Although many eclogites may be upper mantle rocks, their compositions are similar to mantle-derived melts. Primitive mantle compositions can be inferred from such melts, but this approach is addressed in Chapter 4; consequently, I will not discuss petrogenesis of eclogites.

Garnet pyroxenites and clinopyroxenites from Salt Lake Crater, Hawaii, have been studied in the most detail because they have been proposed as sources of Hawaiian basalts. However, as outlined by Frey (1980, p. 429), there has been considerable debate about the large-scale significance of these pyroxenites. Frey (1980) concluded that trace element abundances in these pyroxenites are most consistent with a petrogenesis involving crystal segregation from alkalic basalts within the upper mantle. In particular, their convex-upwards chondrite-normalized REE patterns (Fig. 5.22) are a feature expected in clinopyroxene and garnet-rich solids segregating from alkalic magmas with $(La/Yb)_{cn} \gg 1$. Furthermore, Frey stated that these pyroxenites are unlikely to be the source of Hawaiian tholeiites or alkalic basalts. Similar conclusions are valid for six additional analyses of garnet lherzolites reported by Leeman et al. (1980, fig. 5). Note that most inclusions from Salt Lake Crater designated as garnet lherzolites contain less than 40% olivine and are garnet pyroxenites rather than garnet lherzolites (Fig. 5.1; Frey, 1980, p. 444). Philpotts et al. (1972) also concluded that minerals separated from Kakanui eclogites (garnet pyroxenites) have REE contents reflecting equilibrium with an alkalic basalt magma.

Although characterized by $(La/Yb)_{cn} \leqslant 1$ most garnet pyroxenite inclusions from Salt Lake Crater have higher LREE contents than garnet pyroxenites forming layers in the Ronda and Beni Bouchera alpine peridotite (cf. Figs. 5.9 and 5.22). However, a garnet pyroxenite inclusion from Kapfenstein, Austria, has REE abundances lying within the range of plagioclase-garnet pyroxenites from Ronda, and Kurat et al. (1980) proposed a similar petrogenetic model; that is, this garnet pyroxenite may have formed from crystallization of a liquid, but it subsequently experienced partial melting and now has the REE characteristics of a residual rock.

Clinopyroxene-rich pyroxenites belonging to the Cr-diopside group or Group I of Frey and Prinz (1978) also have convex-upwards chondrite-normalized REE abundances (Frey and Prinz, 1978, fig. 7), consequently,

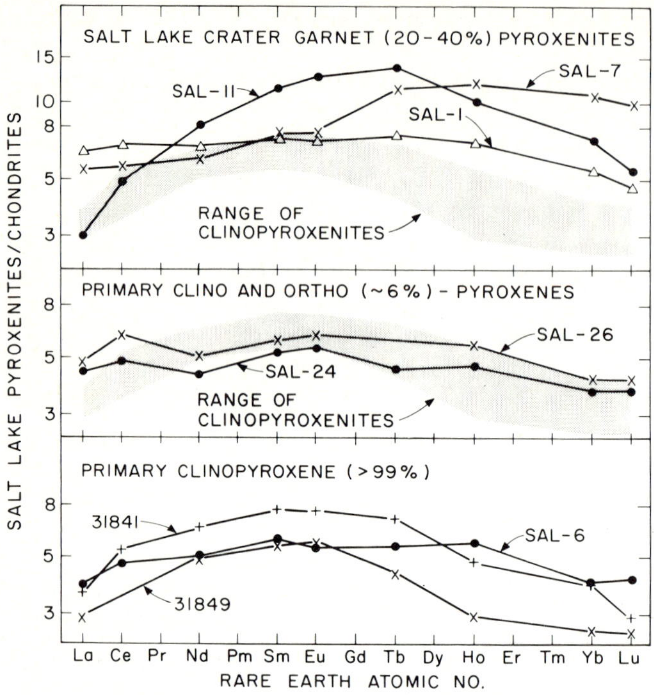

Fig. 5.22. Chondrite-normalized REE abundances for Salt Lake Crater pyroxenites studied by Frey (1980). Lower diagram — three samples whose primary mineralogy was >99% clinopyroxene. Middle diagram — two samples with primary orthopyroxene and clinopyroxene. Upper diagram — three samples with primary garnet and pyroxenes. Shown for comparison in upper and middle sections is the range for the three clinopyroxenites in the lower diagram.

these pyroxenites have also been interpreted as crystalline material segregated from melts within the upper mantle.

REE abundances in Group II pyroxenites, the Al-augite group of Wilshire and Shervais (1975), have been studied in most detail by Frey and Prinz (1978) and Irving (1980). These inclusions vary widely in mineral proportions and major element composition, and typically have $(La/Yb)_{cn} > 1$ (Fig. 5.23). Like other pyroxenites they have convex-upwards, chondrite-normalized REE patterns but with higher LREE contents, and Frey and Prinz concluded that these pyroxenites also formed as pyroxene-rich precipitates segregated from alkali basalts. In particular, at three localities there is close correspondence between the REE abundances in Group II inclusions and the calculated REE contents of pyroxenes and amphibole in equilibrium with the host basalt (Fig. 5.24; Frey and Prinz, 1978 fig. 14). However, in detail, at a single inclusion locality, Group II inclusions do not have coherent variations in modal proportions and mineral compositions. Thus,

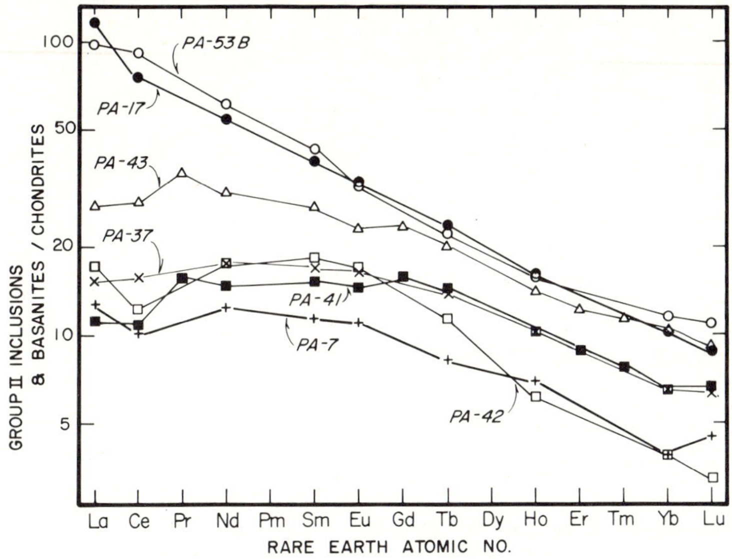

Fig. 5.23. Chondrite-normalized REE abundances in six Group II inclusions from San Carlos, Arizona, from Frey and Prinz (1978). Shown for comparison are two host basanites, PA-53B and PA-17. These inclusions range widely in modal mineralogy from wehrlites to clinopyroxenites to amphibole-olivine rocks (Frey and Prinz, 1978, table 13).

the suite of inclusions apparently formed from several different magmas which may have been precursors to the magma that incorporated the inclusions.

The major important result of REE studies of pyroxenite inclusions within basalts and pyroxenite layers within alpine peridotites is that these pyroxenites are *not* completely crystallized basaltic or proto-basaltic magmas. Instead, their geochemical characteristics reflect crystal segregation from a basaltic magma. Irving (1980) noted that there are several physical mechanisms for separating crystals and liquids, and he postulated that crystallization on narrow conduit walls is the major process creating pyroxenites within the upper mantle, whereas Wilshire et al. (1980) suggested that the pyroxenites represent liquids whose bulk compositions were modified "... by wall rock reactions during their emplacement, filter pressing during their consolidation and partial remelting following their consolidation."

Obviously, a wide variety of pyroxene-rich rocks have been interpreted as crystal segregations within the upper mantle. Their upper mantle origin is not in dispute because there are many examples of composite xenoliths where pyroxene-rich rocks occur as dikes within Group I peridotites or as matrices enclosing Group I peridotites (e.g., Irving, 1980). The diversity of pyroxenite types must reflect differences in parental magma composition plus variations in pressure, temperature and activities of H_2O and CO_2 at

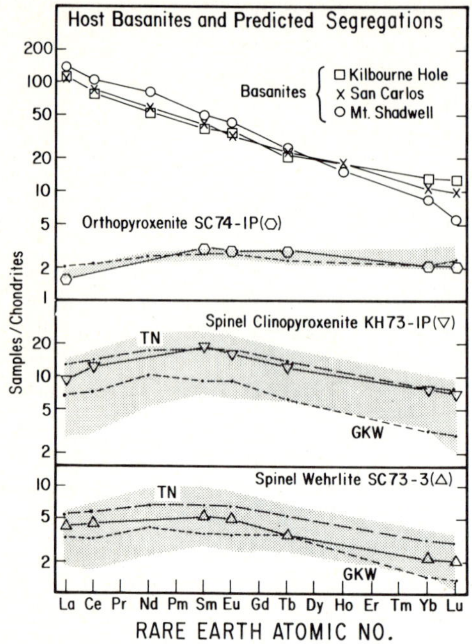

Fig. 5.24. Chondrite-normalized REE abundances for three host basanites and three pyroxene-rich Group II inclusions compared to calculated REE contents for pyroxene-rich solids in equilibrium with the basanites. Stippled areas for models based on megacryst/host REE partition coefficients; dashed lines, labelled "TN" and "GKW", for models using experimentally-determined REE partition coefficients. (Figure modified from Irving, 1980.)

the time of pyroxenite segregation. Moreover, there is abundant evidence that some mantle rocks reflect several generations of melt-crystal segregation and interaction (e.g., Wilshire and Shervais, 1975; Wilshire and Pike, 1975; Irving, 1980; Wilshire et al., 1980).

5.6. Implications of REE data for upper mantle composition and upper mantle processes

REE abundances in upper mantle rocks establish that the upper mantle *is heterogeneous* in REE content, especially in LREE, which range from 0.001 to 25 times the LREE content of ordinary chondrites. The variation in HREE content is much less, ranging only from 0.1 to 5 times ordinary chondrites in alpine peridotites, peridotites from the ocean floor and peridotites occurring as inclusions in alkalic basalts and kimberlites. Moreover, most mantle peridotites exposed in the crust appear to have been derived as residues from partial melting. Prior to this melting event upper mantle spinel and garnet lherzolites had HREE contents ~1.5 to 3 times those of ordinary chondrites (Figs. 5.2 and 5.18).

The observed range in REE content of upper mantle rocks leads to the following conclusions regarding upper mantle processes.

(1) Assuming that the core contains relatively little of the Earth's REE, the mantle + crust are enriched by a factor of 1.5 compared to the bulk Earth. Also, REE abundances vary by about a factor of two between different classes of chondrites. Thus, the inferred upper mantle REE abundances of ~1.5 to 3 times ordinary chondrites requires that the upper mantle has not been created by processes that result in strong enrichment and fractionation of REE.

(2) Almost all CaO- and Al_2O_3-rich (>3%) spinel and garnet lherzolites have $(La/Yb)_{cn} < 1$, HREE contents of 1—3 times ordinary chondrites, and Al_2O_3/CaO ratios similar to chondrites. Given the evidence that the bulk Earth has a chondritic Sm/Nd ratio (Chapter 11), the relative depletion in LREE is most readily interpreted as resulting from loss of a small fraction (<15%) of a fluid relatively enriched in LREE. Consequently, the CaO and Al_2O_3-rich lherzolites occurring in alpine peridotites and as inclusions in alkalic basalts and kimberlites have compositions suitable for the source of mid-ocean ridge basalts (MORB), but they are unlikely to be residues related to MORB generation if MORB are generated by degrees of melting sufficient to create residues with low (<2%) contents of CaO and Al_2O_3.

(3) Compared to lherzolites, upper mantle harzburgites and dunites range more widely in REE content. Some have $(LREE/HREE)_{cn} < 1$ but most have $(LREE/HREE)_{cn} > 1$. Most harzburgites and dunites from alpine peridotites, including the basal sections of ophiolites, have chondrite-normalized REE patterns with well-defined minima in the Sm-Gd region and maximum REE contents of only 0.2—0.5 times ordinary chondrites.

There are probably several processes which create upper mantle harzburgites and dunites with $(LREE/HREE)_{cn} > 1$. For example:

(a) The relative LREE enrichments in highly serpentinized harzburgites and dunites from alpine peridotites, including the basal ultramafic section of ophiolites, and some ocean floor peridotites may result from introduction of LREE during formation of serpentine and other low pressure/low-temperature minerals. These rocks are particularly susceptible to such contamination because of their initially very low REE contents. However, this explanation is not satisfactory for fresh, serpentine-free rocks which have $(LHREE/HREE)_{cn} > 1$.

(b) The relative LREE enrichments in dunites and harzburgites may reflect segregation of olivine from magmas with high LREE/HREE. In this case, ratios of $(LREE/HREE)_{cn} > 1$ in dunites and harzburgites are inherited from the magma even though $D^{ol/melt}_{La/Yb}$ is <1.

(c) The relative LREE enrichment may result from preferential incorporation of LREE relative to HREE in residual minerals formed during partial melting. Residual plagioclase in a harzburgite would have this effect, but there is little evidence, even in plagioclase-bearing ocean floor peridotites, that there was sufficient residual plagioclase to create a bulk peridotite with $(LREE/HREE)_{cn} > 1$. Alternatively, olivine and possibly orthopyroxene may preferentially incorporate LREE under some conditions. Although this interpretation is not consistent with criteria based on ionic radii, such criteria may not be valid for these minerals whose REE contents may not be controlled by substitution of REE at major cation structural sites. The validity of this interpretation requires REE analysis of carefully separated and treated olivine and orthopyroxene crystals.

(4) In contrast to alpine peridotites, harzburgites and low-CaO-Al_2O_3 (<2%) lherzolites from inclusion suites within alkalic basalts and kimberlites have LREE contents *greater* than chondrites with $(La/Yb)_{cn}$ ranging from 1 to 25. The most plausible interpretation of relative LREE enrichment in these upper mantle peridotites is that these peridotites were formed by mixing of at least two geochemically distinct components (Frey and Green, 1974); i.e., component A which reflects the bulk of the peridotite and controls the major mineralogy and major and compatible trace element content, and component B which forms only a small portion of the peridotite but controls the abundances of incompatible trace elements, such as LREE, and is incorporated into the major minerals and accessory minerals such as amphibole. Component A has the geochemical characteristics of a partial melting residue but component B is highly enriched in incompatible elements and is interpreted to be a migratory fluid which is unrelated to component A prior to its infiltration and mixing with component A. This process could also explain the relative LREE enrichment in alpine harzburgites and dunites.

This two- or more, component model has major implications for upper mantle processes and the creation of mantle heterogeneity. Key questions are: (a) What was the composition and mode of formation of component B? (b) What was the timing of the mixing event? and (c) Where did the mixing event occur?

The inferred composition of component B is consistent with a partial melt derived by low (<5%) degrees of melting or a H_2O- and CO_2-rich fluid. Nd isotopic data suggest that spinel peridotite inclusions with $(LREE/HREE)_{cn} > 1$ formed within the last 200 Ma. The location of the mixing event is poorly known, but it is important to note that the mixing event is reflected in peridotites from diverse tectonic regions including not only peridotites in alkalic basalts from continental margins but also garnet peridotite inclusions from kimberlites located within continental cratons and oceanic peridotites exposed on St. Paul's Rocks in the Atlantic and at Salt Lake Crater, Hawaii in the Pacific. If component B formed in the low-velocity zone as a small percent melt which migrated upwards and crystallized, then the mixing zone should be at the low-velocity zone/ lithosphere interface. An important observation is that the mixing process is inferred for spinel peridotites reflecting a possible pressure range of 9—25 kbar and garnet peridotites reflecting higher pressures of 30—50 kbar. Since the garnet peridotites are derived from continental mantle and the spinel peridotites may be oceanic mantle, the different depths inferred for the mixing process may reflect the different depths to the respective low-velocity zones.

In general, CaO- and Al_2O_3-rich lherzolite inclusions and alpine lherzolites have $(La/Yb)_{cn} < 1$ and they contain no evidence for a mixing process. Consequently, the outlined model infers a depleted harzburgite-rich lower lithosphere overlying a fertile lherzolite-rich asthenosphere. This is only a simplistic model and more complex stratigraphy is likely. For example, Stosch and Seck (1980) noted that the amphibole-bearing peridotite suite from Dreiser Weiher, Germany, ranges from harzburgite to lherzolite and all peridotite types of this suite have LREE more than 5 times chondrites. Consequently, they proposed that in this region the upper mantle contains interstratified harzburgite and lherzolite. Stosch and Seck (1980) also found that this amphibole-bearing suite reflects a lower equilibration temperature than the anhydrous suite at Dreiser Weiher, and they concluded that the mixing event occurred in the shallow lithosphere. However, it is not possible to define accurately the pressure-temperature regime of the inferred mixing process for spinel peridotites; in part, because of uncertainties in geobarometry and geothermometry, but also because *P-T* estimates only relate to the last equilibration conditions which are unlikely to reflect the *P-T* regime of the mixing event.

An important observation is that *no* peridotites from high-temperature alpine peridotites (Ronda, Beni Bouchera, Lizard, Tinaquillo, Mount Albert) or alpine peridotites from the Alps have $(La/Yb)_{cn} > 1$. Consequently, there is no evidence for the inferred mixing process in the mantle sampled by these alpine peridotites, and there must be fundamental differences in the petrogenesis of alpine peridotites and ultramafic inclusion suites. The common factor associated with peridotites that have LREE contents greater than chondrites is an intimate association with alkalic volcanism. This is obvious for inclusions in alkalic basalts and kimberlites, but also true for St. Paul's Rocks which represent oceanic mantle. Apparently, these mantle samples derived from regions of alkaline magmatism have been affected *within* the mantle by early phases of the alkalic volcanism. Macroscopic evidence for this possibly multi-stage process are the abundant veining of peridotite by amphibole in several inclusion suites and at St. Paul's Rocks. Alternatively, alkalic volcanism may be limited to mantle regions where the mantle has been enriched in incompatible elements, such as LREE, as suggested by Menzies and Murthy (1980b).

(5) The role of pyroxenites in the upper mantle is a subject of continuing

debate. Important recent proposals are that (a) garnet pyroxenites form the upper mantle transition region (220—670 km) and are the source of MORB (Anderson, 1981), and (b) that veining of olivine-rich peridotite by pyroxenite is an important form of mantle heterogeneity which leads to basalts with diverse trace element abundances and isotopic ratios in a local geographic area (e.g., Hanson, 1977; Wood, 1979; Sun et al., 1979). REE abundances in mantle pyroxenites establish that the studied garnet pyroxenites are not suitable MORB sources, and that mantle pyroxenites are not compositionally equivalent to totally crystallized basaltic magmas. Instead the studied pyroxenites reflect a complex petrogenesis including segregation as solids from basaltic magmas and possibly formation as partial melting residues. Undoubtedly, pyroxenite-rich layers and veins in alpine peridotites and peridotite inclusions reflect magmatism in the mantle; consequently, they provide a mechanism for creating small-scale mantle heterogeneity which will not be readily eliminated by convection. However, the studied pyroxenites are *not representative* of mantle melts and most are *not* suitable sources *for* incompatible element-rich magmas.

Acknowledgements

I thank P. Henderson, A.J. Irving, M. Loubet, G. Ottonello, N.W. Rogers, N. Shimizu and H.G. Stosch for useful critical comments on an early draft of this chapter.

References

Allègre, C.J., Montigny, R. and Bottinga, Y., 1973. Cortège ophiolitique et cortège océanique géochimie comparée et mode de génèse. *Bull. Soc. Géol. Fr.*, 15: 461—477.

Allègre, C.J., Brevart, O., Joron, J.L., Minster, J.F. and Noiret, G., 1980. Trace element geochemistry and the origin of ophiolite complexes. *Colloq. Int. C.N.R.S.*, 272: 125—133.

Anderson, D.L., 1981. Hotspots, basalts and evolution of the mantle. *Science*, 213: 82—89.

Bonatti, E., Hamlyn, P.R. and Ottonello, G., 1981. Upper mantle beneath a young oceanic rift: peridotites from the island of Zabargad (Red Sea). *Geology*, 9: 474—479.

Bowin, C.O., Nalwalk, A.J. and Hersey, J.B., 1966. Serpentinized peridotite from the north wall of the Puerto Rico Trench. *Geol. Soc. Am. Bull.*, 77: 257—270.

Brunfelt, A.O., Roelandts, I. and Steinnes, E., 1974. Determination of rubidium, caesium, barium and eight rare earth elements in ultramafic rocks by neutron activation analysis. *Analyst*, 99: 277—284.

BVSP (Basaltic Volcanism Study Project), 1981. *Basaltic Volcanism on the Terrestrial Planets*. Pergamon, New York, N.Y., pp. 282—310.

Chen, C.-Y. and Frey, F.A., 1980. Geochemistry of lherzolite inclusions from Mt. Leura, Victoria, Australia. *EOS, Trans. Am. Geophys. Union*, 61: 413.

Clarke, D.B. and Loubat, H., 1977. Mineral analyses from the peridotite-gabbro-basalt complex at site 334, DSDP Leg 37. In: F. Aumento, W.G. Melson et al., *Initial Reports of the Deep Sea Drilling Project, 37.* U.S. Government Printing Office, Washington, D.C., pp. 847—855.

Coish, R., Roden, M., Frey, F. and Suen, C.J., 1979. Rare earth abundances in ultramafic rocks from the ocean floor, ophiolites and alpine peridotites. *Geol. Soc. Am. Abstr. Prog.*, 11(7): 403.

Coish, R.A., Hickey, R. and Frey, F.A., 1982. Rare earth element geochemistry of the Betts Cove ophiolite, Newfoundland: complexities in ophiolite formation. *Geochim. Cosmochim. Acta*, 46: 2117—2134.

Coleman, R.G., 1977. Ophiolites. Springer-Verlag, New York, N.Y., 229 pp.

Conquéré, F., 1977. Pétrologie des pyroxenites litées dans les complexes ultramafiques de l'Ariege (France) et autres gisements de lherzolite a spinelle, I. Compositions minéralogiques et chimiques, évolution des conditions d'équilibre des pyroxénites. *Bull. Soc. Fr. Minéral. Cristallogr.*, 100: 42—80.

Dick, H.J.B., 1978. Mineralogy of abyssal peridotites from the far south Atlantic, Caribbean and equatorial Atlantic. *Geol. Soc. Am., Abstr. Prog.*, 10: 388.

Dick, H.J.B. and Bullen, T., 1983. Chromium spinel as a petrogenetic indicator in oceanic environments. *Contrib. Mineral. Petrol.* (in press).

Dickey, J.S., 1970. Partial fusion products in alpine-type peridotites: Serrania de la Ronda and other examples. *Mineral. Soc. Am. Spec. Paper*, 3: 33—49.

Dickey, J.S., Obata, M. and Suen, C.J., 1977. Partial fusion versus fractional crystallization: hypotheses for the differentiation of the Ronda ultramafic massif of southern Spain in magma genesis. *Oreg. Dep. Geol. Miner. Ind. Bull.*, 96: 79—89.

Dickey, J.S., Obata, M. and Suen, C.J., 1979. Chemical differentiation of the lower lithosphere as represented by the Ronda ultramafic massif, Southern Spain. In: L.H. Ahrens (Editor), *Origin and Distribution of the Elements.* Pergamon, Oxford, pp. 587—595.

Dostal, J. and Muecke, G.K., 1978. Trace element geochemistry of the peridotite-gabbro-basalt suite from DSDP Leg 37. *Earth Planet. Sci. Lett.*, 40: 415—422.

Ehrenberg, S.N., 1979. Garnetiferous ultramafic inclusions in minette from the Navajo volcanic field. In: F.R. Boyd and H.O.A. Meyer (Editors), *The Mantle Sample: Inclusions in Kimberlites and other Volcanics.* American Geophysical Union, Washington, D.C., pp. 330—344.

Ehrenberg, S.N., 1982. Rare earth element geochemistry of garnet lherzolite and megacrystalline inclusions from minette of the Colorado Plateau. *Earth Planet. Sci. Lett.*, 57: 191—210.

Ernst, W.G., 1981. Petrogenesis of eclogites and peridotites from the Western and Ligurian Alps. *Am. Mineral.*, 66: 443—472.

Ernst, W.G. and Piccardo, G.B., 1979. Petrogenesis of some Ligurian peridotites, I. Mineral and bulk rock chemistry. *Geochim. Cosmochim. Acta*, 43: 219—237.

Evans, B.W., Goles, G.G. and Tromsdorff, V., 1980. Geochemistry of high-grade metarodingites, Central Alps. *Geol. Soc. Am. Abstr. Prog.*, 12(7): 422.

Flower, M.F.J., 1971. Rare earth element distribution in lavas and ultramafic xenoliths from the Comores Archipelago, western Indian Ocean. *Contrib. Mineral. Petrol.*, 31: 335—346.

Frey, F.A., 1969. Rare earth abundances in a high-temperature peridotite intrusion. *Geochim. Cosmochim. Acta*, 33: 1429—1447.

Frey, F.A., 1970a. Rare earth abundances in alpine ultramafic rocks. *Phys. Earth Planet. Inter.*, 3: 323—330.

Frey, F.A., 1970b. Rare earth and potassium abundances in St. Paul's Rocks. *Earth Planet. Sci. Lett.*, 7: 351—360.

Frey, F.A., 1980. The origin of pyroxenites and garnet pyroxenites from Salt Lake Crater, Oahu, Hawaii: Trace element evidence. *Am. J. Sci.*, 280-A: 427—449.

Frey, F.A. and Green, D.H., 1974. The mineralogy, geochemistry and origin of lherzolite inclusions in Victorian basanites. *Geochim. Cosmochim. Acta*, 38: 1023—1059.

Frey, F.A. and Prinz, M., 1978. Ultramafic inclusions from San Carlos, Arizona: petrologic and geochemical data bearing on their petrogenesis. *Earth Planet. Sci. Lett.*, 38: 129—176.

Frey, F.A. and Suen, C.Y.J., 1983. The Ronda high-temperature peridotite: geochemistry and petrogenesis, *Geochim. Cosmochim. Acta* (submitted).

Frey, F.A., Haskin, L.A. and Haskin, M.A., 1971. Rare earth abundances in some ultramafic rocks. *J. Geophys. Res.*, 75: 2057—2070.

Frey, F.A., Green, D.H. and Roy, S.D., 1978. Integrated models of basalt petrogenesis: a study of quartz tholeiites to olivine melilitites from south eastern Australia utilizing geochemical and experimental petrological data. *J. Petrol.*, 19: 463—513.

Garmann, L.B., Brunfelt, A.O., Finstad, K.G. and Heier, K.S., 1975. Rare-earth elements distribution in basic and ultrabasic rocks from West Norway. *Chem. Geol.*, 15: 103—116.

Green, D.H., 1967. High temperature peridotite intrusions. In: P.J. Wyllie (Editor), *Ultramafic and Related Rocks*. John Wiley and Sons, New York, N.Y., pp. 212—221.

Green, D.H., Hibberson, W.O. and Jaques, A.L., 1979. Petrogenesis of mid-ocean ridge basalts. In: M.W. McElhinny (Editor), *The Earth: Its Origins, Structure and Evolution*. Academic Press, New York, N.Y., 265—299.

Hamlyn, P.R. and Bonatti, E., 1980. Petrology of mantle-derived ultramafics from the Owen fracture zone, northwest Indian Ocean: implications for the nature of the oceanic upper mantle. *Earth Planet. Sci. Lett.*, 48: 65—79.

Hanson, G.N., 1977. Geochemical evolution of the suboceanic mantle. *J. Geol. Soc., London*, 134: 235—253.

Harrison, W.J., 1981. Partition coefficients for REE between garnets and liquids: implications of non-Henry's Law behaviour for models of basalt origin and evolution. *Geochim. Cosmochim. Acta*, 45: 1529—1544.

Harrison, W.J. and Wood, B.J., 1980. An experimental investigation of the partitioning of REE between garnet and liquid with reference to the role of defect equilibria. *Contrib. Mineral. Petrol.*, 72: 145—155.

Haskin, L.A., Frey, F.A., Schmitt, R.A. and Smith, R.H., 1966. Meteoritic, solar and terrestrial rare-earth distributions. *Phys. Chem. Earth*, 7: 169—321.

Haskin, L.A., Haskin, M.A., Frey, F.A. and Wildeman, T.R., 1968. Relative and absolute abundances of the rare earths. In: L.H. Ahrens (Editor), *Origin and Distribution of the Elements*. Pergamon, London, pp. 889—912.

Hickey, R. and Frey, F.A., 1982. Geochemical characteristics of boninite series volcanics: implications for their source. *Geochim. Cosmochim. Acta*, 46: 2099—2115.

Higuchi, H., Tomura, K., Onuma, N. and Hamaguchi, H., 1969. Rare earth abundances in several geochemical standard rocks. *Geochem. J.*, 3: 171—180.

Hodges, F. and Papike, J.J., 1976. DSDP site 334: magmatic cumulates from oceanic layer 3. *J. Geophys. Res.*, 81: 4135—4151.

Hooker, P.J., O'Nions, R.K. and Pankhurst, R.J., 1975. Determination of rare-earth elements in USGS standard rocks by mixed-solvent ion exchange and mass-spectrometric isotope dilution. *Chem. Geol.*, 16: 189—196.

Irving, A.J., 1980. Petrology and geochemistry of composite ultramafic xenoliths in alkalic basalts and implications for magmatic processes within the mantle. *Am. J. Sci.*, 280-A: 389—426.

Jacobsen, S.B. and Wasserburg, G.J., 1979. Nd and Sr isotopic study of the Bay of Islands ophiolite complex and the evolution of the source of midocean ridge basalts. *J. Geophys. Res.*, 84: 7429—7445.

Jagoutz, E., Lorenz, V. and Wänke, H., 1979a. Major trace elements of Al-augites and Cr-diopsides from ultramafic nodules in European alkali basalts. In: F.R. Boyd and H.O.A. Meyer (Editors), *The Mantle Sample: Inclusions in Kimberlites and other Volcanics.* American Geophysical Union, Washington, D.C., pp. 382—390.

Jagoutz, E., Palme, H., Baddenhausen, H., Blum, K., Cendales, M., Dreibus, G., Spettel, B., Lorenz, V. and Wänke, H., 1979b. The abundance of major, minor and trace elements in the earth's mantle as derived from primitive ultramafic nodules. *Proc. 10th Lunar Sci. Conf., Geochim. Cosmochim. Acta, Suppl. II*, 2: 2031—2050.

Jagoutz, E., Carlson, R.W. and Lugmair, G.W., 1980. Equilibrated Nd—unequilibrated Sr isotopes in mantle xenoliths. *Nature*, 286: 708—710.

Jibiki, H. and Masuda, A., 1974. Basalts and serpentinite from the Puerto Rico trench, 2. Rare-earth geochemistry. *Mar. Geol.*, 16: 205—211.

Kay, R.W. and Senechal, R.G., 1976. The rare earth geochemistry of the Troodos ophiolite complex. *J. Geophys. Res.*, 81: 964—970.

Kleeman, J.D., Green, D.H. and Lovering, J.F., 1969. Uranium distribution in ultramafic inclusions from Victorian basalts. *Earth Planet. Sci. Lett.*, 5: 449—458.

Kolesov, G.M., 1976. Determination of rare earth elements in rocks and meteorites by the radioactivation method. *J. Radioanal. Chem.*, 30: 553—560.

Kornprobst, J., 1969. Le massif ultrasique des Beni Bouchera (Rif Interne, Maroc): étude des péridotites de haute température et de haute pression, et des pyroxénolites, à grenat ou sans grenat, qui leur sont associées. *Contrib. Mineral. Petrol.*, 23: 283—322.

Kurat, G., Palme, H., Spettel, B., Baddenhausen, H., Hofmeister, H., Palme, C. and Wänke, H., 1980. Geochemistry of ultramafic xenoliths from Kapfenstein, Austria: evidence for a variety of upper mantle processes. *Geochim. Cosmochim. Acta*, 44: 45—60.

Laul, J.C., 1979. Neutron activation analysis of geological materials. *At. Energy Rev.*, 17: 603—695.

Leeman, W.P., Budahn, J.R., Gerlach, D.C., Smith, D.R. and Powell, B.N., 1980. Origin of Hawaiian tholeiites: trace element constraints. *Am. J. Sci.*, 280-A: 794—819.

Loubet, M. and Allègre, C.J., 1979. Trace element studies in the Alpine type peridotite of Beni-Bouchera (Morocco). *Geochem. J.*, 13: 69—75.

Loubet, M., Shimizu, N. and Allègre, C.J., 1975. Rare earth elements in alpine peridotites. *Contrib. Mineral. Petrol.*, 53: 1—12.

Loubet, M., Bougault, M., Shimizu, N. and Allègre, C.J., 1976. Geochemical study (REE, Ba and partially major and transition elements analysis) of pyroxenolite layers in lherzolite type alpine massives. *EOS, Trans. Am. Geophys. Union*, 57: 1025 (abstract).

Loubet, M., Polvé, M., Richard, P. and Allègre, C.J., 1980. Geochemical studies in orogenic lherzolites: evidence about multiple magmatic events. *Colloq. Int. C.N.R.S.*, 272: 269—277.

Maaløe, S. and Aoki, K., 1977. The major element composition of the upper mantle estimated from the composition of lherzolites. *Contrib. Mineral. Petrol.*, 63: 161—173.

MacGregor, I. and Basu, A.R., 1979. Petrogenesis of the Mount Albert ultramafic massif, Quebec. *Geol. Soc. Am. Bull.*, 90: 1529—1627.

Mattson, P.H., 1964. Petrography and structure of serpentinite from Mayaguez, Puerto Rico. In: C.A. Burk (Editor), A study of Serpentinite. *Natl. Acad. Sci. U.S.A./Natl. Resour. Counc. Publ.*, 1188: 7—24.

McCulloch, M.T., Gregory, R.T., Wasserburg, G.J. and Taylor, H.P., 1981. Sm-Nd, Rb-Sr, and $^{18}O/^{16}O$ isotopic systematics in an oceanic crustal section: Evidence from the Samail Ophiolite. *J. Geophys. Res.*, 86: 2721—2735.

McKay, G.A., 1981. Experimental REE partitioning between olivine and Apollo 12 olivine basaltic liquids. *EOS, Trans. Am. Geophys. Union*, 45: 1070.

Melson, W.G. and Thompson, G., 1971. Petrology of a transform fault zone and adjacent ridge segments. *Philos. Trans. R. Soc. London, Ser. A*, 268: 423—441.

Melson, W.G., Hart, S.R. and Thompson, G., 1972. St. Paul's Rocks, Equatorial Atlantic: petrogenesis, radiometric ages and implications on sea-floor spreading. *Geol. Soc. Am., Mem.*, 132: 241—272.

Menzies, M., 1976. Rare earth geochemistry of fused ophiolitic and alpine lherzolites, I. Othris, Lanzo and Troodos. *Geochim. Cosmochim. Acta*, 40: 645—656.

Menzies, M. and Murthy, V.R., 1980a. Nd and Sr isotope geochemistry of hydrous mantle nodules and their host alkali basalts: implications for local heterogeneties in metasomatically veined mantle. *Earth Planet. Sci. Lett.*, 46: 323—334.

Menzies, M. and Murthy, V.R., 1980b. Mantle metasomatism as a precursor to the genesis of alkaline magmas — isotopic evidence. *Am. J. Sci.*, 280-A: 622—638.

Menzies, M., Blanchard, D., Brannon, J. and Korotev, R., 1977. Rare earth geochemistry of fused ophiolitic and alpine lherzolites. *Contrib. Mineral. Petrol.*, 64: 53—74.

Mitchell, R.H. and Carswell, D.A., 1976. Lanthanium, samarium and ytterbium abundances in some Southern African garnet lherzolites. *Earth Planet. Sci. Lett.*, 31: 175—178.

Miyashiro, A., Shido, F. and Ewing, M., 1969. Composition and origin of serpentinites from the Mid-Atlantic Ridge near 24° and 30° north latitude. *Contrib. Mineral. Petrol.*, 23: 117—127.

Montigny, R. and Allègre, C.J., 1974. A la recherche des océans perdus: les éclogites de Vendée, temoins metamorphises d'une ancienne croûte océanique. *C.R. Acad. Sci, Paris*, 279: 543—545.

Montigny, R., Bougault, H., Bottinga, Y. and Allègre, C.J., 1973. Trace element geochemistry and genesis of the Pindos ophiolite suite. *Geochim. Cosmochim. Acta*, 37: 2135—2147.

Morgan, J.W., Wandless, G.A., Petrie, R.K. and Irving, A.J., 1980. Earth's upper mantle: volatile element distribution and origin of siderophile element content. In: *Lunar and Planetary Science XI.* Lunar and Planetary Institute, Houston, Texas, pp. 740—742.

Morten, L., 1978. REE abundances in spinel-lherzolite nodules and host basalt from S. Giovanni Ilarione quarry, Veneto Region, North Italy. *Neues Jahrb.* Mineral., *Monatsh.*, 159—165.

Morten, L., Brunfelt, A.O. and Mottana, A., 1979. Rare earth abundances in superferrian eclogites from the Voltri Group (Pennidic Belt, Italy). *Lithos*, 12: 25—32.

Mysen, B.O., 1978. Experimental determination of rare earth element partitioning between hydrous silicate melt, amphibole and garnet peridotite minerals at upper mantle pressures and temperatures. *Geochim. Cosmochim. Acta*, 42: 1253—1263.

Mysen, B.O., 1979. Trace-element partitioning between garnet peridotite minerals and water-rich vapor: experimental data from 5 to 30 kbar. *Am. Mineral.*, 64: 274—287.

Nagasawa, H., Wakita, H., Higuchi, H. and Onuma, N., 1969. Rare earths in peridotite nodules: an explanation of the genetic relationship between basalt and peridotite nodules. *Earth Planet. Sci. Lett.*, 5: 377—381.

Nicolas, A. and Jackson, E.D., 1972. Répartition en deux provinces des péridotites des chaînes alpines longeant la Méditerranée: implications géotectoniques. *Schweiz. Mineral. Petrogr. Mitt.*, 52: 479—495.

Nixon, P.H., Rogers, N.W.., Gibson, I.L. and Grey, A., 1981. Depleted and fertile mantle xenoliths from southern African kimberlites. *Annu. Rev. Earth Planet. Sci.*, 8: 285—309.

Noiret, G., Montigny, R. and Allègre, C.J., 1981. Is the Vourinos Complex an island arc ophiolite? *Earth Planet. Sci. Lett.*, 56: 375—386.

Obata, M., Suen, C.J. and Dickey, J.S., 1980. The origin of mafic layers in the Ronda high-temperature peridotite intrusion, S. Spain: an evidence of partial fusion and fractional crystallization in the upper mantle. *Colloq. Int. C.N.R.S.*, 272: 257—268.

Ottonello, G., 1980. Rare earth abundances and distribution in some spinel peridotite xenoliths from Assab (Ethiopia). *Geochim. Cosmochim. Acta*, 44: 1885—1901.

Ottonello, G., Piccardo, G.B., Mazzucotelli, A. and Cimmino, F., 1978. Clinopyroxene-orthopyroxene major and rare earth elements partitioning in spinel peridotite xenoliths from Assab (Ethiopia). *Geochim. Cosmochim. Acta*, 42: 1817—1828.

Ottonello, G., Piccardo, G.B. and Ernst, W.G., 1979. Petrogenesis of some Ligurian peridotites, II. Rare earth element chemistry. *Geochim. Cosmochim. Acta*, 43: 1273—1284.

Pallister, J.S. and Knight, R.J., 1981. Rare-earth element geochemistry of the Samail ophiolite near Ibra, Oman. *J. Geophys. Res.*, 86: 2673—2697.

Philpotts, J.A., Schnetzler, C.C. and Thomas, H.H., 1972. Petrogenetic implications of some new geochemical data on eclogitic and ultrabasic inclusions. *Geochim. Cosmochim. Acta*, 36: 1131—1166.

Potts, M.J. and Condie, K.C., 1971. Rare earth element distributions in a proto-stratiform ultramafic intrusion. *Contrib. Mineral. Petrol.*, 33: 245—258.

Prinz, M., Keil, K., Green, J.A., Reid, A.M., Bonatti, E. and Honnorez, J., 1976. Ultramafic and mafic dredge samples from the equatorial Mid-Atlantic Ridge and fracture zones. *J. Geophys. Res.*, 81: 4087—4103.

Reid, J.B. and Frey, F.A., 1971. Rare earth distributions in lherzolite and garnet pyroxenite xenoliths and the constitution of the upper mantle. *J. Geophys. Res.*, 76: 1184—1196.

Reid, J.B. and Woods, G.A., 1978. Oceanic mantle beneath the Southern Rio Grande rift. *Earth Planet. Sci. Lett.*, 41: 303—316.

Rey, P., Wakita, H. and Schmitt, R.A., 1970. Radiochemical neutron activation analysis of indium, cadium, yttrium and the 14 rare earth elements in rocks. *Anal. Chim. Acta*, 51: 163—178.

Ridley, W.I. and Dawson, J.B., 1975. Lithophile trace element data bearing on the origin of peridotite xenoliths, ankaramite and carbonatite from Lashaine volcano, N. Tanzania. *Phys. Chem. Earth*, 9: 559—569.

Ringwood, A.E., 1975. *Composition and Petrology of the Earth's Mantle*. McGraw-Hill, New York, N.Y., 618 pp.

Roden, M.F., Frey, F.A. and Francis, D.M., 1980. REE, K, Rb, Sr and Sr isotopic geochemistry of peridotite xenoliths in basalt from Nunivak Island, Alaska. *EOS, Trans. Am. Geophys. Union*, 61: 401.

Roden, M.F., Frey, F.A. and Hart, S.R., 1981. The mantle source for the Honolulu Volcanic series: Nd isotope evidence. *EOS, Trans. Am. Geophys. Union*, 62: 423.

Schilling, J.G., 1966. *Rare-earth fractionation in Hawaiian volcanic rocks*. Ph.D. Thesis, Massachusetts Institute of Technology, Cambridge, Mass.

Shaw, D.M., 1970. Trace element fractionation during anatexis. *Geochim. Cosmochim. Acta*, 34: 237—243.

Shih, C.Y., 1972. *The rare earth geochemistry of oceanic igneous rocks*. Ph.D. Thesis, Columbia University, New York, N.Y.

Shih, C. and Gast, P.W., 1971. The distribution of rare earth elements in some dredged ultramafic rocks. *Abstr. Prog., Annu. Meet. Geol. Soc. Am.*, 3(7): 702.

Shimizu, N., 1975. Rare earth elements in garnets and clinopyroxenes from garnet lherzolite nodules in kimberlites. *Earth Planet. Sci. Lett.*, 25: 26—32.

Shimizu, N. and Hart, S.R., 1974. Rare earth element concentrations in clinopyroxene from an ocean-ridge lherzolite. *Carnegie Inst. Washington Yearb.*, 73: 964—967.

Sinton, J.M., 1979a. Ultramafic inclusions and high-pressure xenocrysts in submarine basanitoid, equatorial Mid-Atlantic Ridge. *Contrib. Mineral. Petrol.*, 70: 49—57.

Sinton, J.M., 1979b. Petrology of (Alpine-type) peridotites from site 395, DSDP Leg 45. In: W.G. Melson, P.D. Rabinowitz et al., *Initial Reports of the Deep Sea Drilling Project, 45*. U.S. Government Printing Office, Washington, D.C., pp. 595—601.

Smet, T. and Roelandts, I., 1978. Radiochemical neutron activation trace element analysis of two USGS ultrabasic rock reference samples: Peridotite PCC-1 and Dunite DTS-1. *Geostand. Newslett.*, 2: 61—70.

Stosch, H.G., 1982. Rare earth element partitioning between minerals from anhydrous spinel peridotite xenoliths. *Geochim. Cosmochim. Acta*, 46: 793—811.

Stosch, H.G. and Seck, H.A., 1980. Geochemistry and mineralogy of two spinel peridotite suites from Dreiser Weiher, West Germany. *Geochim. Cosmochim. Acta*, 44: 457—470.

Stosch, H.G., Carlson, R.W. and Lugmair, G.W., 1980. Episodic mantle differentiation: Nd and Sr isotopic evidence. *Earth Planet. Sci. Lett.*, 47: 263—271.

Streckeisen, A., 1976. To each plutonic rock its proper name. *Earth Sci. Rev.*, 12: 1—33.

Strelow, F.W.E. and Jackson, P.F.S., 1974. Determination of trace and ultratrace quantities of rare-earth elements by ion exchange chromatography-mass spectrography. *Anal. Chem.*, 46: 1481—1486.

Suen, C.J., 1978. *Geochemistry of peridotites and associated mafic rocks, Ronda ultramafic complex, Spain.* Ph.D. Thesis, Massachusetts Institute of Technology, Cambridge, Mass., 283 pp.

Suen, C.J. and Frey, F.A., 1977. Rare-earth element geochemistry of the Ronda high-temperature peridotite. *EOS, Trans. Am. Geophys. Union*, 58: 532—533.

Suen, C.J. and Frey, F.A., 1978. Origin of mafic layers in alpine peridotite bodies as indicated by the geochemistry of the Ronda Massif, Spain. *EOS, Trans. Am. Geophys. Union*, 59: 401.

Suen, C.J., Frey, F.A. and Malpas, J., 1979. Bay of Islands Ophiolite Suite, Newfoundland: petrologic and geochemical characteristics with emphasis on rare earth element geochemistry. *Earth Planet. Sci. Lett.*, 45: 337—348.

Sun, S.S. and Nesbitt, R.W., 1978. Petrogenesis of Archean ultrabasic and basic volcanics: evidence from rare earth elements. *Contrib. Mineral. Petrol.*, 65: 301—325.

Sun, S.S., Nesbitt, R.W. and Sharaskin, A.Y., 1979. Geochemical characteristics of mid-ocean ridge basalts. *Earth Planet. Sci. Lett.*, 44: 119—138.

Tanaka, T. and Aoki, K., 1981. Petrogenetic implications of REE and Ba data on mafic and ultramafic inclusions from Itinome-gata, Japan. *J. Geol.*, 89: 369—390.

Varne, R. and Graham, A.L., 1971. Rare earth abundances in hornblende and clinopyroxene of a hornblende lherzolite xenolith: implications for upper mantle fractionation processes. *Earth Planet. Sci. Lett.*, 13: 11—18.

Wass, S.Y. and Rogers, N.W., 1980. Mantle metasomatism-precursor to continental alkaline volcanism. *Geochim. Cosmochim. Acta*, 44: 1811—1823.

Wass, S.Y., Henderson, P. and Elliott, C.J., 1980. Chemical heterogeneity and metasomatism in the upper mantle: evidence from rare earth and other elements in apatite-rich xenoliths in basaltic rocks from eastern Australia. *Philos. Trans. R. Soc. London, Ser. A*, 297: 333—346.

Wendlandt, R.F. and Harrison, W.J., 1979. Rare earth partitioning between immiscible carbonate and silicate liquids and CO_2 vapor: results and implications for the formation of light rare earth-enriched rocks. *Contrib. Mineral. Petrol.*, 69: 409—419.

White, A.J.R., Chappell, B.W. and Jakeš, P., 1972. Coexisting clinopyroxene, garnet and amphibole from an "eclogite", Kakanui, New Zealand. *Contrib. Mineral. Petrol.*, 34: 185—191.

Wilshire, H.G. and Pike, J.E.N., 1975. Upper-mantle diapirism: evidence from analogous features in alpine peridotite and ultramafic inclusions in basalt. *Geology*, 3: 467—470.

Wilshire, H.G. and Shervais, J.W., 1975. Al-augite and Cr-diopside ultramafic xenoliths in basaltic rocks from western United States. *Phys. Chem. Earth*, 9: 257—272.

Wilshire, H.G., Pike, J.N., Meyer, C.E. and Schwarzman, E.C., 1980. Amphibole-rich veins in lherzolite xenoliths, Dish Hill and Deadman Lake, California. *Am. J. Sci.*, 280-A: 576—593.

Wood, D.A., 1979. A variably veined suboceanic upper mantle — genetic significance for mid-oceanic ridge basalts from geochemical evidence. *Geology*, 7: 499—503.

Wyllie, P.J., 1967. Ultramafic and ultrabasic rocks. In: P.J. Wyllie (Editor), *Ultramafic and Related Rocks.* John Wiley and Sons, New York, N.Y., pp. 1—7.

Zindler, A., and Jagoutz, E., 1980. Isotope and trace element systematics in mantle-derived peridotite nodules from San Carlos. *EOS, Trans. Am. Geophys. Union*, 61(197): 374.

Chapter 6

THE RARE EARTH ELEMENT CHARACTERISTICS OF IGNEOUS ROCKS FROM THE OCEAN BASINS

A.D. SAUNDERS

6.1. Introduction

It is well established that the ocean basins are underlain by a thin crust of igneous rocks, predominantly of basaltic composition, formed almost entirely at the mid-ocean ridge system. Normally, the mid-ocean ridges lie at near-abyssal depths, but some segments are elevated to form extensive submarine plateaux (for example, the Azores platform in the North Atlantic Ocean) or even islands (for example, Iceland). In addition to the igneous activity of the essentially linear constructive plate margins, however, more localized magmatism may form volcanic edifices rising from the seabed. Such activity — broadly denoted as "off-axis" or "intra-plate" because it usually occurs away from the constructive plate margins — may be expressed in several diverse forms. On the one hand, intra-plate activity may result in the formation of seamounts or islands, often in parallel linear chains, such as the Hawaiian, Cook and Tuamotou archipelagos in the Pacific Ocean. Alternatively, the intra-plate volcanism can produce extensive submarine lava fields as were recently discovered by drilling in the Nauru Basin.

With so many diverse locations and types of volcanic activity it is not surprising that the magmatic products are also varied, ranging from the tholeiitic basalts typical of ocean ridge activity, to the highly undersaturated magmas and their differentiates found on many ocean islands. Indeed, it is now widely recognized that mid-ocean ridge basalts as a group are chemically heterogeneous. This compositional variability in oceanic basalts has important implications for our understanding of the geochemical evolution of the Earth. All basalts have inherited, to a lesser or greater degree, the chemical characteristics of their mantle source region. The higher the degree of mantle melting, and the more rapidly the melts rise to the surface, thus minimizing the effects of crystal fractionation, the more representative these characteristics are. The importance of oceanic basalts thus becomes clear. Since mid-ocean ridge and oceanic island tholeiites probably represent at least 15% source melting (Green and Ringwood, 1967) and because the melts are not subjected to contamination by continental crust, oceanic basalts are good indicators of the chemistry of their source regions. In effect they provide a window through which we can observe chemical processes within the sub-oceanic mantle. This is, however, a window obscured by a series of events which modifies the composition of the magma from its point of origin to the time that it is sampled as a crystalline rock by the petrologist. These events include fractionation (partial melting and fractional crystallization), flow differentiation, volatile loss and low-temperature alteration; each event must be quantified, evaluated and its effect eliminated before it is possible to see through the window.

To this end, petrologists have utilized a variety of techniques, particularly high-precision analysis of trace elements and radiogenic isotopes. The REE have played an important role, not only because of their coherent behaviour during fractionation events, but also because of their apparent stability during low-temperature alteration of sub-marine basalts. This review will attempt to describe and evaluate the REE geochemistry of oceanic igneous rocks from both the mid-ocean ridge and the intra-plate tectonic setting, within a framework which involves other trace element geochemical, isotope geochemical and experimental petrological data.

6.2. History of research

Prior to the early 1960's, the ocean islands were the only parts of the oceanic crust which had undergone substantial petrological study. The rocks flooring the abyssal plains were largely an unknown quantity until samples were dredged from along sections of the East Pacific Rise (EPR) (e.g., C.G. Engel and Engel, 1963), the Mid-Atlantic Ridge (MAR) (A.E.J. Engel and Engel, 1964; Muir et al., 1966), and the Indian Ocean Ridge (IOR) (C.G. Engel et al., 1965). Subsequent geochemical analysis soon revealed the chemical distinction between abyssal basalts (or low-K tholeiites) dredged from the mid-ocean ridge system, and the alkaline basalts and their differentiates recovered from seamounts and ocean islands. Abyssal basalts were found to have low concentrations of many incompatible elements (K, Rb, Ba, Nb, Th, Pb, U and Zr), low $^{87}Sr/^{86}Sr$ ratios, and to exhibit relative LREE depletion (Frey and Haskin, 1964; A.E.J. Engel et al., 1965; Tatsumoto et al., 1965). Samples from ocean islands and seamounts, on the other hand, usually follow alkaline rather than tholeiitic differentiation trends, contain significantly higher abundances of many incompatible elements, and have higher $^{87}Sr/^{86}Sr$ and $(La/Yb)_{cn}$ ratios than abyssal basalts (e.g., Frey and Haskin, 1964; A.E.J. Engel et al., 1965). It was soon realized that this scenario is a gross oversimplification; tholeiitic rocks occur on several ocean islands (e.g. Iceland, Hawaii), and Muir et al. (1964) described nepheline-normative basalts from latitude 45°N on the MAR, a topographically normal segment of ridge. Nevertheless, this broad two-fold distribution of rock types continued to influence petrological debate for several years.

Initially, attempts were made to relate the chemistry of alkaline basalts and abyssal tholeiites by simple crystal fractionation (A.E.J. Engel et al., 1965) but Gast (1968a) demonstrated that the magma types are derived from chemically distinct mantle sources. Gast indicated that the chemical characteristics of abyssal basalts probably reflect those of their source and, in order to explain the incompatible element depletion, suggested that this source must have undergone previous episodes of partial fusion and melt removal. Indeed, Gast (1965, 1968a, b) proposed that the very low Cs, Rb and K contents of abyssal basalts indicate that the Earth as a whole is depleted in these volatile elements relative to chondrites. The source of abyssal basalts, as defined by Gast, is therefore considered as being "doubly depleted" — once prior to Earth formation, and again during melt extraction at some earlier time(s) during Earth history.

Subsequent studies have been considerably enhanced by the advent of the Deep Sea Drilling Project, and the general acceptance of sea-floor spreading. Systematic, co-ordinated dredging, drilling and submersible operations, particularly in the North Atlantic, have led to the realization that different segments of ridge are erupting chemically distinct magma types. There is an apparent, broad correlation between the degree of incompatible element and radiogenic isotope enrichment, topography and heat flow which is difficult to explain by simple, low-pressure fractional crystallization (e.g., Schilling, 1975a). Whereas some differences between magma types may be explained by disequilibrium partial melting, dynamic partial melting and so forth, it is now generally accepted that mantle heterogeneity is the principal controlling factor. It is the nature of the heterogeneity which is now disputed, some workers invoking hot, enriched mantle plumes beneath the elevated "anomalous" ridge segments (e.g., Iceland, Azores platform) and triple junctions (e.g., Afar region, Ethiopia) (Schilling, 1973a, b; 1975a, b; Hart et al., 1973; White et al., 1976; White and Schilling, 1978), while others prefer the concept of variable-scale regional heterogeneities which reflect the complex history of the Earth's mantle (Langmuir et al., 1978; Tarney et al., 1979, 1980; Wood et al., 1979a, b; Dupré and Allègre, 1980; Bougault et al., 1979; 1980). These points will be returned to in the final sections of this review.

Before proceeding with an evaluation of the chemical characteristics of oceanic igneous rocks, it is necessary to briefly introduce the terminology used in trace element systematics. The term *incompatible* will be used in the standard way, i.e., a trace element having a bulk crystal/liquid coefficient (D) significantly less than 1. Of course, the D value of any given element may vary according to the physical and chemical conditions of the system in which it occurs; Eu for example has a high D value when the melt is in equilibrium with feldspar. It will be apparent, therefore that some elements will be more incompatible than others, again depending on the P-T-X condition of the system. Thus, we may define *highly incompatible* elements as those elements with estimated D values less than 0.01. In ocean ridge basalt petrogenesis, these elements include Cs, Rb, K, U, Th, Ba, Ta, Nb, La and Ce (see Bougault et al., 1979; Wood et al., 1979b). Other less incompatible elements ($D \lesssim 0.1$) are the remaining REE, Y, Ti, Zr, Hf, and P, and include, if feldspar is not a major fractionating phase, Sr and Eu. Measured ratios between two highly incompatible elements (e.g. La/Ce, La/Ta) *or* between two incompatible elements with similar chemical properties (Tb/Y, Zr/Hf, Ta/Nb) are thought not to be significantly affected by known fractionation processes occurring during basalt genesis, and are thought to reflect the ratios of their source (see discussion by Bougault et al., 1980). The REE are of particular interest in this respect, because it is possible to measure not only the ratios between adjacent REE (La/Ce, Sm/Nd, Yb/Lu), but also the ratios between REE which have slightly different ionic properties and distribution coefficients (Ce/Yb). The former allow assessment of the *source* characteristics (cf. La/Ta, La/Th ratios) whereas the latter ratios are more readily changed during fractionation processes and allow assessment of magma genesis and evolution.

6.3. Ocean ridge basalts

The predominant magma erupted at mid-ocean ridges is a tholeiitic basalt which, on the basis of mineralogy and major element chemistry alone, appears remarkably uniform in composition. Detailed studies have revealed, however, that such uniformity does not extend to the trace elements and isotope ratios; indeed, mid-ocean ridge basalts are extremely variable in their composition, and these variations reflect fundamental differences in mantle source compositions.

Chondrite-normalized REE distribution patterns of representative mid-ocean ridge basalts serve to illustrate some of these variations in trace element chemistry (Figs 6.1—6.3). All of the REE patterns plotted in Figs. 6.1—6.3 represent tholeiitic (or very mildly nepheline-normative) basalts erupted at constructive plate margins, but it is evident that the shape of the patterns varies considerably, with $(Ce/Yb)_{cn}$ ratio ranging from less than 0.5 to more than 10.

It is, however, convenient to place the REE patterns into three groups on the basis of the chemistry and of the site of eruption of the basalts; this has been done in Figs. 6.1—6.3. The reasoning behind this classification is that it is possible, using topographic, heat flow, gravity and geochemical criteria, to subdivide the mid-ocean ridge system into three different types of ridge segment: "normal", "anomalous", and "transitional" (cf. Schilling, 1975a).

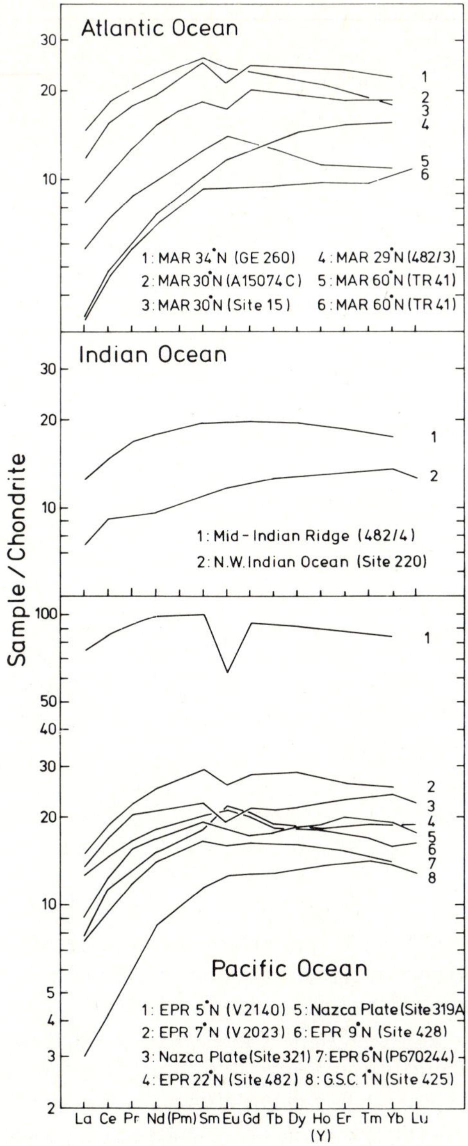

Fig. 6.1. Chondrite-normalized REE distributions in tholeiitic N-type ocean ridge basalts. Data sources: *Atlantic Ocean* — GE-260, A-1507-4C from Kay et al. (1970); Site 15, TR41 from Schilling (1975a); 482/3 from Sun et al. (1979). *Indian Ocean* — 482/4 from Sun et al. (1979); Site 220 from Frey et al. (1980). *Pacific Ocean* — V21-40, V20-23, P6702-44 from Kay et al. (1970); Site 321, Site 319A from Thompson et al. (1976); Site 482 from Saunders (1983); Site 428 and Site 425 from Srivastava et al. (1980). V21-40 is a high-silica glass (Kay et al., 1970).

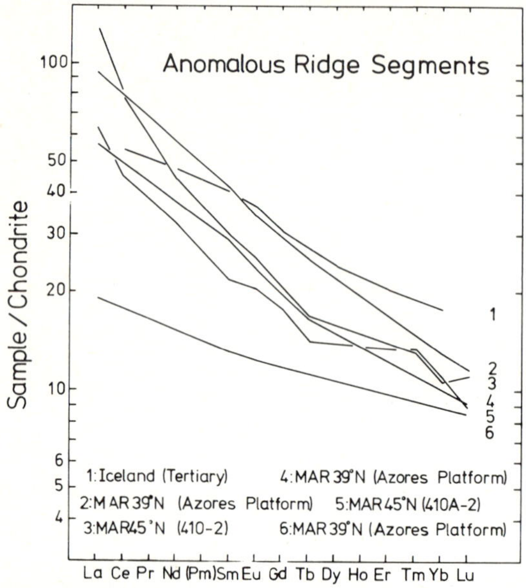

Fig. 6.2. Chondrite-normalized REE distributions in E-type basalts from "anomalous" ridge segments. Data sources: Iceland (Tertiary) from O'Nions et al. (1976); Azores platform samples from Schilling (1975b); MAR 45°N from Wood et al. (1979b). Sample 410-2 contains minor normative nepheline; the remaining samples are hypersthene normative.

"Normal" ridge segments

"Normal" ridge segments (e.g., MAR 20—30°N, 50—60°N; most sampled portions of the EPR) are characterized by normal ridge elevations (i.e., water depths greater than 3500 m), well-developed magnetic lineations, and typical oceanic crustal thickness and structure (Vogt et al., 1969). Along normal ridge segments, it is assumed that the rising mantle peridotite undergoes adiabatic decompression, variable but extensive partial melting (15—30%: Green and Ringwood, 1967; Green 1970, 1971; Sun et al., 1979) and essentially passive eruption in response to crustal extension.

Basalts erupted at normal ridge segments are by far the most abundant type of ocean ridge basalt. They are invariably LREE-depleted, whether recovered from the MAR, EPR or IOR (Fig. 6.1), a feature which reflects the very low absolute abundances of other highly incompatible elements (Kay et al., 1970; Schilling, 1971; Frey et al., 1974; Kay and Hubbard, 1978; Wood et al., 1979b). In addition, they have low $^{87}Sr/^{86}Sr$ ratios and high $^{143}Nd/^{144}Nd$ ratios (Hart, 1971; DePaolo and Wasserburg, 1976; O'Nions

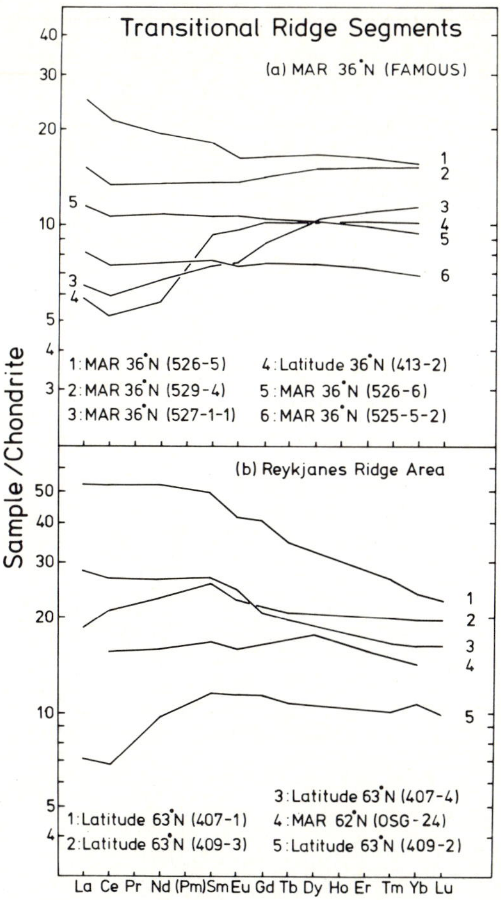

Fig. 6.3. Chondrite-normalized REE distributions in tholeiitic basalts from transitional ridge segments. Data sources: FAMOUS area from Langmuir et al. (1977), apart from sample 413-2 which is from Wood et al. (1979b); Reykjanes Ridge area from Wood et al. (1979b), apart from sample OSG-34 which is from O'Nions et al. (1976). Compare the FAMOUS data with Fig. 6.2, and the Reykjanes Ridge data with Fig. 6.5.

et al., 1977; Cohen et al., 1980), indicating derivation from a mantle source which has suffered a previous melt extraction (Gast, 1968a; O'Nions et al., 1979). The high $^{143}Nd/^{144}Nd$ ratios (higher than bulk Earth values) necessitate derivation of the basalts from a source which has been LREE-depleted (i.e., having high Sm/Nd ratios) for several billion years on a time-integrated basis.

Basalts erupted at normal ridge segments exhibit essentially constant highly incompatible element ratios (La/Ta ~ 18; Th/Ta ~ 0.7; Ba/La ~ 3;

Th/Hf ~ 0.05) (Fig. 6.4), implying derivation from a similar, world-wide source. They have been variously termed low-K tholeiites (A.E.J. Engel et al., 1965), "mid-ocean ridge basalts" or "MORB" (Hart et al., 1972) and, more specifically, "normal", or "N-type MORB" (Sun et al., 1979; Tarney et al., 1980). The latter term will be used in this account.

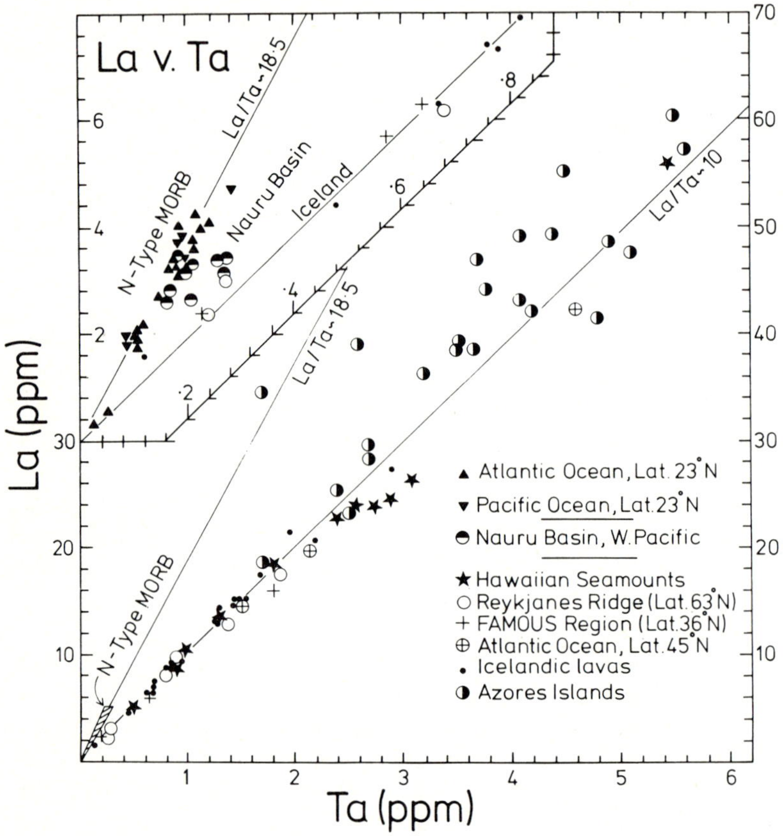

Fig. 6.4. La versus Ta relationships in igneous rocks from the ocean basins. Note the two distinct trends exhibited by samples from normal ridge segments (N-type MORB) (inset) and anomalous and transitional ridge segments (E- and T-type MORB). Data from islands and seamounts tend to fall on the E-type MORB trend (La/Ta ~ 10) although there is a tendency for La/Ta ratio to increase at higher SiO_2 levels, possibly due to removal of ilmenite (e.g., Wood, 1978). Note also the very low values of La and Ta in N-type MORB, and in basalts from the Nauru Basin. Data sources: N-type MORB, Atlantic Ocean — Bougault et al. (1978); N-type MORB, Pacific Ocean — Saunders (1983); Nauru Basin (ocean plateau tholeiites) — Seifert (1981); Hawaiian seamounts — Cambon et al. (1980); T- and E-type MORB from the Reykjanes Ridge FAMOUS region, and lat. 45°N — Wood et al. (1979b); Iceland — Wood et al. (1979a); Azores Islands — White et al. (1979).

"Anomalous" ridge segments

"Anomalous" (or "plume": Schilling, 1973a, b; 1975a) ridge segments are usually characterized by elevated ridge topography and often lie over large plateaux (e.g., Iceland, Azores platform, Afar region) with positive gravity anomalies (Anderson et al., 1973). The crustal structure is markedly thickened, and the geothermal gradient is higher than at normal ridge segments (Talwani et al., 1971), consistent with more extensive eruption of basaltic magma. Tholeiitic basalts are again the predominant rock type, but alkaline basalts also occur. However, both basalt types show far greater chemical variation than those erupted at normal ridge segments, with variable enrichment of the LREE (Fig. 6.2); $(Ce/Yb)_{cn}$ may be as high as 10. Other highly incompatible elements, particularly Nb and Ta, also have much higher abundances than in N-type MORB; thus, La/Ta (Fig. 6.4) and La/Nb ratios are much lower than in N-type MORB which, together with the more radiogenic Sr, Pb and Nd ratios, has been used as evidence to demonstrate that the two basalt types are derived from distinct mantle sources (Hart et al., 1973; Schilling, 1975a; White et al., 1976; O'Nions et al., 1977; Wood et al., 1979b; Bougault et al., 1979, 1980). The isotopic ratios are, however, still much lower (higher for Nd) than calculated bulk Earth values, suggesting that these basalts, as well as those from normal ridge segments, are derived from mantle sources that have suffered previous melt extraction (Gast, 1968a; O'Nions et al., 1979).

Basalts recovered from "anomalous" ridge segments have been variously termed enriched (or "E-type") MORB (Tarney et al., 1980), because of their chemical characteristics, or plume- or "P-type" basalts (Sun et al., 1979) following Schilling's (1973a, b) suggestion that mantle plumes or blobs are responsible for the anomalous heat flow, topography and chemistry of these regions. It is worth noting, however, that E-type MORB have been recovered from MAR 45°N, a region which does not have a pronounced topographic high (see Tarney et al., 1979; Wood et al., 1979b; and Fig. 6.2). This implies that the correlation between topography and geochemistry is not sharply defined.

Interestingly, E-type MORB have not been recovered from along the EPR, where all the analysed basalts are depleted, N-type MORB, although LREE-enriched basalts are erupted in some fracture zones (e.g., Siqueiros fracture zone: Batiza et al., 1977; Batiza and Johnson, 1980).

"Transitional" ridge segments

"Transitional" ridge segments represent intermediate zones — topographical, geophysical and geochemical — between normal and anomalous ridge segments. Basalts from transitional ridge segments — "T-type" MORB — may be either LREE-depleted or enriched (Fig. 6.3), although highly incompatible element ratios (La/Ta, Nb/Zr) and isotope ratios are clearly different from those in N-type MORB (Fig. 6.4). For example, T-type MORB from the Reykjanes Ridge, a transitional area between the MAR and Iceland, have a La/Ta ratio of 10, which is lower than in N-type MORB (~18) (Wood et al., 1979b). Again, note that T-type MORB from the FAMOUS area (MAR 36°N, on the southern slope of the Azores platform) exhibit both LREE depletion and enrichment, although $(La/Ce)_{cn}$ ratios are consistently greater than 1 (Langmuir et al., 1977; and Fig. 6.3); $(La/Ce)_{cn}$ ratios in N-type MORB are less than 1.

Many of the above observations were made possible by intensive dredging operations along the mid-ocean ridge system, particularly in the North Atlantic (e.g., Schilling, 1973b; White et al., 1976; Bougault and Treuil, 1980). However, deep-sea drilling in areas inaccessible to dredging has recovered samples from transects along mantle flow lines, thus allowing an assessment of variations in basalt chemistry with respect to *time*. Data are now available from several transects in the North Atlantic (e.g. Legs 37, 45, 46, 49, 51, 52, 53 of the Deep Sea Drilling Project) and across the EPR (Legs 34, 64 and 65), and they demonstrate that the fundamental variations in chemistry observed in basalts recovered from along the ridge system (i.e., those variations which may be due to source heterogeneity) have been maintained for tens of millions of years. In other words, each ridge segment is erupting its own distinctive basalt type, and has been for a long period of time (Tarney et al., 1980). This in turn implies that source heterogeneities are large-scale phenomena.

From the foregoing it is evident that the isotope data effectively preclude the derivation of N- and E-type MORB from a common primary magma, or indeed from a common mantle source. Before discussing these variations in source chemistry in detail, however, it is necessary to review the chemistry of ocean island igneous rocks because in several respects ocean island basalts resemble, or indeed are more enriched versions of, E-type MORB.

6.4. Ocean islands

Unlike the mid-ocean ridge system, where tholeiitic basalts are the predominant rock type, ocean islands comprise tholeiitic, alkalic and even highly undersaturated magma series. The relative abundances of saturated and undersaturated magma types varies from island to island: tholeiitic activity is abundant on Iceland, for example, whereas it is rare on the Azores Islands. As well as the broad variations in magma type between different provinces, however, there are important, subtle variations in trace element and isotope geochemistry. Thus, although ocean island basalts — tholeiitic and alkalic — are *generally* enriched in the highly incompatible elements (particularly Nb, Ta, LREE, Th, U, Cs, Rb, Ba, K) and resemble E-type MORB, ratios of these elements, and of the isotopes of Sr, Pb and Nd, vary between different islands. Indeed, studies of the larger island groups (Hawaii, Azores) have demonstrated isotopic and trace element variations between suites from the same island (Peterman and Hedge, 1971; Sun and Hanson, 1975; Hawkesworth et al., 1979; Sun, 1980; Wood et al., 1981).

In the following condensed account, attention will concentrate on five island groups which have been reasonably well studied and which are representative of the various tectonic settings of ocean islands: Iceland, astride the MAR; the Azores Islands, which represent intra-plate activity but which rise from a broad plateau intersecting the MAR; Reunion and Gough Islands, both representing intra-plate activity in the Indian and South Atlantic Oceans, respectively; and the Hawaiian Islands, perhaps the most famous example of intra-plate activity. Iceland is in fact an elevated portion of the MAR but is included in this section because it forms an effective link between mid-ocean ridge tholeiitic activity and the magmatism characteristic of most intra-plate activity.

Iceland

Iceland has been intensively studied over the last decade, with particular emphasis on the regional and temporal variation in magma composition and the nature of the underlying mantle (O'Nions and Grönvold, 1973; O'Nions and Pankhurst, 1973; Schilling, 1973b; Hart et al., 1973; O'Hara, 1973, 1975; Sigvaldasson, 1974; Sigvaldasson et al., 1974; O'Nions et al., 1976; Wood, 1978, Wood et al., 1979a). The presently active rift zones are erupting a wider variety of rock types — tholeiites, olivine tholeiites and alkali olivine basalts — than are usually found along the Reykjanes Ridge and MAR, although tholeiitic magmas still predominate.

Icelandic basalts have variable REE distribution patterns (Fig. 6.5). Tholeiites from the southern end of the Reykjanes Ridge have $(Ce/Yb)_{cn}$ and $(Ce/Sm)_{cn}$ ratios less than 1, whereas tholeiites erupted at fissures along the Reykjanes Peninsula and many other areas on Iceland have flat to LREE-enriched patterns (Schilling, 1973b; O'Nions et al., 1976). Absolute abundances of Fe, Ti, K and P, and $^{87}Sr/^{86}Sr$ ratios also increase with Ce/Sm and Ce/Yb ratios northwards along the Reykjanes Ridge (Schilling, 1973b; Hart et al., 1973), although O'Nions et al. (1976) have demonstrated that the strong positive correlation between $^{87}Sr/^{86}Sr$ (0.7029—0.7035) and $(Ce/Yb)_{cn}$ ratios (0.3—3.4) is a feature of Tertiary to Recent tholeiites collected from many areas of Iceland. Alkali basalts tend towards high $(Ce/Yb)_{cn}$ ratios — over 7 — but form a separate trend of lower slope on the $^{87}Sr/^{86}Sr$ versus $(Ce/Yb)_{cn}$ correlation diagram. The strong co-variance between degree of LREE-enrichment and $^{87}Sr/^{86}Sr$ in Icelandic tholeiites cannot be adequately explained by simple partial melting of a homogeneous source,

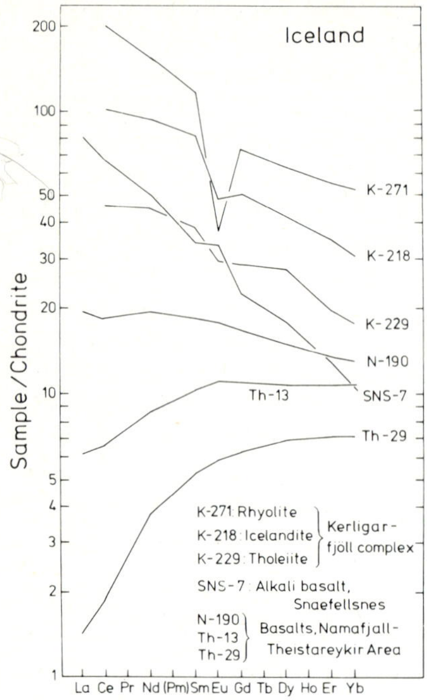

Fig. 6.5. Chondrite-normalized REE patterns for Iceland igneous rocks. Data sources: Kerlingarfjöll (Pleistocene/Recent) — O'Nions and Grönvold (1973); Snaefellsnes (post-glacial), Theistareykir (Th-13 and Th-29) and Namafjall (N-190) (post-glacial) — O'Nions et al. (1976).

nor by subsequent fractionation or contamination processes; rather, it must reflect the geochemical characteristics of either a heterogeneous source or at least the primary magmas supplying the Icelandic basalts.

However, some variations in $(Ce/Yb)_{cn}$ ratio observed in suites of basalts with constant $^{87}Sr/^{86}Sr$ ratio may be due to partial melting processes. Shield-building volcanism often supersedes fissure eruptions in Iceland, and olivine tholeiites and picritic basalts erupted at these centres are often LREE-depleted, or have flat chondrite-normalized REE patterns (Fig. 6.5; O'Nions et al., 1976; Wood, 1978). Wood (1979) has proposed that these variations may be caused by dynamic partial melting processes (Langmuir et al., 1977), the LREE-depleted shield tholeiites and picrites representing later remelts of the mantle source which produced the fissure basalts. Indeed, as will be shown in the next section, it is theoretically possible to explain all the observed variations in $(Ce/Yb)_{cn}$ ratio in Icelandic tholeiites and alkali basalts by dynamic partial melting of a single homogeneous source, but of course the isotopic considerations preclude this possibility.

Fractional crystallization has been proposed to account for the abundant silicic differentiates found on Iceland (Carmichael, 1964; Walker, 1966;

Wood, 1978). For example, the REE patterns and abundances of intermediate and rhyolitic volcanic rocks in eastern Iceland are consistent with derivation from a LREE-enriched basaltic parent by fractional crystallization (Wood, 1978). O'Nions and Grönvold (1973), however, proposed that the rhyolites found in the Torfajökull complex were derived from partially melted hydrous gabbroic layer 3 crust of Tertiary age. Such a mechanism appears necessary to expalin the observed increase in $^{87}Sr/^{86}Sr$ ratio in the rhyolites.

Azores Islands

The nine islands of the Azores archipelago rise from the Azores Plateau, an elevated section of the Atlantic floor which intersects the MAR at about 39°N. The islands are composed predominantly of alkali basalt and its differentation products: hawaiites, trachytes and, less frequently, comendites and pantellerites, but there is considerable inter-island variation in the relative abundances and chemistry of these eruptive products (Schmincke, 1973; Schmincke and Weibel, 1973; Flower et al., 1976; White et al., 1976; 1979; Marriner et al., 1982).

Most analysed samples of basalts from the Azores Islands are strongly LREE-enriched (Fig. 6.6), with $(Ce/Yb)_{cn}$ ratio up to 14 (e.g., White et al., 1979; Marriner et al., 1982), and with strong enrichment of the highly incompatible elements (particularly K, Rb, Ba, Th, Nb, Ta) (Wood et al., 1981; Marriner et al., 1982) — more so than is observed in the Icelandic rocks (Fig. 6.4). Basalts and hawaiites from the same eruptive centre usually have similar chondrite-normalized REE patterns, despite variations in absolute REE contents, but systematic differences in the patterns are observed when samples from different islands are compared. For instance, samples from the islands of Corvo and Flores have the highest Ce/Yb ratios, whereas Saõ Miguel lavas tend to have higher REE abundances (White et al., 1979). Fractionation of alkali feldspars in lavas from Agua de Pau volcano, Saõ Miguel has produced negative Eu anomalies in trachytes, and complementary positive anomalies in feldspar—phyric hawaiites (Fig. 6.6; Marriner et al., 1982); although the Ce/Yb ratio remains approximately constant throughout the series.

In several respects — degrees of LREE-enrichment, incompatible element ratios, and isotopic ratios — the alkali basalts of the Azores Islands resemble the enriched mid-ocean ridge tholeiites recovered from the MAR/Azores Plateau intersection (Bougault and Treuil, 1980; White et al., 1979), although the Azores basalts generally exhibit higher absolute abundances of the incompatible elements. However, isotope ratios (particularly $^{87}Sr/^{86}Sr$) and incompatible element ratios (e.g., Nb/Zr, Ba/Zr) vary between the different islands, implying variation in the chemistry of the primary magmas, and possibly the source beneath the Azores Province (White et al., 1979; Hawkesworth et al., 1979; Wood et al., 1981).

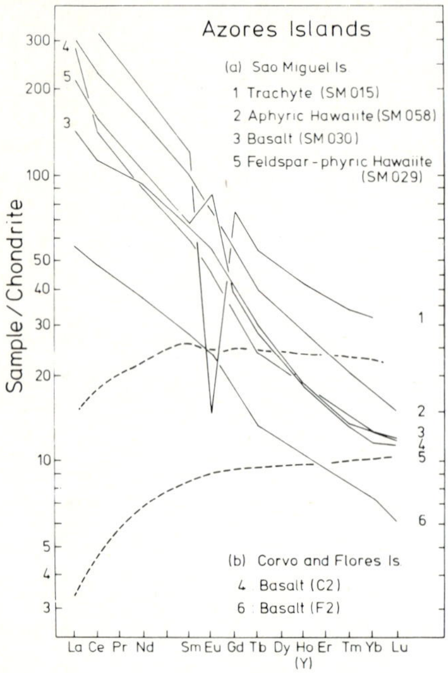

Fig. 6.6. Chondrite-normalized REE distribution in igneous rocks from the Azores Islands, Atlantic Ocean. Data sources: São Miguel Island from Marriner et al. (1982, and personal communication); Corvo (C2) and Flores (F2) Islands from White et al. (1979). Compare with REE data on Azores platform tholeiites (Fig. 6.2). Broken lines delimit field of N-type MORB data in Fig. 6.1.

Reunion Island

Reunion Island, in the western Indian Ocean, comprises two volcanic centres. The older shield (Piton des Neiges) is composed of hypersthene-normative olivine basalt (2—0.5 Ma: McDougall, 1971) capped by a series of more silicic volcanics ranging in composition from hawaiite to trachyte (Upton and Wadsworth, 1965, 1966, 1972). The younger centre is currently at the basaltic shield-building stage. The volcanic rocks are, as with the Azores Islands, strongly LREE-enriched, with high levels of Th, U and the LREE, and $^{87}Sr/^{86}Sr$ ratios of between 0.7040 and 0.7046 (Zielinski, 1975; McDougall and Compston, 1965).

On the basis of their REE geochemistry, Zielinski (1975) divided the Piton des Neiges series into two groups (Fig. 6.7). Group 1 (basalt through to trachyte) has sub-parallel chondrite-normalized REE patterns ($(Ce/Yb)_{cn}$ ~ 6), increasing REE contents with fractionation, and development of negative Eu anomalies in the trachytes. These observations can be explained by fractional crystallization of a mantle-derived olivine basalt melt involving observed phenocryst phases (Zielinski, 1975). The three analysed rocks of

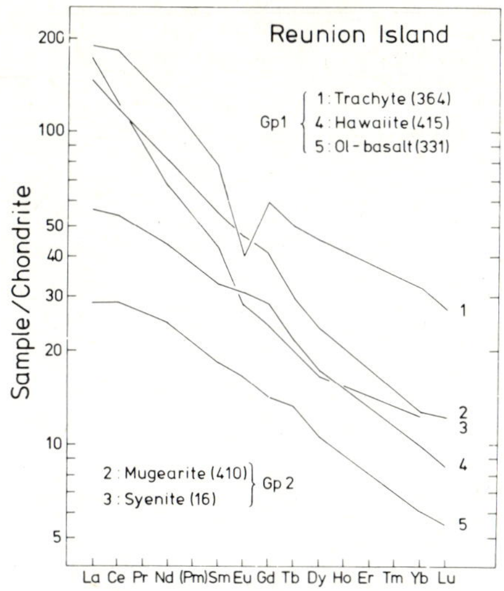

Fig. 6.7. Chondrite-normalized REE distribution in igneous rocks from Reunion Island, Indian Ocean. Data from Zielinski (1975).

group 2, however, have variable, steeper REE patterns ($(Ce/Yb)_{cn} > 10$), and there is no increase in total REE content with differentiation index. One of the rocks of group 2, a syenite, may be derived by partial melting of source rich in clinopyroxene and/or amphibole, whereas the other rocks (benmoreite and mugearite) may represent hybrids between group 1 magmas and the group 2 syenite (Zielinski, 1975).

Gough Island

Gough Island lies near the MAR in the South Atlantic Ocean. Approximately 50% of its surface consists of trachyte, apparently erupted in cyclical fashion within an alkali basalt—trachyte series (Le Maitre, 1960). The rocks are again LREE-enriched, with increasing total REE content and Ce/Yb ratio with increasing fractionation, and relative depletion of Eu, Sr and Ba in the trachytes (Zielinski and Frey, 1970). As with the Reunion Island lavas, the observed rock types could have been generated by fractional crystallization of parental alkali olivine basalts involving olivine, pyroxene, feldspar and apatite (Zielinski and Frey, 1970).

The Hawaiian Islands and Emperor Seamounts

The Hawaiian Islands and Emperor Seamounts form a continuous chain of volcanic edifices stretching from Hawaii to the Aleutian Trench, a distance of over 6000 km. Present-day volcanism is restricted to the southeast tip of the chain, with magmatism progressively decaying towards the northeast as the Pacific plate carries each volcanic pile away from the magma- or heat-source (McDougall, 1964). It would appear that the heat source (or hot spot) is stationary relative to the deep mantle (Morgan, 1972; Wilson, 1963), and hence originates from beneath the lithospheric plate, perhaps ascending as a vertical column or plume.

Each Hawaiian volcano tends to erupt lavas of distinct chemical composition during its history, although most volcanoes pass through similar evolutionary stages (Macdonald and Katsura, 1964; MacDonald, 1968). During the initial stage, voluminous outpourings of predominantly tholeiitic basalt build the main shields within only 1 Ma or so (Jackson et al., 1972). Following the shield-building stage, caldera collapse occurs. Next, both tholeiitic and alkalic basalts, together with their differentiates, are erupted. After a prolonged period of quiescence volcanism may resume, producing a minor capping of highly alkaline basic rocks — nephelinites, basanites and melilite basalts. Leg 55 of the Deep Sea Drilling Project recovered similar sequences of rocks from four of the Emperor Seamounts, so it is reasonable to assume that the style of magmatic activity at the Hawaiian "hot spot" has not changed significantly during the last 63 Ma (Jackson et al., 1980).

Representative chondrite-normalized REE patterns of Hawaiian and Emperor lavas are given in Fig. 6.8. Note the distinctive sigmoidal patterns, with low $(Ce/Sm)_{cn}$ ratios compared with Azorean and Reunion lavas, and increasing $(Ce/Yb)_{cn}$ ratio with increasing silica undersaturation. Schilling and Winchester (1969) argued that the alkalic series (basalts, hawaiites and mugearites) could be related by fractional crystallization, and that the nephelinites may be linked to the alkali olivine basalts by either fractional crystallization or partial melting processes. Indeed, ratios of highly incompatible elements (e.g., La/Ta and Th/Ta) are virtually constant throughout the alkalic *and* tholeiitic lava suites from the Emperor Seamounts, indicating derivation of both saturated and undersaturated magmas from similar mantle sources (Cambon et al., 1980) (Fig. 6.4).

6.5. Ocean plateau volcanism

Seismic surveys and drilling operations in the central and western Pacific have demonstrated the existence of thick, extensive basalt provinces which, on the basis of palaeontological and magnetic evidence, apparently formed in an intra-plate setting. Three examples of this type of activity may be cited: the Ontong-Java Plateau, the Manihiki Plateau (Winterer, 1975) and the Nauru Basin (Larson et al., 1981). The thick ($>$500 m) complex of lavas and sills discovered in the Nauru Basin during Leg 61 of the Deep Sea Drilling Project essentially comprises altered, quartz and olivine-hypersthene-normative tholeiites of at least Barremian age, overlying Pacific plate of

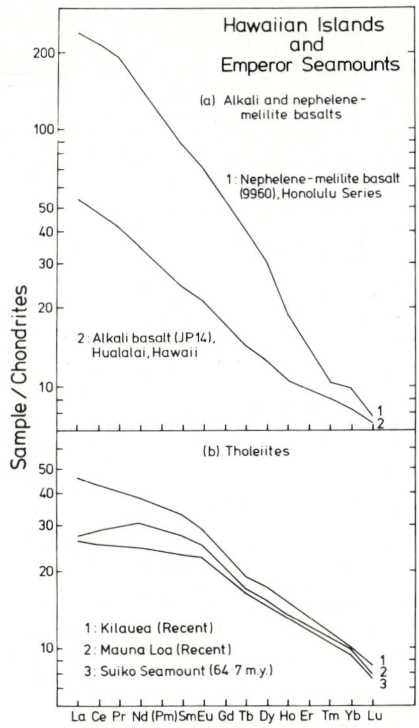

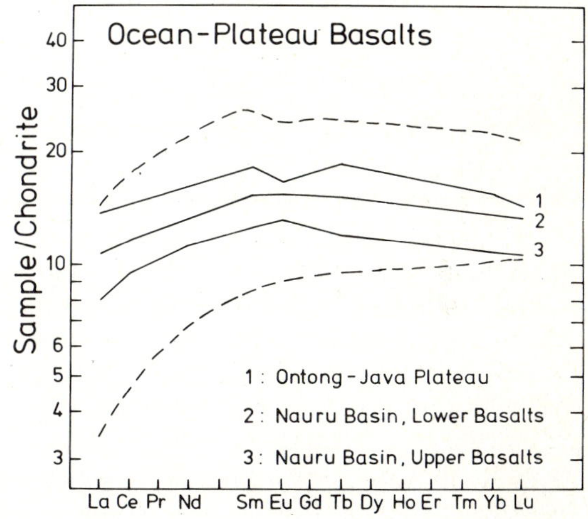

Fig. 6.8 (left). Chondrite-normalized REE distribution in igneous rocks from the Hawaiian Island—Emperor Seamount chain. Data sources: nepheline-melilite basalt and alkali basalt from Schilling and Winchester (1969); Kilauea (average of 10 samples) and Mauna Loa (average of 13 samples) tholeiites from Leeman et al. (1980); Suiko Seamount tholeiite from DSDP Site 433C (433C-39-5, 87—94) (Clague and Frey, 1980).

Fig. 6.9 (right). Chondrite-normalized REE patterns for basalts from the Nauru Basin (Seifert, 1981) and Ontong-Java Plateau (Batiza, 1981). Dashed lines indicate limits of N-type MORB in Fig. 6.1.

estimated Oxfordian age (Larson et al., 1981). These lavas have been termed ocean plateau tholeiites by Tokuyama and Batiza (1981).

In several respects the trace element geochemistry of ocean plateau tholeiites resembles that of N-type MORB: flat to LREE-depleted REE patterns (Fig. 6.9), La/Ta ratios of about 16 (Fig. 6.4), and low absolute abundance of Sr and Th (Seifert, 1981; Batiza, 1981). However, the predominance of quartz-normative varieties is atypical of N-type MORB. In addition, the Nauru Basin basalts have significantly higher $^{87}Sr/^{86}Sr$ ratios (0.7036—0.7040) than N-type MORB (Fujii et al., 1981) and slightly higher

$(La/Sm)_{cn}$ ratios (Batiza, 1981), although both of these observations could in part be due to the effects of secondary alteration. However, the Nauru Basin samples have consistently lower Zr/Nb ratios (less than 20) when compared with N-type MORB (generally greater than 20), and in this respect they more closely resemble Hawaiian tholeiites (A.D. Saunders, unpublished data). Clearly, further studies of ocean plateau tholeiites are required before detailed comparisons with other provinces can be made.

6.6. Factors controlling the distribution of REE in oceanic basalts

Three major factors control the distribution of REE in igneous rocks: the composition of the source; the degree and conditions of partial melting of the source; and fractional crystallization. Studies of isotopes and highly incompatible trace elements effectively preclude derivation of all oceanic basalts from a single, compositionally homogeneous source, and indeed it would appear that source composition exerts the dominant control on basalt composition — at least in terms of isotope and trace element contents. Nevertheless, before attempting to evaluate the chemical variability of the mantle, it is necessary to assess first the effect of partial melting and fractional crystallization processes.

Modelling of these processes has in fact really only been made possible by the properties of the REE (Chapter 4). They exhibit systematic, coherent behaviour in magmatic systems, the smaller ions of the HREE substituting more readily than the LREE into the crystal structures of most ferromagnesian minerals (Chapter 1). Equally important, however, is the large body of experimentally determined natural and synthetic crystal/liquid partition coefficient (D) data (similar data for other trace and minor elements, particularly Ti, P, Ta, Nb, Zr and Hf, are still sparse). Although the REE are less mobile than the alkaline and alkaline earth elements during alteration of basalt by seawater, it is realized that recent studies have demonstrated variable mobility of the REE in altered basalts and altered glassy selvages (Ludden and Thompson, 1979; Terrell et al., 1978). S.E. Humphris discusses REE mobility in Chapter 9.

Fractional crystallization

Most MORB have Fe/Mg ratios too high, and Ni and Cr contents too low, for them to represent primary melts in equilibrium with mantle compositions (O'Hara, 1968, Kay et al., 1970; Green et al., 1979). It is, therefore, generally assumed that the melt has undergone fractionation en route to the surface, probably in sub-ridge magma chambers, and that this low-pressure fractionation is controlled by precipitation of olivine + plagioclase, and at a later stage of olivine + plagioclase + clinopyroxene + iron oxides

(Miyashiro et al., 1970; Wood, 1978). Although the cumulate sequences expected to be formed from such a process are rarely sampled in the ocean basins, studies of ophiolite complexes do reveal that fractional crystallization must play an important role in MORB genesis (e.g., Elthon, 1979). In addition, silicic differentiates are particularly common on oceanic islands, possibly reflecting the formation of long-lived, sub-volcanic magma chambers.

Fractional crystallization may occur in *closed* or *open* magmatic systems. Closed-system fractionation describes the evolution of a single batch of cooling magma in an isolated chamber and may be modelled by the Rayleigh equation (Neumann et al., 1954; Gast, 1968a). Although closed-system fractional crystallization involving olivine, plagioclase and clinopyroxene increases the absolute abundance of the incompatible elements in the liquid phase, it does not significantly fractionate the REE; thus, the Ce/Yb and Ce/Sm ratios remain essentially constant within the basaltic range of low-pressure fractionation. This is illustrated by vector *A* in Fig. 6.10.

Open-system fractional crystallization (O'Hara, 1977; O'Hara and Mathews, 1981), which involves continual replenishment of a magma chamber with primitive basaltic melt and partial extraction of the fractionated liquid, has been proposed as a mechanism of fractionating the light from the heavy REE. However, Pankhurst (1977) has shown that under conditions of low pressure, significant fractionation between light and heavy REE occurs only when the fraction of liquid removed from the magma chamber is very small (less than 0.01). Indeed, as illustrated by vector *B* in Fig. 6.10, the trend is similar to that predicted by closed-system fractionation.

Low-pressure fractional crystallization, either in closed or open magma chambers can, therefore, provide an explanation for the increasing total REE abundances (with constant Ce/Sm and Ce/Yb ratios), observed in many consanguineous MORB suites. In addition, significant removal of plagioclase will produce the negative Eu anomalies often seen in evolved basalts and in silicic differentiates (Philpotts and Schnetzler, 1970) (e.g. Fig. 6.1).

Whereas low-pressure fractionation involving plagioclase, olivine and clinopyroxene is unable to alter Ce/Yb ratios significantly, high-pressure crystal fractionation involving garnet (eclogite fractionation: O'Hara, 1975) may do so. This is because garnet has high partition coefficients for the HREE ($\sim$4: Shimizu and Kushiro, 1975; and see Chapter 1). Indeed, by invoking eclogite fractionation in an open-system magma chamber — assuming for the moment that conditions in the mantle facilitate such a mechanism — it is theoretically possible to produce most of the variation in Ce/Yb ratios (line *C* in Fig. 6.10) observed in most MORB and ocean island basalts (Tarney et al., 1980). This mechanism cannot, however, account for the variation of Ce/Sm ratios in oceanic basalts, nor does it explain systematic variations in isotopic and highly incompatible element ratios (e.g., La/Ta, Zr/Nb, etc.) observed between N- and E-type MORB.

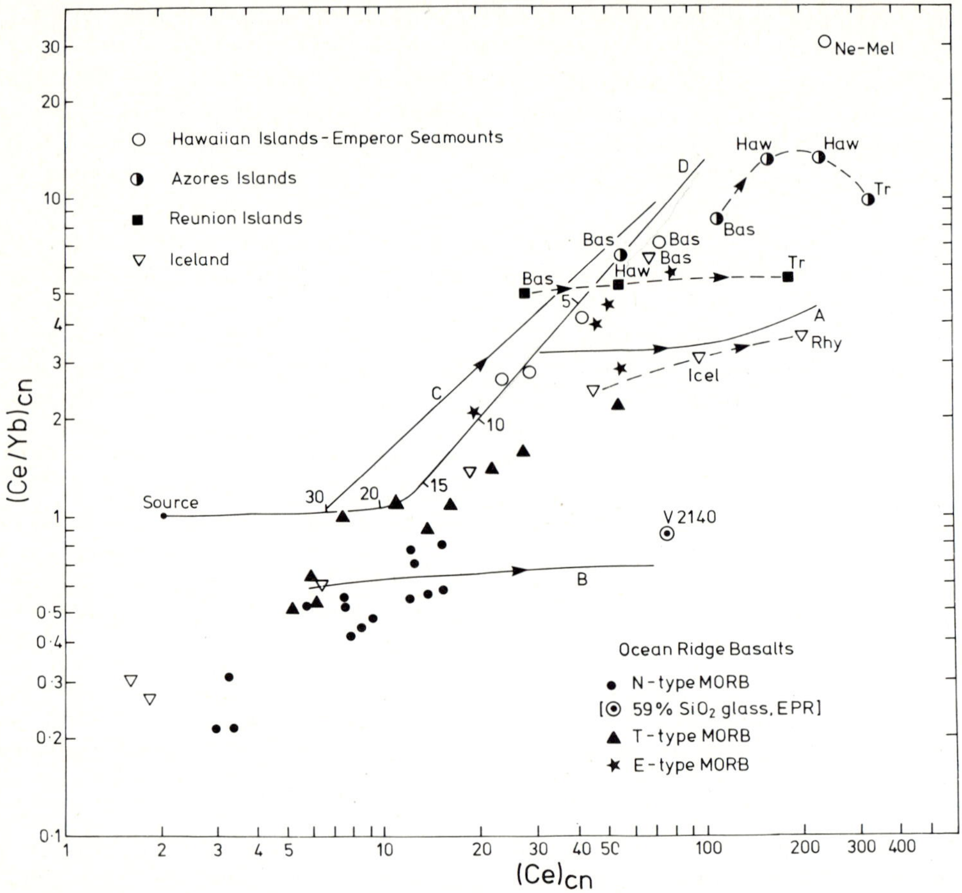

Fig. 6.10. Representative oceanic igneous rocks plotted on a log-log plot of $(Ce/Yb)_{cn}$ ratio versus Ce_{cn}. All the data used in this diagram are from Fig. 6.1—6.8. Fractionation trends, modified from Tarney et al. (1980), *A*: closed-system fractional crystallization (low pressure); *B*: open-system fractional crystallization (O'Hara, 1977) (low pressure); *C*: open-system fractional crystallization (high-pressure eclogite fractionation); *D*: equilibrium batch partial melting (garnet lherzolite source, 55% ol, 25% opx, 15% cpx, 5% garnet, melting mode of 10:20:40:30, respectively, and 2 times chondritic REE abundances). Numbers on batch melting curve indicate percentage melt. (Note that because this is a log-log plot, the vectors may be moved without affecting their shape. For example, curve *D* could be moved to originate from a source with $(Ce/Yb)_{cn} \sim 0.5$, and $Ce_{cn} \sim 1$.). Broken lines indicate rock types from common volcanic centres, but do not necessarily indicate consanguinuity. All samples are tholeiitic basalts except where indicated: Bas = alkali or alkali olivine basalt; Ice = icelandite; Haw = hawaiite; Rhy = rhyolite; Tr = trachyte; Ne-Mel: nepheline-melilite basalt. Full data sources are given in the captions to Figs. 6.1—6.8.

Equilibrium partial melting

Several models describing the behaviour of trace elements during partial melting are available (Gast, 1968a; Shaw, 1970; Arth, 1976), but the two most geologically acceptable models are (a) *batch melting* which involves continuous equilibrium of the liquid phase with the residual solid until removal of the melt from the solid (Shaw, 1970), and (b) *continuous melting*, which involves multi-stage melting of the source, but with a small fraction of liquid retained in the residue throughout and after melting (Langmuir et al., 1977).

A more complete evaluation of these models is presented in Chapter 4, and only the following salient conclusions are given here:

(1) Significant fractionation of the light from the heavy REE will occur only if garnet is a residual phase during melting (see curve *D*, Fig. 6.10).

(2) If melting exceeds 15% of the source, all the garnet is likely to be consumed, and hence the Ce/Yb ratio of the melt will approach that of the source. Since most ocean ridge tholeiites (N- and E-type MORB) represent at least 15% melting (Green and Ringwood, 1967; Sun et al., 1979), it is reasonable to assume that their REE patterns resemble those of their source, and hence reflect existing heterogeneities (Gast, 1968a; Schilling, 1975a).

(3) Very small degrees of melting of a garnet-bearing source will produce the high Ce/Yb ratios found in E-type MORB and in ocean island nephelinites, but only if the source has at least chondritic REE abundances (Gast, 1968a; Kay and Gast, 1973) or, more preferably, is LREE-enriched (Sun and Hanson, 1975; Frey et al., 1978).

(4) Continuous partial melting can produce melts with lower Ce/Yb ratios than their initial source. A combination of continuous melting and batch partial melting (dynamic melting) is capable of producing a wide range of Ce/Yb ratios in basalts derived from compositionally similar sources — as demonstrated by Langmuir et al. (1977) in their study of LREE-enriched and LREE-depleted basalts from the FAMOUS area (MAR 37°N). It is unlikely, however, that dynamic partial melting is responsible for all variations in MORB REE patterns. For example, most of the analysed basalts from the FAMOUS area study have consistently high La/Ce ratios — "dished" REE patterns — distinct from N-type MORB (see Figs. 6.1 and 6.3). This implies that whereas dynamic partial melting is capable of producing a wide range of La/Yb and Ce/Yb ratios, probably on a local scale, it cannot fractionate La and Ce. The same argument applies to other highly incompatible elements (e.g., La/Ta, La/Th ratios) and of course to the isotope ratios.

Mantle heterogeneity

From the previous section it is apparent that whilst REE abundance data facilitate semi-quantitative modelling of partial melting and fractional crystallization processes, they do not, if taken in isolation, provide *unequivocal* evidence of chemical heterogeneity in the source regions of oceanic basalts. Nevertheless, it is difficult to envisage how both LREE-enriched ($(La/Sm)_{cn} > 1$) and depleted ($(La/Sm)_{cn} < 1$) tholeiites could be derived from sources with similar REE contents, if geologically reasonable processes are involved (e.g., Schilling, 1975a). Evidence from other incompatible elements and isotope ratios supports this view. LREE-enriched tholeiites and alkali basalts from ocean islands and anomalous ridge segments exhibit systematic enrichment of highly incompatible elements (Cs, U, K, Rb, Th, U, Nb and Ta), when compared with most LREE-depleted basalts. Absolute abundance variations can, of course, be explained by fractionation processes; it is the variation in *ratios* of highly incompatible elements and isotopes which is of significance here. Thus, observed differences in La/Ta, La/Ce, La/Th, $^{87}Sr/^{86}Sr$, $^{143}Nd/^{144}Nd$ and Pb isotope ratios between ocean island, E-type MORB and N-type MORB can be satisfactorily explained only by invoking mantle heterogeneity (O'Nions et al., 1977; Sun et al., 1979; Bougault et al., 1980; Tarney et al., 1980). It follows that the source for the ocean island basalts and E-type MORB is probably generally LREE-enriched, whereas the N-type MORB source is LREE-depleted.

Indeed there are now many models which attempt to describe the size and causes of mantle heterogeneity; an exhaustive evaluation of these is beyond the scope of this chapter. The proposed scale of the heterogeneities ranges from the mineral grain size (e.g., Allègre et al., 1980) to tens of kilometers (Schilling, 1975a, b; Machado et al., 1982); although several recent models incorporate variable-scale heterogeneities (e.g., Sun and Hanson, 1975; Hanson, 1977; Tarney et al., 1980). Most models envisage a broad division of the upper mantle into chemically "enriched" and "depleted" components, but the relative disposition of these components is poorly understood (in the following account, the term "enriched" is used in a wide sense, implying source enrichment in those incompatible elements and radiogenic isotopes with high concentrations in E-type MORB and most ocean island basalts, e.g., K, Rb, Ba, Th, U, LREE, Ta and Nb. Conversely, "depleted" implies relative depletion in these elements, as found in N-type MORB). Gast (1968a) suggested that ocean island basalts come from an "enriched" source, equivalent to the low-velocity zone (LVZ) whereas normal MORB are derived from overlying, "depleted" mantle. Schilling (1973a, b, 1975a, b) and co-workers subsequently proposed that deeper, "enriched" mantle rises as vertical columns or plumes (Morgan, 1972) through a chemically-depleted LVZ, the latter being the world-wide source for N-type MORB. In this model, Schilling (1973a, b) and Hart et al. (1973) envisage that Iceland is underlain by a plume of "enriched" mantle material, and that mixing of plume material and LVZ mantle occurs along a transitional ridge segment, the Reykjanes Ridge. The model is able to explain some geochemical features of MAR basalts, in particular the apparent increase in incompatible elements concentrations, $^{87}Sr/^{86}Sr$ and $(Ce/Yb)_{cn}$ ratios as Iceland is approached along the Reykjanes Ridge (the "geochemical gradient"). However, whereas there well may be "plumes" underlying certain oceanic islands (for example, Hawaii, where the heat source for magmatic activity is demonstrably stationary in the deeper mantle), the model does not concur with several observations made on the geochemistry of North Atlantic basalts.

Schilling's (1975a) proposal may be tested by a binary mixing equation since his model is essentially one involving the mixing of two end-members. Such a process should produce a linear relationship between data plotted on a graph of two ratios which have a common denominator (e.g., Hf/Th versus Ta/Th, or $^{87}Sr/^{86}Sr$ versus $Ce/^{86}Sr$). Icelandic basalt and N-type MORB data do not all fall on such a straight line (Langmuir et al., 1978, Dupré and Allègre, 1980), and thus do not completely support either mixing of magmas or simple binary mixing of an "enriched" source with a "depleted" ocean ridge source. Rather, Langmuir et al. (1978) suggest that either (a) there are two separate mixing trends, one beneath Iceland with the enriched alkali basalt source and a depleted Iceland source as end-members, the second along the Reykjanes Ridge with a heterogeneous MORB source and intermediate Icelandic basalt source as end-members (the depleted Iceland source and the MORB source are not the same); or (b) the basalts are derived from a multiplicity of sources which have been isolated for hundreds of millions of years. The multiple source model is difficult to evaluate with available geochemical data, but clearly the real situation is more complex than that suggested by the plume hypothesis.

More recently, it has been proposed that the chemical (isotopic and trace element) heterogeneity observed in both oceanic and continental basalts may result from successive depletion and enrichment events which have affected their source regions (e.g., Hanson, 1977; Frey et al., 1978). Detailed observations of mantle-derived xenoliths by Frey and Green (1974) indicate that the mantle may be variably metasomatized or veined. These veins, which are commonly rich in clinopyroxene, hydrous phases and the highly incompatible elements (including the LREE), may be either in equilibrium with the host mantle (e.g., Dick, 1977), or derived from an isotopically and chemically distinct mantle source (Frey and Green, 1974; Frey and Prinz, 1978). Frey et al. (1978) suggested that the metasomatizing veins may have a bulk chemistry similar to highly undersaturated olivine melilitite ($(Ce/Yb)_{cn} \sim 26$) and, by using binary mixing equations, Tarney et al. (1980) showed that it is possible to generate a wide spectrum of mantle sources, ranging from LREE-depleted bulk source (insignificant vein component) to strongly LREE-enriched bulk source (high percentage of veins). In this model, the composition of the final liquid depends on (a) the composition of the host mantle, (b) the composition of the veins, and whether they are in equilibrium with their host, (c) the percentage of veins and (d) the percentage of melting of the bulk source. A small degree of melting could feasibly involve only the vein component, leaving behind refractory host mantle and producing an undersaturated melt enriched in LREE and other highly incompatible elements.

The metasomatized or veined-mantle model has the advantage of being able to resolve — theoretically, at least — several apparently conflicting geochemical characteristics of oceanic basalts. LREE-enriched tholeiites from ocean ridges and ocean islands (e.g., Hawaii) probably represent more than 15—20% melting of their source regions, and therefore it is likely that their *bulk* source is also LREE-enriched (Green, 1970; Clague and Frey, 1980; Leeman et al., 1977, 1980). This conflicts with available Nd isotope data which clearly imply that the bulk source for most alkali and tholeiitic ridge and island basalts is LREE-*depleted* (i.e., having a high Sm/Nd ratio) relative to the bulk Earth (DePaolo and Wasserburg, 1976; O'Nions et al., 1977; Zindler et al., 1979; Leeman et al., 1980). This apparent paradox may be resolved by metasomatizing and re-enriching a mantle source or host which has been LREE-depleted for billions of years. Thus, any subsequent melts from such a heterogeneous source would contain a significant portion of the incompatible elements from the metasomatizing veins, and thus be LREE-enriched, but would retain the high $^{143}Nd/^{144}Nd$ ratio of the depleted host.

The nature of both the depletion and metasomatizing events is imperfectly understood. The depletion of the LREE within the sub-oceanic mantle probably occurred in early Earth history, if our understanding of isotope systematics is correct. Available isotope data indicate that a significant portion of the Earth's basalt-producing mantle is LREE-depleted relative to the bulk Earth on a time-integrated basis, so the event(s) which produced the depletion must have been on a major scale. Gast (1968a) suggested that silica-undersaturated and incompatible-element-enriched fluids (perhaps similar to contemporaneous metasomatic fluids?) migrated upwards through the mantle leaving behind a depleted, refractory residue. In addition, extraction of significant amounts of material from the mantle must have occurred during the formation of the continental crust which, on average, has an andesitic composition enriched in those elements (particularly K, Rb, Ba, Th, Sr, and the LREE) which have low concentrations in N-type MORB (Ringwood, 1974; Saunders et al., 1980; Tarney et al., 1980).

Re-enrichment of this depleted mantle may then occur when undersaturated fluids, enriched in highly incompatible elements, percolate from deeper portions of the mantle. Studies of continental basalt suites and associated xenoliths and volcanic gases indicate that such fluids are CO_2-enriched, and may well be related to major degassing of the mantle (cf. Lloyd and Bailey, 1975; Frey et al., 1978; Wass et al., 1980). Possibly, the metasomatism associated with present-day ocean island and ridge volcanism is a relatively recent event, perhaps even assisting (or causing?) mantle melting (cf. Wass and Rodgers, 1980), although the longevity of chemically distinct mantle sources feeding the MAR (Tarney et al., 1980) and Pacific island chains (e.g., Hawaiian, Cook and Tuamotou archipelagos) implies that they are large-scale phenomena.

Finally, it is worth briefly mentioning the chemical differences between MAR and EPR basalts. All analysed basalts from the EPR are LREE-depleted (i.e., N-type MORB), whereas enriched tholeiites are not uncommon along the MAR, particularly in the North Atlantic. One explanation may be that prior to the development of the North Atlantic, the underlying mantle was in a sub-continental setting, where the effects of mantle metasomatism may be expected to be particularly significant. The Pacific Basin, however, has been of oceanic character for several hundred million years, so the potential effects of mantle metasomatism may be reduced as a result of mantle convection. Of course, localized mantle metasomatism may be responsible for enriched basalts found on *off-axis* ocean islands (Hawaiian Islands, Cook Islands, etc.). Regional studies of Pacific Ocean basalts from a variety of settings are presently underway, and it is hoped that comparisons with similar data from the North Atlantic (e.g., Wood et al., 1981) will yield further insights into the petrogenesis of oceanic basalts and into the chemical evolution of the sub-oceanic mantle.

6.7. Summary and conclusions

(1) Igneous rocks of the ocean basins exhibit considerable diversity in their mode of eruption (ocean ridges or intraplate seamounts and islands) and in their chemistry.

(2) Mid-ocean ridge basalts (MORB) range in composition from tholeiites with marked depletion of the highly incompatible elements, low $^{87}Sr/^{86}Sr$ ratios and high $^{143}Nd/^{144}Nd$ ratios (N-type MORB), to enriched tholeiitic and alkali basalts (E-type MORB). There appears to be a complete spectrum of compositional types between N- and E-type MORB, although each ridge segment appears to erupt its own basalt type(s).

(3) Ocean island basalts are also variable in composition, although LREE-enriched varieties predominate. In most respects, ocean island basalts resemble E-type MORB, although the degree of LREE and incompatible element enrichment is often greater in the former.

(4) Most of the observed variations in REE distributions in oceanic igneous rocks (total REE contents, $(Ce/Yb)_{cn}$ ratios, etc.) can *theoretically* be explained by fractional crystallization and partial melting processes, particularly if garnet is a residual mineral phase. Whereas the REE have been successfully used to model these processes, it is apparent, however, that they are of lesser value in providing unequivocal evidence of source heterogeneity. Nonetheless, if reasonable petrological models are used (i.e., that tholeiitic magmas represent more than 20% melting of the source), and if highly

incompatible element and isotope ratios are considered, then derivation of all oceanic basalts from a single compositionally homogeneous source becomes untenable.

(5) It is now generally accepted that the sub-oceanic mantle is chemically heterogeneous. Most models involve chemically "depleted" and "enriched" components. For example, the plume model (Schilling, 1973a, b, 1975a) envisages rising columns of hot, enriched mantle peridotite beneath anomalous ridge segments and some oceanic island provinces (e.g., the Azores Islands and Iceland). Mixing of the enriched plume material with a depleted, world-wide mantle source produces the geochemical gradients recorded by some workers along transitional ridge segments (e.g. Reykjanes Ridge and FAMOUS area).

(6) The plume model does not satisfactorily account for *all* geochemical variations in Icelandic and Reykjanes Ridge basalts; the available data do not accord with a simple two-end-member mixing model (Langmuir et al., 1978).

(7) Recent models which invoke intimate, fine-scale metasomatism of the source of "enriched" oceanic basalts more satisfactorily explain the available trace element, isotopic and petrological data (Frey et al., 1978; Tarney et al., 1980). Re-enrichment of a previously depleted mantle host by undersaturated fluids enriched in CO_2 and highly incompatible elements (including the LREE) can produce observed trace element distribution in E-type MORB and LREE-enriched ocean island tholeiites without recourse to garnet fractionation. More important, if the re-enrichment is a relatively recent event, the bulk source could still generate the $^{143}Nd/^{144}Nd$ ratios observed in the basalts, but which on a time integrated basis were derived from a LREE-*depleted* source.

(8) If the sub-oceanic mantle is variably metasomatised, then the composition of parental (or primary) basaltic liquids will depend on the proportion and composition of the metasomatising component and on the conditions and extent of partial melting of the bulk source. Theoretically, the model can account for many of the variations in isotopic and highly incompatible element ratios observed in oceanic basalts.

Acknowledgements

I am very grateful to the following people who commented upon and improved an earlier version of this chapter: P.H. Banham, R.C.O. Gill, R.J. Pankhurst, J. Tarney and B.L. Weaver. Also, many thanks to Giz Marriner for making available REE data from lavas from the Azores Islands, Mrs. Sheila Bishop for typing the manuscript, and N. Sinclair-Jones for drawing the text figures.

References

Allègre, C.J., Brévart, O., Dupré, B. and Minster, J.F., 1980. Isotopic and chemical effects produced in a continuously differentiating convecting Earth mantle. *Philos. Trans. R. Soc. London, Ser. A*, 297: 447—477.

Anderson, R.N., McKenzie, D.P. and Slater, J.G., 1973. Gravity, bathymetry and convection in the earth. *Earth Planet. Sci. Lett.*, 18: 391—407.

Arth, J.G., 1976. Behaviour of trace elements during magmatic processes — a summary of theoretical models and their applications. *J. Res. U.S. Geol. Surv.*, 4: 41—47.

Batiza, R., 1981. Trace-element characteristics of Leg 61 Basalts. In: R.L. Larson, S.O. Schlanger et al., *Initial Reports of the Deep Sea Drilling Project, 61.* U.S. Government Printing Office, Washington, D.C., pp. 689—695.

Batiza, R. and Johnson, J.R., 1980. Trace element and isotopic evidence for magma mixing in alkalic and transitional basalts near the East Pacific Rise. In: B.R. Rosendahl, R. Hekinian et al., *Initial Reports of the Deep Sea Drilling Project, 54.* U.S. Government Printing Office, Washington, D.C., pp. 63—70.

Batiza, R., Rosendahl, B.R. and Fisher, R.L., 1977. Evolution of the oceanic crust, 3. Petrology and chemistry of basalts from the East Pacific Rise and the Siqueiros transform fault. *J. Geophys. Res.*, 82: 265—276.

Bougault, H. and Treuil, M., 1980. Mid-Atlantic Ridge: zero-age geochemical variations between Azores and 22°N. *Nature (London)*, 286: 209—212.

Bougault, H., Treuil, M. and Joron, J.-L., 1978. Trace elements in basalts from 23°N and 36°N in the Atlantic Ocean: fractional crystallization, partial melting, and heterogeneity of the upper mantle. In: W.G. Melson, P.D. Rabinowitz et al., *Initial Reports of the Deep Sea Drilling Project, 45.* U.S. Government Printing Office, Washington, D.C., pp. 493—506.

Bougault, H., Cambon, P., Corre, O., Treuil, M. and Joron, J.-L., 1979. Evidence for variability of magmatic processes and upper mantle heterogeneity in the axial region of the Mid-Atlantic Ridge near 22° and 36°N. *Tectonophysics*, 55: 11—34.

Bougault, H., Joron, J.-L. and Treuil, M., 1980. The primordial chondritic nature and large-scale heterogeneities in the mantle: evidence from high and low partition coefficient elements in oceanic basalts. *Philos. Trans. R. Soc. London, Ser. A*, 297 203—213.

Cambon, P., Joron, J.-L., Bougault, H. and Treuil, M., 1980. Leg 55, Emperor Seamounts: trace elements in transitional tholeiites, alkali basalts and hawaiites — mantle homogeneity and magmatic processes. In: E.D. Jackson, I. Koisumi et al., *Initial Reports of the Deep Sea Drilling Project, 55.* U.S. Government Printing Office, Washington, D.C., pp. 585—598.

Carmichael, I.S.E., 1964. The petrology of Thingmuli, a Tertiary volcano in eastern Iceland. *J. Petrol.*, 5: 453—460.

Clague, D.A. and Frey, F.A., 1980. Trace-element geochemistry of tholeiitic basalts from Site 433C, Suiko Seamount. In: E.D. Jackson, I. Koisumi et al., *Initial Reports of the Deep Sea Drilling Project, 55.* U.S. Government Printing Office, Washington, D.C., pp. 559—570.

Cohen, R.S., Evensen, N.M., Hamilton, P.J. and O'Nions, R.K., 1980. U-Pb, Sm-Nd and Rb-Sr systematics of mid-ocean ridge basalt glasses. *Nature (London)*, 283: 149—153.

DePaolo, D.J. and Wasserburg, G.J., 1976. Inferences about magma sources and mantle structure from variations of $^{143}Nd/^{144}Nd$. *Geophys. Res. Lett.*, 3: 249—253.

Dick, H.J.B., 1977. Partial melting in the Josephine Peridotite, 1. The effect of mineral composition and its consequence for geobarometry and geothermometry. *Am. J. Sci.*, 277: 801—832.

Dupré, B. and Allègre, C.J., 1980. Pb-Sr-Nd isotopic correlation and the chemistry of the North Atlantic mantle. *Nature (London)*, 286: 17—22.

Elthon, D., 1979. High magnesia liquids as the parental magma for ocean floor basalts. *Nature (London)*, 278: 514–518.

Engel, A.E.J. and Engel, C.G., 1964. Composition of basalts from the Mid-Atlantic Ridge. *Science*, 144: 1130—1333.

Engel, A.E.J., Engel, C.G. and Havens, R.G., 1965. Chemical characteristics of oceanic basalts and the upper mantle. *Geol. Soc. Am. Bull.*, 76: 719–734.

Engel, C.G. and Engel, A.E.J., 1963. Basalts dredged from the northeastern Pacific Ocean. *Science*, 140: 1321–1324.

Engel, C.G., Fisher, R.L. and Engel, A.E.J., 1965. Igneous rocks from the Indian Ocean floor. *Science*, 150: 605—609.

Flower, M.F.M., Schmincke, H.-U. and Bowman, H., 1976. Rare earth and other trace elements in historic Azorean lavas. *J. Volcanol. Geotherm. Res.*, 1: 127—147.

Frey, F.A. and Green, D.H., 1974. The mineralogy, geochemistry and origin of lherzolite inclusions in Victorian basanites. *Geochim. Cosmochim. Acta*, 38: 1023—1059.

Frey, F.A. and Haskin, L., 1964. Rare earths in oceanic basalts. *J. Geophys. Res.*, 69: 775—779.

Frey, F.A. and Prinz, M., 1978. Ultramafic inclusions from San Carlos, Arizona: petrologic and geochemical data bearing on their petrogenesis. *Earth Planet. Sci. Lett.*, 38: 129—176.

Frey, F.A., Bryan, W.B. and Thompson, G., 1974. Atlantic Ocean floor: geochemistry and petrology of basalts from Legs 2 and 3 of the Deep Sea Drilling Project. *J. Geophys. Res.*, 79: 5507—5527.

Frey, F.A., Green, D.H. and Roy, S.D., 1978. Integrated models of basalt petrogenesis: a study of quartz tholeiites to olivine melilitites from southeastern Australia utilizing geochemical and experimental petrological data. *J. Petrol.*, 19: 463—513.

Frey, F.A., Dickey, J.S., Thompson, G.S., Bryan, W.B. and Davies, H.L., 1980. Evidence for heterogeneous primary MORB and mantle sources, N.W. Indian Ocean. *Contrib. Mineral Petrol.*, 74: 387—402.

Fujii, N., Notsu, K. and Onuma, N., 1981. Chemical compositions and Sr-isotopes of Deep Sea Drilling Project Leg 61 basalts. In: R.L. Larson, S.O. Schlanger et al., *Initial Reports of the Deep Sea Drilling Project, 61.* U.S. Government Printing Office, Washington, D.C., pp. 697—700.

Gast, P.W., 1965. Terrestrial ratio of potassium to rubidium and the composition of the Earth's mantle. *Science*, 147: 858—860.

Gast, P.W., 1968a. Trace element fractionation and the origin of tholeiitic and alkaline magma types. *Geochim. Cosmochim Acta*, 32: 1057—1086.

Gast, P.W., 1968b. Upper mantle chemistry and evolution of the Earth's crust. In: R. Phinney (Editor), *The History of the Earth's Crust*, Princeton University Press, Princeton, N.J., p. 15.

Green, D.H., 1970. The origin of basaltic and nephelinitic magmas. *Trans. Leicester Lit. Philos. Soc.*, 64: 26—45.

Green, D.H., 1971. Composition of basaltic magmas as indicators of conditions of origin: application to oceanic volcanism. *Philos. Trans. R. Soc. London, Ser. A*, 268: 707—725.

Green, D.H. and Ringwood, A.E., 1967. The genesis of basaltic magmas. *Contrib. Mineral. Petrol.*, 15: 103—190.

Green, D.H., Hibberson, W.O. and Jacques, A.L., 1979. Petrogenesis of mid-ocean ridge basalts. In: M.W. McElhinny (Editor), *The Earth: Its Origins, Structure and Evolution.* Academic Press, New York, N.Y., pp. 265—269.

Hanson, G.N., 1977. Geochemical evolution of the suboceanic mantle. *J. Geol. Soc. London*, 134: 235—253.

Hart, S.R., 1971. K, Rb, Cs, Sr and Ba contents and Sr isotope ratios of ocean floor basalts. *Philos. Trans. R. Soc. London, Ser. A*, 268; 573—578.

Hart, S.R., Glassley, W.E. and Karig, D.E., 1972. Basalts and sea-floor spreading behind the Mariana island arc. *Earth Planet. Sci. Lett.*, 15: 12—18.

Hart, S.R., Schilling, J.-G. and Powell, J.L., 1973. Basalts from Iceland and along the Reykjanes Ridge: Sr isotope geochemistry. *Nature Phys. Sci.*, 246: 104—107.

Hawkesworth, C.J., Norry, M.J., Roddick, J.C. and Vollmer, R., 1979. $^{143}Nd/^{144}Nd$ and $^{87}Sr/^{86}Sr$ ratios from the Azores and their significance in LIL-element-enriched mantle. *Nature (London)*, 280: 28—31.

Jackson, E.D., Silver, E.A. and Dalrymple, G.B., 1972. Hawaiian-Emperor chain and its relation to Cenozoic circum-Pacific tectonics. *Geol. Soc. Am. Bull.*, 83: 601—618.

Jackson, E.D., Koisumi, I. et al., 1980. *Initial Reports of the Deep Sea Drilling Project, 55.* U.S. Government Printing Office, Washington, D.C., 868 pp.

Kay, R.W. and Gast, P.W., 1973. The rare earth content and origin of alkali-rich basalts. *J. Geol.*, 81: 653—682.

Kay, R.W. and Hubbard, N.J., 1978. Trace elements in ocean ridge basalts. *Earth Planet. Sci. Lett.*, 38: 95- 116.

Kay, R.W., Hubbard, N.J. and Gast, P.W., 1970. Chemical characteristics of oceanic ridge volcanic rocks. *J. Geophys. Res.*, 75: 1585—1613.

Langmuir, C.H., Bender, J.F., Bence, A.E., Hanson, G.N. and Taylor, S.R., 1977. Petrogenesis of basalts from the FAMOUS area: Mid-Atlantic Ridge. *Earth Planet. Sci. Lett.*, 36: 133—156.

Langmuir, C.H., Vocke, R.D. and Hanson, G.N., 1978. A general mixing equation with applications to Icelandic basalts. *Earth Planet. Sci. Lett.*, 37: 380—392.

Larson, R.L., Schlanger, S.O. et al., 1981. *Initial Reports of the Deep Sea Drilling Project, 61.* U.S. Government Printing Office, Wasington, D.C., 885 pp.

Leeman, W.P., Murali, A.V., Ma, M.S. and Schmitt, R.A., 1977. Mineral constitution of mantle source regions for Hawaiian basalts — rare earth evidence for mantle heterogeneity. In: H.J.B. Dick (Editor), *Magma Genesis. Oreg., Dep. Geol. Miner. Ind., Bull.*, 96: 169—183.

Leeman, W.P., Budahn, J.R., Gerlach, D.C., Smith, D.R. and Powell, B.N., 1980. Origin of Hawaiian tholeiites: trace element constraints. *Am. J. Sci.*, 280-A: 794—819.

Le Maitre, R.W., 1960. The geology of Gough Island, South Atlantic. *Overseas Geol. Miner. Resour.*, 7: 371—380.

Lloyd, F.E. and Bailey, D.K., 1975. Light element metasomatism of the continental mantle: the evidence and consequences. In: L.H. Ahrens, F. Press, S.K. Runcorn and H.C. Urey (Editors), *Physics and Chemistry of the Earth, 9.* Pergamon, Oxford, pp. 389—416.

Ludden, J.N. and Thompson, G., 1979. An evaluation of the behaviour of rare earth elements during weathering of sea floor basalt. *Earth Planet. Sci. Lett.*, 43: 85—92.

Macdonald, G.A., 1968. Composition and origin of Hawaiian lavas. *Geol. Soc. Am., Mem.*, 116: 477—522.

Macdonald, G.A. and Katsura, T., 1964. Chemical composition of Hawaiian lavas. *J. Petrol.*, 5: 82—113.

Machado, N., Ludden, J.N., Brooks, C. and Thompson, G., 1982. Fine-scale isotopic heterogeneity in the sub-Atlantic mantle. *Nature (London)*, 295: 226—228.

Marriner, G.F., Norry, M.J. and Gibson, I.L., 1982. The petrology and geochemistry of the Agua de Pau volcano, Sao Miguel, Azores. In: *Proceedings, Symposium on the Activity of Oceanic Volcanoes, Azores 1981. Arquipélago (Rev. Univ. Azores)*, 3: 159—173.

McDougall, I., 1964. Potassium-argon ages from lavas of the Hawaiian Islands. *Geol. Soc. Am. Bull.*, 75: 107—128.

McDougall, I., 1971. The geochronology and evolution of the young volcanic island of Reunion, Indian Ocean. *Geochim. Cosmochim. Acta*, 35: 261—268.

McDougall, I. and Compston, W., 1965. Strontium isotope composition and potassium-rubidium ratios in some rocks from Reunion and Rodriguez, Indian Ocean. *Nature (London)*: 207: 252—253.

Miyashiro, A., Shido, F. and Ewing, M., 1970. Crystallization and differentiation in abyssal tholeiites and gabbros from mid-ocean ridges. *Earth Planet. Sci. Lett.*, 7: 361—365.

Morgan, W.J., 1972. Deep mantle convection plumes and plate motions. *Bull. Am. Assoc. Pet. Geol.*, 56: 203—213.

Muir, I.D., Tilley, C.E. and Scoon, J.H., 1964. Basalts from the northern part of the rift zone of the Mid-Atlantic Ridge. *J. Petrol.*, 5: 409—434.

Muir, I.D., Tilley, C.E. and Scoon, J.H., 1966. Basalts from the northern part of the Mid-Atlantic Ridge, II. The Atlantis Collections near 30°N. *J. Petrol.*, 7: 193—201.

Neumann, H., Mead, J. and Vitaliano, C.J., 1954. Trace element variation during fractional crystallization as calculated from the distribution law. *Geochim. Cosmochim. Acta*, 6: 90—99.

O'Hara, M.J., 1968. Are ocean floor basalts primary? *Nature (London)*, 220: 683—686.

O'Hara, M.J., 1973. Non-primary magmas and dubious mantle plume beneath Iceland. *Nature (London)*, 243: 507—508.

O'Hara, M.J., 1975. Is there an Icelandic mantle plume? *Nature (London)*, 253: 708—710.

O'Hara, M.J., 1977. Geochemical evolution during fractional crystallization of a periodically refilled magma chamber. *Nature (London)*, 266: 503—507.

O'Hara, M.J. and Mathews, R.E., 1981. Geochemical evolution in an advancing, periodically replenished, periodically tapped, continuously fractionated magma chamber. *J. Geol. Soc. London*, 138: 237—278.

O'Nions, R.K. and Grönvold, K., 1973. Petrogenic relationships of acid and basic rocks in Iceland: Sr-isotopes and rare-earth elements in late and postglacial volcanics. *Earth Planet. Sci. Lett.*, 19: 379—409.

O'Nions, R.K. and Pankhurst, R.J., 1973. Secular variations in Sr-isotope composition of Icelandic volcanic rocks. *Earth Planet. Sci. Lett.*, 21: 13—21.

O'Nions, R.K., Pankhurst, R.J. and Grönvold, K., 1976. Nature and development of basalt magma sources beneath Iceland and the Reykjanes Ridge. *J. Petrol.*, 17: 315—338.

O'Nions, R.K., Hamilton, P.J. and Evensen, N.M., 1977. Variations in $^{143}Nd/^{144}Nd$ and $^{87}Sr/^{86}Sr$ ratios in oceanic basalts. *Earth Planet. Sci. Lett.*, 34: 13—22.

O'Nions, R.K., Evensen, N.M. and Hamilton, P.J., 1979. Geochemical modelling of mantle differentiation and crustal growth. *J. Geophys. Res.*, 84: 6093—6101.

Pankhurst, R.J., 1977. Open system crystal fractionation and incompatible element variation in basalts. *Nature (London)*, 268: 36—38.

Peterman, Z.E. and Hedge, C.E., 1971. Related strontium isotopic and chemical variations in oceanic basalts. *Geol. Soc. Am. Bull.*, 82: 493—499.

Philpotts, J.A. and Schnetzler, C.C., 1970. Phenocryst-matrix partition coefficients for K, Rb, Sr and Ba with applications to anorthosite and basalt genesis. *Geochim. Cosmochim. Acta*, 34: 307—322.

Ringwood, A.E., 1974. The petrological evolution of island arc systems. *J. Geol. Soc. London*, 130: 183—204.

Saunders, A.D., 1983. The Gulf of California: geochemistry of basalts recovered during Leg 65 of the Deep Sea Drilling Project. In: B.T.R. Lewis, P. Robinson et al., *Initial Reports of the Deep Sea Drilling Project, 65.* U.S. Government Printing Office, Washington, D.C., pp. 591—621.

Saunders, A.D., Tarney, J. and Weaver, S.D., 1980. Transverse geochemical variations across the Antarctic Peninsula: implications for the genesis of calc-alkaline magmas. *Earth Planet. Sci. Lett.*, 46: 344—360.

Schilling, J.-G., 1971. Sea floor evolution: rare-earth evidence. *Philos. Trans. R. Soc. London, Ser. A*, 268: 663—706.

Schilling, J.-G., 1973a. Afar mantle plume: rare-earth evidence. *Nature Phys. Sci. (London)*, 242: 2—5.

Schilling, J.-G., 1973b. Iceland mantle plume: geochemical study of Reykjanes Ridge. *Nature (London)*, 242: 565—571.

Schilling, J.-G., 1975a. Rare-earth variations across "normal segments" of the Reykjanes Ridge, 60—53°N, Mid-Atlantic Ridge, 29°S, and East Pacific Rise, 2—19°S, and evidence of the composition of the underlying low-velocity layer. *J. Geophys. Res.*, 80: 1459—1473.

Schilling, J.-G., 1975b. Azores mantle blob: rare earth evidence. *Earth Planet. Sci. Lett.*, 25: 103—115.

Schilling, J.-G. and Winchester, J.W., 1969. Rare earth contribution to the origin of Hawaiian lavas. *Contrib. Mineral. Petrol.*, 23: 27—37.

Schmincke, H.-U., 1973. Magmatic evolution and tectonic regime in the Canary, Madeira, and Azores island groups. *Geol. Soc. Am. Bull.*, 84: 633—648.

Schmincke, H.-U. and Weibel, H., 1973. Magmatic evolution and tectonic evolution in the Canary, Madeira and Azores Islands groups. *Geol. Soc. Am. Bull.*, 84: 633—648.

Seifert, K.E., 1981. Geochemistry of Nauru Basin basalts from the lower portion of Hole 462A, Deep Sea Drilling Project Leg 61. In: R.L. Larson, S.O. Schlanger et al., *Initial Reports of the Deep Sea Drilling Project, 61.* U.S. Government Printing Office, Washington, Washington, D.C., pp. 705—708.

Shaw, D.M., 1970. Trace element fractionation during anatexis. *Geochim. Cosmochim. Acta*, 34: 237—243.

Shimizu, N. and Kushiro, I., 1975. The partitioning of rare-earth elements between garnet and liquid at high pressures: preliminary experiments. *Geophys. Res. Lett.*, 2: 413—416.

Sigvaldasson, G.E., 1974. Basalts from the centre of the assumed Icelandic mantle plume. *J. Petrol.*, 15: 497—525.

Sigvaldasson, G.E., Steinthorsson, S. and Oskarsson, N., 1974. Compositional variation in recent Icelandic tholeiites and the Kverkfjöll hot spot. *Nature, (London)*, 251: 579—582.

Srivastava, R.K., Emmermann, R. and Puchelt, H., 1980. Petrology and geochemistry of basalts from Deep Sea Drilling Project Leg 54. In: B.R. Rosendahl, R. Hekinian et al., *Initial Reports of the Deep Sea Drilling Project, 54.* U.S. Government Printing Office, Washington, D.C., pp. 671—694.

Sun, S.-S., 1980. Lead isotope study of young volcanic rocks from mid-ocean ridges, ocean islands and island arcs. *Philos. Trans. R. Soc. London, Ser. A*, 297: 409—445.

Sun, S.-S. and Hanson, G.N., 1975. Evolution of the mantle: geochemical evidence from alkali basalt. *Geology*, 3: 297—302.

Sun, S.-S., Nesbitt, R.W. and Sharaskin, A.Ya., 1979. Geochemical characteristics of mid-ocean ridge basalts. *Earth Planet. Sci. Lett.*, 44: 119—138.

Talwani, M., Windisch, C.C. and Langseth, M.J., Jr., 1971. Reykjanes Ridge crest: a detailed geophysical study. *J. Geophys. Res.*, 76: 473—517.

Tarney, J., Wood, D.A., Varet, J., Saunders, A.D. and Cann, J.R., 1979. Nature of mantle heterogeneity in the North Atlantic: evidence from Leg 49 basalts. In: M. Talwani (Editor), *Implications of Deep Sea Drilling Results in the Atlantic Ocean. Am. Geophys. Union, Maurice Ewing Ser.*, 2: 285—301.

Tarney, J., Wood, D.A., Saunders, A.D., Cann, J.R. and Varet, J., 1980. Nature of mantle heterogeneity in the North Atlantic: evidence from deep sea drilling. *Philos. Trans. R. Soc. London, Ser. A*, 297: 179—202.

Tatsumoto, M., Hedge, C.E. and Engel, A.E.J., 1965. Potassium, rubidium, strontium, thorium, uranium and the ratio of strontium-87 to strontium-86 in oceanic tholeiitic basalts. *Science*, 150: 886—888.

Terrell, D.J., Pal, S., López, M.M. and Pérez, J.R., 1979. Rare-earth elements in basalt samples, Gulf of California. *Chem. Geol.*, 26: 267—275.

Thompson, G., Bryan, W.B., Frey, F.A., Dickey, J.S. and Suen, C.J., 1976. Petrology and geochemistry of basalts from DSDP Leg 34, Nazca Plate. In: S.R. Hart, R.S. Yeats et al. *Initial Reports of the Deep Sea Drilling Project, 34.* U.S. Government Printing Office, Washington, D.C., pp. 215—226.

Tokuyama, H. and Batiza, R., 1981. Chemical composition of igneous rocks and origin of the sill and pillow-basalt complex of Nauru Basin, southwest Pacific. In: R.L. Larson, S.O. Schlanger et al., *Initial Reports of the Deep Sea Drilling Project, 61.* U.S. Government Printing Office, Washington, D.C., pp. 673—683.

Upton, B.G.J. and Wadsworth, W.J., 1965. The geology of Reunion Island, Indian Ocean. *Nature (London)*, 207: 151—154.

Upton, B.G.J. and Wadsworth, W.J., 1966. The basalts of Reunion Island, Indian Ocean. *Bull. Volcanol.*, 29: 7—22.

Upton, B.G.J. and Wadsworth, W.J., 1972. Aspects of magmatic evolution on Reunion Island. *Philos. Trans. R. Soc. London, Ser. A*, 271: 105—130.

Vogt, P.R., Schneider, E.D. and Johnson, G.L., 1969. The crust and upper mantle beneath the sea. In: P.J. Hart (Editor), *The Earth's Crust and Upper Mantle. Am. Geophys. Union, Geophys. Monogr. Ser.*, 13: 556—617.

Walker, G.P.L., 1966. Acid volcanic rocks in Iceland. *Bull. Volcanol.*, 29: 375—402.

Wass, S.Y. and Rodgers, N.W., 1980. Mantle metasomatism, precursor to continental alkaline volcanism. *Geochim. Cosmochim. Acta*, 44: 1811—1824.

Wass, S.Y., Henderson, P. and Elliot, C.J., 1980. Chemical heterogeneity and metasomatism in the upper mantle: evidence from rare-earth and other elements in apatite-rich xenoliths in basaltic rocks from eastern Australia. *Philos. Trans. R. Soc. London, Ser. A*, 297: 333—346.

White, W.M. and Schilling, J.-G., 1978. The nature and origin of geochemical variation in Mid-Atlantic Ridge basalts from the Central North Atlantic. *Geochim. Cosmochim. Acta*, 42: 1501—1516.

White, W.M., Schilling, J.-G. and Hart, S.R., 1976. Evidence for the Azores mantle plume from strontium isotope geochemistry of the central North Atlantic. *Nature (London)*, 263: 659—663.

White, W.M., Tapia, M.D.M. and Schilling, J.-G., 1979. The petrology and geochemistry of the Azores Islands. *Contrib. Mineral. Petrol.*, 69: 201—213.

Wilson, J.T., 1963. A possible origin of the Hawaiian Islands. *Can. J. Phys.*, 41: 863—870.

Winterer, E.L., 1975. Bathymetry and regional tectonic setting of the Line Islands Chain. In: S.O. Schlanger, E.D. Jackson et al., *Initial Reports of the Deep Sea Drilling Project, 33.* U.S. Government Printing Office, Washington, D.C., pp. 731—747.

Wood, D.A., 1978. Major and trace element variations in the Tertiary lavas of eastern Iceland with respect to the Iceland geochemical anomaly. *J. Petrol.*, 19: 393—436.

Wood, D.A., 1979. Dynamic partial melting: its application to the petrogenesis of basalt lava series from Iceland, the Faeroe Islands, the Isle of Skye (Scotland) and the Troodos Massif (Cyprus). *Geochim. Cosmochim. Acta*, 43: 1031—1046.

Wood, D.A., Joron, J.-L., Treuil, M., Norry, M. and Tarney, J., 1979a. Elemental and Sr-isotope variations in basic lavas from Iceland and the surrounding ocean floor. *Contrib. Mineral. Petrol.*, 70: 319—339.

Wood, D.A., Tarney, J., Varet, J., Saunders, A.D., Bougault, H., Joron, J.-L., Treuil, M. and Cann, J.R., 1979b. Geochemistry of basalts drilling in the North Atlantic by IPOD Leg 49: implications for mantle heterogeneity. *Earth Planet. Sci. Lett.*, 42: 77—97.

Wood, D.A., Tarney, J. and Weaver, B.L., 1981. Trace element variations in Atlantic Ocean basalts and Proterozoic dykes from northwest Scotland: their bearing upon the nature and geochemical evolution of the upper mantle. *Tectonophysics*, 75: 91—112.

Zielinski, R.A., 1975. Trace element evaluation of a suite of rocks from Reunion Island, Indian Ocean. *Geochim. Cosmochim. Acta*, 39: 713—734.

Zielinski, R.A. and Frey, F.A., 1970. Gough Island: evaluation of a fractional crystallization model. *Contrib. Mineral. Petrol.*, 29: 242—254.

Zindler, A., Hart, S.R., Frey, F.A. and Jakobsson, S.P., 1979. Nd and Sr isotope ratios and rare earth element abundances in Reykjanes Peninsula basalts: evidence for mantle heterogeneity beneath Iceland. *Earth Planet. Sci. Lett.*, 45: 249—262.

Chapter 7

RARE EARTH ELEMENTS IN IGNEOUS ROCKS OF THE CONTINENTAL CRUST: PREDOMINANTLY BASIC AND ULTRABASIC ROCKS

ROBERT L. CULLERS and JOSEPH L. GRAF

7.1. Introduction

Studies that interpret the petrogenesis of basic and ultrabasic rocks are best when REE abundances are used as only one of many lines of evidence including field relations, major element and trace element compositions of the rocks and minerals, isotopes, and experimental petrology (e.g., Frey et al., 1978). In such studies the REE can be quite useful in limiting the number of hypotheses. One major problem is that many papers seem to propose a unique hypothesis to explain the data when in reality several hypotheses could explain the data equally well.

This can be illustrated using the REE to understand the petrogenesis of basic and ultrabasic rocks. The results of experimental petrology suggest that small degrees of melting (less than about 10%) of peridotite in the mantle yield alkali-rich basalt or ijolites, while greater degrees of melting (about 15—30%) yield tholeiitic basalts. The REE are concentrated in the magma during melting of peridotite unless garnet is present to concentrate the HREE. As described more fully in Chapter 4, LREE contents and $(La/Lu)_{cn}$ ratios are the highest in the magma at the smallest degrees of melting, and these values decrease rapidly with increased melting up to about 20—30 percent. With further melting the REE contents of the melts do not change very much.

Alkali-rich basalts, ijolites, and kimberlites have very high LREE contents and $(La/Lu)_{cn}$ ratios (Table 7.1), consistent with their formation by small degrees of melting of garnet peridotite. One problem is that only garnet peridotite more enriched in the LREE than most peridotite is required to melt to produce such LREE-enriched melts. Even then, the predictions tend to be lower than the observed contents. Therefore, many theoretical models also include other non-unique models such as subsequent crystallization of garnet and clinopyroxene, volatile transport of the REE, or crustal contamination to increase the LREE contents and La/Lu ratios in the magmas. Every effort needs to be made to limit the range of possible models by explaining all data available. Authors need to strive to emphasize what range of possible hypotheses are acceptable to explain the data and which hypotheses are not acceptable.

TABLE 7.1

Range of REE contents in basic and ultrabasic rocks (associated more silicic rocks are included for the alkali-rich group)

Rock type	Σ REE[a] (ppm)	$(La/Lu)_{cn}$	Eu/Sm
Kimberlites	115—4613	15.8—216	0.035—0.29
Carbonatites	72—15,515	7.1—1240	0.15—0.50
Ijolite series (urtite-melteigite)	145—1334	2.9—64.3	0.25—0.39 (one is 2.25)
Lamprophyres	261—1033	20.2—55	0.22—0.29
Komatiites	10.1—59.1	0.24—4.8	0.25—0.55
peridotitic komatiites	10.1—24.9	0.24—1.2	0.25—0.45
basaltic komatiites	16.7—59.1	0.57—4.8	0.25—0.55
Associated tholeiites	21.6—146	0.42—3.8	0.29—0.46
All alkaline rocks			
mafic	69—1453	2.7—262	0.24—0.40
intermediate and felsics	92—1750	0.78—114	0.13—0.47
Alkali basalt group ($Na_2O > K_2O$; $K_2O + Na_2O \approx 3.4$; $K_2O/Na_2O \approx 0.5$)			
mafics	90—610	3.6—34	0.24—0.40
transitional basalts	90—243	3.6—12.2	0.27—0.40
alkali basalts	130—250	5.35—28.4	0.27—0.35
basanites, nepheline basalts, nephelinites	202—597	13.7—34	0.24—0.33
intermediate (trachytes, phonolites, syenites)	154—1750	0.78—42	0.13—0.47
associated with transitional basalts	159—390	5.8—9.2	0.28—0.32
associated with alkali basalts	235—988	4.8—41.8	0.13—0.47
associated with basanites, etc.	380—1750	0.73—8.7	0.13—0.18
felsic rocks			
associated with alkali or transitional basalts	316—738	0.73—8.7	0.13—0.18
Shoshonite group ($Na_2O \approx K_2O$; $K_2O + Na_2O \approx 5—6$)			
shoshonites	69—240	2.7—18.3	0.25—0.34
intermediate and felsic rocks associated with shoshonites	124—387	3.2—23.1	0.16—0.28
K-rich basalt group ($K_2O \gg Na_2O$; $K_2O + Na_2O \approx 7—10$; $K_2O/Na_2O \approx 7—10$)			
K-rich mafic rocks	255—1453	19.5—262	0.24—0.31
intermediate rocks associated with K-rich mafics	92—688	3.8—114	0.06—0.27
Continental tholeiites	15.2—322	0.5—7.6	0.16—0.55
Island arc and back-arc tholeiites	10—262	0.4—7.3	0.22—0.54
Tholeiitic intrusive complexes	2.7—547	0.31—19.3	0.23—1.55

[a]This is the total REE content from La to Lu, including estimates of REE not analyzed from chondrite-normalized curves.

The greater percent melting of peridotite needed to produce tholeiites compared to alkali basalts results in lower REE contents and La/Lu ratios in the tholeiites compared to the alkali basalts. At such large degrees of melting, REE contents of the magmas are more sensitive to variation of REE content in the source and less sensitive to variations in degree of melting. Relatively undepleted peridotite is required to melt and produce the REE contents of continental tholeiites (Table 7.1). Peridotites depleted in the LREE from a previous melting episode may undergo a similar percent melting as for continental tholeiites to produce LREE-depleted tholeiites at subduction zones or at oceanic rises.

This review summarizes the range of REE contents in mainly basic and ultrabasic rocks in the order kimberlites, carbonatites and associated alkalic silicate rocks, lamprophyres, komatiites, alkali-rich basic rocks and associated more silicic rocks, and tholeiitic basalts. Also the major interpretations of the origin of these rock types using the REE will be summarized.

7.2. Kimberlite

General

Kimberlite is an ultramafic rock containing phenocrysts of ilmenite, phlogopite, olivine, and pyrope in a groundmass of olivine, phlogopite, serpentine, chlorite, carbonate, magnetite, and apatite. Kimberlites are often crowded with xenoliths of crustal rocks and upper mantle rocks (garnet peridotite and eclogite). The hydrous and carbonate mineral assemblages indicate a high fluid pressure during their emplacement. Kimberlite usually occurs in small volume as pipes or dikes.

Chemically they contain larger contents of the alkalies Ca, Ti, Al, and Fe, and less Si and Mg than peridotites (Dawson, 1972). Many trace elements such as Sc, Th, Ba, Li, Rb, Sr, Y, Ga, Zr, V, Mo, Pb, and the LREE are enriched compared to other ultramafic rocks (e.g., Dawson, 1962; 1972).

REE content

Kimberlites are characterized by their large absolute REE content (ΣREE = 115—4613 ppm), large LREE/HREE ratios ($(La/Lu)_{cn}$ = 15.8—216), and linear REE patterns, with or without Eu anomalies in chondrite-normalized abundance plots (Fig. 7.1; Burkov and Podporina, 1966; Haskin et al., 1966; Frey et al., 1971; Philpotts et al., 1972; Mitchell and Brunfelt, 1974; Fesq et al., 1974; Ilupin et al., 1974; Paul et al., 1975; Reitz and Cullers, 1980; Cullers et al., 1982). Two types of kimberlite are sometimes distinguished: micaceous kimberlite and "basaltic" kimberlite (with little mica). "Basaltic" kimberlite is a misnomer; it should be called kimberlite (Mitchell, 1970) or

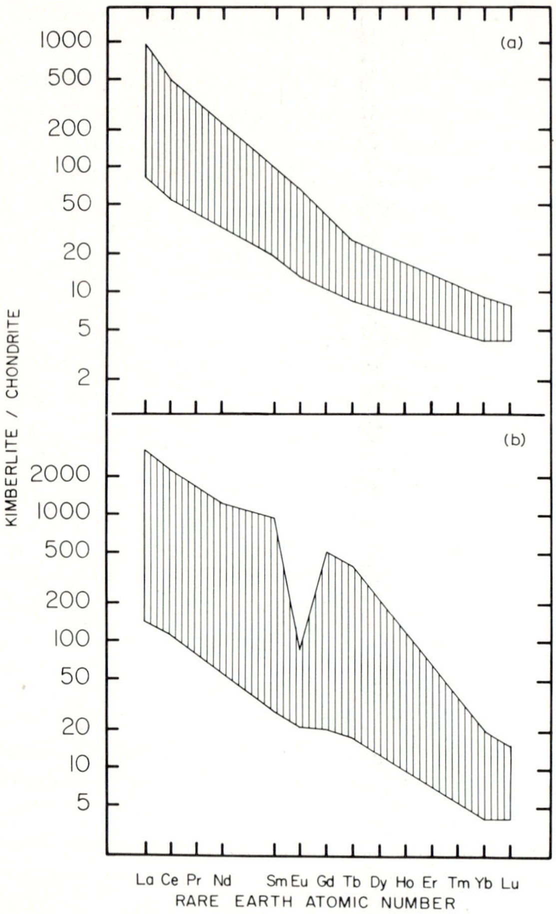

Fig. 7.1. (a) The range of REE content found in kimberlites that contain no Eu anomaly. REE contents of the sample are divided element by element by those from chondritic meteorites (Wakita et al., 1971) and are plotted vs. REE atomic number in this and in most of the subsequent figures. (b) The range of REE content found in kimberlites that contain Eu anomalies. Large, negative Eu anomalies such as found in the top sample are rare since only 4 out of 50 analyses thus far contain such large anomalies. Most anomalies, if present, are small like those of the lower sample.

micaceous-poor kimberlite. Although data are limited, micaceous kimberlites tend to have small to large negative Eu anomalies (some contain no Eu anomalies; Eu/Sm = 0.035—0.28); Eu anomalies tend to be absent in non-micaceous kimberlites (Eu/Sm = 0.20—0.29; Fesq et al., 1974, Mitchell and Brunfelt, 1974). The presence of negative Eu anomalies in the micaceous kimberlites has been attributed to accumulation of phlogopite or perovskite containing negative Eu anomalies (Fesq et al., 1974; Mitchell and Brunfelt, 1974). Certainly feldspar/melt equilibria during partial melting or crystallization could not have produced the anomalies since feldspar is not believed to be present during kimberlite petrogenesis.

Abundances of the LREE correlate well with P content in kimberlites. This suggests phosphate minerals (e.g., apatite, monazite, or perovskite) may be predominant host minerals for the LREE, and that carbonated, phosphate minerals may be the source of the LREE in the upper mantle (Fesq et al., 1974). The problems involved with this interpretation will be discussed later.

General petrogenesis

The petrogenesis of kimberlites with regard to the REE has been reviewed most extensively by Fesq et al. (1974), Mitchell and Brunfelt (1974), and Paul et al. (1975). A major problem is to determine if the extremely large LREE content and LREE/HREE ratios were produced solely by partial melting or whether processes subsequent to partial melting were dominant (Mitchell and Brunfelt, 1974). Theories of kimberlite petrogenesis will be reviewed in order of partial melting, zone refining, fractional crystallization, volatile transport, and contamination.

Partial melting

The extreme REE contents in rocks such as kimberlites (also ijolites, carbonatites, and lamprophyres) are difficult or impossible to produce using upper mantle mineralogy, known D values, REE content of upper mantle peridotite, and theoretical melting models (Gast, 1968; Kay and Gast, 1973; Cullers and Medaris, 1977; Mysen, 1977). For example, Mitchell and Brunfelt (1974) calculated that less than 1% melting of a garnet lherzolite could produce the La/Yb ratios of kimberlites, but not the REE abundances. REE abundances predicted in the models are lower than those observed. The large LREE content in rocks like kimberlite can only be approached by using the lowest D values, the largest REE contents found in peridotite, and the very smallest degrees of melting (Cullers and Medaris, 1977; Cullers et al., 1982). Such conditions are probably unrealistic. For example, Nd isotopic anomalies suggest that widespread LREE-enriched mantle was not present through geologic time (DePaolo and Wasserburg, 1976). The extreme REE content of kimberlite must be influenced by processes in addition to partial melting.

A possible exception to this conclusion would be possible if the mantle contained a LREE-rich accessory phase which could undergo disequilibrium melting (Beswick and Carmichael, 1978; Campbell and Gorton, 1980). This accessory phase would have to be the first to melt completely, thereby enriching the small amounts of first-formed melt with its large LREE content. REE contents in monazite are at the appropriate level for this model. The thermodynamic stability of fluorapatite or whitlockite in the mantle allow these minerals to be likely candidates. The correlation of the abundances of LREE with phosphorus in kimberlites and in mafic rocks in general does, according to Beswick and Carmichael (1978), support the presence of a phosphate accessory in the mantle source.

One deficiency in this argument is that such LREE-enriched phases have not been found in upper mantle assemblages. The major problem with this interpretation, however, is the conclusion that the more P_2O_5-rich and REE-rich basic magmas are formed from progressively more refractory sources. In other words, the greater the degree of melting the greater is the REE and P_2O_5 contents of the melt. This interpretation is fundamentally different to the usual interpretation that phosphorus and REE behave as incompatable elements so that the concentrations of these elements decrease with increased melting of the source. For this reason and other reasons, this possible interpretation has been severely criticized by Frey et al. (1980). This problem is further discussed by Beswick and Carmichael (1980).

Zone refining

The large REE content of kimberlites might also be produced by a zone refining process. The effect of this process would be quite similar to that produced by a very small degree of partial melting so that the LREE enrichment would be maximized (Paul et al., 1975).

Fractional crystallization

REE contents of kimberlites may be produced by high-pressure fractionation of eclogite from a less REE-enriched basic melt, consistent with models suggested by O'Hara and Yoder (1967) and MacGregor (1970). For example, 96% crystallization of a solid with equal amounts of garnet and clinopyroxene from Hawaiian basic lavas could produce the REE content of kimberlites (Mitchell and Brunfelt, 1974). However, on the basis of evidence other than the REE, models of partial melting are favored over those of fractional crystallization.

Volatile transport

Fesq et al. (1974) believe the enrichment of LREE relative to HREE in kimberlites is principally due to volatile transport of stable REE complexes. At low pressure, HREE complexes are more stable than LREE complexes in carbonate-rich fluids separating from silicate melts (Kosterin, 1959; Mineyev, 1963) hence, LREE/HREE ratios would increase in the melts during volatile removal (Fesq et al., 1974).

An alternative model of volatile transport is suggested by experimental studies at mantle and crustal pressure (Mysen, 1979; Wendlandt and Harrison, 1979). In contrast to previous experimental studies, Wendlandt and Harrison (1979) find the LREE are more soluble than the HREE in CO_2- and H_2O-rich fluids in fluid/melt equilibria at 5 and 20 kbar. For example, the REE (especially the LREE) become very soluble in H_2O-rich fluids relative to silicate melts or solids at mantle pressure. A large amount of the REE could be contained in the H_2O-rich fluids in contact with silicate melt and solid during partial melting if the mass of the H_2O was large enough. At lower pressure, the REE become less soluble in the H_2O fluids relative to silicate melt (Cullers et al., 1973; Zielinski and Frey, 1974; Mysen, 1979) so the REE may partition back into the silicate melt as these fluids rise to shallow depth. This model might explain the enriched LREE content of such rocks as kimberlite which could have formed in contact with such abundant volatiles.

Contamination

Kimberlites that are of small volume and have pierced large volumes of continental crust are suspect to crustal contamination. Many kimberlites contain swarms of crustal xenoliths. Major element contents (e.g., low silica) of most kimberlites suggest they have not been extensively contaminated, but those kimberlites enriched in silica might be the products of extensive crustal contamination.

Average crustal rocks contain somewhat smaller LREE content and larger HREE content compared to kimberlites, so assimilation of crustal rocks would tend to lower LREE/HREE ratios. Intermediate crustal igneous rocks and shales have small, negative Eu anomalies (e.g., Cullers et al., 1979), and most granites contain negative Eu anomalies (Chapter 8). Assimilation of these rocks could result in some of the negative Eu anomalies observed in some kimberlites. For example, the Koffyfontein kimberlites, South Africa, have small Eu anomalies, and they contain small REE contents, small LREE/HREE ratios, and large silica contents compared to most kimberlites (Fesq et al., 1974).

7.3. Carbonatites and associated alkalic silicate rocks

General

Carbonatites are carbonate-rich rocks that appear to be magmatically derived (cf. Heinrich, 1966; Tuttle and Gittins, 1966). Intrusive carbonatites commonly occur as dikes. Extrusives occur as pyroclastics and flows. The carbonate minerals are usually calcite (sövite), dolomite (rauhaugite), or ankerite, but many other minerals may occur.

Associated alkalic silicate rocks are commonly ultramafic-mafic rocks of the urtite-ijolite-melteigite-pyroxenite series (varied amounts of nepheline and pyroxene) and nepheline syenite. Many other rock types may be present. The size of the complex is usually small; most are less than 20 square miles.

Chemically, carbonatites contain large contents of Nb, Ti, Zr, Ba, Sr, Th, and the LREE (Heinrich, 1966). These concentrations, as well as isotopic and experimental constraints, preclude carbonatites being remobilized limestones (e.g., Tuttle and Gittins, 1966; Loubet et al., 1972).

REE contents

Carbonatites contain the largest REE contents (ΣREE = 72—15,515 ppm) and LREE/HREE ratios ($(La/Lu)_{cn}$ = 7.1—1240) of any rock type (Fig. 7.2; Vainshtein et al., 1961; Schofield and Haskin, 1964; Kapustin, 1966; Balashov and Pozharitskaya, 1968; Loubet et al., 1972; Barber, 1974, Eby,

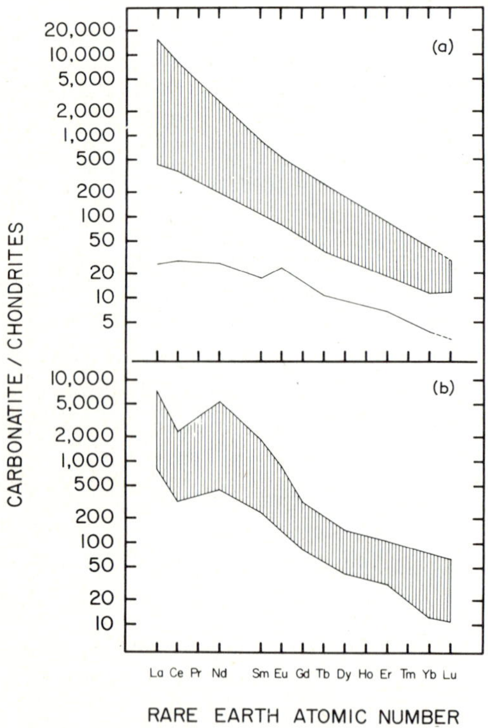

Fig. 7.2. (a) The range of REE content found in most carbonatites that contain no Ce anomaly is normalized to chondrites. Most carbonatite concentrations range within the vertical striped lines, but one carbonatite that is unusually low in REE content and may contain a positive Eu anomaly is also included (single, solid line). (b) The range of REE content found in carbonatites that contain a negative Ce anomaly is normalized to chondrites.

1975; Mitchell and Brunfelt, 1975; Cullers and Medaris, 1977; Armbrustmacher, 1979). Rauhaugites and sövites do not appear to contain significantly different REE contents or patterns. Carbonatites have no Eu anomalies, but some have negative Ce anomalies (Fig. 7.2). Samples with no Ce anomalies may have formed under less oxidizing conditions than those with Ce anomalies.

Correlation of REE content with other parameters have been found in some studies. REE contents and LREE/HREE ratios of carbonatites increase with decreasing age in some complexes (Vainshtein et al., 1961; Kapustin, 1966; Balashov and Pozharitskaya, 1968; Barber, 1974). Also in a given complex, the REE are scattered in the earliest-formed carbonatites in minerals like calcite, apatite, and pyrochlore, but they are concentrated in the late-stage carbonatites in REE minerals.

Armbrustmacher (1979) emphasizes the difference in REE content between replacement and primary, magmatic carbonatites in the Wet Mountains, Colorado. Replacement carbonatites show textures indicating replacement of porphyritic or hypidiomorphic-granular igneous rocks by carbonates, but primary magmatic carbonatites show no evidence of replacement. The replacement carbonatites have lower REE contents than primary magmatic carbonatites. REE are dispersed in the rock-forming minerals in replacement carbonatites, but are concentrated in REE minerals in the primary carbonatites.

Alkalic silicate rocks associated with carbonatites contain lower REE content and lower to similar LREE/HREE ratios than associated carbonatites (Fig. 7.3). REE distributions in the common ijolite series associated with the carbonatites contain REE contents quite similar to kimberlites. Most of the alkalic silicate rocks do not contain Eu anomalies although a few samples at the Oka complex contain either positive or negative Eu anomalies (Eby, 1975).

General petrogenesis

The use of the REE for the petrogenesis of carbonatites and associated alkaline silicate rocks has been reviewed most extensively by Eby (1975) at Oka, Quebec, by Mitchell and Brunfelt (1975) at Fen, Norway, and by Cullers and Medaris (1977) at Seabrook Lake and Callander Bay, Canada. As with kimberlite petrogenesis, the processes of melting, crystallization, and volatile transport are believed to be important in affecting REE distributions in carbonatites and associated alkalic silicate rocks. In addition, immiscibility of carbonate melts from basic silicate melts is also an important process. A major deficiency in the use of REE contents for petrogenetic models in these systems is the lack of appropriate D values for mineral/melt/fluid equilibria in these systems as they cool and crystallize.

Partial melting

As with kimberlites, it is difficult or impossible to produce the large REE contents and LREE/HREE ratios of carbonatites and associated alkalic silicate melts by partial melting of upper mantle peridotite. In concert with the pioneering models of Gast (1968), an alkali-rich, basaltic parent melt of lower REE content than rocks of the ijolite series is presumed to form by less than about 5% melting of a garnet peridotite (Cullers and Medaris, 1977; Eby, 1975; Mitchell and Brunfelt, 1975). Fractional crystallization of this parent can then produce the varied silicate rock types in a complex. A major criticism of the proposal for such a parent melt is that no direct evidence of its occurrence in many complexes is available. Also, partial melting of volatile-free spinel peridotites, rather than garnet peridotites, may produce alkali olivine basalts (Mysen, 1977). Melting of spinel peridotites does not produce the required REE contents in the melt. Perhaps the alternative models discussed with regard to kimberlite petrogenesis (e.g., volatile transport of REE in the mantle) may be more realistic explanations of the large REE contents in these rocks.

At about 1200°C and 25 kbar, carbonatite melts can be produced by a small percent melting of peridotite enriched in CO_2 (Eggler, 1976; Wyllie and Huang, 1975). *D* values for the REE between such carbonate melts and residual minerals are not available to evaluate such models quantitatively, but recent experiments by Wendlandt and Harrison (1979) suggest the REE would be more concentrated in carbonate melts than silicate melts, as required during small degrees of melting.

Fractional crystallization

Fractional crystallization of minerals from a presumed alkali basalt parent at varied pressure can produce the REE content of many silicate rocks in these complexes. At Seabrook Lake, about 50% high-pressure crystallization of clinopyroxene (50—99%), olivine (⩽50%), and garnet (~1%) from an alkali basalt parent could have produced the REE content of the ijolite (Cullers and Medaris, 1977). The REE contents of the carbonatite there could not be produced by further crystallization of minerals in the ijolite since residual melts would contain lowered REE contents due to sphene and apatite crystallization, instead of increased REE contents as required. A similar conclusion was reached to explain the sequence of alkali basalt (not observed)-ijolite-okaite at Oka (Eby, 1975). About 50% crystallization of augite and olivine in a ratio of 60/40 from a presumed alkali basalt parent could produce the REE content of the ijolite. Then about 85% crystallization of nepheline, pyroxene and garnet in a ratio of 49/49/2 from the ijolites could produce the REE contents of the okaites. Further crystallization of minerals from the okaites (melilite:

biotite:perovskite = 16/3/1) could not generate the REE content of the carbonatites.

The large degree of fractional crystallization to produce the required REE contents in such models may be unrealistic. Balashov and Krigman (1975) emphasize that lower mineral/melt *D* values for the REE in alkaline magmas may be more realistic to use than the higher values used in most studies. Using these lower *D* values could result in the extreme REE enrichment for more reasonable degrees of fractional crystallization. At Fen, the varied REE distributions of the alkaline rocks were not evaluated quantitatively, but some generalizations were summarized (Mitchell and Brunfelt, 1975). Low-pressure fractional crystallization could produce the REE distributions within the ijolite series, but further fractionation of nephelinite could not produce the REE contents of the sövites and rauhaugites.

The effect of fractional crystallization of minerals on REE contents of evolving carbonatite melts has not been evaluated to any extent. Melts approximating carbonatite compositions are very fluid (Wyllie and Tuttle, 1960) so rapid crystal settling might be expected. Crystallization of calcite from carbonatites in Kenya appeared to concentrate the REE in residual magmas since younger carbonatites had larger REE contents than older ones (Barber, 1974). The increase in REE content of carbonatites with decreasing age discussed previously is also consistent with this conclusion.

Liquid immiscibility

Since fractional crystallization of minerals from alkalic silicate magmas associated with carbonatites could not produce the REE content of carbonatites, a process of liquid immiscibility has been proposed to explain the REE content of the presumably cogenetic magmas (Eby, 1975; Mitchell and Brunfelt, 1975; Cullers and Medaris, 1977). Clearly the REE tend to concentrate in carbonatite melts relative to silicate melts that are often of ijolitic or lamprophyric composition in several petrographic provinces and in experimental systems (Fig. 7.3). Ratios of LREE in carbonatite to ijolite (lamprophyre) in a given natural complex tend to be similar or larger than the HREE (Fig. 7.3). Ratios of LREE in the carbonate ocelli relative to lamprophyric host rock measured at Callander Bay are similar or smaller than for the HREE (Cullers and Medaris, 1977). Carbonate-rich ocelli in lamprophyric host rock from Callander Bay have been interpreted as having formed by liquid immiscibility in ultrabasic, alkaline magma (Ferguson and Currie, 1971).

Contrary to results for most natural rocks, *D* values for the LREE in the experimental system of carbonate (K_2CO_3) relative to silicate (sanidine) melt are smaller than for the HREE (Wendlandt and Harrison, 1979). The difference between natural rocks and the experimental system could be due to several causes. One is the large difference in composition between the

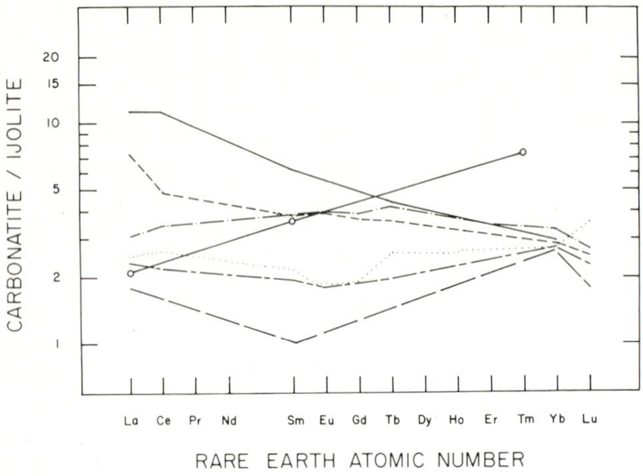

Fig. 7.3. The ratio of the REE content of carbonatite to associated ijolite found in different complexes. Averages are used if more than one analysis of the complex is compared. The key for the complexes is as follows: Oka (Eby, 1975) — solid line; Seabrook Lake (Cullers and Medaris, 1977) — dash, dot; Callander Bay (Cullers and Medaris, 1977) — dots; Fen (Mitchell and Brunfelt, 1975) — long dash, short dash; Magnet Cove (R.L. Cullers, unpublished) — short dash; Iron Hill, Colorado (R.L. Cullers, unpublished) — short dash; experimental sanidine melt/K_2CO_3 melt, 5 kbar, 1300°C (Wendlandt and Harrison, 1979) — solid line with circles.

natural systems and the experimental system. Also, the carbonatite magma could have separated from silicate magmas that were not of ijolitic or lamprophyric composition so rocks of the correct composition may not be compared in Fig. 7.3. As pointed out by Wendlandt and Harrison, estimated *D* values of REE between sövite and a lamprophyre at the Fen complex are similar to their experimental results. Finally, subsequent fenitization may wipe out the REE distributions produced by liquid immiscibility. Carbonatites with high LREE/HREE abundance ratios relative to those of coexisting silicate melts could be due to addition of the LREE to the carbonatite during or after solidification. The higher potential mobility of the LREE relative to that of the HREE in CO_2-rich fluids at crustal pressure is consistent with this possibility.

A process of liquid immiscibility to produce two silicate melts has been proposed for some alkaline rocks, but the lack of experimental *D* values for the REE between the melts makes it impossible to evaluate the role of this process. For example, a lamprophyric melt at Fen is presumed to have separated into a syenite liquid and an ijolite liquid during petrogenesis of the suite (Mitchell and Brunfelt, 1975); REE data are available, but cannot be evaluated.

Volatile transfer

As discussed previously, volatile transport of REE in H_2O-rich fluids at mantle pressures and in CO_2-rich fluids at crustal pressures may be an important process. Volatile transport of the REE is considered to be extremely important in the petrogenesis of carbonatites or associated alkaline rocks (Kapustin, 1966, Balashov and Pozharitskaya, 1968, and Loubet et al., 1972). At Fen, the extreme LREE enrichment of a hematitic sövite (Fig. 7.2a) is attributed to LREE volatile transport during hydrothermal alteration of a lamprophyre (Mitchell and Brunfelt, 1975).

If REE are transported in CO_2-rich fluids, LREE enrichment of the fenites compared to the unaltered rock may occur. LREE enrichment was noted in fenitized quartzite in the Borralan complex, Scotland (Martin et al., 1978). In contrast, little difference in REE content between fenitized and unfenitized country rock occurs around a lamprophyric dike at Seabrook Lake (Cullers and Medaris, 1977). There, the fenitized parts are depleted in HREE compared to unfenitized parts taken only 2.85 m apart. One extremely fenitized sample at another location at Seabrook Lake is very much depleted in REE compared to unaltered country rock. The depletion of REE in fenites might be explained by the availability of REE in fenitizing solutions. Solutions greatly undersaturated in REE would have a large dissolving capacity for the REE; hence, such solutions might leach the REE from the fenite and move them elsewhere.

7.4. Lamprophyres

General

Lamprophyres are dark rocks which occur as dikes, and they may contain two generations of euhedral minerals. Phenocrysts include olivine, hornblende, biotite, and augite. The groundmass may include considerable felsic minerals such as plagioclase, sanidine, or melilite as well as varied secondary minerals. Chemically, lamprophyres contain small silica contents and large alkali and incompatible trace element contents. Minettes (mica-alkali feldspar lamprophyres), for example, are chemically similar to kimberlites (Bachinski and Scott, 1979).

REE Contents

REE contents (ΣREE = 261—1033 ppm) and LREE/HREE ratios are large $(La/Lu)_{cn}$ = 20—55 and similar to the REE content of kimberlites; there are no Eu anomalies (Eu/Sm = 0.22—0.29) in the few analyzed samples (Fig. 7.4; Kay and Gast, 1973; Faerseth et al., 1976; Cullers and Medaris,

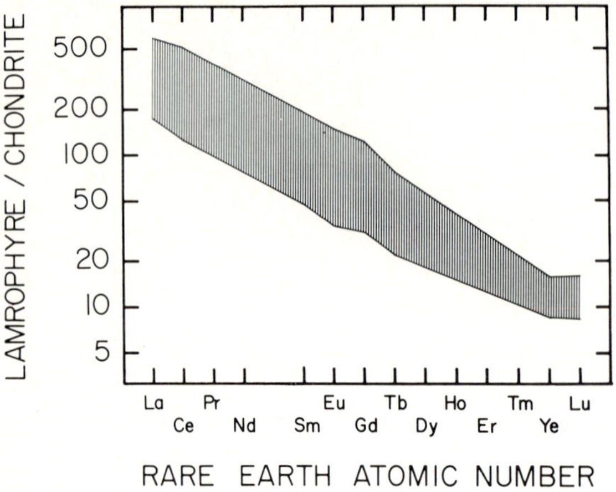

Fig. 7.4. The range of REE content found in lamprophyres is normalized to chondrites.

1977; Bachinski and Scott, 1979; Jahn et al., 1979; Cullers and Arnold, 1981). There may be a correlation of increased LREE content in lamprophyres with increased alkalinity although too few samples have been analyzed to make a definite statement.

Petrogenesis

Because the REE contents of lamprophyres are very similar to those of kimberlites, theoretical models of the petrogenesis of lamprophyres, using the REE, tend to be similar to those models for kimberlites. Preferred models to explain incompatible trace element contents thus include less than 5% melting of undepleted, upper mantle peridotite followed by possible fractional crystallization or volatile transport (Kay and Gast, 1973; Bachinski and Scott, 1979; Jahn et al., 1979; Cullers and Arnold, 1981).

Models for the formation of K-rich lamprophyres by the reaction of a basic magma with crustal sedimentary rocks, granite liquid, or granite solid have been rejected on the basis of REE abundances, as well as major element and other trace element data. For example, at the Spanish Peaks igneous complex, Colorado (U.S.A.) the alkali gabbro dikes and all analyzed granitic rocks contain lower LREE contents than the minettes (Jahn et al., 1979). Most sedimentary rocks also contain lower LREE contents compared to the lamprophyres (e.g., Bachinski and Scott, 1979; Cullers et al., 1979). Mixing of any combination of the gabbroic melts with granites or sedimentary rocks with low LREE contents could not produce the lamprophyric melts with larger LREE contents. There are also problems of mixing most crustal materials that tend to contain negative Eu anomalies to produce lamprophyres with no Eu anomalies (Bachinski and Scott, 1979).

Details of melting, crystallization or volatile transport models will not be elaborated on here. The few studies that consider these models argue in an identical fashion to arguments discussed in connection with kimberlites (see section 7.2).

7.5. Komatiites and associated basic rocks

Introduction

Although the terminology of komatiites is not yet standardized, komatiites are usually defined as non-cumulative rocks with ≥9% MgO and $CaO/Al_2O_3 \geq 1$ (e.g., Sun and Nesbitt, 1978; Jahn and Sun, 1979). Komatiites usually contain a quench, spinifex texture. The higher-Mg types (MgO ≥ 20%) of komatiites may be called spinifex-textured peridotitic komatiites, and the lower-Mg types (MgO = 9—20%) may be called basaltic komatiites. The basaltic komatiites do not necessarily contain spinifex, quench textures.

Komatiites usually occur in the lower part of some Precambrian greenstone belts as interlayered basic and ultrabasic rocks with pillow structures (e.g., Windley, 1977). Recently, a possible Tertiary or Mesozoic komatite complex has been found (Echeverria, 1980). The basic and ultrabasic rocks of the lower part of Precambrian greenstone belts, if present, grade upward into a calc-alkaline sequence containing basalts, andesites, and rhyolites. Only the REE geochemistry of the peridotitic komatiites, basaltic komatiites, and associated basalts will be summarized in this section.

REE contents

Komatiites contain fairly low REE contents (Σ REE = 10.1—59.1 ppm) and LREE/HREE ratios ($(La/Lu)_{cn}$ = 0.24—4.8; Herrmann et al., 1976; Arth et al., 1977; Hawkesworth and O'Nions, 1977; McGregor and Mason, 1977; Sun and Nesbitt, 1977, 1978; O'Nions and Pankhurst, 1978; Whitford and Arndt, 1978; Echeverria, 1980; Jahn et al., 1980). REE contents and La/Lu ratios tend to increase in the order peridotitic komatiite, basaltic komatiite, and tholeiite (Fig. 7.5). Certainly within a given komatiite suite REE contents and La/Lu ratios increase with decreasing MgO. Komatiites may or may not contain Eu anomalies (range Eu/Sm = 0.25—0.55). Such Eu anomalies may be the result of extensive alteration of some of these samples rather than due to the feldspar/melt equilibria (Sun and Nesbitt, 1978; Jahn and Sun, 1979).

General petrogenesis

Komatiites and associated basic rocks are often highly altered; therefore, many elements that are mobile during rock alteration cannot be used for the

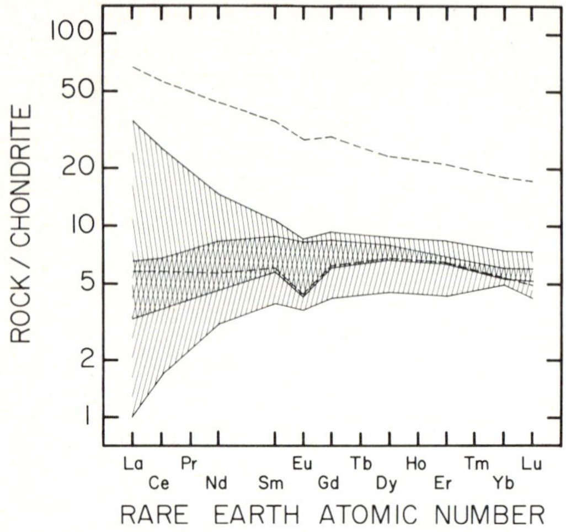

Fig. 7.5. The range of REE content found in peridotitic komatiites (northeast-southwest ruled lines), basaltic komatiites (northwest-southeast ruled lines), and associated tholeiites (between the two dashed lines) is normalized to chondrites.

interpretation of the igneous processes. The REE, however, exhibit minimal mobility compared to other elements. Therefore, they have been among the most important elements used to elucidate the petrogeneses of these rocks. Alteration effects in basic and ultrabasic rocks are discussed in more detail in Chapter 9 (also see Frey et al., 1974; Condie et al., 1977, or Sun and Nesbitt, 1978). Of special interest is the possibility that the LREE contents and Eu anomaly size may be changed in komatiites and basalt during alteration. Thus, Eu anomalies in komatiite sequences have not been used extensively to interpret igneous processes in altered samples.

The petrogenesis of peridotitic komatiites, basaltic komatiites, or associated basalts using the REE or other trace elements has been reviewed most extensively by Condie (1976), Hawkesworth and O'Nions (1977), O'Nions and Pankhurst (1978), Sun and Nesbitt (1978), and Jahn and Sun (1979). Basaltic komatiites and tholeiites associated with peridotitic komatites cannot in general be formed by fractional crystallization of more Mg-rich komatiites, so models for the formation of basaltic komatiites and tholeiites focus on incremental melting from heterogeneous sources with undepleted REE contents. Much larger degrees of partial melting (>50%) are required to produce peridotitic komatiites than to produce tholeiites or basaltic komatiites (e.g., D.H. Green, 1975) so the melt of peridotitic komatiites closely reflects the REE content of the source peridotite. Since most peridotitic komatiites in the Archean are depleted in the LREE relative to HREE in chondrite-normalized plots, the peridotite source of the peridotitic komatiites must have been depleted in the LREE in the same manner as in the source for modern oceanic-rise tholeiites (e.g., Sun and Nesbitt, 1977). Nd-Sm isotope studies suggest that the source for the peridotitic komatiites should be chondritic (no depletion of the LREE relative to the HREE; Hamilton et al., 1977) which means that the depletion of the LREE may have occurred a short time before melting or by a multi-stage (incremental) melting (Arth et al., 1977; Sun and Nesbitt, 1978; Echeverria, 1980). Thus, many models focus on an initial formation of basalts or basaltic komatiites by a relatively small percent melting of undepleted peridotite followed by extensive melting of the LREE-depleted peridotite. Melting of the LREE-depleted peridotite could produce the depleted peridotitic komatiites.

Partial melting

As summarized above, extensive melting of a source with varied LREE depletion can explain the varied LREE depletion of peridotitic komatiites (Arth et al., 1977; Hawkesworth and O'Nions, 1977; Sun and Nesbitt, 1978; Whitford and Arndt, 1978; Echeverria, 1980; Jahn et al., 1980). A few peridotitic komatiites have relatively flat chondrite-normalized REE patterns, so they appear to have been derived from mantle peridotite with fairly undepleted REE contents (Herrmann et al., 1976; Jahn et al., 1980). One peridotitic komatiite with relatively enriched LREE to HREE could be explained if the melt equilibrated with garnet (Sun and Nesbitt, 1978). One example is appropriate. Herrmann et al. (1976) believe that a peridotitic komatiite with a slight LREE depletion from the Barberton greenstone belt could have formed by about 60—80% melting of an upper mantle source with REE contents about 2—2½ times chondrite (residue is assumed to be olivine and pyroxene). The LREE content of the source would be slightly LREE depleted, but, in this case, it would not be as depleted as the source for most modern oceanic rise tholeiites.

Basaltic komatiites and tholeiitic basalts contain correspondingly larger REE contents and LREE/HREE ratios than peridotitic komatiites so these basaltic rocks are believed to have formed by a lesser degree of melting often from a less depleted source than the peridotitic komatiites (e.g., Arth et al., 1977; Jahn et al., 1980). For example, 8—19% partial melting of fairly undepleted peridotite (2—2½ times chondrite) concentrated in olivine and pyroxene will produce the REE content of the basaltic komatiites in the Barberton greenstone belt (Herrmann et al., 1976). In another study, garnet peridotite needed to have undergone a small percent melting to produce the tholeiites in Munro Township, Ontario (Arth et al., 1977). In contrast to some of the above melting models, a melting model for African greenstone belts, using a homogeneous source which underwent progressive melting at different pressure, could have produced the continuum of element contents in calc-alkaline andesites, tholeiites, and komatiites (Hawkesworth and O'Nions, 1977).

Although Eu anomalies in highly altered samples may be produced by the alteration effects, some relatively unaltered samples contain Eu anomalies. Basaltic komatiites with the largest REE contents and negative Eu anomalies within the suite from Munro Township, Ontario, could have formed by a relatively small percent melting of plagioclase peridotite at less than 40 km (Arth et al., 1977). Further melting of the source could produce melts with lower REE contents and no Eu anomalies as all the plagioclase would melt.

Fractional crystallization

Peridotitic komatiites cannot usually be shown to undergo fractional crystallization to produce the LREE-enriched basaltic komatiites and tholeiites associated with the peridotitic komatiites (Herrmann et al., 1976; Sun and Nesbitt, 1978; Echeverria, 1980; Jahn et al., 1980). In the Barberton greenstone belt mentioned above, the peridotitic komatiites would have to undergo 70—85% fractional crystallization of minerals that reject the REE (including olivine) to produce the basaltic komatiites. This model was rejected in favor of a melting model, since Ni should have decreased drastically during crystallization due to its concentration in the olivine, but, in fact, it showed little decrease. In the Gorgona Island complex, Colombia, the komatiitic magmas would have needed to precipitate very large amounts of olivine and clinopyroxene to produce the very much larger REE contents and LREE/HREE ratios of the associated basalts. Thus, melting models are favored over crystallization models.

Fractional crystallization models to produce some basalts or basaltic komatiites from more Mg-rich melts can nevertheless be allowed in a few cases (e.g., Sun and Nesbitt, 1978; Whitford and Arndt, 1978). For example, some peridotitic komatiites in the Barberton greenstone belt may have undergone some garnet crystallization to produce residual melts with large CaO/Al_2O_3 ratios and low HREE and Sc. Also, a large komatiite flow (parent has 19.6% MgO) in Munro Township, Ontario, has been shown to have undergone internal fractional crystallization of mainly olivine and clinopyroxene to produce progressively REE-enriched melts with lower MgO contents (MgO $\leqslant$ 8%). This latter example thus demonstrates that such crystallization processes are possible.

7.6. Alkaline mafic rocks and associated intermediate and felsic rocks

General

Alkaline magmatic rocks are defined as those rocks that have high levels of Na_2O and K_2O relative to their SiO_2 contents (e.g., Macdonald, 1968, fig. 1). Although this chapter is mainly concerned with basic rocks, some consideration will also be given in this section to the more silicic varieties of alkaline magma such as syenites and intermediate and felsic lavas associated with alkali mafic lavas.

While making up a small proportion of basaltic rocks, the alkali basic and associated lavas are found in a wide variety of geologic settings such as island arcs, oceanic islands, continental interiors, and rift zones. One question which must be addressed by any petrogenetic model is how a magmatic process, seemingly oblivious to the overlying crustal material, can generate magmas which are similar with respect to both major and trace element contents in such a range of plate tectonic settings.

Any attempt to review alkaline rocks soon enters on a collision course with the nomenclature of this diverse group of rocks. Even an abbreviated list of all the rock types is beyond the scope of this chapter (Sørensen, 1974). Therefore, no attempt will be made to treat all of the many varieties of alkaline rocks. Instead, a very general scheme will be used to divide basalts which classify as alkaline based upon $K_2O + Na_2O$ content into the following three groups:

(1) Alkali basalt group — $Na_2O > K_2O$, $K_2O + Na_2O \approx 3$–4, $K_2O/Na_2O \approx 0.5$.

(2) Shoshonite group — $Na_2O \approx K_2O$, $K_2O + Na_2O \approx 5$–6.

(3) Potassic basalt group — $K_2O \gg Na_2O$, $K_2O + Na_2O \approx 7$–10, $K_2O/Na_2O \approx 7$–10.

Felsic and intermediate rocks are classified with the associated basaltic rocks.

REE contents

All the alkaline rocks are characterized by a large enrichment of total REE and LREE relative to chondrites (Gast, 1968; Herrmann, 1968; Jakeš and Gill, 1970; Philpotts et al., 1971; Capaldi et al., 1972; Kay and Gast, 1973; Barberi et al., 1975; Goles, 1975; Masuda et al., 1975; Sun and Hanson, 1975a, b; 1976; Ewart et al., 1976; Faerseth et al., 1976; Kyle and Rankin, 1976; Lo and Goles, 1976; Mitchell and Bell, 1976; Baker et al., 1977; Kay, 1977; Neumann et al., 1977; Frey et al., 1978; T.H. Green, 1978; Johnson et al., 1978; Nicholls and Whitford, 1978; Rock, 1978; Gest and McBirney, 1979; Griffiths and Gibson, 1980; Morrison, 1980; Pe-Piper, 1980; Price and Taylor, 1980; Thompson et al., 1980). The range of REE contents of about 140 mafic, alkaline rocks and about 120 associated intermediate and felsic rocks are as follows (see also Table 7.1 and Fig. 7.6):

mafic rocks: ΣREE = 69–1453 ppm, $(La/Lu)_{cn}$ = 2.7–262, Eu/Sm = 0.24–0.40

intermediate and felsic rocks: ΣREE = 92–1750 ppm, $(La/Lu)_{cn}$ = 0.78–114, Eu/Sm = 0.13–0.47

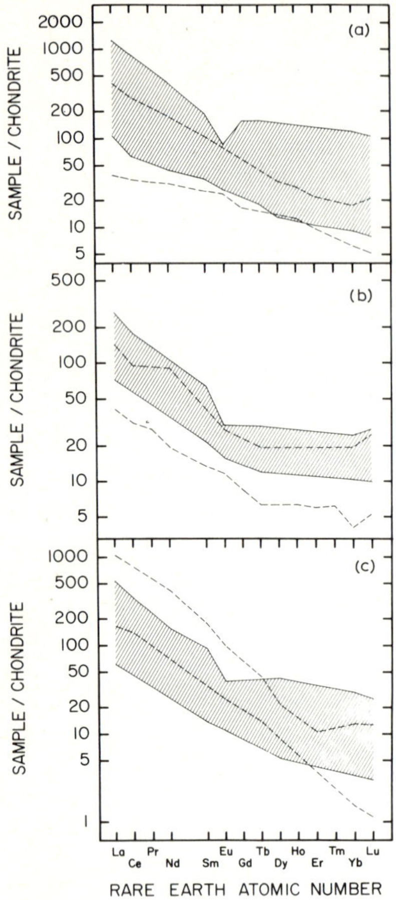

Fig. 7.6. (a) The range of REE content in the alkali basalt group (between the two dashed lines) and associated intermediate and felsic rocks (between the two solid lines in the ruled area) is normalized to chondrites. (b) The range of REE content in the shoshonite basalts (between the two dashed lines) and associated intermediate and felsic rocks (between the two solid lines and in the ruled area) is normalized to chondrites. (c) The range of REE content in the K-rich group of basalts (between the two dashed lines) and associated intermediate and felsic rocks (between the two solid lines and in the ruled area) is normalized to chondrites.

Among the mafic rocks, the largest ΣREE contents and La/Lu ratios belong to potassic lavas. Shoshonitic and alkali basalts have smaller ΣREE contents and La/Lu ratios. None of these mafic, alkaline rocks has even a moderate Eu anomaly. All of the Eu/Sm ratios are close to or slightly below the chondritic value with the lower ratios due to the steep slopes of the REE patterns. The more undersaturated rocks in each of the major groups (e.g., nephelinites and basanites as opposed to alkali basalts) generally have larger

ΣREE and La/Lu ratios. The more evolved rocks in each association show a tendency toward smaller La/Lu ratios and toward negative Eu anomalies. With the exception of potassic lavas, the more evolved rocks also tend to have larger ΣREE than associated mafic rocks.

Alkali basalt group. This group is the most diverse of those treated here, consisting of rocks ranging from transitional basalts through highly undersaturated nephelinites. The ranges of REE contents for all the mafic rocks and associated intermediate and felsic rocks are as follows (Fig. 7.6a):

mafic rocks: ΣREE = 90—610 ppm, $(La/Lu)_{cn}$ = 3.6—34, Eu/Sm = 0.24—0.4

intermediate and felsic rocks: ΣREE = 154—1750 ppm, $(La/Lu)_{cn}$ = 0.78—42, Eu/Sm = 0.13—0.47 (felsic rocks all have negative Eu anomalies while most intermediate rocks do not)

For the purposes of this discussion, the alkali basalt group will be further subdivided into the transitional (and alkali olivine basalt) association, the alkali basalt association, and the highly undersaturated basalt association (basanites, nepheline basalts and nephelinites). Transitional basalts have the lowest REE contents and the lowest LREE/HREE ratios of the three subgroups (ΣREE = 90—243 ppm, $(La/Lu)_{cn}$ = 3.6—12.2). Alkali basalts have marginally higher total REE and LREE/HREE ratios than transitional basalts (ΣREE = 130—250 ppm, $(La/Lu)_{cn}$ = 5.4—28.4). Undersaturated basalts have the highest REE content and LREE enrichments (ΣREE = 202—597 ppm, $(La/Lu)_{cn}$ = 13.7—34). While total REE are essentially identical for nephelinites and other undersaturated mafic rocks, the LREE enrichment is strongest in the nephelinites ($(La/Lu)_{cn}$ = 13.7—29.6 for other undersaturated rocks and 23—34 for nephelinites). None of the subgroups has a Eu anomaly, and Eu/Sm ratios are very similar for all three (transitional basalt = 0.27—0.4; alkali basalt = 0.27—0.35; undersaturated basalt = 0.24—0.33).

As is the case with the mafic rocks, intermediate rocks associated with transitional basalts have the lowest total REE contents and the lowest LREE enrichment (Fig. 7.6; ΣREE = 159—390 ppm; $(La/Lu)_{cn}$ = 5.8—9.2). Intermediate rocks associated with undersaturated alkali basalts have the highest REE contents (ΣREE = 380—1750 ppm), and those associated with alkali basalts are between the two extremes (ΣREE = 235—988 ppm). Both of the latter subgroups show wide ranges of LREE/HREE ratios ($(La/Lu)_{cn}$ = 4.8—41.8 for alkali basalt association and 13.5—30.4 for the undersaturated basalt association). Syenites span the entire range of La/Lu ratios and total REE.

A few of the intermediate rocks in the alkali basalt group have negative Eu anomalies. Intermediate rocks in the transitional basalt association have no Eu anomalies (Eu/Sm = 0.28—0.32) while some of those in the undersaturated association may contain negative Eu anomalies (Eu/Sm = 0.13—0.35). Intermediate rocks associated with alkali basalts and syenites range from slight, positive anomalies to moderate negative anomalies (Eu/Sm = 0.13—0.47). Of the syenitic rocks, only one analysis (Eu/Sm = 0.13) shows a moderate anomaly, and this sample also has the highest total REE content of the syenites (Neumann et al., 1977). The other syenite analyses have Eu/Sm ratios no lower than 0.22.

Only twelve analyses were found of felsic rocks associated with alkali or transitional basalts. Their total REE range of 316—738 ppm falls within the range of intermediate rocks, but the felsic rocks are distinguished by their low LREE/HREE ratios and consistent negative Eu anomalies $(La/Lu)_{cn}$ = 0.73—8.7. Eu/Sm = 0.13—0.18).

Shoshonite group. Mafic rocks of the shoshonite association have REE contents and distributions very similar to those of alkali and transitional basalts (Fig. 7.6b; 12 shoshonites; ΣREE = 69—240 ppm; $(La/Lu)_{cn}$ = 2.7—18.3; Eu/Sm = 0.25—0.34). The lower patterns belong to either intrusive varieties (absarokites) or to lavas in oceanic island arcs. The higher patterns are for continental shoshonites (e.g., Wyoming) or for continental island arc volcanic rocks.

Analyses of six felsic and intermediate rocks of the shoshonite association have slightly higher total REE and LREE/HREE ratios (Fig. 7.6b; ΣREE = 124—387 ppm; $(La/Lu)_{cn}$ = 3.2—23.1) than do the mafic shoshonites. They also show weak to moderate Eu anomalies (Eu/Sm = 0.16—0.28) with the lowest ratios belonging to the felsic rocks.

Potassic basalt group. K-rich volcanic rocks have by far the highest REE contents and LREE/HREE ratios of any of the alkaline, mafic rocks (Fig. 7.6c; 30 mafic rocks; ΣREE = 255—1453 ppm; $(La/Lu)_{cn}$ = 19.5—262). Like all the other mafic rocks in this section, there are no Eu anomalies in the K-rich basalts (Eu/Sm = 0.24—0.31).

Twenty intermediate K-rich rocks have a lower range of ΣREE (92—688 ppm) and a somewhat lower range of LREE/HREE ($(La/Lu)_{cn}$ = 3.8—114) than the mafic rocks. The intermediate rocks range from being without Eu anomalies to having large negative anomalies (Eu/Sm = 0.06—0.27).

Partial melting

Most petrogenetic models for alkaline basaltic magmas focus on partial melting of mantle peridotite because fractionation schemes cannot be generated which are consistent with all trace element data (Gast, 1968; Kay and Gast, 1973; Sun and Hanson, 1975a, 1975b; Frey et al., 1978). The

REE characteristics which must be explained by any model are the large LREE/HREE ratios and the strong enrichment of all REE relative to chondrites. A related problem (Chapter 6) is the generation of both unfractioned, or LREE-depleted, ocean floor basalts and highly fractionated (LREE-enriched) alkaline basalts by partial melting of the mantle.

Partial melting of any peridotite will enrich the melt in all REE and increase its LREE/HREE ratio relative to the original peridotite. Most of the REE reside in clinopyroxene and its mineral/melt D values are larger for HREE than for LREE (Gast, 1968; Kay and Gast, 1973; Sun and Hanson, 1975a). While some transitional and alkali basalts with low La/Lu ratios may be produced by melting of spinel peridotite, the predicted LREE/HREE ratios are much lower than those of most alkali basalts and all nephelinites and potassic basalts (Gast, 1968). In order to reproduce the higher La/Lu ratios, garnet, with its very large mineral/melt D values for HREE (Sun and Hanson, 1975a), must also be present in the residue left by partial melting. Kay and Gast (1973) suggested that variations in the garnet to clinopyroxene ratio of mantle peridotite could occur due to different P-T regimes within the mantle and that partial melting of these different zones would give rise to magmas with varying La/Lu ratios. Since the trace element content of the melt is controlled by mineral/melt partitioning, the amount of each mineral which melts will only affect the trace element content of the melt when a phase melts completely. Model calculations show that as the percent of partial melting increases, the REE content and LREE/HREE ratio of the melt decreases (Kay and Gast, 1973; Gast, 1968; Sun and Hanson, 1975a).

By varying the garnet/clinopyroxene ratio and percent of melting, Kay and Gast (1973) were able to generate theoretically the relative REE abundances of potassic basalts, nephelinites, and alkali basalts from a peridotite with chondritic REE pattern. For a source peridotite containing 20% clinopyroxene, 55% olivine, and 25% orthopyroxene, and varied amounts of garnet, they found the following parameters to be necessary to produce the corresponding rock types:

alkali basalt — 4.5—5% garnet and 0.8—2.9% partial melting
nephelinite — 8—10% garnet and 0.65—1.5% partial melting
potassic basalt — 11—17% garnet and 0.3—0.8% partial melting

Although the relative REE concentrations (i.e., shape of the REE patterns) can be produced by partial melting of a garnet peridotite, the absolute concentrations cannot be produced unless the mantle itself is enriched in REE relative to chondrites (e.g., Gast, 1968; Mitchell and Bell, 1976). The calculations of Kay and Gast (1973) and of Frey et al. (1978) for unfractionated peridotite required the mantle to be enriched over chondrites by a factor of 2—5. Sun and Hanson (1975a) estimated from covariance plots of REE for Ross Island alkali basalts that the source mantle must have been not only enriched in all REE (enrichment factor of about 3 for HREE) but also fractionated ($(La/Lu)_{cn}$ of about 4.5). Frey et al. (1978) also suggested from model calculations that the distributions of REE and of other trace elements in alkaline, mafic rocks are best explained by an enriched and fractionated mantle with HREE 2.5—3 times and LREE 7—9 times chondrite values. A discussion of mantle REE patterns can be found in Chapter 5.

Accepting the suggestion that the mantle may be, at least locally, enriched and fractionated allows for a higher and presumably more realistic degree of partial melting. Sun and Hanson (1975a) estimated 3—7% partial melting of mantle peridotite to produce nephelinite and 7—15% to produce alkali basalts. Frey et al. (1978) estimated partial melts of 4—6% for olivine melilite, 5—7% for basanite, and 11—15% for alkali olivine basalt. Also, garnet would no longer be essential to produce those transitional and alkali basalt magmas with relatively low La/Lu ratios. If partial melting of spinel peridotite increases the La/Lu of the melt to 2—4 times that of the source rocks (Gast, 1968) and that source rock has a La/Lu of 4.5, the resulting melt could then have a $(La/Lu)_{cn}$ of up to 18. Only nephelinites and potassic basalts have their entire ranges of $(La/Lu)_{cn}$ above this value (Table 7.1).

The most commonly proposed mechanism to enrich locally the mantle in REE (especially LREE) and in other incompatible elements is a two-stage process in which limited partial melting takes place in one part of the mantle to produce a LREE-enriched fluid or melt phase which intrudes another region of the mantle (Gast, 1968; Kay and Gast, 1973; Sun and Hanson, 1975a). This second region, which contains both peridotite and enriched material, is then subjected to partial melting to produce a highly fractionated and enriched alkali basic magma.

Generation of K-rich alkaline magmas such as potassic basalts and possibly shoshonites, requires the presence in the mantle peridotite of a K-bearing mineral such as phlogopite (Mitchell and Bell, 1976). In addition, local metasomatism prior to or during partial melting has been proposed to explain the very large LREE/HREE ratios and the trace element character of potassic basalts (e.g., Mitchell and Bell, 1976). Such a situation is nearly impossible to model because of lack of data regarding initial REE contents and D values for the minerals in a metasomatized mantle.

To our knowledge, no one has proposed a specific petrogenetic model for the production of shoshonite magmas, although Gest and McBirney (1979) suggested that Wyoming shoshonites were derived from an absarokite magma by fractionation (see below). The possibility that shoshonites are derived by greater degrees of melting from mantle sources similar to those for potassic basalts has not been evaluated but would face the same modelling problems as the genesis of potassic basalts.

Fractionation mechanisms

Intermediate and felsic, alkaline magmas associated with alkaline basic rocks may be produced by crystal fractionation from the more basic magmas (Barberi et al., 1975; Ewart et al., 1976; Kyle and Rankin, 1976; Sun and Hanson, 1976; Baker et al., 1977; Rock, 1978; Griffiths and Gibson, 1980; Thompson et al., 1980). Gest and McBirney (1979) suggested that shoshonites and absarokites from Wyoming are related to one another by crystal fractionation. Absarokites would represent either the parent magma which gave rise to shoshonites by fractionation of olivine and pyroxene (the interpretation preferred by Gest and McBirney, 1979) or alternatively, the absarokites may be a zone of accumulation of olivine and pyroxene within a system of shoshonite magmas.

Fractionation models for generation of intermediate and felsic, alkaline magmas center around crystallization of olivine, pyroxene, and feldspar, which increases the REE and SiO_2 contents of the residual melt. Crystallization of feldspar also reduces the LREE/HREE ratio of the melt and causes a negative Eu anomaly in the more evolved melts. Starting with a basaltic or basanitic magma, the general sequence of crystallization is as follows:

(1) Fractionation dominated by olivine and clinopyroxene to produce an evolved basaltic magma (hawaiite and ferrobasalt).

(2) Further fractionation of clinopyroxene accompanied by plagioclase to produce an intermediate magma (benmoreite and trachyte).

(3) Increased fractionation of plagioclase to drive the magma to more silicic compositions (phonolite).

(4) Fractionation of alkali feldspar to drive the magma toward peralkaline rhyolitic compositions (commendites and pantellerites).

Different variations on this theme are proposed for different volcanic suites. Crystallization of Fe-Ti oxides, kaersutite, apatite, and zircon sometimes needs to be invoked. In the McMurdo volcanic group (Kyle and Rankin, 1976; Sun and Hanson, 1976) original basanite or basanitoid magma differentiated along two fractionation paths depending upon whether early kaersutite formed. Barberi et al. (1975) traced the differentiation process from transitional basalt to pantellerite in the Afar Rift region. At roughly 45% residual liquid, fractionation of Fe-Ti oxides became important along with clinopyroxene and plagioclase. At about 30% liquid, Eu, Gd, and Tb contents and Fe^{2+}/Fe^{3+} ratio in the melt increased sharply, then dropped off to resume a trend similar to that of the other REE. Barberi et al. (1975) attributed this behavior to a f_{O_2} minimum at 30% liquid with both later and earlier magmas being more oxidized.

The general observation that the more evolved rocks within a given sequence have large, negative Eu anomalies has some exceptions. Thompson et al. (1980) found strong evidence for plagioclase fractionation in more felsic rocks of the Skye lava series in Scotland but found no Eu anomalies. They suggested that the lack of Eu fractionation was due in part to the high f_{O_2} of many of the magmas which precluded significant reduction of Eu^{3+} to Eu^{2+}. But, other lavas which apparently formed from reduced magmas (below the Ni-NiO buffer) are also without Eu anomalies, possibly due to some bulk compositional effect on Eu reduction or partitioning (Thompson et al., 1980).

7.7. Tholeiitic basalts — continental, back-arc basins, and island arcs

General

Tholeiitic basalts contain normative orthopyroxene and no normative nepheline while this is reversed in alkali basalts (Yoder and Tilley, 1962). Tholeiitic basalts can be distinguished from calc-alkali basalts or alkali basalts by lower K_2O content at a given SiO_2 (Macdonald, 1968; Jakěs and Gill, 1970).

Continental tholeiites (including intrusive complexes), island arc tholeiites, and back-arc basin tholeiites are included in this section. Island arc tholeiites tend to have a higher percent more silicic rocks associated with them than do oceanic tholeiites (Jakeš and Gill, 1970). Island arc tholeiites contain greater K, Rb, Ba, Cs, Pb, and Sr and lesser Fe, Mg, Ni, and Cr contents than oceanic rise tholeiites (Jakeš and Gill, 1970). Island arc tholeiites have similar Th, U, and REE contents as oceanic rise tholeiites.

Those tholeiites associated with komatiites are discussed in section 7.5. Ocean ridge and ocean floor tholeiites are examined in Chapter 6.

REE contents

The REE contents and La/Lu ratios of tholeiites tend to be lower than the alkali-rich basalts discussed in the previous section (Table 7.1). Tholeiitic sequences will be summarized in the order continental tholeiites, island arc and back-arc basin tholeiites, and intrusive complexes. The REE characteristics of continental tholeiites (not including intrusive complexes) are as follows (Fig. 7.7a): ΣREE = 15.2—322 ppm, $(La/Lu)_{cn}$ = 0.5—7.6, Eu/Sm = 0.16—0.55 (Balashov and Nesterenko, 1966; Frey et al., 1968; Herrmann and Jung, 1970; O'Nions and Clarke, 1972; Helmke and Haskin, 1973; Nathan and Fruchter, 1974; Mark et al., 1975; Leeman, 1976; Alexander and Gibson, 1977; Frey et al., 1978). Two lower patterns are shown in Fig. 7.7a to distinguish the lowest LREE-enriched pattern from the lowest LREE-depleted pattern. The largest LREE depletions belong to some tholeiites along continental margins which may have been related to continental breakup (O'Nions and Clarke, 1972), but not all of these basalts have

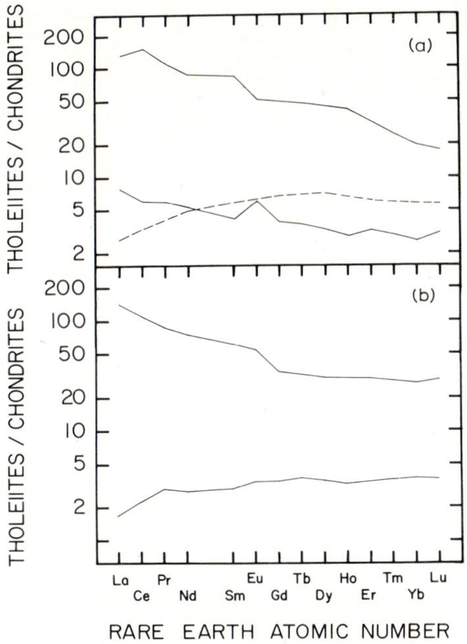

Fig. 7.7. (a) The range of REE content in most continental tholeiites (between the two solid lines) is normalized to chondrites. The lower pattern has an atypical positive Eu anomaly so a pattern with a depleted LREE content relative to the HREE and no Eu anomaly (dashed line) is also shown. (b) The range of REE content in island arc and back-arc tholeiites is normalized to chondrites.

depleted patterns. Although there is a fairly wide variation of Eu/Sm ratios, most continental tholeiites do not have Eu anomalies. Within a given suite the negative Eu anomaly size tends to increase with increasing REE content if plagioclase fractionation is important. For convenience, island arc, back-arc basin, and a few oceanic island tholeiites will be treated as one group. The REE contents of these rocks are as follows (Fig. 7.7b): ΣREE = 10—262 ppm, $(La/Lu)_{cn}$ = 0.4—7.3, Eu/Sm = 0.22—0.54 (Frey et al., 1968; Jakeš and Gill, 1970; Jakeš and White, 1972; Ewart and Bryan, 1973; Ewart et al., 1973, 1977; Tanaka and Sugisaki, 1973; Tanaka, 1974, 1975, 1977, Fujimaki, 1975; Ewart, 1976; Lo and Goles, 1976; Gorton, 1977; Hawkesworth et al., 1977; Tarney et al., 1977; T.H. Green, 1978; Masuda and Aoki, 1978; Dixon and Batiza, 1979; Heming and Rankin, 1979; Saunders and Tarney, 1979; Weaver et al., 1979; Whitford et al., 1979). Both LREE-depleted and LREE-enriched patterns exist for these tholeiites (Fig. 7.7b). Depleted patterns tend to have lower REE contents (Fig. 7.7b), but they are apparently not restricted to any one geologic setting. Both island arc and back-arc tholeiites may be either depleted or enriched, even within the same rock suite. High REE contents and greatly LREE-enriched patterns tend to be most common in island arc and oceanic island tholeiites. These tholeiites like continental tholeiites, have a wide variation of Eu/Sm ratio which is related more to the degree of LREE depletion than a reflection of Eu anomalies of varying size.

The REE contents of intrusive complexes are as follows (Fig. 7.8): ΣREE = 2.7–547 ppm, $(La/Lu)_{cn}$ = 0.31–19.3, Eu/Sm = 0.23–1.55 (Frey et al., 1968; Haskin and Haskin, 1968; Balashov et al., 1970; Potts and Condie, 1971; Balashov and Suschevskaya, 1973; Kosiewicz, 1973; Paster et al., 1974; Paul et al., 1977; Kuo and Crocket, 1979). Also shown in Fig. 7.8 are the ranges of REE patterns of chill or border zones. Such chill zones are believed to represent unfractionated melt that was quickly cooled at the margins of the intrusions and the country rock.

General petrogenesis

The results of experimental petrology suggest that tholeiites may form by about 15–30% melting of upper mantle peridotite (e.g., T.H. Green and Ringwood, 1968; D.H. Green, 1973). The *D* values for REE between tholeiitic melt and peridotite are less than one. At the relatively large percent melting needed to produce tholeiites (greater than about 15%), the REE concentrations in the melt are relatively insensitive to the degree of melting compared to the smaller degree of melting needed to produce alkali-rich

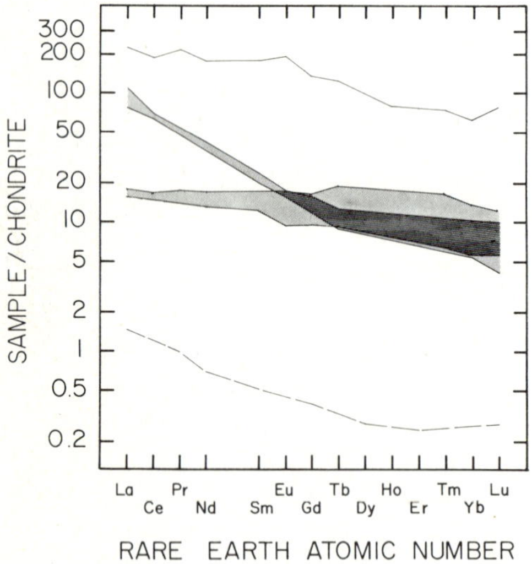

Fig. 7.8. Range of REE content in intrusive complexes is normalized to chondrites (range of intrusive complexes fall between the upper solid line and the lower dashed line; most chill zones: lower shaded lines for the REE; the Sudbury complex: upper shaded lines for the LREE).

basalts (Chapter 4). Thus, large variations in especially the LREE in tholeiites are usually attributed to variations in LREE content of the source. Certainly fractional crystallization of mainly plagioclase, pyroxene, or olivine from tholeiites does not result in much change in REE content unless the degree of crystallization is large.

The first study that used a successful quantitative approach to understanding trace element distributions in tholeiites was by Gast (1968). Good recent examples of applying trace elements to the understanding of tholeiitic petrogenesis can be found in Paster et al. (1974; intrusive rocks) and in Frey et al. (1978; extrusive rocks).

Partial melting

Most models for the formation of tholeiites assume about 20—30% melting of upper mantle peridotite with varied REE contents to explain the range of REE contents of the tholeiites. Tholeiites with the greater REE contents ($(La)_{cn}$ about 30—80; Lu about 5—20) and $(La/Lu)_{cn}$ ratios are assumed to melt from peridotite with REE contents enriched over chondrites ($(La)_{cn}$ about 5—20; $(Lu)_{cn}$ about 2—7) and with $(La/Lu)_{cn}$ ratios greater than one (e.g., D.H. Green, 1973; Frey et al., 1978). Some tholeiites from Norfolk Island which are abnormally enriched in the REE (La_{cn} about 100; Lu_{cn} about 35) and have unusual REE patterns may have inherited their REE contents from accessory minerals enriched in these elements (D.H. Green, 1973).

Tholeiites with low REE contents and flat to depleted REE distributions in chondrite-normalized curves are assumed to melt from primitive mantle with little or no enrichment to depletion of the LREE relative to the HREE (e.g., Jakeš and White, 1972; Saunders and Tarney, 1979; Masuda and Aoki, 1979; Whitford et al., 1979). Island arc tholeiites tend to have chondritic to LREE-depleted contents so they may come from similar sources as for oceanic ridge tholeiites, but at least some island arc tholeiites tend to come from less depleted sources compared to oceanic rise tholeiites (e.g., Jakeš and Gill, 1970; Dixon and Batiza, 1979).

A few tholeiites contain negative Ce anomalies in chondrite-normalized curves (Heming and Rankin, 1979; Dixon and Batiza, 1979). These anomalies may be inherited from the source since no phases which abnormally concentrate Ce relative to the REE have been found in mineral separates from such rocks.

Several examples of the use of REE in partial melting models are summarized below. Dixon and Batiza (1979) determined the petrology of the northern Marianas island arc, an example of an immature island arc. Least fractionated tholeiites in this sequence have larger K, Rb, Ba, U, and $^{87}Sr/^{86}Sr$ and smaller Ti, Zr, Nb, and Hf contents than for oceanic rise tholeiites. The peridotite source for the island arc tholeiites was interpreted to be similar to the source for the oceanic rise tholeiites but distinguishable in the above elements. The REE patterns are most decisive in precluding a subducted oceanic crust as the source. Only peridotite above the subduction zone and not subducted oceanic crust could melt and produce the REE content of these tholeiites. The subducted oceanic crust is too depleted in the LREE relative to the HREE to produce the REE patterns of the tholeiites.

Wood (1979) notes gradual depletion of the LREE relative to HREE from the lower to upper portions of volcanic piles in a number of tensional environments (Iceland, Faeroe Islands, Scotland, Troodos). He proposes a dynamic partial melting model to explain these changes in which some melt remains in the residue as increments of melt are continuously removed. Initial increments are enriched in the LREE and subsequent increments gradually become depleted in the LREE as the source becomes depleted. The melt remaining in the residue is a necessary feature of this model in order to keep the LREE from becoming depleted too rapidly during fractional melting.

Fractional crystallization

Crystallization in tholeiitic melts is dominated by plagioclase, pyroxene, or olivine (Haskin and Haskin, 1968; Jakeš and White, 1972; Helmke and Haskin, 1973; Kosiewicz, 1973; Potts and Condie, 1971; Paster et al., 1974; Tanaka, 1975; Hawkesworth et al. 1977; Frey et al., 1978; Kuo and Crocket, 1979; Saunders and Tarney, 1979; Weaver et al., 1979). Plagioclase-dominated crystallization will tend to produce noticeable negative Eu anomalies in the melt after about 15—20% crystallization, but much larger degrees of crystallization are required to produce very large negative Eu anomalies. Plagioclase-rich cumulates may contain positive Eu anomalies and enriched Al contents compared to melts not enriched in such cumulates. Olivine-dominated crystallization will cause REE contents to increase slowly with fractionation, but will not produce much change in REE patterns or produce a Eu anomaly. Clinopyroxene crystallization from tholeiitic melts at low pressure causes REE contents to increase with little fractionation resulting in normal tholeiitic trends (Tanaka, 1975). Clinopyroxene crystallization from tholeiitic melts at high pressure causes a large fractionation of the LREE from the HREE, producing large LREE relative to HREE in residual melts. The tholeiites differentiate into the alkali basalt field during this high-pressure differentiation.

In the large intrusive complexes from the Skaergaard, Muskox, Bushveld, Stillwater, and Sudbury intrusions, the early cumulates tend to have positive Eu anomalies and smaller La/Lu ratios than the presumed, parent melt of the chill zone gabbros (Haskin and Haskin, 1968; Kosiewicz, 1973; Paster et al., 1974; Kuo and Crocket, 1979). REE contents and La/Lu ratios tend to increase in the stratigraphic sequence so that the late-stage granophyres also contain negative Eu anomalies. These trends are caused by mainly plagioclase, pyroxene, or olivine accumulation or removal as discussed above.

7.8. Summary

Although there may be considerable overlap in the REE content of groups of ultrabasic and basic rocks, the abundances tend to decrease in the following order: carbonatites > ijolites and associated alkalic rocks, kimberlites, and lamprophyres > alkali basalts > continental tholeiites > island arc tholeiites > basaltic komatiites > komatiites (Table 7.1). There is a tendency for the LREE/HREE ratio or $(La/Lu)_{cn}$ ratio to decrease with decreasing REE content. Most basic and ultrabasic rocks do not contain Eu anomalies, in marked contrast, to more silicic rocks (Chapter 8). A few carbonatites showing a negative Ce anomaly have also been found.

Most models to explain REE contents of basic and ultrabasic rocks have focused on a varied percent melting of upper mantle peridotite with or without having undergone a previous melting episode (depleted vs. undepleted peridotite, respectively) possibly followed by low- or high-pressure fractional crystallization, crustal contamination, or volatile transport. The most successful studies include the REE as parts of comprehensive studies in which the model for the formation of a suite of rocks is found to be consistent with field, petrographic, major element, trace element (especially Rb, Ba, Sr, Ni), and isotopic data.

Undepleted or even relatively enriched peridotite is required to undergo a very small percent melting (less than 5%) to produce the large content of REE (especially LREE) and La/Lu abundance ratios of ijolites, kimberlites, alkali basalts, or some carbonatites. Even then, the predictions tend to be lower than the observed contents so many theoretical models also include subsequent crystallization of garnet and clinopyroxene at high pressure or volatile transport of the REE, in order to enrich the magmas in the LREE and to increase the La/Lu ratios. Also, the REE abundances in some carbonatites appear to be consistent with a petrogenesis involving liquid immiscibility from associated alkalic, silicate magmas. Small degrees of partial melting of a garnet peridotite which contains some K-bearing phase (phlogopite) are required to produce kimberlites, lamprophyres, and K-rich basalts. The large contents of LREE, K, and other incompatible elements in these rocks may also require that the peridotite be affected by some enrichment process such as alkali metasomatism. Intermediate and felsic, alkaline magmas are derived from parent basaltic magmas by crystal fractionation.

Relatively undepleted peridotite is required to undergo a larger percent melting (15—30%) to produce the lower REE contents and La/Lu ratios for continental tholeiites compared to ijolites, kimberlites, or alkali basalts. Depleted peridotite is required to melt to produce tholeiites at subduction zones or oceanic rises that are depleted in the LREE. Fractional crystallization of plagioclase, for example, from such tholeiitic magmas at low pressure may produce differentiated magmas with negative Eu anomalies.

Depleted peridotites undergoing greater than about 50% melting produce peridotitic komatiites depleted in the LREE. The depletion of the source peridotite may have been produced during a previous partial melting episode or by incremental melting in which progressive melting produces tholeiites, basaltic komatiites, and peridotitic komatiites with progressively depleted REE contents (especially the LREE).

Acknowledgements

Reviews by Fred Frey, J. Lawford Anderson, Karl Seifert, and John Fountain have improved this chapter. We thank Renée Hambleton and Monique Santilli for doing the typing and Bob Barnett for doing the drafting of the figures.

References

Alexander, P.O. and Gibson, I.L., 1977. Rare earth abundances in Deccan trap basalts. *Lithos*, 10: 143—147.

Armbrustmacher, T.J., 1979. Replacement and primary magmatic carbonatites from the Wet Mountains area, Fremont and Custer Counties, Colorado. *Econ. Geol.*, 74: 888—901.

Arth, J.G., Arndt, N.T. and Naldrett, A.J., 1977. Genesis of Archean komatiites from Munro Township, Ontario: trace element evidence. *Geology*, 5: 590—594.

Bachinski, S.W. and Scott, R.B., 1979. Rare-earth and other trace element contents and the origin of minettes (mica-lamprophyres). *Geochim. Cosmochim. Acta*, 43: 93—100.

Baker, B.H., Goles, G.G., Leeman, W.P. and Lindstrom, M.M., 1977. Geochemistry and petrogenesis of a basalt-trachyte suite from the southern part of Gregory rift. Kenya. *Contrib. Mineral. Petrol.*, 64: 303—332.

Balashov, Y.A. and Krigman, L.D., 1975. The effects of alkalinity and volatiles on rare-earth separation in magmatic systems. *Geochem. Int.*, 12: 165—170.

Balashov, Y.A. and Nesterenko, K.V., 1966. Distribution of the rare earths in the traps of the Siberian platform. *Geochem. Int.*, 3: 672—679.

Balashov, Y.A. and Pozharitskaya, L.K., 1968. Factors governing the behavior of rare-earth elements in the carbonatite process. *Geochem. Int.*, 5: 271—288.

Balashov, Y.A. and Suschevskaya, N.M., 1973. The rare earths in stratified ultrabasic intrusions. *Geokhimiya*, 12: 1823—1830.

Balashov, Y.A., Nadareyshvili, D.G. and Kekekeiya, M.A., 1970. Rare-earth balance in the gabbroids of the Kviran intrusion. *Geokhimiya*, 10: 1204—1214.

Barber, C., 1974. The geochemistry of carbonatites and related rocks from two carbonatite complexes, south Nyonza, Kenya. *Lithos*, 7: 53—63.

Barberi, F., Ferrara, G., Santacroce, R., Treuil, M. and Varet, J. 1975. A transitional basalt-pantellerite sequence of fractional crystallization, the Boina Centre (Afar Rift, Ethiopia). *J. Petrol.*, 16: 22—56.

Beswick, A.E. and Carmichael, I.S.E., 1978. Constraints on mantle source compositions imposed by phosphorus and the rare-earth elements. *Contrib. Mineral. Petrol.*, 67: 317—330.

Burkov, V.V. and Podporina, Y.K., 1966. First data on rare earths in kimberlites. *Dokl. Akad. Nauk S.S.S.R.*, 171: 215—219.

Campbell, I.H. and Gorton, M.P., 1980. Accessory phases and the generation of LREE-enriched basalts — a test for disequilibrium melting. *Contrib. Mineral. Petr.*, 72: 157—163.

Capaldi, G., Gasparini, P., Moauro, A., Salvia, E. and Travaglione, O., 1972. Rare earth abundances in the alkaline volcanic rocks from Campania, South Italy. *Earth Planet. Sci. Lett.*, 17: 247—257.

Condie, K.C., 1976. Trace-element geochemistry of Archean greenstone belts. *Earth-Sci. Rev.*, 12: 393—417.

Condie, K.C., Viljoen, M.J. and Kable, E.J.D., 1977. Effects of alteration on element distributions in Archean tholeiites from the Barberton greenstone belt, South Africa. *Contrib. Mineral. Petrol*, 64: 75—89.

Cullers, R.L. and Arnold, B., 1981. The petrogenesis of the Tertiary Spanish Peaks igneous complex, Colorado, U.S.A. (submitted).

Cullers, R.L. and Medaris, L.G., 1977. Rare earth elements in carbonatite and cogenetic alkaline rocks: Examples from Seabrook Lake and Callander Bay, Ontario. *Contrib. Mineral. Petrol.*, 65: 143—153.

Cullers, R.L., Medaris, L.G. and Haskin, L.A., 1973. Experimental studies of the distribution of rare earths as trace elements among silicate minerals and liquids and water. *Geochim. Cosmochim. Acta*, 37: 1499—1512.

Cullers, R.L., Chaudhuri, S., Kilbane, N. and Koch, R., 1979. Rare-earths in size fractions and sedimentary rocks of Pennsylvanian-Permain age from the mid-continent of the U.S.A. *Geochim. Cosmochim. Acta*, 43: 1285—1301.

Cullers, R.L., Mullenax, J., Demarco, M. and Nordeng, S. 1982. Trace element content and petrogenesis of kimberlites, Riley County, Kansas. *Am. Mineral.*, 67: 223—233.

Dawson, J.B., 1962. Basutoland kimberlites. *Geol. Soc. Am. Bull.*, 73: 545—560.

Dawson, J.B., 1972. Kimberlites and their relation to the mantle. *Philos. Trans. R. Soc. London, Ser. A*, 271: 297—311.

DePaolo, D.J. and Wasserburg, G.J., 1976. Nd isotopic variations and petrogenetic models. *Geophys. Res. Lett.*, 3: 249.

Dixon, T.H. and Batiza, R., 1979. Petrology and chemistry of recent lavas in the northern Marianas: implications for the origin of island arc basalts. *Contrib. Mineral. Petrol.*, 70: 167—181.

Eby, G.N., 1975. Abundance and distribution of the rare-earth elements and yttrium in the rocks and minerals of the Oka carbonatite complex, Quebec. *Geochim. Cosmochim. Acta*, 39: 597—620.

Echeverria, L.M., 1980. Tertiary or Mesozoic komatiites from Gorgona Island, Columbia: field relations and geochemistry. *Contrib. Mineral. Petrol.*, 73: 253—266.

Eggler, D.H., 1976. Does CO_2 cause partial melting in the low-velocity layer of the mantle? *Geology*, 4: 69—72.

Emmermann, R., Daicva, L. and Schneider, J., 1975. Petrologic significance of rare earth distribution in granites. *Contrib. Mineral. Petrol.*, 52: 267—283.

Ewart, A., 1976. A petrological study of the younger Tongan andesites and dacites, and the olivine tholeiites of Niua Fo'ou Island, S.W. Pacific. *Contrib. Mineral. Petrol.*, 58: 1—21.

Ewart, A. and Bryan, W.B., 1973. The petrology and geochemistry of the Tongan Islands: the western Pacific. In: P.J. Coleman (Editor), *Island Arcs, Marginal Seas, Geochemistry*, pp. 503—522.

Ewart, A., Bryan, W.B. and Gill, J.B., 1973. Mineralogy and geochemistry of the younger volcanic islands of Tonga, S.W. Pacific. *J. Petrol.*, 14: 429—465.

Ewart, A., Mateen, A. and Ross, J.A., 1976. Review of mineralogy and chemistry of Tertiary central volcanic complexes in southeast Queensland and northeast New South Wales. In: R.W. Johnson (Editor), *Volcanism in Australasia*. Elsevier, Amsterdam, pp. 21—40.

Ewart, A., Brothers, R.N. and Mateen, A., 1977. An outline of the geology and geochemistry, and the possible petrogenetic evolution of the volcanic rocks of the Tonga-Kermadec-New Zealand island arc. *J. Volcanol. Geotherm. Res.*, 2: 205—250.

Faerseth, R.B., MacIntyre, R.M. and Naterstad, J., 1976. Mesozoic alkaline dykes in the Sunnhordland region, western Norway: ages, geochemistry and regional significance. *Lithos*, 9: 331—345.

Ferguson, J. and Currie, K.L., 1971. Evidence of liquid immiscibility in alkaline ultrabasic dikes at Callander Bay, Ontario. *J. Petrol.*, 12: 561—585.

Fesq, H.W., Kable, E.J.D. and Gurney, J.J., 1974. Aspects of the geochemistry of kimberlites from the Premier mine, and other selected South African occurences with particular reference to the rare earth elements. *Phys. Chem. Earth*, 9: 687—707.

Frey, F.A., Haskin, M.A., Poetz, J.A. and Haskin, L.A., 1968. Rare earth abundances in some basic rocks. *J. Geophys. Res.*, 73: 6085—6098.

Frey, F.A., Haskin, L.A. and Haskin, M.A., 1971. Rare-earth abundances in some ultramafic rocks. *J. Geophys. Res.*, 76: 2057—2070.

Frey, F.A., Bryan, W.B. and Thompson, G., 1974. Atlantic Ocean floor: geochemistry and petrology of basalts from Legs 2 and 3 of the Deep-Sea Drilling Project. *J. Geophys. Res.*, 79: 5507—5527.

Frey, F.A., Green, D.H. and Roy, S.D., 1978. Integrated models of basalt petrogenesis: a study of quartz tholeiites to olivine melilitites from southeastern Australia utilizing geochemical and experimental petrological data. *J. Petrol.*, 19: 463—513.

Frey, F.A., Roden, M.F. and Zindler, A., 1980. Constraints on mantle source compositions imposed by phosphorus and the rare-earth elements. *Contrib. Mineral. Petrol.*, 75: 165—173.

Fujimaki, H., 1975. Rare-earth elements in volcanic rocks from Hakone volcano and northern Izu peninsula, Japan. *J. Fac. Sci., Univ. Tokyo, Sec. II*, 19: 81—93.

Gast, P.W., 1968. Trace-element fractionation and the origin of tholeiitic and alkaline magma types. *Geochim. Cosmochim. Acta*, 32: 1057—1086.

Gest, D.E. and McBirney, A.R., 1979. Genetic relations of shoshonitic and absarokitic magmas, Absaroka Mountains, Wyoming. *J. Volcanol. Geotherm. Res.*, 6: 85—104.

Goles, G.G., 1975. Basalts of unusual composition from the Chyulu Hills, Kenya. *Lithos*, 8: 47—58.

Gorton, M.P., 1977. The geochemistry and origin of quaternary volcanism in the New Hebrides. *Geochim. Cosmochim. Acta*, 41: 1257—1270.

Green, D.H., 1973. Experimental melting studies on a model upper mantle composition under water-saturated and water-undersaturated condition. *Earth Planet. Sci. Lett.*, 19: 37—53.

Green, D.H., 1975. Genesis of Archean peridotitic magmas and constraints on Archean geothermal gradients and tectonics. *Geology*, 3: 15—18.

Green, T.H., 1978. Rare earth geochemistry of basalts from Norfolk Island, and implications for mantle inhomogeneity in the rare earth elements. *Geochem. J.*, 12: 165—172.

Green, T.H. and Ringwood, A.E., 1968. Genesis of the calc-alkaline igneous rock suite. *Contrib. Mineral. Petrol.*, 18: 105—162.

Griffiths, P.S. and Gibson, I.L., 1980. The geology and petrology of the Hannington trachyphonolite formation, Kenya Rift Valley. *Lithos*, 13: 43—63.

Haskin, L.A. and Haskin, M.A., 1968. Rare-earth elements in the Skaergaard intrustion. *Geochim. Cosmochim. Acta*, 32: 433—447.

Haskin, L.A., Frey, F.A., Schmitt, R.A. and Smith, R.H., 1966. Meteoritic, solar, and terrestrial rare-earth distributions. *Phys. Chem. Earth*, 7: 167—321.

Hawkesworth, C.J. and O'Nions, R.K., 1977. The petrogenesis of some Archean volcanic rocks from southern Africa. *J. Petrol.*, 18: 487—520.

Hawkesworth, C.J., O'Nions, R.K., Pankhurst, R.J., Hamilton, P.J. and Evensen, N.M., 1977. A geochemical study of island-arc and back-arc tholeiites from the Scotia Sea. *Earth Planet. Sci. Lett.*, 36: 253—262.

Heinrich, E.W., 1966. *The Geology of Carbonatites.* Rand McNally Co., Chicago, Ill., 555 pp.

Helmke, P.A. and Haskin, L.A., 1973. Rare-earth elements, Co, Sc, and Hf in the Steens Mountain basalts. *Geochim. Cosmochim. Acta*, 37: 1513—1529.

Heming, R.F. and Rankin, P.C., 1979. Ce-anomalous lavas from Rabaul Caldera, Papua New Guinea. *Geochim. Cosmochim. Acta*, 43: 1351—1355.

Herrmann, A.G., 1968. Die Verteilung der Lanthaniden in basaltischen Gesteinen. *Contrib. Mineral. Petrol.*, 17: 275—314.

Herrmann, A.G. and Jung, D., 1970. Die Verteilung der Lanthaniden im Tholeyit von Tholey (Saar) und in Platatiniten, Pseudopegmatiten und Apliten des permischen Vulkanismus in Saar-Nake-Pflaz-gebiet. *Contrib. Mineral. Petrol.*, 29: 33—42.

Herrmann, A.G., Blanchard, D.P., Haskin, L.A., Jacobs, J.W., Knake, D., Korotev, R.L. and Brannon, J.C., 1976. Major, minor, and trace element compositions of peridotitic and basaltic komatiites from the Precambrian crust of southern Africa. *Contrib. Mineral. Petrol.*, 59: 1—12.

Ilupin, I.P., Varshal, G.M., Pavlutskaya, V.I. and Kalenchuk, G.E., 1974. Rare earth elements in Yakutian kimberlites. *Geokhimiya*, 1: 126—131.

Jahn, B. and Sun, S.S., 1979. Trace element distribution and isotopic composition of Archean greenstones. In: L.H. Ahrens (Editor), *Origin and Distribution of the Elements.* Pergamon, Oxford, pp. 597—618.

Jahn, B., Sun, S.S. and Nesbitt, R.W., 1979. REE distribution and petrogenesis of the Spanish Peaks igneous complex, Colorado. *Contrib. Mineral. Petrol.*, 70: 281—298.

Jahn, B., Auvray, B., Blais, S., Capdevila, R., Cornichet, J., Vidal, R. and Hameurt, J., 1980. Trace element geochemistry and petrogenesis of Finnish greenstone belts. *J. Petrol.*, 21: 201—244.

Jakeš, P. and Gill, J., 1970. Rare earth elements and the island or tholeiitic series. *Earth Planet. Sci. Lett.*, 9: 17—28.

Jakeš, P. and White, A.J.R., 1972. Major and trace element abundances in volcanic rocks or orogenic areas. *Geol. Soc. Am. Bull.*, 83: 29—40.

Johnson, R.W., MacKenzie, D.E. and Smith, I.E.M., 1978. Delayed partial melting of subduction-modified mantle in Papua, New Guinea. *Tectonophysics*, 46: 197—216.

Kapustin, Y.L., 1966. Geochemistry of rare earth elements in carbonatites. *Geochem. Int.*, 3: 1054—1054.

Kay, R.W., 1977. Geochemical constraints on the origin of Aleutian magmas. In: M. Talwani, C.G. Harrison and W.C. Pitman III (Editors), *Island Arcs, Deep Sea Trenches, and Back-arc Basins. Am. Geophys. Union, Maurice Ewing Ser.*, 1: 229—242.

Kay, R.W. and Gast, P.W., 1973. The rare-earth content and origin of alkali-rich basalts. *J. Geol.*, 81: 653—682.

Kosiewicz, S.T., 1973. *Rare-earth elements in U.S.G.S. rocks SCo-1 and STM-1, basalts from the Servilleta and Hindale formations, and rocks from the Stillwater and Muskox intrusions.* Ph.D. Thesis, University of Wisconsin, Madison, Wisc. (unpublished).

Kosterin, A.V., 1959. The possible modes of transport of the rare earths by hydrothermal solutions. *Geochemistry (U.S.S.R.)*, 4: 381—387.

Kuo, H.Y. and Crocket, J.H., 1979. Rare earth elements in the Sudbury nickel irruptive: comparison with layered gabbros and implications for nickel irruptive petrogenesis. *Econ. Geol.*, 74: 590—605.

Kyle, P.R. and Rankin, P.C., 1976. Rare earth element geochemistry of Late Cenozoic alkaline lavas of the McMurdo volcanic group, Antarctica. *Geochim. Cosmochim. Acta*, 40: 1497—1507.

Leeman, W.P., 1976. Petrogenesis of McKinney (Snake River) olivine tholeiite in light of rare-earth element and Cr/Ni distributions. *Geol. Soc. Am. Bull.*, 87: 1582—1586.

Lo, H.H. and Goles, G.G., 1976. Compositions of Formosan basalts and aspects of their petrogenesis. *Lithos*, 9: 149—159.

Loubet, M., Bernat, M., Javoy, M., and Allègre, C.J., 1972. Rare earth contents in carbonatites. *Earth Planet. Sci. Lett.*, 14: 226—232.

Macdonald, G.A., 1968. Composition and origin of Hawaiian Lavas. *Geol. Soc. Am., Mem.*, 116: 477—522.

MacGregor, I.D., 1970. A hypothesis for the origin of kimberlite. *Mineral Soc. Am., Spec. Paper*, 3: 51—62.

Mark, R.K., Hu, C.L., Bowman, H.R., Asaro, F., McKee, E.H. and Coats, R.R., 1975. A high $^{87}Sr/^{86}Sr$ mantle source for low alkali tholeiite, northern Great Basin. *Geochim. Cosmochim. Acta*, 39: 1671—1678.

Martin, R.F., Whitley, J.E. and Woolley, A.R., 1978. An investigation of rare-earth mobility: fenitized quartzites, Borralan Complex, N.W. Scotland. *Contrib. Mineral. Petrol.*, 66: 69—73.

Masuda, Y. and Aoki, K., 1978. Two types of island arc tholeiites in Japan. *Earth Planet. Sci. Lett.*, 39: 298—302.

Masuda, Y. and Aoki, K., 1979. Trace element variation in the volcanic rocks from the Nasu zone, northeast Japan. *Earth Planet. Sci. Lett.*, 44: 139—149.

Masuda, Y., Nishimura, S., Ikeda, T. and Katsui, Y., 1975. Rare-earth and trace elements in the Quaternary volcanic rocks of Hokkaido, Japan. *Chem. Geol.*, 15: 251—271.

McGregor, V.R. and Mason, B., 1977. Petrogenesis and geochemistry of metabasaltic and metasedimentary enclaves in the Amitsoq gneisses, West Greenland. *Am. Mineral.*, 62: 887—904.

Mineyev, D.A., 1963. Geochemical differentiation of the rare earths. *Geochemistry (U.S.S.R.)*, 12: 1129—1149.

Mitchell, R.H., 1970. Kimberlite and related rocks — a critical reappraisal. *J. Geol.*, 78: 686—704.

Mitchell, R.H. and Bell, K., 1976. Rare earth element geochemistry of potassic lavas from the Birunga and Toro-Ankole regions of Uganda, Africa. *Contrib. Mineral. Petrol.*, 58: 293—303.

Mitchell, R.H. and Brunfelt, A.O., 1974. Rare earth element geochemistry of kimberlite. *Phys. Chem. Earth*, 9: 671—686.

Mitchell, R.H. and Brunfelt, A.O., 1975. Rare earth geochemistry of the Fen alkaline complex, Norway. *Contrib. Mineral. Petrol.*, 52: 247—259.

Morrison, G.W., 1980. Characteristics and tectonic setting of the shoshonite rock association. *Lithos*, 13: 97—108.

Mysen, B.O., 1977. Experimental determination of rare earth fractionation patterns in partial melts from peridotite in the upper mantle. *Earth Planet. Sci. Lett.*, 34: 231—237.

Mysen, B.O., 1979. Trace element partitioning between garnet peridotite minerals and water-rich vapor: experimental data from 5 to 30 kb. *Am. Mineral.*, 64: 274—287.

Nathan, S. and Fruchter, J.S., 1974. Geochemical and paleomagnetic stratigraphy of the Picture Gorge and Yakima Basalts (Columbia River Group) in central Oregon. *Geol. Soc. Am. Bull.*, 85: 63—76.

Neumann, E.R., Brunfelt, A.O. and Finstad, K.G., 1977. Rare earth elements in some igneous rocks in the Oslo rift, Norway. *Lithos*, 10: 311—319.

Nicholls, I.A. and Whitford, D.J., 1978. Geochemical zonation in the Sundra volcanic arc, and the origin of K-rich lavas. *Bull. Aust. Soc. Explor. Geophys.*, 9: 93—98.

O'Hara, M.J. and Yoder, H.S., 1967. Formation and fractionation of basic magma at high pressure. *Scott. J. Geol.*, 3: 67—117.

O'Nions, R.K. and Clarke, D.B., 1972. Comparitive trace element geochemistry of Tertiary basalts from Baffin Bay. *Earth Plant. Sci. Lett.*, 15: 436—446.

O'Nions, R.K. and Pankhurst, R.J., 1978. Early Archean rocks and geochemical evolution of the earth's crust. *Earth Planet. Sci. Lett.*, 38: 211—236.

Paster, T.P., Schauwecker, D.S. and Haskin, L.A., 1974. The behavior of some trace elements during solidification of the Skaergaard layered series. *Geochim. Cosmochim. Acta*, 38: 1549—1577.

Paul, D.K., Potts, P.J., Gibson, I.L. and Harris, P.G., 1975. Rare-earth abundances in Indian kimberlite. *Earth Planet. Sci. Lett.*, 25: 151—158.

Paul, D.K., Potts, P.J., Rex, D.C. and Beckinsale, R.D., 1977. Geochemical and petrogenetic study of the Girnar igneous complex Deccan volcanic province, India. *Geol. Soc. Am. Bull.*, 88: 227—234.

Pe-Piper, G., 1980. Geochemistry of Miocene shoshonites, Lesbos, Greece. *Contrib. Mineral. Petrol.*, 72: 387—396.

Philpotts, J.A., Martin, W. and Schnetzler, C.C., 1971. Geochemical aspects of some Japanese lavas. *Earth Planet. Sci. Lett.*, 12: 89—96.

Philpotts, J.A., Schnetzler, C.C. and Thomas, H.H., 1972. Petrogenetic implications of some new geochemical data on eclogitic and ultrabasic inclusions. *Geochim. Cosmochim. Acta*, 36: 1131—1166.

Potts, M.J. and Condie, K.C., 1971. Rare earth element distributions in a protostratiform ultramafic intrusion. *Contrib. Mineral. Petrol.*, 33: 245—258.

Price, R.C. and Taylor, S.R., 1980. Petrology and geochemistry of the Banks Peninsula volcanoes, South Island, New Zealand. *Contrib. Mineral. Petrol.*, 72: 1—18.

Reitz, B. and Cullers, R.L., 1980. Petrogenesis of kimberlites, Riley County, Kansas. *Geol. Soc. Am. Abstr. Prog.*, 12: 16 (abstract).

Rock, N.M.S., 1978. Petrology and petrogenesis of the Monchique alkaline complex, southern Portugal. *J. Petrol.*, 19: 171—214.

Saunders, A.D. and Tarney, J., 1979. The geochemistry of basalts from a backarc spreading centre in the East Scotia Sea. *Geochim. Cosmochim. Acta*, 43: 555—572.

Schofield, A. and Haskin, L., 1964. Rare-earth distribution patterns in eight terrestrial materials. *Geochim. Cosmochim. Acta*, 28: 437—446.

Sørensen, H., 1974. *The Alkaline Rocks*. John Wiley and Sons, New York, N.Y.

Sun, S.S. and Hanson, G.N., 1975a. Evolution of the mantle: geochemical evidence from alkali basalt. *Geology*, 3: 297—302.

Sun, S.S. and Hanson, G.N., 1975b. Origin of Ross Island basanitoids and limitations upon the heterogeneity of mantle sources for alkali basalts and nephelinites. *Contrib. Mineral. Petrol.*, 52: 77—106.

Sun, S.S. and Hanson, G.N. 1976. Rare earth element evidence for differentiation of McMurdo volcanics, Ross Island, Antarctica. *Contrib. Mineral. Petrol.*, 54: 139—155.

Sun, S.S. and Nesbitt, R.W., 1977. Chemical heterogeneity of the Archaean mantle, composition of the Earth and mantle evolution. *Earth Planet. Sci. Lett.*, 35: 429—448.

Sun, S.S. and Nesbitt, R.W., 1978. Petrogenesis of Archean ultrabasic and basic volcanics: evidence from rare earth elements. *Contrib. Mineral. Petrol.*, 65: 301—325.

Tanaka, T., 1974. Rare earth elements in a gabbroic body of the Japanese Paleozoic geosyncline. *Geochem. J.*, 8: 47—60.

Tanaka, T., 1975. Geological significance of rare earth elements in Japanese geosynclinal basalts. *Contrib. Mineral. Petrol.*, 52: 233—246.

Tanaka, T., 1977. Rare earth abundances in Japanese Paleozoic geosynclinal basalts and their geological significance. *Bull., Geol. Surv. Jpn.*, 28: 27—57.

Tanaka, T. and Sugisaki, R., 1973. Successive eruption of alkaline and tholeiitic magmas in a Japanese geosynclinal basalt body with special reference to rare earth element features. *J. Petrol.*, 14: 489—507.

Tarney, J., Saunders, A.D. and Weaver, S.D., 1977. Geochemistry of volcanic rocks from the island arcs and marginal basins of the Scotia Arc region. In: M. Talwani, C.G. Harrison and W.C. Pitman III (Editors), *Island Arcs, Deep Sea Trenches, and Back-arc Basins, Am. Geophys. Union, Maurice Ewing Ser.*, 1: 367—377.

Thompson, R.H., Gibson, I.L., Marriner, G.F., Mattey, D.P. and Morrison, M.A., 1980. Trace-element evidence of multistage mantle fusion and polybasic fractional crystallization in the Palaeocene lavas of Skye, N.W. Scotland. *J. Petrol.*, 21: 265—293.

Tuttle, O.F. and Gittins, J., 1966. *Carbonatites.* John Wiley and Sons, New York, N.Y., 591 pp.

Vainshtein, E.E., Pozharitskaya, L.K. and Turanskaya, N.V., 1961. Behavior of the rare earths in the process of formation of carbonatites. *Geochemistry (U.S.S.R.)*, 11: 1151—1154.

Wakita, H., Rey, P. and Schmitt, R.A., 1971. Abundances of the 14 rare earth elements and 12 other rare elements in Apollo 12 samples: five igneous and one breccia rocks and four soils. *Proc. 2nd Lunar Sci. Conf., Geochim. Cosmochim. Acta, Suppl.*, 2, 2: 1319.

Weaver, S.D., Saunders, A.D., Pankhurst, R.J. and Tarney, J., 1979. A geochemical study of magmatism associated with the initial stages of back-arc spreading. *Contrib. Mineral. Petrol.*, 68: 151—169.

Wendlandt, R.F. and Harrison, W.J., 1979. Rare earth partitioning between immiscible carbonate and silicate liquids and CO_2 vapor: results and implications for the formation of light rare earth-enriched rocks. *Contrib. Mineral. Petrol.*, 69: 409—419.

Whitford, D.J. and Arndt, N.T., 1978. Rare earth element abundances in a thick, layered komatiite lava flow from Ontario, Canada. *Earth Planet. Sci. Lett.*, 41: 188—196.

Whitford, D.J., Nicholls, I.A. and Taylor, S.R., 1979. Spatial variations in the geochemistry of Quaternary lavas across the Sunda arc in Java and Bali. *Contrib. Mineral. Petrol.*, 70: 341—356.

Windley, B.F., 1977. *The Evolving Continents.* John Wiley and Sons, New York, N.Y., 385 pp.

Winkler, H.G.F., 1979. *Petrogenesis of Metamorphic Rocks.* Springer-Verlag, Berlin.

Wood, D.A., 1979. Dynamic partial melting: its application to the petrogenesis of basalts erupted in Iceland, the Faeroe Islands, the Isle of Skye (Scotland) and the Troodos Massif (Cyprus). *Geochim. Cosmochim. Acta*, 43: 1031—1046.

Wyllie, P.J. and Huang, W.L., 1975. Peridotite, kimberlite, and carbonatite explained in the system CaO-MgO-SiO_2-CO_2. *Geology*, 3: 621—624.

Wyllie, P.J. and Tuttle, O.F., 1960. The system CaO-CO_2-H_2O and the origin of carbonatites. *J. Petrol.*, 1: 1—46.

Yoder, H.S. and Tilley, C.E., 1962. Origin of basalt magmas: an experimental study of natural and synthetic rock systems. *J. Petrol.*, 3: 342—532.

Chapter 8

RARE EARTH ELEMENTS IN IGNEOUS ROCKS OF THE CONTINENTAL CRUST: INTERMEDIATE AND SILICIC ROCKS – ORE PETROGENESIS

ROBERT L. CULLERS and JOSEPH L. GRAF

8.1. Introduction

Models to explain the petrogenesis of basic and ultrabasic rocks are simple compared to those required to explain the petrogenesis of the intermediate and silicic rocks discussed in this chapter. Andesites and granitic rocks can be derived from a multitude of possible source materials and by a variety of crystallization paths so that it is even more difficult to limit possible hypotheses for the origin of these rock types compared to basic and ultrabasic rocks. In addition, small amounts of accessory minerals that concentrate the REE, rather than major rock-forming minerals, may control the REE contents during melting or crystallization, and thereby can limit severely the use of these elements for the interpretation of igneous processes in intermediate and silicic rocks. Often, good D values are not even available for these accessory minerals to evaluate models in these systems.

At least two-stage models are required to explain the origin of these rocks. For example, eclogite in an oceanic subduction zone should be able to melt and produce dacite and andesite magmas, but the predicted REE contents and other incompatible element contents of the dacites and andesites are too low compared to those observed in a simple one-stage model. Thus, some second stage of enrichment of the incompatible elements is required. In some models volatile transport in H_2O-rich fluid is assumed to occur. Alternatively, the silicic magmas derived from the melting of eclogite may react with overlying mantle peridotite to produce a garnet pyroxenite. The garnet pyroxenite may in turn rise and melt to generate the calc-alkaline andesites with the required incompatible element content in this second stage of enrichment. Furthermore, detailed studies of small areas using a variety of data in addition to the REE often demand very complex models to explain all the data. These complex models include combinations of melting, crystallization, magma mixing, assimilation, and volatile transport to explain the variety of magmas coming from even just one volcanic vent (Luhr and Carmichael, 1980). Such complex systems demand detailed studies integrating field work, petrography, isotope, major elements, and trace elements in order to limit possible models for the petrogenesis of these systems.

8.2. Andesites and associated rocks

General

Large volumes of andesite or basaltic andesite occur with lesser amounts of basalt, dacite, rhyodacite, and rhyolite in calc-alkaline sequences along island arc or continental subduction zones. Siliceous volcanic rocks associated with andesites seem to be more voluminous over continental subduction zones than over island arc ones. Also volcanic rocks over continental subduction zones contain large FeO/Fe_2O_3, K_2O/Na_2O ratios, and higher contents of incompatible elements than corresponding rocks over island arc subduction zones (e.g., Jakeš and White, 1972).

Basalts and andesites of the calc-alkaline association may contain phenocrysts of plagioclase, olivine, augite, or hypersthene. Hornblende phenocrysts may also occur in andesites to rhyolites, and biotite phenocrysts may occur in rhyolite. The basalts and basaltic andesites tend to have high Al_2O_3 and low TiO_2 contents. The abundances of the alkalies are intermediate between those of alkali olivine basalts and tholeiitic basalts. In contrast to tholeiitic associations, more silicic rocks in this association do not concentrate Fe relative to Mg and alkalies.

Small volumes of tholeiitic andesite and dacite and small volumes of shoshonitic andesite and dacite may be associated with larger volumes of tholeiitic basalts and alkali-rich basalt respectively (Jakeš and White, 1972). These small volumes of andesites may form by differentiation processes from the large volumes of more basic melts.

REE contents

Andesites contain low to moderate REE contents (ΣREE = 25—341 ppm) and LREE/HREE ratios ($(La/Lu)_{cn}$ = 1.0—21.5; Fig. 8.1, Table 8.1); andesites seldom have Eu anomalies (Eu/Sm = 0.15—0.49), and if anomalies are present, they are not large*.

As summarized in a review of andesite petrogenesis by Gill (1978), REE contents increase with increasing K for a given Si content in andesite suites

---

*Sources of data are: Taylor, 1968, 1969; Taylor et al., 1969; Jakeš and Gill, 1970; Yajima et al., 1972; Condie and Swenson, 1973; Ewart et al., 1973, 1977; López-Escobar et al., 1974, 1977; Condie and Baragar, 1974; Gill, 1970, 1974, 1976, 1978; Condie and Hayslip, 1975; Fujimaki, 1975; Masuda et al., 1975; Condie and Harrison, 1976; Noble et al., 1976; Thorpe et al., 1976, 1979; Zielinski and Lipman, 1976; Davis and Condie, 1977; Dostal et al., 1977a, 1977b; Hawkesworth and O'Nions, 1977; Kay, 1977, 1978; Mertzman, 1977; White and McBirney, 1977; Dostal and Zerbi, 1978; Fryer and Jenner, 1978; Johnson et al., 1978; Dixon and Batiza, 1979; Dupuy et al., 1979; Fountain, 1979; Jahn et al., 1979, 1980; Masuda and Aoki, 1979; Smith et al., 1979; Thorpe and Francis, 1979; Whitford et al., 1979; Leo et al., 1980; Luhr and Carmichael, 1980; Cullers and Arnold, 1981.

TABLE 8.1

Range of REE contents in intermediate and silicic igneous rocks (anorthosites are included since they are often associated with these rocks)

Rock type	ΣREE (ppm)	$(La/Lu)_{cn}$	Eu/Sm
Andesite	25—341	1.0- 21.5	0.15—0.51
oceanic	25—178	1.0—8.6	0.19—0.51
continental	67—341	1.5—21.5	0.15—0.35
Anorthosite	1.7—148	0.13--58	0.30—5.0
Associated jotunite, mangerite, and anorthositic gabbro	34.5—312	3.4—14.9	0.34—0.73
Quartz diorite, tonalite, granodiorite, and trondhjemite	10.5—499	0.34—413	0.041—1.76
Those with negative Eu anomalies	60—499	8.9—66	0.041—0.26
Those with negative Eu anomalies from young island arcs or ophiolites	34.4—131	0.34—1.7	0.13—0.29
Those with positive Eu anomalies	10.5—144	5.0—77.5	0.39—1.76
Those with little or no Eu anomalies	12—273	2.8—413	0.24—0.38
Monzogranites and syenogranites	8—1977	0.54—137	0.0009—1.07
Those with small to moderate negative Eu anomalies	106—877	2.5—50	0.09—0.23
Those with moderate to large negative Eu anomalies	40—1977	1.1—22	0.0009—0.074
Those with positive Eu anomalies	40—210	4.8—97	0.41—1.07
Those with little or no Eu anomalies	8—1426	0.54—137	0.20—0.36

so REE contents correlate to increasing vertical distance from the subduction zone. Within a given suite REE content increases with increasing Si.

General petrogenesis

Simple one-stage melting or crystallization models to explain trace element distributions in andesites do not explain all the data. There are several reasons for this problem. Experimentally, there are a variety of sources which can melt and produce andesites. A variety of magmas less silicic than andesites can crystallize several kinds of minerals and can still produce andesites. Trace element modelling is hindered by possible changing D values during magma evolution. Also processes such as volatile transport or assimilation of some elements may produce complex patterns of trace element distributions superimposed on effects caused by crystal/melt equilibria which make the data difficult to interpret.

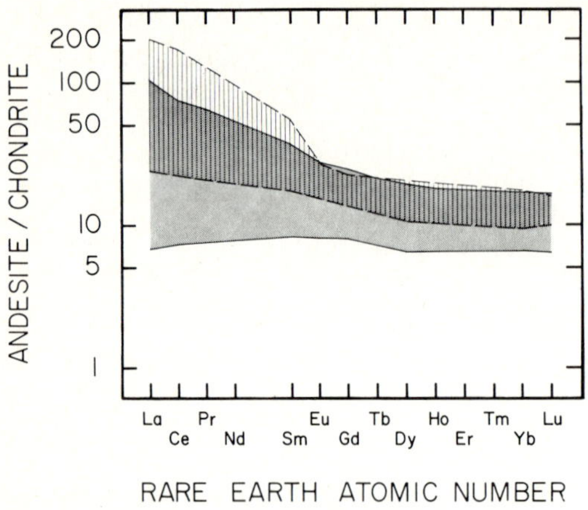

Fig. 8.1. The range in chondrite-normalized REE abundances of most continental andesites associated with subduction zones (vertical ruled lines) compared to the range of most island arc andesites (shaded portion). Continental andesites tend to contain greater REE contents than island arc andesites.

Gabbros, amphibolites, or eclogites can melt to produce dacites and andesites of the calc-alkaline suite (T.H. Green and Ringwood, 1968) although eclogite is usually taken as the main source from this group. It is not clear if hydrous melting of garnet peridotite produces andesites, since quench reactions of melts during these experiments may obscure the composition of the original melt. One group of experimentalists believe that it is possible to produce andesites by hydrous melting of peridotite at mantle pressure (Kushiro, 1972; Mysen and Boettcher, 1975); Another group believes that andesites can only be produced from the hydrous melting of peridotite at less than 10 kbar (Nicholls and Ringwood, 1973; D.H. Green, 1973, 1976). The latter group finds that tholeiites are produced instead by the hydrous melting of peridotite at pressures greater than about 10 kbar.

Most subducted, low-K eclogite and perhaps garnet peridotite are too depleted in incompatible elements (U, Th, Rb, Ba, Sr, LREE) to generate the typical contents of these elements in andesites (Gill, 1974; López-Escobar et al., 1977). Thus, at least two-stage models such as proposed by Ringwood (1974) have been used recently to explain the trace element constraints of andesites. Small degrees of melting of the slab of subducted eclogite produce silicic magmas, enriched in incompatible elements, which react with olivine in the overlying mantle to produce garnet pyroxenite. Melting of this garnet pyroxenite can produce andesites which are further enriched in the incompatible elements.

The crystallization of amphibole or magnetite and plagioclase has been postulated to produce andesites from less silicic melts. These models are the most consistent of the simple models to explain all the trace element variations (Gill, 1978). They can be used to explain small volumes of tholeiitic andesite with low REE content, associated with large volumes of tholeiitic basalt.

Thus, the petrogenesis of andesites is complex and must involve complex processes. The detailed studies mentioned in the introduction are needed to evaluate these processes. The following discussion is organized into predominantly partial melting and fractional crystallization processes. More complex processes are discussed in the examples within each of these sections.

Predominately partial melting

The earliest studies of REE in andesites suggested that the rocks might have formed by melting of the mantle, subducted oceanic crust (e.g., Taylor, 1968, 1969), or of metagraywacke-argillites (Ewart et al., 1968). These studies were done before D values and source contents of trace elements were available. When these data became available, it was soon realized that simple melting of subducted oceanic crust or overlying peridotite could not produce the incompatible element content of the calc-alkaline suite (Gill, 1974; López-Escobar et al., 1974). For example, 40% melting of a LREE-depleted oceanic tholeiite, changed to eclogite mineralogy, would produce a bowed REE pattern in chondrite-normalized plots in which abundances of the intermediate REE would be too high and those of the LREE and HREE would be too low compared to recorded values for andesites in the Andes.

Most recent models for the production of calc-alkali or high-K andesites with fairly large REE contents and LREE/HREE ratios focus on melting a LREE-enriched eclogite, garnet peridotite, amphibolite, or garnet pyroxenite involving at least two stages. For example, Dostal et al. (1977a,b) suggest a garnet peridotite could become enriched in the LREE by ascending hydrous fluids above the subduction zone. Models of these kinds gain support from experimental studies that suggest the LREE become soluble in H_2O-rich fluids at mantle pressure (Mysen, 1979; Wendlandt and Harrison, 1979). The garnet peridotite source for andesite need only be enriched by 2—6 times chondritic abundances if small degrees of hydrous melting of peridotite are allowed. Experimental petrologists need to resolve the problem of whether or not andesites can indeed form by melting of hydrous peridotite before such models can gain general acceptance.

The Brokeoff volcano in California is mainly pyroxene andesite with lesser basalt, basaltic andesite, dacite, and rhyodacite (Fountain, 1979). The voluminous andesites have large LREE/HREE ratios which could be produced by 4—5% melting of a hydrous peridotite containing about 2—3% garnet, 17% clinopyroxene, 20% orthopyroxene, and 60% olivine. Smaller volumes of rhyodacites of flank eruptions could have been produced by a smaller degree of melting of the peridotite than that required by the andesites followed by 20% fractional crystallization of hornblende. This model would produce higher contents of the LREE and lower contents of the HREE in the rhyodacites compared to the andesites. Some andesite of the flank eruptions contain intermediate chemistry and mineralogy between the most voluminous andesites and the rhyodacites. This andesite of the flank eruptions may have been produced by mixing of andesitic and rhyodacitic magmas. Ringwood's previously discussed two-stage model involving eclogite melting in the first stage has been used in some recent papers (e.g., Thorpe et al, 1976; Thorpe and Francis, 1979). This attractive model will not be discussed further.

Eclogite enriched in the LREE (similar to continental tholeiites) has been considered as a source that could produce the REE content of andesites or syenodiorites (e.g., Zielinski and Lipman, 1976; Davis and Condie, 1977; López-Escobar et al., 1977; Jahn et al., 1980; Cullers and Arnold, 1981). For example, an eclogite containing 0.85 clinopyroxene and 0.15 garnet and melting in a ratio of clinopyroxene: garnet = 0.7/0.3 yields the predicted range of REE contents in andesite in the Summer Coon volcano, Colorado. A smaller degree of melting of a similar eclogite is required to produce the more K-rich and LREE-rich syenodiorites in the hypabyssal Spanish Peaks complex to the east of the Summer Coon volcano. How such LREE-rich eclogites may form at the correct depth is unknown, but Brooks et al. (1976) propose a model to explain the varied Sr isotopic composition of Tertiary volcanism in the western U.S.A. In this model, peridotite is assumed to upwell from the mantle periodically through geologic time and produce basaltic melts which crystallize to eclogite relatively rich in Rb and presumably the LREE. The thermal or dehydration event that occurred during the Tertiary could cause melting of eclogite of varied ages and produce the varied Sr isotopic composition and LREE-enriched patterns of the intermediate melts.

Andesitic rocks at Medicine Lake, California are believed to have formed by melting of altered oceanic rise tholeiite (Condie and Hayslip, 1975). Crystallization and removal of plagioclase and Ti-rich magnetite could explain the lowered Sr, Eu, and Ti of some of the volcanic rocks.

Finally, andesites along the west coast of Sumatra, containing fairly large $^{87}Sr/^{86}Sr$ initial ratios and negative Eu anomalies, are assumed to have formed by melting of lower crustal material with residual plagioclase. Further fractionation of plagioclase, hornblende, and pyroxene may have produced some of the differentiates.

Fractional crystallization

Small volumes of andesites associated with large volumes of basalt or basaltic andesites could form by fractional crystallization of plagioclase, amphibole, pyroxene, or magnetite from the more basic melts (Taylor, 1968;

Yajima et al., 1972; Ewart et al., 1973; Fujimaki, 1975; Masuda et al., 1975; Gill, 1976; Noble et al., 1976; Ewart et al., 1977; Gorton, 1977; Kay, 1977; Mertzman, 1977; Dixon and Batiza, 1979; Masuda and Aoki, 1979). Such crystallization models are common to explain tholeiitic trends. Tholeiitic andesites formed by fractional crystallization tend to have low LREE contents (higher than the parent tholeiite) and low La/Lu ratios, and they may have small, negative Eu anomalies (Fig. 8.2). The last feature is surprising in view of the large amount of plagioclase crystallization needed in some of these models. Perhaps the oxygen fugacity was large enough to prevent much Eu^{2+} from forming and producing the Eu anomalies.

Small volumes of dacites, which are more silica-rich than andesite may be formed by fractional crystallization from andesite (Ewart et al., 1973, 1976; Gill, 1976; Dostal and Zerbi, 1978; Thorpe et al., 1979). For example, fractional crystallization of plagioclase and hornblende from andesitic parent melts were used to explain the formation of small volumes of some of the dacites and rhyodacites of the Brokeoff volcano, California (Fountain, 1979).

8.3. Anorthosites and associated jotunites, mangerites, and charnockites

General

Anorthosites range in size from small bodies tens of meters across to very large bodies thousands of square kilometers in size. Two types of anorthosites are often distinguished: massif-type anorthosites and layered anorthosites (Griffin et al., 1974). The layered anorthosites occur in layered gabbroic

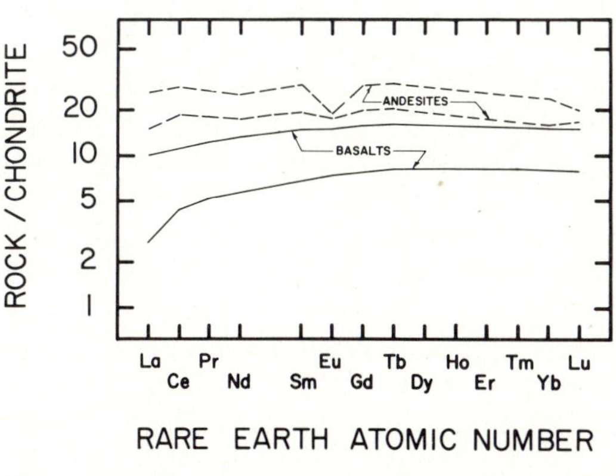

Fig. 8.2. REE content of basalts (solid line) and andesites (dashed) of the pigeonitic rock series found in the Izu-Hakone series, Japan (Yajima et al., 1972). More differentiated rocks contain larger REE contents and higher LREE/HREE ratios in chondrite-normalized curves. The basalts contain no Eu anomalies and the andesites contain negative Eu anomalies.

complexes or in anorthosite-mangerite complexes in which the anorthosite surely results from crystal accumulation. Small inclusions of anorthosite may also occur in basic rocks such as dolerite.

Massif-type anorthosites are largely absent in the Phanerozoic although some smaller bodies are present. The massif-type are dominately clustered between 1.0—1.8 Ga before the present although some are greater than 2.5 Ga before the present (Duchesne and Demaiffe, 1978). Plagioclase may range from An_{35} to An_{65}. Thus, massif-type anorthosites are sometimes divided into labradorite and andesine types. Plagioclase may be surrounded by more mafic, interstitial minerals of pyroxene, magnetite-ilmenite, and plagioclase. The anorthosite may grade gradually into gabbroic borders. Such gabbros can be related by comagmatic processes to the anorthosites (e.g., Simmons and Hanson, 1978, Ashwal and Seifert, 1980). Less clear is the relation of associated more silicic jotunites (pyroxene monzodiorites), mangerites (hypersthene monzonites), and charnockites (hypersthene granites) since they and the massif-type anorthosites are rarely in contact. When the more silicic rocks are in contact with the anorthosites, the boundaries can be sharp to gradational.

Another important problem is the nature of the parent melt of the anorthosite. Melts ranging in composition from that of gabbroic anorthosite (70—80% normative plagioclase) through gabbro to granodiorite have been suggested. Trace elements and most especially the REE have helped to limit possible hypotheses.

REE contents

The REE contents (ΣREE = 1.7—148) and LREE/HREE ratios ($(La/Lu)_{cn}$ = 0.13—58) of anorthosites are fairly low, and most anorthosites contain positive Eu anomalies (Eu/Sm = 0.30—5.0; Fig. 8.3; Philpotts et al., 1966; T.H. Green et al., 1969, 1972; O'Nions and Pankhurst, 1974; Henderson et al., 1976; Morgan et al., 1976; Demaiffe et al., 1979; Seifert et al., 1977; Anderson and Cullers, 1978; Seifert, 1978; Duchesne and Demaiffe, 1978; Simmons and Hanson, 1978; Ashwal and Seifert, 1980). Plagioclase separates in anorthosites contain lower REE contents (ΣREE = 0.7—57.6), larger LREE/HREE ratios ($(La/Lu_{cn}$ = 0.84—37), and larger positive Eu anomalies (0.75—10.0) than the anorthosite from which the plagioclase has been removed (Fig. 8.3). This is presumably because the mafic minerals in anorthosites contain larger REE contents, smaller LREE/HREE ratios, and smaller Eu anomalies than the coexisting plagioclase. The positive Eu anomaly generally shown by anorthosites suggests they are cumulates.

REE contents of more silicic rocks such as jotunites and mangerites associated with anorthosites usually contain larger REE contents and smaller Eu anomaly size than the associated anorthosites (Fig. 8.4; e.g.,

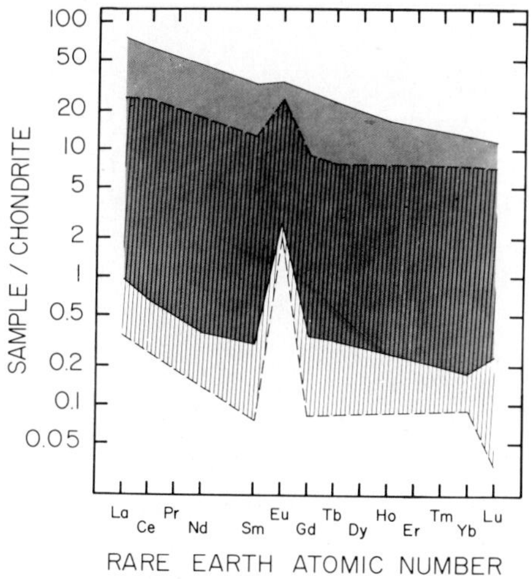

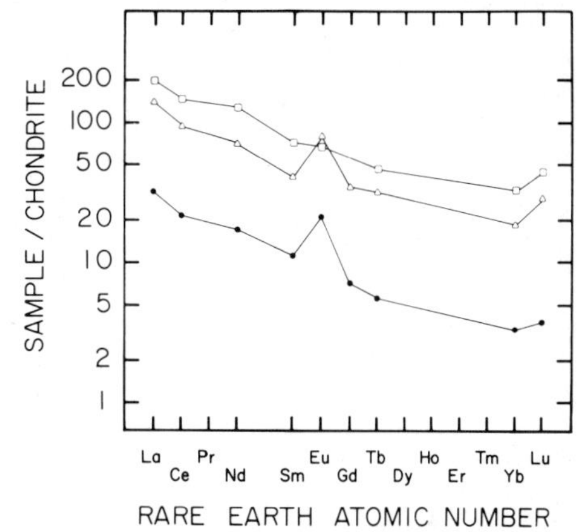

Fig. 8.3 (left). The range of REE contents found in anorthosites (solid curves — dark shading) and in plagioclase separates from some anorthosites (dashed curves — vertical ruled lines).

Fig. 8.4 (right). REE contents found in an anorthosite (solid circles) and associated jotunite (open squares) and magnerite (open triangles) at Baker Mountain, New York, U.S.A. (Seifert, 1978).

Anderson and Cullers, 1978; Simmons and Hanson, 1978; Ashwal and Seifert, 1980).

General petrogenesis

Most studies test models in which the anorthosite is assumed to be a solid residue from melting or, more often, a cumulate from a silicate melt. Thus, the trace element content of the magma can be predicted using the following relation (see Chapter 4):

Trace element content in magma = (distribution coefficient for bulk solid) × (trace element content in the solid)

Predicted trace element contents in magmas can then be compared to those observed in rocks associated with anorthosites to see if these rocks could have been magmas which precipitated the plagioclase. The problems with this kind of simplistic model are outlined in Haskin and Korotev (1977). For example, the separation of the cumulate minerals from the parent magma is likely to be incomplete. Any magma actually incorporated into the cumulate anorthosite would lead to high concentrations of the incompatible elements in this rock and, thus, the predictions of the incompatible element concentrations in the coexisting magma would be too high.

Partial melting

Only two studies of REE abundances have indicated how primary melts might form in anorthosite complexes. Simmons and Hanson (1978) considered that anorthositic melts may form by 5—15% melting of a tholeiite at a shallower depth than the basalt-eclogite transition leaving a pyroxene-rich residue. Fractionation of mainly plagioclase from this anorthositic melt could produce the trace element characteristics of the anorthosite. This model is also consistent with the low Mg/(Mg + Fe^{2+}) ratios of these rocks. The predicted trace element contents of the melt using this melting model agree with predicted trace element contents of parent melts predicted from anorthositic cumulates.

T.H. Green et al. (1969, 1972) consider how mangerite with positive Eu anomalies might form by a two-stage process. During the first stage of melting the source rock melts leaving a residue with a positive Eu anomaly. This residue could undergo a large percent melting during the second stage to produce a melt with a positive Eu anomaly.

Crystallization models

Predicted REE contents of magmas that equilibrated with anorthositic plagioclase or with anorthosites which are assumed to be adcumulates can be compared to the REE contents of associated rocks to see if they could be related as parent magmas or differentiates to the anorthosites. D values of REE (except Eu) for plagioclase/melt are low (see Chapter 1), hence melts from which plagioclase precipitated can be expected to contain much larger REE contents than the plagioclase. Since Eu concentrates in plagioclase relative to the other REE as it precipitates from the magma, the differentiated magma may develop larger negative Eu anomalies from an original magma that may have little or no Eu anomaly.

By using such reasoning, andesine anorthosites at Rogaland, Norway, could have equilibrated with associated jotunites as the parent melt, but labradorite anorthosites could not have equilibrated with the jotunites (Duchesne and Demaiffe, 1978). In the Grenville Province of the northeastern U.S.A., calculated contents of REE predicted in the melts from presumed anorthositic cumulates have similar REE contents as associated anorthositic gabbros so the anorthosites could have equilibrated with the anorthositic gabbros (Simmons and Hanson, 1978; Ashwal and Seifert, 1980). Varied amounts of presumed parent liquid of the anorthositic gabbros and solid anorthosite could mix and produce the spectrum of REE contents intermediate between these two extremes. By similar reasoning, plagioclase inclusions in dolerites from Iceland could have equilibrated with oceanic tholeiites with low LREE contents (Griffin et al., 1974). Plagioclase in some complexes contains large LREE contents and small HREE contents.

Such plagioclase yields a predicted parent with large LREE/HREE ratios that may have equilibrated with garnet (Griffin et al., 1974; Henderson et al., 1976).

Some mangerites and charnockites contain positive Eu anomalies and large REE contents. The anomalies are not consistent with rocks being related, as differentiates, to the anorthosite parent melts since they should contain negative Eu anomalies (e.g., T.H. Green et al., 1972; Ashwal and Seifert, 1980). At Rogaland, Norway, however, an increase in both REE content and negative Eu anomaly size in the order andesine anorthosite, leuconorite, norite, and charnockite allows a fractional crystallization model involving mainly plagioclase, to be consistent with the formation of this suite (Duchesne and Demaiffe, 1978). Finally, in the Wolf River Batholith, Wisconsin (U.S.A.), a parent anorthositic melt could have undergone 75—85% fractional crystallization of plagioclase:orthopyroxene: clinopyroxene:apatite = 85/10/5/0.5 to produce the REE, Ba, Rb, and Sr contents of the associated monzonite (Anderson and Cullers, 1978).

8.4. Granitic rocks — quartz diorite, tonalite, granodiorite, and trondhjemite

General

The granitic rocks discussed are those that contain abundant plagioclase relative to alkali feldspar; those that contain abundant alkali feldspar, e.g., monzogranite (admellite) and syenogranite (granite), are discussed in the next section. Granodiorites and quartz diorites are often the most abundant rocks in large batholiths; they are often accompanied by lesser amounts of monzogranite or syenogranite.

REE contents

The REE content of quartz diorites, tonalites, granodiorites, and trondhjemites from a wide variation in age and setting exhibit a large range in composition (Fig. 8.5). The range of about 120 samples are as follows: ΣREE = 10.5—499 ppm; $(La/Lu)_{cn}$ = 0.34—413, and Eu/Sm = 0.041—1.76*.

---

*Sources of data are: Arth and Hanson, 1972; Hanson and Goldich, 1972; Yajima et al., 1972; O'Nions and Pankhurst, 1974, 1978; Emmermann et al., 1975; Barker et al., 1976; Condie, 1976, 1978; Condie and Hunter, 1976; Glickson, 1976; Lambert and Holland, 1976; McCarthy and Hasty, 1976; Albuquerque, 1977, 1978; Hawkesworth and O'Nions, 1977; Arth et al., 1978; Compton, 1978; Drury, 1978; Frey et al., 1978; Fryer and Jenner, 1978; Hunter et al., 1978; Simmons and Hedge, 1978; Arth, 1979; Barker and Millard, 1979; Barker et al., 1979a, b; Birk et al., 1979; Collerson and Bridgwater, 1979; López-Escobar et al., 1979; Peccerillo et al., 1979; Phelps, 1979; Saunders et al, 1979; Perfit et al., 1980.

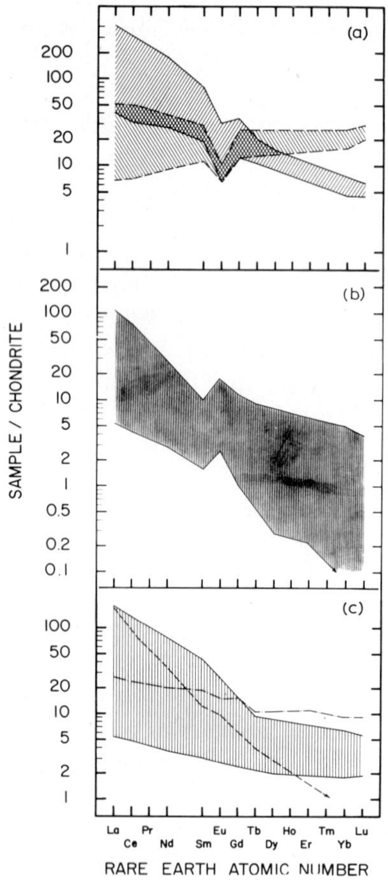

Fig. 8.5. The range of REE contents found in trondhjemites, tonalites, granodiorites, and quartz diorites (solid curves — northeast-southwest ruled lines) that contain negative Eu anomalies compared to rocks of similar composition in island arcs (dashed curves — northwest-southeast ruled lines). The island arc samples contain smaller LREE and ΣREE and larger HREE contents than samples from other tectonic environments. (b) The range of REE content of trondhjemites, tonalites, and granodiorites that contain positive Eu anomalies (this figure) have lower REE contents (especially HREE) compared to most samples with negative Eu anomalies that are not from island arcs (Fig. 8.5a). (c) The range of REE content of most trondhjemites, tonalites; and granodiorites with little or no Eu anomalies lie in the ruled area between the solid curves. The dashed curves represent the extreme ranges of slopes of two samples which lie outside the range of most samples for the HREE.

The granitic rocks discussed in this section are arbitrarily divided into groups containing negative Eu anomalies (range Eu/Sm = 0.041—0.27), positive Eu anomalies (Eu/Sm = 0.39—1.76), and little or no Eu anomalies (Eu/Sm = 0.23—0.38; Fig. 8.5). The reason for this division is that the presence or absence of the Eu anomalies usually is related to the amount of equilibration of the magma with plagioclase. For example, increasingly larger, negative Eu anomalies in a series of volcanic rocks may indicate the progressive removal of plagioclase from the magma if this mineral is observed as a phenocryst. The group containing negative Eu anomalies may be divided into two subgroups (Fig. 8.5a). One subgroup has a large REE content (range ΣREE = 60—499 ppm) and LREE/HREE ratios (range $(La/Lu)_{cn}$ = 8.9—66). Samples of the first subgroup are from continental or continent-margin settings and are of any geologic age. The second subgroup contains low REE contents (ΣREE = 34—131 ppm) and LREE/HREE ratios ($(La/Lu)_{cn}$ = 0.34—1.7), and are only obtained from trondhjemites and dacites of young island arcs or plagiogranites from ophiolites. The rocks from island arcs or ophiolites seem to contain different REE contents than in young continental margins (Arth, 1979; Barker et al., 1979b). Oceanic or island arc trondhjemites have less than 14.5—15% Al_2O_3, flat REE chondrite-normalized curves and negative Eu anomalies (Fig. 8.5a). In contrast, continental trondhjemites contain more than 14.5% Al_2O_3 and steep REE curves, and they may or may not contain Eu anomalies (Fig. 8.5a and c). This conclusion should be considered tenative since few oceanic or island arc samples have been analyzed.

Trondhjemites, tonalites, quartz diorites, and granodiorites that have positive Eu anomalies have low absolute REE contents (ΣREE = 10.5—144 ppm) more similar to island arc abundances, but they have LREE/HREE ratios ($(La/Lu)_{cn}$ = 5.0—77.5) that are fairly large (Fig. 8.5b). All rocks with this distribution are Precambrian, and most are in Archean high-grade gneisses. Several of these suites have a negative correlation of REE content and SiO_2 (e.g., Arth et al., 1978).

Rocks with little or no Eu anomalies contain low to moderate REE contents (ΣREE = 12—273 ppm) and quite variable LREE/HREE ratios (Fig. 8.5c; $(La/Lu)_{cn}$ = 2.8—413). Most of these are Precambrian, but some are Phanerozoic in age.

General petrogenesis

The application of trace elements to the petrogenesis of granitic rocks has been reviewed most extensively by Hanson (1978). Important conceptual models to explain trace element changes during crystallization of granitic melts were reviewed by McCarthy and Hasty (1976). Arth (1979) and Tarney et al. (1979) reviewed models for the petrogenesis of trondhjemites or tonalites.

The use of trace elements to help unravel the petrogenesis of basaltic rocks has been quite successful due to the limited source that can melt and give basaltic magmas, and also because we now have a fairly clear understanding of how basic melts undergo crystallization at different pressures. In contrast, the range of granitic melts can be produced by melting a variety of sources. These sources include basaltic rocks (amphibolite to gabbro at low pressure or eclogite at high pressure), metagraywacke, meta-arkose, meta-shales, granitic gneisses of other igneous rocks, or even possibly wet peridotite. Other problems with partial melting models include lack of knowledge on the precise melting relations of the sources, and how mineral/melt *D* values change as a function of temperature, pressure, or composition.

Several problems arise during fractional crystallization. Most models developed to explain trace element changes during fractional crystallization of instrusives assume complete separation of crystals and liquid. Yet granitic magmas are very viscous and assumed complete separation of crystals and magma is not realistic. A petrographic distinction between minerals and coexisting melt in coarse-grained, plutonic rocks is not possible in practice, and this would be important to determine since the rocks may be mixes of any ratio of magma and solid. In addition, petrography and experimental petrology indicate a wide range in mineral parageneses due to differences in magma composition, temperature, pressure, and contents of dissolved volatile species (e.g., H_2O, O_2, and CO_2). Finally some *D* values of the REE for some minerals in silicic magmas are not known very well even though some minerals like sphene or allanite drastically concentrate the REE. In spite of these difficulties, many possible sources or models can be eliminated and narrowed to a few using the reasoning described in this section.

Almost all models using REE to help explain the petrogenesis of granitic melts argue in terms of melting or crystallization processes because these are relatively easy to model. The influence of such processes as magma mixing, liquid immiscibility, assimilation of wall rock, or evolution of volatiles on REE evolution are seldom addressed. An exception is a study of the Bishop Tuff in which all usual models to explain element variation in a zoned magma chamber are convincingly eliminated (Hildreth, 1979). A little understood thermogravitational diffusion model was required to explain the element distributions. A potentially important model to evaluate trace element distributions in such evolving solid-melt-fluid systems was developed by Shaw (1978).

Many rocks discussed in this section occur in high-grade metamorphic terrain of Precambrian age. Most studies assume that metamorphism has not affected REE distributions although many other elements have been shown to be mobile. The immobility of the REE during metamorphism of Archean rocks has been challenged (Collerson and Fryer, 1978; Collerson and Bridgwater, 1979). HREE depletions observed in many Archean rocks could result from volatile transport of the HREE as Earth degassed CO_2 during its early history. This conclusion was partially based on experiments by Mineyev (1963) in which the HREE were transported more readily in some fluids than the LREE. Recent experiments that suggest the LREE are more soluble than HREE in H_2O- and CO_2-rich fluids (Mysen, 1979; Wendlandt and Harrison, 1979) are not so easy to reconcile with this argument.

Nevertheless, these experiments suggest the REE become very soluble in CO_2- or H_2O-rich fluids at certain pressures. Detailed studies of the influence of metamorphism on REE distributions in meta-igneous rocks during the Archean in which extensive CO_2 may have been evolved are required before REE models of igneous processes can be used with confidence. In addition, further studies of the effects of metamorphism on REE distributions of any age should be done on meta-igneous rocks if the igneous processes forming these rocks are to be interpreted.

In the melting and crystallization models summarized below, plagioclase/melt or alkali feldspar/melt equilibria result in negative Eu anomalies in evolving magmas and positive Eu anomalies in feldspar-rich residues. Most studies using fractional crystallization models argue in terms of evolving magmas crystallizing feldspar to explain increases in negative Eu anomaly size, and it is curious that few residual feldspar-rich rocks with positive Eu anomalies have been found. Also many studies suggest that a source rock may melt leaving a feldspar-rich residue and produce a melt with a negative Eu anomaly. Few feldspar-rich rocks with the positive Eu anomalies expected in these residue rocks have been found. The Scourion complex in Scotland, however, has been metamorphosed to the granulite facies and has been interpreted to be a residue left after a melting event that produced granites (Pride and Muecke, 1980).

Precambrian granodiorites or trondhjemites with positive Eu anomalies are not believed to be a result of feldspar accumulation. Rather these may result from hornblende/melt equilibria in which Eu concentrates in evolving melts relative to the other REE as it is rejected from separating hornblende. High D values for REE between hornblende and melt, also cause HREE contents of residual melts to decrease with increased crystallization.

The crystallization of hornblende, however, does not always appear to produce positive Eu anomalies in residual magmas as hornblende precipitation is required to explain HREE depletion in melts which have no Eu anomalies. The behavior of Eu in hornblende could partially be a result of differing oxygen fugacity. Under more reducing conditions more Eu^{2+} is present and presumably would be partitioned more strongly than the other REE into the melt. Under more oxidizing conditions most of the Eu would be in the 3+ oxidation state so it would behave like the rest of the REE. Garnet is also commonly used in models to explain HREE depletion in evolving melts.

Partial melting

Granodiorites, tonalites, quartz diorites, and trondhjemites containing negative Eu anomalies and fairly large REE contents and LREE/HREE ratios (Fig. 8.5a) require sources with abundant plagioclase (to produce the negative Eu anomalies in the melt) and small amounts of garnet, amphibole, or pyroxene (to reduce the HREE relative to LREE in the melts). Presumably little K-rich phases are present in these source rocks because of the lack of alkali feldspar in the melts. Such sources are garnet-bearing siliceous granulite (Condie, 1976; Condie and Hunter, 1976; Fryer and Jenner, 1978; Peccerillo et al., 1979), hornblende-garnet-clinopyroxene-plagioclase basic rock (Barker et al., 1976), and metagraywacke (Albuquerque, 1977, 1978). For example, 50% melting of a garnet-bearing siliceous granulite containing mostly plagioclase, orthopyroxene, quartz, and some K-feldspar and garnet are required to produce REE distributions in Dalmein-type granodiorites in the Barberton region, South Africa. Rocks in this group with somewhat higher HREE contents may not require garnet or amphibole in the source. For example, 50% melting of a siliceous granulite (30—40% plagioclase, 15% orthopyroxene + clinopyroxene, less than 8% K-feldspar, 10—40% quartz) produces the correct REE distribution in high-Ca granodiorites in New Mexico (Condie, 1978).

Low-Al dacites and quartz diorites in island arcs with negative Eu anomalies and low LREE/HREE ratios (Fig. 8.5a) require a basic source with low LREE/HREE ratios and abundant plagioclase. Melting of such a source will produce the negative Eu anomalies and low LREE/HREE ratios observed in the dacitic melts (Barker et al., 1979b; Phelps, 1979; Tarney et al., 1979). No garnet or amphibole are allowed because of the large HREE content. For example, a basaltic source with LREE depletion of oceanic tholeiites is required to melt and produce the REE contents in the Sparta quartz diorite—trondhjemite complex, Oregon.

The Precambrian rocks with positive Eu anomalies, small REE contents, and fairly large LREE/HREE ratios (Fig. 8.5b) require either an amphibolite, eclogite, or garnet amphibolite source (Arth and Hanson, 1972, 1975; Hanson and Goldich, 1972; O'Nions and Pankhurst, 1974, 1978; Glickson, 1976; Lambert and Holland, 1976; Compton, 1978; Drury, 1978; Tarney et al., 1979). Amphibole and garnet residues keep total REE contents of melts low and LREE/HREE ratios large. Both minerals are presumed to produce positive Eu anomalies in the melts. Unfortunately, only two sets of REE distribution coefficients between hornblende phenocrysts and coexisting silicic groundmass have been determined (Arth and Barker, 1976; Zielinski and Lipman, 1976). Much more work needs to be done on this problem.

Rocks containing no Eu anomalies, low to moderate REE contents, and variable LREE/HREE ratios (Fig. 8.5c) require little or no residual plagioclase in the source. Alternatively, subequal amounts of plagioclase and amphibole could have opposing effects on Eu anomalies so that no net anomaly in the melt would result. Eclogitic or mafic granulite sources of basic rocks are used for rocks in this group containing the larger LREE/HREE ratios (e.g., Condie and Harrison, 1976; Glickson, 1976; Frey et al., 1978; Simmons and Hedge, 1978; Barker and Millard, 1979). Some tonalite and trondhjemite gneisses of granulite facies mineralogy have also been used as sources to melt and produce the largest REE contents in the suite (Collerson and Bridgwater, 1979).

Fractional crystallization

Reconnaissance studies can be valuable in giving a first approximation to the origin of a suite of igneous rocks. The most successful studies, however, are those in which detailed field relations are integrated with mineralogic and major element changes geographically so that probable primary melts and differentiates may be identified. Trace element and isotopic studies may further limit possible sources of primary melts and fractional crystallization schemes. Many of the fractional crystallization models below result from the need in such detailed studies to explain continuous major and trace element trends in variation diagrams and systematic mineralogic changes in igneous plutons that are closely related in space and time.

Trondhjemites, granodiorites, quartz diorites, and tonalites containing large REE contents and negative Eu anomalies (Fig. 8.5a) require fractional crystallization models in which much plagioclase is removed (Drury, 1978; Perfit et al., 1980). For example, the Captains Bay pluton in Alaska is zoned from a gabbro rim to a central granodiorite and quartz monzodiorite. Mineralogic and trace element changes require a model in which a high-Al basaltic parent melt undergoes fractional crystallization of Ca-rich plagioclase and lesser pyroxene, olivine, and Fe-Ti oxides at less than 6 kbar to produce the more intermediate rocks. During crystallization the plagioclase becomes more Na-rich and in the latter stages, hornblende crystallization becomes important so that dacite liquids are the last product of the differentiation. This model is consistent with the gradually increasing negative Eu anomaly size and increased LREE fractionation with differentiation.

Some rocks in the group containing large REE contents and negative anomalies have very large anomalies. Rocks with large negative anomalies may require two stage models. These stages include melting with much residual plagioclase and crystallization of much plagioclase (e.g., Condie, 1978).

Small volumes of low-Al trondhjemites containing negative Eu anomalies and low LREE/HREE ratios (Fig. 8.5a) require formation by abundant plagioclase crystallization from less differentiated melts (Barker et al., 1976, 1979a; Arth and Barker, 1976; Bruhn et al., 1978; Phelps, 1979; Saunders et al., 1979). Plagioclase crystallization produces the negative Eu anomalies and low Al content in the differentiated liquids. A fractional crystallization model was considered in the Sparta complex in Oregon to relate a quartz diorite parent by fractional crystallization to a trondhjemite differentiate. This model explained the increased Eu anomaly size but not the increased Ce/Yb ratio of the differentiate compared to the parent. Thus, a melting model was favored over the crystallization model. In a similar problem in a Devonian island arc complex in California, sphene (may concentrate the LREE over the HREE) and plagioclase crystallization explain increasing negative Eu anomaly size and decreasing LREE and intermediate REE depletion with increased differentiation. The problems associated with the REE *D* values for sphene/melt will be discussed later.

Trondhjemite melts with positive Eu anomalies, low REE content, and fairly large LREE/HREE ratios could have been produced by hornblende or perhaps garnet crystallization from less differentiated melts (O'Nions and Pankhurst, 1974; Arth and Barker, 1976; Arth et al., 1978; Compton, 1978; Hunter et al., 1978; Tarney et al., 1979). In southwest Finland, intermediate rocks (60% SiO_2) have gradually decreasing REE content and increasing positive anomaly size to more silicic rocks (72.4% SiO_2; Fig. 8.6). This may be explained by fractional crystallization of mainly hornblende, and lesser plagioclase and biotite. Partial melting models discussed previously are also consistent with these data, but the presence of cumulates and rocks with the continuous range of compositions tend to support the crystallization models (Arth et al., 1978). Tarney et al. (1979) prefer partial melting models over ones involving fractional crystallization for similar sequences in Scotland and East Greenland, because of the lack of intermediate rocks or cumulates.

The range of REE contents in some suites of tonalites to trondhjemites containing little or no Eu anomalies (Fig. 8.5c) have been explained using fractional crystallization. A negative correlation of Si to REE content in the

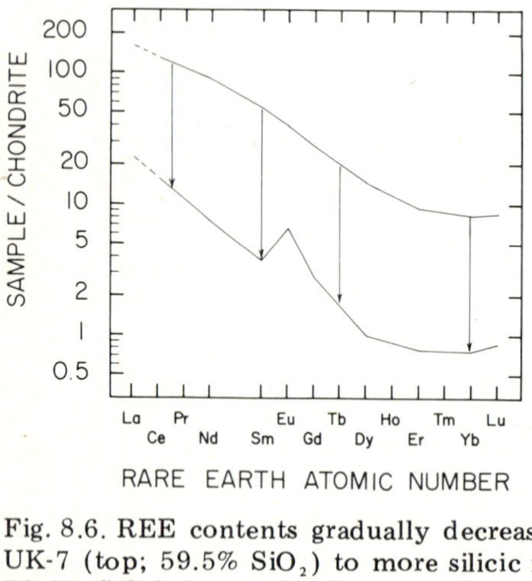

Fig. 8.6. REE contents gradually decrease and Eu anomaly size increases from tonalite UK-7 (top; 59.5% SiO_2) to more silicic members such as trondhjemite UK-12 (bottom; 72.4% SiO_2) in a suite from southwest Finland (Arth et al., 1978). Other rocks are intermediate between these two extremes. The arrows indicate crystallization trends to lowered REE contents and larger positive Eu anomaly size due to hornblende crystallization.

Wabigoon volcanic-plutonic belt was related to crystallization of hornblende (concentrates REE; Birk et al., 1979). The large concentrations of Sr and Ba in these rocks were presumed to prevent the Eu anomalies. Alternatively conditions may have been oxidizing enough to stabilize Eu in the 3+ oxidation state. Hornblende crystallization was also tentatively attributed to cause HREE depletion without changing Eu anomaly size in the Tuolumne Intrusive Series of the Sierra Nevada Batholith (Frey et al., 1978). A model of fractional crystallization of plagioclase, hornblende, and biotite was proposed to produce a trondhjemite magma without Eu anomalies and lower REE content from a parent diorite (Barker and Millard, 1979). Presumably the plagioclase and hornblende compensate one another so that no net Eu anomaly is produced in the residual magma.

8.5. Granitic rocks — monzogranites and syenogranites

General

Monzogranites (adamellites) and syenogranites (granites) contain more abundant alkali feldspar and less plagioclase than the granitic rocks discussed in section 8.4. Some corresponding silicic extrusives are included in this section.

REE contents

The range of REE content of about 260 monzogranites and syenogranites are as follows: Σ REE = 8—1977 ppm; $(La/Lu)_{cn}$ = 0.54—137, and Eu/Sm = 0.0009—1.07*.

These alkali feldspar-rich granitic rocks are grouped into those containing small to moderate negative Eu anomalies (range Eu/Sm = 0.09—0.23), moderate to large negative Eu anomalies (Eu/Sm = 0.0009—0.074), positive Eu anomalies (Eu/Sm = 0.41—10.7), and little or no Eu anomalies (Eu/Sm =

---

*Sources of data are: Mineyev, 1963; Ewart et al., 1968; Taylor et al., 1968; Allksiyev, 1970; Buma et al., 1971; Bowman et al., 1973; Bowden and Whitley, 1974; Duchesne et al., 1974; Hunter, 1974; Jahn et al., 1974; Koljonen and Rosenberg, 1974; O'Nions and Pankhurst, 1974; Dostal, 1975; Emmerman et al., 1975; Condie and Hunter, 1976; Glickson, 1976; Lambert and Holland, 1976; Albuquerque, 1977; Baker and Henage, 1977; Lenthall and Hunter, 1977; McCurry and Wright, 1977; Price and Taylor, 1977; Thorpe et al., 1977; Zielinski et al., 1977; Albuquerque, 1978; Anderson and Cullers, 1978; Bruhn et al., 1978; Condie, 1978; Coulon et al., 1978; Drury, 1978; Fryer and Jenner, 1978; McCarthy and Kable, 1978; Simmons and Hedge, 1978; Bowden et al., 1979; Ewing, 1979; Hildreth, 1979; López-Escobar et al., 1979; Noble et al., 1979; Peccerillo et al., 1979; Rose et al., 1979; Thorpe et al., 1979; Walsh et al., 1979; Anderson et al., 1980; Cullers et al., 1981; Jahn et al., 1980; Perfit et al., 1980; Weaver, 1980; Cullers and Arnold, 1981.

0.20—0.36; Figs. 8.7 and 8.8) to aid discussion as in section 8.4. Rocks containing small to moderate Eu anomalies have a limited range of REE content (ΣREE = 106—877 ppm) and LREE/HREE ratios ($(La/Lu)_{cn}$ = 2.5—50; Fig. 8.7a). Rocks that have moderate to large negative Eu anomalies have a wider range of REE content (ΣREE = 40–1977 ppm) and lower LREE/HREE ratios ($(La/Lu)_{cn}$ = 1.1—22) than those with small to moderate negative Eu anomalies (Fig. 8.7b). Rocks in the group with positive Eu anomalies contain low REE content (ΣREE = 40—210 ppm) and variable LREE/HREE ratios ($(La/Lu)_{cn}$ = 4.8–97; Fig. 8.8). Most rocks with little

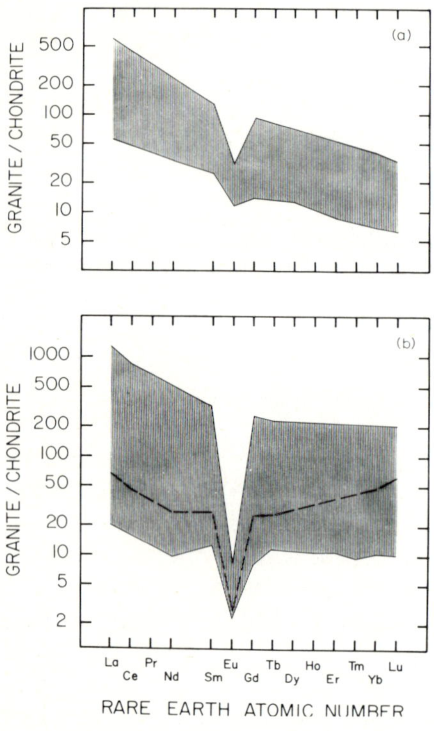

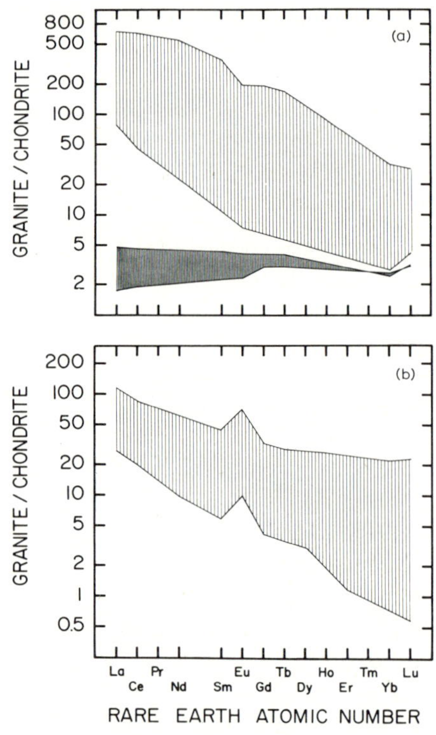

Fig. 8.7 (left). (a) The range of REE content found in monzogranites (adamellites) and syenogranites (granites) containing small to moderate, negative Eu anomalies. (b) The range of REE content found in monzogranites and syenogranites containing moderate to large, negative Eu anomalies. One concave-up pattern (dashed line) is also included to illustrate the variability of REE content in granitic rocks.

Fig. 8.8 (right). (a) The range of REE contents found in monzogranites and syenogranites containing little or no Eu anomalies (upper, wide-spaced vertical lines). Microgranites from the Spanish Peaks igneous complex, U.S.A., contain unusually low LREE and total REE contents and flat REE distributions (lower, dark pattern; Cullers and Arnold, 1981). (b) The range of REE in monzogranites and syenogranites containing positive Eu anomalies.

or no Eu anomalies have REE contents (ΣREE = 108—1426 ppm) and LREE/HREE ratios ($(La/Lu)_{cn}$ = 4.5—137) that are similar to the range of rocks with negative Eu anomalies (Fig. 8.8). One group of high-silica granitic rocks with no Eu anomalies from the Spanish Peaks igneous complex contain very low REE contents (ΣREE = 8—15 ppm) and flat REE patterns ($(La/Lu)_{cn}$ = 0.54—1.5; Fig. 8.8; Cullers and Arnold, 1981).

There does not seem to be a clear correlation of these groups of rocks with geologic age although those with large, negative Eu anomalies are not abundant in the Archean. However, moderate to large negative Eu anomalies are much more common and positive Eu anomalies are less common in monzogranites and syenogranites than in granodiorites, quartz diorites, tonalites, or trondhjemites.

General petrogenesis

Many of the review papers and general comments in section 8.4 concerning plagioclase-rich granitic rocks apply to alkali feldspar-rich granitic rocks. These comments will not be repeated here.

Examples of how the REE results may be integrated with detailed field, petrographic, major element, other trace elements, and isotopic studies for intrusive, granitic rocks may be found in Anderson and Cullers (1978) and in Cullers et al. (1981). Most studies of these alkali feldspar-rich granitic rocks again focus on melting and crystallization models. The thermogravitational diffusion process discussed below may be important at the top of some very silicic magma chambers (Hildreth, 1979).

In contrast to models for plagioclase-rich granitic rocks, models for alkali feldspar-rich granitic rocks may require a more K-rich source and a corresponding source mineralogy that includes potassic phases such as K-rich feldspar, muscovite, or biotite. The large number of rocks in this group with negative Eu anomalies require that much feldspar is involved in the mineral/melt equilibria during melting or crystallization. Fractional crystallization models requiring much feldspar are often used to explain increased negative Eu anomaly size in presumably related intrusive rocks. Rocks with positive Eu anomalies that would presumably represent a large amount of cumulus minerals relative to magma are seldom found in these sequences (e.g., Cullers et al., 1981). Probably the separation of crystals from magma are incomplete in such viscous magmas so that few rocks with enough cumulus feldspar form to develop a positive Eu anomaly. A melt with a large, negative Eu anomaly mixed with cumulus plagioclase may never produce a rock with a positive anomaly unless the cumulus plagioclase is predominant. Alternatively, many studies lack the detailed field work needed to find the small pods of cumulate-rich material.

Also the REE contents often decrease with increased differentiation especially in the most SiO_2-rich rocks of a given suite. The decrease in REE is probably due to fractional crystallization of accessory minerals such as allanite that contain high abundances of the REE.

The samples with large LREE/HREE ratios and with no Eu anomalies require garnet in the source without much feldspar. Thus, as with plagioclase-rich granitic rocks, eclogitic sources are postulated for alkali feldspar-rich granitic rocks. It is not clear why eclogites might give varied K contents to granitic melts while still producing similar REE patterns.

Partial melting

Monzogranites, syenogranites, and volcanic rocks of similar composition with small to moderate negative Eu anomalies (Fig. 8.7a) require abundant residual feldspar in the source. Varied amounts of minerals like garnet, amphibole, or pyroxenes are included in sources to explain varied LREE/HREE ratios and absolute REE contents. Most models do not have alkali feldspar or muscovite in the residuum. Presumably the alkali feldspar or muscovite completely melt to provide the large K content for the syenogranites or monzogranites. Sources include metasediments such as metagraywackes or metapelitics (Arth and Hanson, 1975; Albuquerque, 1977, 1978; Drury, 1979; Cullers et al., 1981), metaquartz diorites or tonalites (Anderson and Cullers, 1978; Coulon et al., 1978; Anderson et al., 1980; Cullers et al., 1981), and siliceous granulites (Freyer and Jenner, 1978; Weaver, 1980). For example, a quartz diorite or granodiorite source melted (20%) to produce undifferentiated monzogranite and syenogranite melts of the Wolf River Batholith, Wisconsin, yielding a model residue of plagioclase (40—65%), biotite (10—15%), apatite (0.1—0.2%), hornblende (0.5%), pyroxene (0—20%), and olivine or quartz (0—10%). Some studies emphasize that plagioclase must be in the source residue, but do not necessarily specify a particular rock (e.g., Bowden and Whitley, 1974; Bruhn et al., 1978; Walsh et al., 1979).

At least two-stage melting or crystallization models involving much feldspar are required to explain rocks with the largest negative Eu anomalies (Fig. 8.7b). A source with a negative Eu anomaly produced originally by igneous processes could re-melt and produce a larger negative Eu anomaly in the melt. Further crystallization of feldspar (discussed in the next section) could further enhance the negative Eu anomalies. Garnet is seldom included as a possible phase in the source because of the presence of relatively large abundances of HREE compared to the LREE in most of these rocks.

In Archean rocks of southern India, silicic granulites (quartz:plagioclase:hypersthene:ore = 30/60/5/5) were presumed to melt producing residues with positive Eu anomalies, and lower REE contents and larger LREE/HREE ratios than the melts (Weaver, 1980). This is one of the few studies in the alkali feldspar-rich group that suggests metasomatism may affect REE distributions.

Monzogranites or syenogranites containing no Eu anomalies (Fig. 8.8a) often have large LREE/HREE ratios that require a quartz eclogite or eclogite source (e.g., Buma et al., 1971; Jahn et al., 1974, 1980; Ewing, 1979; Cullers and Arnold, 1981). Granitic rocks with a slight negative Eu anomaly allow plagioclase in the source. These sources include ancient tonalites (Glickson, 1976), quartz-feldspar-hornblende-biotite gneiss (Drury, 1978), or graywacke (Emmerman et al., 1975). One model assumes that kaersutite in the upper mantle melts to produce quartz monzonites or monzonorites in the South Rogaland complex, Norway (Duchesne et al., 1974).

High-silica and K-rich granitic rocks in the Spanish Peaks igneous complex, Colorado (U.S.A.), contain very small REE contents and LREE/HREE ratios, and no Eu anomalies (Fig. 8.8a). Most possible crustal sources are eliminated since these cannot have residual feldspar, and they must have residual phases that concentrate all the REE. Melting of a hydrous mafic granulite with residual sphene at 12 kbar as suggested by Hellman and Green (1979) is a possible source although the HREE tend to be too abundant for this model. Alternatively, a biotite-plagioclase-quartz gneiss (metagraywacke?) may undergo about 50% melting with a complex reaction over a temperature interval of only 670—690°C. This produces a granitic melt and leaves a mafic residue of hornblende, sphene, garnet, and either quartz or plagioclase (Winkler, 1979). The REE content of the predicted melt produced by 50% fusion of such a source and leaving a residue of quartz:hornblende:garnet: sphene = 0.31/0.60/0.01/0.08 matches closely the range of REE observed in these rocks (Cullers and Arnold, 1981). These models assume D values of sphene/melt are larger for the LREE than for the HREE. However, only one D value has been estimated with the LREE being concentrated over the HREE, and this value is speculative (Hellman and Green, 1979). Other D values determined for sphene/melt have the intermediate REE higher than the LREE or HREE (Simmons and Hedge, 1978; Henderson, 1980). Such D values would not work for this model. We emphasize that the D values for accessory minerals, like sphene relative to silicic magma, need to be determined carefully before further significant progress can be made in the interpretation of melting and crystallization in these systems.

Fractional crystallization

Fractional crystallization models focus on the production of higher silica granitic rocks by crystallization of feldspar to increase the negative Eu anomaly during differentiation (Taylor et al., 1968; Bowden and Whitley, 1974; Emmerman et al., 1975; Condie and Hunter, 1976; McCurry and Wright, 1977; Thorpe et al., 1977; Zielinski et al., 1977; Anderson and Cullers, 1978; Condie, 1978; Coulon et al., 1978; Fryer and Jenner, 1978; McCarthy and Kable, 1978; Rose et al., 1979; Walsh et al., 1979; Anderson et al., 1980; Cullers et al., 1981). It is important to realize that absolute REE contents may be controlled by the fractional crystallization of small amounts of minerals such as sphene, allanite, hornblende, or apatite. REE contents may thus decrease in residual melts due to crystallization of these minerals. Also models of hornblende crystallization have been precluded in favor of partial melting models in some studies. For example, in an Archean high-grade meta-igneous suite in southern India (Weaver, 1980), there was an increase in Ce/Yb ratios and Eu anomaly size with decreasing REE content similar to that observed by Arth et al. (1978) in southwest Finland. The trend could not be explained by hornblende crystallization since there was no correlation of REE content to percent SiO_2 in the Indian rocks. Also few intermediates of the right composition were present in this bimodal suite to be consistent with a fractional crystallization model. Fractional crystallization of amphibole was also precluded as an explanation for the increasing REE content and negative Eu anomaly size with increasing SiO_2 content in the Ben Ghnema batholith, North Africa (Rogers et al., 1980). The K content increased too rapidly relative to Si to be consistent with a fractional crystallization model.

Several examples of crystallization models illustrate how the variation of REE content of the alkali feldspar-rich granitic rocks with Eu anomalies may be produced. Field relations, mineral composition changes, major and trace element changes in the Wolf River Batholith suggest the large volume of undifferentiated Wolf River Granite is a parent, some portions of which underwent fractional crystallization to produce smaller volumes of Belongia Coarse and Belongia Fine Granites (Anderson and Cullers, 1978; Fig. 8.9). Small pods of rocks presumed to represent a large amount of cumulate were also found in this study. About 55% fractional crystallization of alkali feldspar, plagioclase, hornblende, biotite, apatite, zircon, and allanite in the ratio of 62/31/3/3/0.5/0.05/0.03 from the Wolf River Granite magma will produce the REE content, other trace element, and major element contents of the Belongia Coarse Granite (Fig. 8.9a). The REE content and negative Eu anomaly size increase during this differentiation. Also the Belongia Coarse Granite had to contain about 90% melt and 10% cumulus crystals. Otherwise, predicted negative Eu anomaly size, Sr, and Ba contents would be too low. Further crystallization of Belongia Coarse-type magmas could produce the Belongia Fine Granite assuming the same major phases continued crystallizing (Fig. 8.9b). More allanite needed to be crystallized to account for a decrease in LREE content from the Belongia Coarse to Belongia Fine Granites. The rocks presumed to represent a large amount of cumulate contained larger Sr, Ba, Eu (positive anomaly) and lower Rb and REE content (except Eu) compared to the presumed Wolf River Granite parent (Fig. 8.9c).

A second example of a study in an alkali feldspar-rich granitic system is provided by a comparison of REE contents of interlayered residual parent gneisses and presumed granitic granulite melts in Namaqualand, South Africa

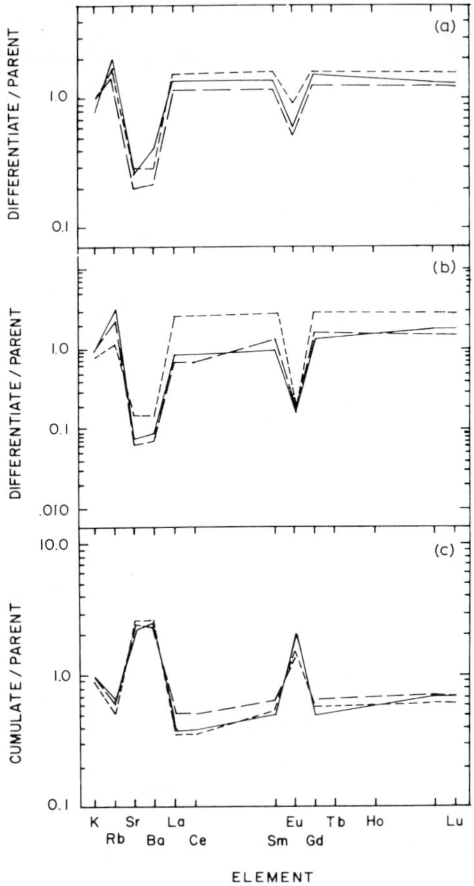

Fig. 8.9. Comparison of actual (solid lines) to model (dashed lines) data for differentiated and cumulate samples relative to the presumed parent (Wolf River Granite). (a) Belongia Coarse Granite: upper dashed line represents the composition of melt derived from the Wolf River by 45% fractionation of feldspar with K-feldspar/plagioclase ratio of 4:1. Lower dashed line represents 55% fractionation of 62% K-feldspar, 31% plagioclase, 3% biotite, 3% hornblende, 0.5% apatite, 0.05% zircon, and 0.03% allanite. (b) Belongia Fine Granite: upper and lower dash lines represent 70% and 75% fractionation of minerals in the same proportions of the upper and lower dashed lines in (a), respectively, with the exception of having 0.06% allanite as a cumulate phase (allows late depletion of LREE). (c) Cumulate portion of the Wolf River Granite (most Si-depleted sample (GR24A): both upper and lower lines represent 30% accumulation of 60% K-feldspar, 30% plagioclase, 7% biotite, 2% hornblende, 0.5% apatite, and 0.1% zircon. Upper dashed model contains 25% interstital melt while the lower dashed model is pure solid.

(McCarthy and Kable, 1978). The melt was not enriched or depleted in REE relative to the residue during melting. This was attributed to changing D values of the REE during progressive melting. Crystallization of minor minerals from the primary melt controlled REE distributions resulting in a decrease of the LREE and increase in HREE that produced concave up REE patterns in the residual magmas (Fig. 8.7b). Large, negative Eu anomalies in the residual magmas were produced by feldspar crystallization. McCarthy and Kable emphasize that the major control of minor minerals on REE distributions restricts severely the use of the REE as petrogenetic indicators in granitic rocks.

Alkali feldspar-rich melts with no Eu anomalies have models for their formation that focus on partial melting (see previous section) rather than fractional crystallization. Thus, such melts are often taken as primary and are assumed to undergo feldspar fractionation in order to produce associated rocks with Eu anomalies (e.g., Dostal, 1975; Emmerman et al., 1975).

Thermogravitational diffusion

Some very silicic magma chambers appear to become vertically zoned in element concentration prior to eruption. The Bishop Tuff, California, U.S.A., represents a continuous outpouring of more than 170 km^3 of rhyolitic ash-falls and outflows from a zoned magma chamber (Hildreth, 1979). Higher portions of the zoned magma chamber were tapped and extruded first. Deeper portions of the chamber were gradually tapped and extruded. Temperature gradually increased during the eruption process as deeper portions of the chamber were tapped. Mineral compositions changed systematically with progressive eruption. Major element contents changed modestly, but many trace element concentrations changed drastically with eruption. The LREE contents increased and the HREE and negative Eu anomaly size decreased with progressive eruption (Fig. 8.10).

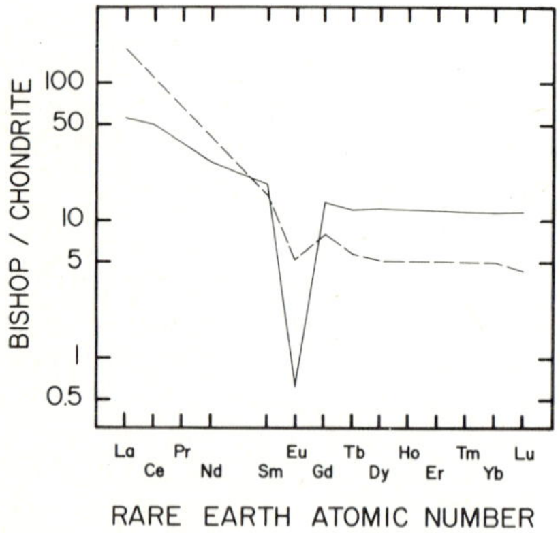

Fig. 8.10. Compositional zonation of the REE in the Bishop Tuff was observed in which the initial eruption from the top of the magma chamber (solid line) contained smaller LREE and larger HREE and negative Eu anomaly size compared to the final part of the eruption (dashed line) from lower in the chamber (Hildreth, 1979). Samples intermediate between the two extremes changed progressively with eruption.

Models of progressive fusion, fractional crystallization, large-scale assimilation, liquid immiscibility, and contamination by underplating mafic magma are excluded as possible causes for the variations. Instead a little understood convection-diffusion process driven by a temperature gradient and gravity is proposed in which all fractionation takes place in an all fluid state. Compositional gradients of elements are presumably produced at the top of the magma chamber by combinations of temperature gradient, gravity, and changes in the structure of the melt. The main problem is that exact mechanisms of transfer are little understood as yet. It is fairly easy to visualize that the HREE may form more stable complexes than the LREE so the HREE concentrate at the top of the chamber due to the HREE complexes moving with upward migrating volatiles. It is difficult, however, to visualize how the LREE decrease upward in the chamber by this mechanism.

8.6. REE in hydrothermal systems

General

A hydrothermal system is considered here to be any system in which heated water interacts with rocks or magmas. Direct or indirect involvement of igneous rocks or magmas may occur but is not necessarily implied by the term hydrothermal. The REE distribution in hydrothermal solutions will be controlled by the partitioning of REE between the solution and rock phases. However, solution/mineral or solution/melt partition coefficients are not available for the conditions prevalent in most hydrothermal systems. In addition, most sulfide and oxide ore minerals are not good hosts for REE, so REE analyses are usually restricted to non-sulfide gangue minerals.

The limited amount of REE data suggests that REE contents and distributions vary widely depending upon the mineral analyzed, the geologic setting of the deposit, and the position of the sample in the paragenetic sequence. Therefore, ranges of REE contents or patterns have little meaning except where there are a large number of analyses from a single deposit or district. The purpose of this section is to describe the REE distributions observed in hydrothermal minerals and to explore the potential of REE studies to provide information concerning ore formation processes.

REE contents

The only analyses of sulfide minerals were made on galenas from Creede, Colorado (Morgan and Wandless, 1980). Only La and Sm were present at concentrations above their detection limits, and these had concentrations of under 1 ppb.

Analyses are available for calcite and dolomite from Pb-Zn deposits in

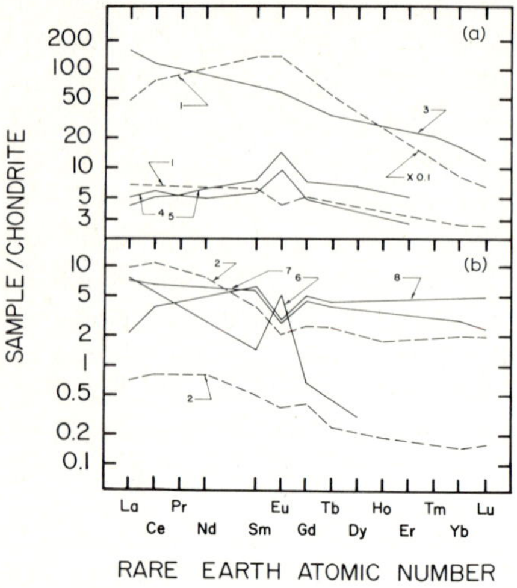

Fig. 8.11. REE patterns of hydrothermal minerals: *1* = range of REE patterns of hydrothermal calcites, western Harz (dashed lines; Moller et al., 1979); *2* = range of REE patterns of hydrothermal dolomites, Leadville limestone, Colorado (dashed lines; Jarvis et al., 1975); *3* = anhydrite (Morgan and Wandless, 1980); *4* = quartz-carbonate vein (Kerrich and Fryer, 1979); *5* = quartz-tourmaline vein (Kerrich and Fryer, 1979); *6* = barite (Morgan and Wandless, 1980); *7* = hydrothermal dolomite, eastern Missouri (Graf and Richards, 1981); *8* = hydrothermal calcite eastern Missouri (Graf and Richards, 1981).

Colorado, Missouri, and Germany (Fig. 8.11; Jarvis et al., 1975; Moller et al., 1979; Graf and Richards, 1981). Both Colorado and Missouri carbonates have low REE contents (ΣREE $<$ 25 ppm) while ΣREE of the German calcites range up to 2500 ppm. Missouri dolomites and calcites (Fig. 8.11, patterns 7 and 8, respectively) have fairly flat REE patterns and negative Eu anomalies although calcites are depleted in LREE relative to dolomite patterns. Dolomites from the Colorado mineral belt (Fig. 8.11, pattern 2) have similar ΣREE contents to the Missouri carbonates but are enriched in LREE relative to HREE and have no significant Eu anomalies. As shown by the two patterns marked 1 in Fig. 8.11, calcites from the western Harz, Germany, display two different REE patterns. The first (upper pattern 1) is characterized by very large ΣREE, strong depletion of HREE, and depletion of La and Ce relative to Sm and Eu. The second (lower pattern 1) is characterized by lower ΣREE and flatter REE patterns with negative Eu, and sometimes Ce, anomalies. A third pattern type with very low ΣREE was observed in late stage calcites (Moller et al., 1979).

Two siderite patterns are reported in Morgan and Wandless (1980). Both are depleted in all REE relative to Yb and Lu ($(La/Lu)_{cn}$ = 0.02). One has a negative Eu anomaly while the other does not (Eu/Sm = 0.58, 0.06). Hydrothermal quartz-carbonate and quartz-tourmaline veins from the Dome Mine, Ontario, Canada, have positive Eu anomalies and fairly flat REE patterns (Fig. 8.11, patterns 4 and 5; Kerrich and Fryer, 1979).

Typical patterns of hydrothermal barite and anhydrite are also shown in Fig. 8.11 (patterns 6 and 3, respectively; Guichon et al., 1979; Morgan and Wandless, 1980). Barites have a large range of Eu/Sm ratios (0.25—12) but most have positive Eu anomalies. These anomalies probably result from substitution of Eu^{2+} for Ba^{2+} (Morgan and Wandless, 1980). Two anhydrite samples from porphyry copper deposits have $(La/Lu)_{cn}$ of about 13 and little or no Eu anomaly (Eu/Sm = 0.18—0.38). Either Eu^{2+} does not substitute as readily for Ca^{2+} in the anhydrite structure as for Ba^{2+} in the barite structure or the conditions of deposition of these anhydrites were too oxidizing for an appreciable amount of Eu^{2+} to have been present. REE patterns of chemical sediments associated with volcanic activity are shown in Fig. 8.12b (Laajoki, 1975; Fryer, 1977; Graf, 1977, 1978). A common characteristic of these rocks is a positive Eu anomaly (Eu/Sm $\geqslant$ 0.35). Massive sulfide samples from New Brunswick, Canada, have a range of REE patterns as shown in Fig. 8.12b (patterns 9). Only copper sulfide zones do not have positive Eu anomalies (Graf, 1977). Iron formations associated with the New Brunswick sulfide deposits have REE patterns similar to those of the

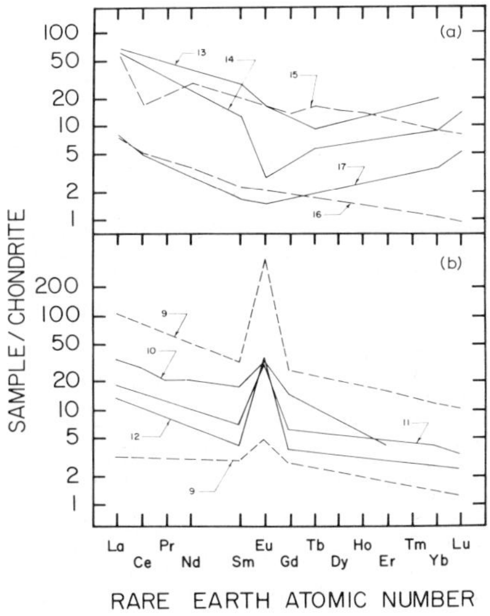

Fig. 8.12. (a) REE patterns of oxide facies of iron formations not directly associated with volcanic rocks: *13* = Mn layer, Urucum region, Brazil (Van Leeuwen and Graf, 1981); *14* = hematite ore, Urucum region, Brazil (Van Leeuwen and Graf, 1981); *15* = magnetite iron formation, Finland (Laajoki, 1975); *16* = oxide facies, Sokoman Iron Formation, Canada (Fryer, 1977); *17* = magnetite iron formation, Hamersley, Australia (Graf, 1978). (b) REE patterns of some chemical sedimentary ores associated with volcanic rocks: *9* = range of REE patterns for massive sulfide ores, New Brunswick (dashed lines; Graf, 1977); *10* = stratiform carbonate, Dome Mine, Ontario (Kerrick and Fryer, 1979); *11* = banded magnetite-quartz, Adams Mine, Ontario (Graf, 1978); *12* = magnetite iron formation, New Brunswick (Graf, 1977).

massive sulfides (Fig. 8.12b, pattern 12). Iron formations and stratiform carbonates within Archean greenstone belts have REE patterns with positive Eu anomalies very similar to the patterns of the younger, New Brunswick rocks (Fig. 8.12b, patterns 10 and 11; Graf, 1978; Kerrich and Fryer, 1979). Iron formations with no demonstrable volcanic association do not, as a rule, have positive Eu anomalies (Fig. 8.12a, Laajoki, 1975; Fryer, 1977; Graf, 1978; Van Leeuwen and Graf, 1981).

REE and ore genesis

The use of REE as hydrothermal tracers requires knowledge of the partitioning behavior of REE between hydrothermal solutions and minerals either during precipitation or during interactions with host or source rocks. Many of these data are lacking for conditions appropriate to hydrothermal systems. Hydrothermal temperatures range from as low as 50°C up to granite melting temperatures. Most hydrothermal systems also operated at upper crustal, sometimes almost surficial, pressures. However, partitioning experiments have been concerned with magmatic systems and carried out at high temperatures and often high pressures as well (Cullers et al., 1973; Flynn and Burnham, 1978; Wendlandt and Harrison, 1979).

Melt/vapor experiments have shown that REE do not partition favorably into aqueous vapor phases at low pressure but become mobile at pressures of around 20 kbar (Wendlandt and Harrison, 1979). CO_2 vapors tend to concentrate REE (especially LREE) at lower pressures but the vapor/melt D values decrease with increasing pressure (Wendlandt and Harrison, 1979). Melt REE partition more favorably into a chloride-bearing vapor phase than into pure water (Flynn and Burnham, 1978). The only mineral/water D values available are those of Cullers et al. (1973).

The experimental behavior of Eu is intriguing and illustrates the need for more data. Cullers et al. (1973) found little Eu fractionation during partitioning. However, Flynn and Burnham (1978) found Eu to fractionate into a chloride brine relative to silicate melts. Even though all their experiments were carried out at an f_{O_2} buffered by the quartz-fayalite-magnetite assemblage, the extent of Eu fractionation depended upon pressure and melt composition. At 4 kbar Eu partitioned preferentially into the vapor phase, but at 1.25 kbar the vapor became depleted in Eu relative to the melt. A jadeite-nepheline melt had much lower D values for all REE and produced a lower vapor Eu anomaly than did a pegmatite melt under identical conditions. A fluoride-bearing vapor phase showed little or no Eu fractionation relative to the pegmatite melt (Flynn and Burnham, 1978).

In contrast to the partitioning data, experiments on the Eu^{2+}-Eu^{3+} transition (Shul'gin and Koz'min, 1963) and REE complexes (Peppard et al., 1962; Choppin and Unrein, 1963) have all been at temperatures below or at the low end of the range of interest for hydrothermal systems. The conditions of Eu reduction are such that either Eu^{2+} or Eu^{3+} might dominate in a given hydrothermal system but Eu^{2+} is probably favored by higher temperatures (Shul'gin and Koz'min, 1963). At low temperatures the Eu^{2+}-Eu^{3+} transition occurs within the Fe^{2+} field (Shul'gin and Koz'min, 1963), so many ore solutions which are precipitating ferrous minerals should have significant contents of reduced Eu as well. At low temperature and pressure, HREE form stronger complexes with chloride ions which are the most likely complexing ligands in hydrothermal solutions.

The above data do not permit sophisticated modelling of hydrothermal systems but some generalizations are probably valid. If the water/rock volume ratio of the system is low, the REE distribution in the solution may be similar to that of the rock. The LREE/HREE ratio of the solution may be affected by slightly higher D values for LREE or by complexing of HREE. When the solution is altering certain rock phases but only exchanging REE with the other phases, REE from the altered phase may dominate the solution REE pattern. The amount of REE fractionation which occurs during precipitation of hydrothermal minerals is difficult to assess but may affect only Eu and the LREE/HREE ratio.

Changes in LREE/HREE ratios within a deposit may be used to infer changes in the extent of complexing in the solution which will be related to the concentration of the hydrothermal solution. In general, very concentrated brines would be expected to become enriched in HREE (Graf, 1977) and if precipitation occurs due to mixing with less saline solutions, the resulting precipitates should also reflect this HREE enrichment. Such changes have been observed in many hydrothermal systems (Jarvis et al., 1975; Graf, 1977; Moller et al., 1979).

Behavior of Eu is very complex, but it does lend itself to some interpretation. Eu anomalies in hydrothermal minerals may have been caused either by fractionation of the REE during precipitation of the minerals or by precipitation of the minerals from solutions which had already developed Eu anomalies. Based upon experimental separation of Eu from other REE (Kazimirz and Madjar, 1958) Graf (1977) suggested that the large Eu/Sm ratios of some New Brunswick sulfides could not be explained solely by precipitation fractionation. An Eu anomaly could have been imparted to the solution either by fractionation of REE during partitioning between the solution and host or source rocks or by preferential alteration of an anomalous rock phase such as feldspar (Graf, 1977). If the low pressure data of Flynn and Burnham (1978) can be extended to lower temperatures, the latter option seems more likely for positive anomalies.

The above considerations plus the association of the highest Eu anomalies with Pb-Zn sulfides led Graf (1977) to propose that the alteration of feldspar was responsible for mobilization of Pb and Zn and at the same time produced a positive Eu anomaly in the New Brunswick hydrothermal solutions. The lack of an Eu anomaly in copper sulfides was attributed to alteration of non-anomalous, ferromagnesian minerals and volcanic glass (Graf, 1977). The lack of a positive Eu anomaly or the negative Eu anomaly observed in some carbonate minerals from Pb-Zn deposits (Fig. 8.11) may have resulted from host rock REE dominating in the solution or from interaction of the solution with other minerals than feldspar (e.g., amphiboles). Alternatively, the carbonate minerals may have been deposited from different solutions than those which deposited the sulfide minerals.

In summary, REE studies can provide information regarding solution histories but only in a very general way. More sophisticated modelling of the behavior of REE in water-rock systems will require more detailed knowledge of the partitioning behavior and solution properties of the REE than is presently available.

8.7. Summary

Andesites contain REE contents similar to the range found in basalts (Table 8.1), and they rarely contain Eu anomalies. Gabbro, amphibolite, eclogite, garnet pyroxenite, and hydrous garnet peridotite have been used as sources which can melt and produce the observed REE contents. Generally one-stage melting models do not account for the major and trace element characteristics of the melt; models involving at least two-stages are required. For example, the melting of eclogite in subducted oceanic slabs produces silicic melts, enriched in the LREE and incompatible elements (e.g., Rb, Ba, Sr, Th), which may react with overlying peridotite to produce a garnet pyroxenite. In turn, the garnet pyroxenite may melt to produce andesites with a second stage enrichment of these elements. The incompatible element contents predicted by this model are similar to those of calc-alkaline andesites produced at subduction zones. Some tholeiitic andesites, however, may be produced by fractional crystallization of plagioclase, amphibole, pyroxene, or magnetite from tholeiitic basalt magmas.

Anorthosites are often associated with fairly silicic rocks so they are included here. Anorthosites contain low REE contents and LREE/HREE ratios, and they usually have positive Eu anomalies, while the associated, more silicic rocks usually have larger REE contents and smaller Eu anomalies (positive to negative). Most theoretical models focus on the anorthosite being a cumulate; it is then possible to estimate the REE contents of the parent melt and compare these to contents in associated rocks to see if the rocks are related.

Granitic rocks contain variable REE contents, LREE/HREE ratios, and Eu anomaly size. Negative Eu anomalies are more frequent and positive Eu anomalies are less frequent in K-rich granitic rocks compared to K-poor granitic rocks. Granitic magmas with large REE contents, large LREE/HREE ratios, and large negative Eu anomalies can be generated by melting of a source rich in residual feldspar to produce the negative Eu anomaly. In addition, this source needs to have residual phases like garnet, amphibole, or pyroxene to increase the LREE/HREE ratios (siliceous granulites, hornblende-garnet-clinopyroxene-plagioclase basic rocks, metagraywackes, and metatonalites). Alternatively, crystallization of minerals like feldspar, garnet, or amphibole from a more basic magma could produce similar REE contents in residual, silicic magmas. Those silicic rocks with the largest negative Eu anomalies require at least two stages of melting and crystallization involving much feldspar. A convection-diffusion process driven by a temperature gradient and gravity in the upper portions of some very silicic magma chambers may cause the LREE to increase and the HREE and negative Eu anomaly size to decrease from the top downward in the chamber.

Those low-Al dacites and quartz diorites in island arcs which contain negative Eu anomalies and small LREE/HREE ratios require the melting of a basic source with small LREE/HREE ratios and abundant plagioclase. Alternatively, some of these dacites and quartz diorites may form by fractional crystallization of plagioclase from less differentiated melts with low LREE/HREE ratios.

Granitic rocks with positive Eu anomalies, small REE, and large LREE/HREE ratios may form by melting of amphibolite, eclogite, or garnet amphibolite. Alternatively, crystallization of garnet and amphibole from less differentiated melts may produce the positive Eu anomalies and large fractionation of LREE from the HREE.

Since granitic rocks with no Eu anomalies arise from sources with no significant plagioclase, eclogites or mafic granulites are likely to be common source rocks if their LREE/HREE ratios are fairly large. One group of silicic rocks with no Eu anomalies and very low REE contents and La/Lu ratios may form by melting a biotite-plagioclase-quartz gneiss leaving a residue of hornblende, sphene, garnet, and quartz.

REE patterns may be useful to interpret the formation history of mineral deposits because the behavior of REE in hydrothermal systems can be treated in a manner analagous to that used for magmatic systems. REE distributions in aqueous solutions will be controlled by the partitioning of REE between the solution and the rock phases and by additional release of REE during alteration of the rock phases most probably in open systems. REE patterns of ore minerals tend to show variation in Eu/Sm and LREE/HREE ratios. Eu anomalies may be due to reduction of Eu^{3+} to Eu^{2+} or simply to interaction of the solution with minerals such as feldspar. LREE/HREE ratios in solutions can be changed by increasing or decreasing complexing of REE in the solution. REE studies of ore deposits are hampered by such problems as the need to assess the degree to which wall rocks can mask the REE pattern of the initial solution and the lack of good data for mineral/water *D* values over a wide range of temperatures and fluid compositions. As the data base increases, REE studies should contribute more to our understanding of ore-forming processes.

Acknowledgements

Reviews by Fred Frey, J. Lawford Anderson, Karl Seifert, and John Fountain have improved this chapter. We thank Renée Hambleton and Monique Santilli for doing the typing and Bob Barnett for doing the drafting of the figures.

The papers presented in Chapters 7 and 8 represent the literature available to us up to the summer of 1980. A few additional references were inserted at the last minute up to April, 1981.

References

Albuquerque, C.A.R. de, 1977. Geochemistry of the tonalitic and granitic rocks of the Nova Scotia southern plutons. *Geochim. Cosmochim. Acta*, 41: 1—13.

Albuquerque, C.A.R. de, 1978. Rare earth elements in "younger" granites, northern Portugal. *Lithos*, 11: 219—229.

Aleksiyev, E.I., 1970. Genetic significance of the rare earth elements in the younger granites of northern Nigeria and the Cameroons. *Geochem. Int.*, 7: 127—132.

Anderson, J.L. and Cullers, R.L., 1978. Geochemistry and evolution of the Wolf River batholith, a Late Precambrian rapakivi massif in north Wisconsin, U.S.A. *Precambrian Res.*, 7: 287—324.

Anderson, J.L., Cullers, R.L. and VanSchmus, W.R., 1980. Post-orogenic metaluminous and peraluminous granite plutonism in the Mid-Proterozoic of Wisconsin, U.S.A. *Contrib. Mineral. Petrol.*, 74: 311—328.

Arth, J.G., 1979. Some trace elements in trondhjemites — their implications to magma genesis and paleotectonic setting. In: F. Barker (Editor), *Trondhjemites, Dacites, and Related Rocks.* Elsevier, Amsterdam, pp. 123—132.

Arth, J.G. and Barker, F., 1976. Rare-earth partitioning between hornblende and dacitic liquid and implications for the genesis of trondhjemitic-tonalitic magmas. *Geology*, 4: 534—536.

Arth, J.G. and Hanson, G.N., 1972. Quartz diorites derived by partial melting of eclogite or amphibolite at mantle depths. *Contrib. Mineral. Petrol.*, 37: 161—174.

Arth, J.G. and Hanson, G.N., 1975. Geochemistry and origin of the early Precambrian crust of northeastern Minnesota. *Geochim. Cosmochim. Acta*, 39: 325—362.

Arth, J.G., Barker, F., Peterman, Z.E. and Friedman, I., 1978. Geochemistry of the gabbro-diorite-tonalite-trondhjemite suite of southwest Finland and its implications for the origin of tonalitic and trondhjemitic magmas. *J. Petrol.*, 19: 289—316.

Ashwal, L.D. and Seifert, K.E., 1980. Rare-earth-element geochemistry of anorthosite and related rocks from the Adirondocks, New York, and other massif-type complexes, Parts I and II. *Geol. Soc. Am. Bull.*, 91: 105—107 (Part I), 91: 659—684 (Part II).

Baker, B.H. and Henage, L.F., 1977. Compositional changes during crystallization of some peralkaline silicic lavas of the Kenya rift valley. *J. Volcanol. Geotherm. Res.*, 2: 17—28.

Barker, F. and Millard, H.T., Jr., 1979. Geochemistry of the type trondhjemite and three associated rocks, Norway. In: F. Barker (Editor), *Trondhjemites, Dacites, and Related Rocks.* Elsevier, Amsterdam, pp. 517—529.

Barker, F. Arth, J.G., Peterman, Z.E. and Friedman, I., 1976. The 1.7- to 1.8-b.y.-old trondhjemites of southwestern Colorado and northern New Mexico: geochemistry and depths of genesis. *Geol. Soc. Am. Bull.*, 87: 189—198.

Barker, F., Arth, J.G. and Millard, H.T., Jr., 1979a. Archean trondhjemites of the southwestern Big Horn Mountains, Wyoming: a preliminary report. In: F. Barker (Editor), *Trondhjemites, Dacites, and Related Rocks.* Elsevier, Amsterdam, pp. 401—414.

Barker, F., Millard, H.T., Jr. and Knight, R.J., 1979b. Reconnaissance geochemistry of Devonian island-arc volcanic and intrusive rocks, West Shasta district, California. In: F. Barker (Editor), *Trondhjemites, Dacites, and Related Rocks.* Elsevier, Amsterdam, pp. 531—545.

Birk, D., Koljonen, J. and Rosenberg, R.J., 1979. Rare earth distribution in Archean granitoid plutons of the Wabigoon volcanic-plutonic belt, northwestern Ontario. *Can. J. Earth Sci.*, 16: 270—289.

Bowden, P. and Whitley, J.E., 1974. Rare-earth patterns in peralkaline and associated granites. *Lithos*, 7: 15—21.

Bowman, H.R., Asaro, F. and Perlman, I., 1973. On the uniformity of composition in obsidians and evidence for magmatic mixing. *J. Geol.*, 81: 312—327.

Brooks, C., James, D.E. and Hart, S.R., 1976. Ancient lithosphere: its role in young continental volcanism. *Science*, 193: 1086—1094.

Bruhn, R.L., Stern, C.R. and DeWit, M.J., 1978. Field and geochemical data bearing on the development of a Mesozoic volcano-tectonic rift zone and back arc basin in southernmost South America. *Earth Planet. Sci. Lett.*, 41: 32—46.

Buma, G., Frey, F.A. and Wones, D.R., 1971. New England granites: trace element evidence regarding their origin and differentiation. *Contrib. Mineral. Petrol.*, 31: 300—320.

Choppin, G.R. and Unrein, P.J., 1963. Halide complexes of the lanthanide elements. *J. Inorg. Nucl. Chem.*, 25: 387—393.

Collerson, K.D. and Bridgwater, D., 1979. Metamorphic development of early Archean tonalitic and trondhjemitic gneisses: Saglek area, Labrador. In: F. Barker (Editor), *Trondhjemites, Dacites, and Related Rocks.* Elsevier, Amsterdam, pp. 205—273.

Collerson, K.D. and Fryer, B.J., 1978. The role of fluids in the formation and subsequent development of early continental crust. *Contrib. Mineral. Petrol.*, 67: 151—167.

Compton, P., 1978. Rare earth evidence for the origin of the Nûk gneisses Buksefjorden region, southern West Greenland. *Contrib. Mineral. Petrol.*, 66: 283—293.

Condie, K.C., 1976. Trace-element geochemistry of Archean greenstone belts. *Earth-Sci. Rev.*, 12: 393—417.

Condie, K.C., 1978. Geochemistry of Proterozoic granitic plutons from New Mexico, U.S.A. *Chem. Geol.*, 21: 131—149.

Condie, K.C. and Baragar, W.R.A., 1974. Rare-earth element distributions in volcanic rocks from Archean greenstone belts. *Contrib. Mineral. Petrol.*, 45: 237—246.

Condie, K.C. and Harrison, N.M., 1976. Geochemistry of the Archean Bulawayan Group, Midlands greenstone belt, Rhodesia. *Precambrian Res.*, 3: 253—271.

Condie, K.C. and Hayslip, D.L., 1975. Young bimodal volcanism at Medicine Lake volcanic center, northern California. *Geochim. Cosmochim. Acta*, 39: 1165—1178.

Condie, K.C. and Hunter, D.R., 1976. Trace element geochemistry of Archean granitic rocks from the Barberton region, South Africa. *Earth Planet. Sci. Lett.*, 29: 389—400.

Condie, K.C. and Swenson, D.J., 1973. Composition variation in three Cascade stratovolcanoes: Jefferson, Rainier, and Shasta. *Bull. Volcanol.*, 37: 205—230.

Coulon, C., Dostal, J. and Dupuy, C., 1978. Petrology and geochemistry of the ignimbrites and associated lava domes from N.W. Sardinia. *Contrib. Mineral. Petrol.*, 68: 89—98.

Cullers, R.L. and Arnold, B., 1981. The petrogenesis of the Tertiary Spanish Peaks igneous complex, Colorado, U.S.A. (submitted).

Cullers, R.L., Medaris, L.G. and Haskin, L.A., 1973. Experimental studies of the distribution of rare earths as trace elements among silicate minerals and liquids and water. *Geochim. Cosmochim. Acta*, 37: 1499—1512.

Cullers, R.L., Koch, R. and Bickford, M.E., 1981. Chemical evolution of magmas in the igneous terrane of the St. Francois Mts., Mo., II. Trace element evidence. *J. Geophys. Res.*, 86: 10365—10387.

Davis, P.A., Jr. and Condie, K.C., 1977. Trace element model studies of Nyanzian greenstone belts, western Kenya. *Geochim. Cosmochim. Acta*, 41: 271—277.

Demaiffe, D., Duchesne, J.C. and Hertogen, J., 1979. Trace element variations and isotopic composition of charnockitic acidic rocks related to anorthosites (Rogaland, S.W. Norway). In: L.H. Ahrens (Editor), *Origin and Distribution of the Elements.* Pergamon, New York, N.Y., pp. 417—429.

Dixon, T.H. and Batiza, R., 1979. Petrology and chemistry of recent lavas in the northern Marianas: implications for the origin of island arc basalts. *Contrib. Mineral. Petrol.*, 70: 167—181.

Dostal, J., 1975. Geochemistry and petrology of the Loon Lake pluton, Ontario. *Can. J. Earth Sci.*, 12: 1331—1345.

Dostal, J. and Zerbi, M., 1978. Geochemistry of the Savalan volcano (northwestern Iran). *Chem. Geol.*, 22: 31—42.

Dostal, J., Dupuy, C. and Lefevre, C., 1977a. Rare earth element distribution in Plio-Quaternary volcanic rocks from southern Peru. *Lithos*, 10: 173—183.

Dostal, J., Zentilli, M., Caelles, J.C. and Clark, A.H., 1977b. Geochemistry and origin of volcanic rocks of the Andes (26°—28°S). *Contrib. Mineral. Petrol.*, 63: 113—128.

Drury, S.A., 1978. REE distributions in a high-grade Archean gneiss complex in Scotland: implications for the genesis of ancient sialic crust. *Precambrian Res.*, 7: 237—257.

Drury, S.A., 1979. Rare-earth and other trace element data bearing on the origin of Archean granitic rocks from Yellowknife, Northwest Territories. *Can. Earth Sci.*, 16: 809—815.

Duchesne, J.C. and Demaiffe, D., 1978. Trace elements and anorthosite genesis. *Earth Planet. Sci. Lett.*, 38: 249—272.

Duchesne, J.C., Roelandts, I. and Demaiffe, D., 1974. Rare-earth data on monzonoritic rocks related to anorthosites and their bearing on the nature of the parental magma of the anorthosite series. *Earth Planet. Sci. Lett.*, 24: 325—335.

Dupuy, C., Dostal, J. and Coulon, C., 1979. Geochemistry and origin of andesitic rocks from northwestern Sardinia. *J. Volcanol. Geotherm. Res.*, 6: 375—389.

Emmermann, R., Daieva, L. and Schneider, J., 1975. Petrologic significance of rare earth distribution in granites. *Contrib. Mineral. Petrol.*, 52: 267—283.

Ewart, A., Taylor, S.R. and Capp, A.C., 1968. Trace and minor element geochemistry of the rhyolitic volcanic rocks, Central North Island, New Zealand. *Contrib. Mineral. Petrol.*, 18: 76—104.

Ewart, A., Bryan, W.B. and Gill, J.B., 1973. Mineralogy and geochemistry of the younger volcanic islands of Tonga, S.W. Pacific. *J. Petrol.*, 14: 429—465.

Ewart, A., Mateen, A. and Ross, J.A., 1976. Review of mineralogy and chemistry of Tertiary central volcanic complexes in southeast Queensland and northeast New South Wales. In: R.W. Johnson (Editor), *Volcanism in Australasia*, Elsevier, Amsterdam, pp. 21—40.

Ewart, A., Brothers, R.N. and Mateen, A., 1977. An outline of the geology and geochemistry, and the possible petrogenetic evolution of the volcanic rocks of the Tonga-Kermadec-New Zealand island arc. *J. Volcanol. Geotherm. Res.*, 2: 205—250.

Ewing, T.E., 1979. Two calc-alkaline volcanic trends in the Archean: trace element evidence. *Contrib. Mineral. Petrol.*, 71: 1—7.

Flynn, R.T. and Burnham, C.W., 1978. An experimental determination of rare earth partition coefficients between a chloride containing vapor phase and silicate melts. *Geochim. Cosmochim. Acta*, 42: 685—701.

Fountain, J.C., 1979. Geochemistry of Brokeoff volcano, California. *Geol. Soc. Am. Bull.*, 90: 294—300.

Frey, F.A., Chappell, B.W. and Roy, S.D., 1978. Fractionation of rare-earth elements in the Tuolumne Intrusive Series, Sierra Nevada batholith, California. *Geology*, 6: 239—242.

Fryer, B.J., 1977. Trace element geochemistry of the Sokoman Iron Formation. *Can. J. Earth Sci.*, 14: 1598—1610.

Fryer, B.J. and Jenner, G.A., 1978. Geochemistry and origin of the Archean Prince Albert Group volcanics, western Melville Peninsula, Northwest Territories, Canada. *Geochim. Cosmochim. Acta*, 42: 1645—1654.

Fujimaki, H., 1975. Rare earth elements in volcanic rocks from Hakone volcano and northern Izu peninsula, Japan. *J. Fac. Sci., Univ. Tokyo, Sec. II*, 19: 81—93.

Gill, J.B., 1970. Geochemistry of Viti Levu, Fiji, and its evolution as an island arc. *Contrib. Mineral. Petrol.*, 27: 179—203.

Gill, J.B., 1974. Role of underthrust oceanic crust in the genesis of a Fijian calc-alkaline suite. *Contrib. Mineral. Petrol.*, 43: 29—45.

Gill, J.B., 1976. Composition and age of Lau Basin and Ridge volcanic rocks: implications for evolution of an interarc basin and remnant arc. *Geol. Soc. Am. Bull.*, 87: 1384—1395.

Gill, J.B., 1978. Role of trace element partition coefficients in models of andesite genesis. *Geochim. Cosmochim. Acta*, 42: 709—724.

Glickson, A.Y., 1976. Trace element geochemistry and origin of early Precambrian acid igneous series, Barberton Mountain Land, Transvaal. *Geochim. Cosmochim. Acta*, 40: 1261—1280.

Gorton, M.P., 1977. The geochemistry and origin of quaternary volcanism in the New Hebrides. *Geochim. Cosmochim. Acta*, 41: 1257—1270.

Graf, J.L., Jr., 1977. Rare earth elements as hydrothermal tracers during the formation of massive sulfide deposits in volcanic rocks. *Econ. Geol.*, 72: 527—548.

Graf, J.L., Jr., 1978. Rare earth elements, iron formations and sea water. *Geochim. Cosmochim. Acta*, 42: 1845—1850.

Graf, J.L. and Richards, B.D., 1981. Rare earth elements in hydrothermal carbonates associated with breccia-filling and replacement, Pb-Zn-Cu sulfide deposits in dolomite host rock, Viburnum Trend, Missouri. *Geol. Soc. Am., Abstr.*, 13(6): 279.

Green, D.H., 1973. Experimental melting studies on a model upper mantle composition under water saturated and water undersaturated condition. *Earth Planet. Sci. Lett.*, 19: 37—53.

Green, D.H., 1976. Experimental testing of "equilibrium" partial melting of peridotite under water-saturated, high-pressure conditions. *Can. Mineral.*, 14: 255—268.

Green, T.H. and Ringwood, A.E., 1968. Genesis of the calc-alkaline igneous rock suite. *Contrib. Mineral. Petrol.*, 18: 105—162.

Green, T.H., Brunfelt, A.O. and Heier, K.S., 1969. Rare earth element distribution in anorthosites and associated high grade metamorphic rocks, Lofoten-Vesteraalen, North Norway. *Earth Planet. Sci. Lett.*, 7: 93—98.

Green, T.H., Brunfelt, A.O. and Heier, K.S., 1972. Rare-earth element distribution and K/Rb ratios in granulites, mangerites, and anorthosites, Lofoten-Vesteraalen, Norway. *Geochim. Cosmochim. Acta*, 36: 241—257.

Griffin, W.L., Sundvoll, B. and Kristmannsdottir, H., 1974. Trace element composition of anorthosite plagioclase. *Earth Planet. Sci. Lett.*, 24: 213—223.

Guichon, F., Church, T.M., Treuil, M. and Jaffrezic, H., 1979. Rare earths in barites: distribution and effects on aqueous partitioning. *Geochim. Cosmochim. Acta*, 43: 983—997.

Hanson, G.N., 1978. The application of trace elements to the petrogenesis of igneous rocks of granitic composition. *Earth Planet. Sci. Lett.*, 38: 26—43.

Hanson, G.N. and Goldich, S.S., 1972. Early Precambrian rocks in the Saganaga Lake—

Northern Light Lake Area, Minnesota-Ontario, II. Petrogenesis. *Geol. Soc. Am., Mem.*, 135: 179—192.

Haskin, L.A. and Korotev, R.L., 1977. Test of a model for trace element partition during closed-system solidification of a silicate liquid. *Geochim. Cosmochim. Acta*, 41: 921—939.

Hawkesworth, C.J. and O'Nions, R.K., 1977. The petrogenesis of some Archean volcanic rocks from southern Africa. *J. Petrol.*, 18: 487—520.

Hellman, P.L. and Green, T.H., 1979. The role of sphene as an accessory phase in the high-pressure partial melting of hydrous mafic compositions. *Earth Planet. Sci. Lett.*, 42: 191—201.

Henderson, P., 1980. Rare earth element partition between sphene, apatite, and other coexisting minerals of the Kangerdlugaauaq intrusion, E. Greenland. *Contrib. Mineral. Petrol.*, 72: 81—85.

Henderson, P., Fishlock, S.J., Laul, J.C., Cooper, T.D., Conard, R.L., Boynton, W.V. and Schmitt, R.A., 1976. Rare earth element abundances in rocks and minerals from the Fiskenaesset complex, West Greenland. *Earth Plant. Sci. Lett.*, 30: 37—49.

Hildreth, W., 1979. The Bishop Tuff: evidence for the origin of compositional zonation in silicic magma chambers. In: C.E. Chapin and W.E. Elston (Editors), *Ash Flow Tuffs. Geol. Soc. Am., Spec. Paper*, 180: 43—75.

Hunter, D.R., 1974. Crustal development in the Kaapvaal Craton, II. The Proterozoic. *Precambrian Res.*, 1: 295—326.

Hunter, D.R., Barker, F. and Millard, H.T., Jr., 1978. The geochemical nature of the Archean ancient gneiss complex and granodiorite suite, Swaziland: a preliminary study. *Precambrian Res.*, 7: 105—127.

Jahn, B., Shih, C. and Murthy, V.R., 1974. Trace element geochemistry of Archean volcanic rocks. *Geochim. Cosmochim. Acta*, 38: 611—627.

Jahn, B., Sun, S.S. and Nesbitt, R.W., 1979. REE distribution and petrogenesis of the Spanish Peaks igneous complex, Colorado. *Contrib. Mineral. Petrol.*, 70: 281—298.

Jahn, B., Auvray, B., Blais, S., Capdevila, R., Cornichet, J., Vidal, R. and Hameurt, J., 1980. Trace element geochemistry and petrogenesis of Finnish greenstone belts. *J. Petrol.*, 21: 201—244.

Jakeš, P. and Gill, J., 1970. Rare earth elements and the island or tholeiitic series. *Earth Planet. Sci. Lett.*, 9: 17—28.

Jakeš, P. and White, A.J.R., 1972. Major and trace element abundances in volcanic rocks or orogenic areas. *Geol. Soc. Am. Bull.*, 83: 29—40.

Jarvis, J.C., Wildeman, T.R. and Banks, N.G., 1975. Rare earths in the Leadville Limestone and its metamorphic derivatives. *Chem. Geol.*, 16: 27—37.

Johnson, R.W., MacKenzie, D.E. and Smith, I.E.M., 1978. Delayed partial melting of subduction-modified mantle in Papua, New Guinea. *Tectonophysics*, 46: 197—216.

Kay, R.W., 1977. Geochemical constraints on the origin of Aleutian magmas. In: M. Talwani, C.G. Harrison and W.C. Pitman III (Editors), *Island Arcs, Deep Sea Trenches, and Back-arc Basins. Am. Geophys. Union, Maurice Ewing Ser.*, 1: 229—242.

Kay, R.W., 1978. Aleutian magnesian andesites: melts from subducted Pacific Ocean crust. *J. Volcanol. Geotherm. Res.*, 4: 117—132.

Kerrich, R. and Fryer, B.J., 1979. Archaean precious-metal hydrothermal systems, Dome Mine, Abitibi Greenstone Belt, II. REE and oxygen isotope relations. *Can. J. Earth Sci.*, 16: 440—458.

Koljonen, T. and Rosenberg, R.J., 1974. Rare earth elements in granitic rocks. *Lithos*, 7: 249—261.

Kushiro, I., 1972. Effect of water on the composition of magmas formed at high pressures. *J. Petrol.*, 13: 311—334.

Laajoki, 1975. Rare earth elements in Precambrian iron formations in Väyrylänkylä, south Puolanka area, Finland. *Bull. Geol. Soc. Finl.*, 47: 93—107.
Lambert, R.St.J. and Holland, J.G., 1976. Amîtsoq gneiss geochemistry: preliminary observations. In: B.F. Windley (Editor), *The Early History of the History of the Earth.* Wiley, New York, N.Y., pp. 191—201.
Lenthall, D.H. and Hunter, D.R., 1977. The geochemistry of the Bushveld granites in the Potgietersaus tin-field. *Precambrian Res.*, 5: 359—400.
Leo, G.W., Hedge, C.E. and Marvin, R.F., 1980. Geochemistry, strontium isotope data, and potassium-argon ages of the andesite-rhyolite association in the Padang area, west Sumatra. *J. Volcanol. Geotherm. Res.*, 7: 139—156.
López-Escobar, L., Frey, F.A. and Vergara, M., 1974. Andesites from central-south Chile: trace element abundances and petrogenesis. *Proc. Symp. on Andean and Antarctic Volcanology Problems, Santiago*, pp. 725—761.
López-Escobar, L., Frey, F.A. and Vergara, M., 1977. Andesites and high-alumina basalts from the central-south Chile High Andes: geochemical evidence bearing on their petrogenesis. *Contrib. Mineral. Petrol.*, 63: 199—228.
López-Escobar, L., Frey, F.A. and Oyarzún, J., 1979. Geochemical characteristics of Central Chile (33°—34°S) granitoids. *Contrib. Mineral. Petrol.*, 70: 439—450.
Luhr, J.F. and Carmichael, I.S.E., 1980. The Colima volcanic complex, Mexico. *Contrib. Mineral. Petrol.*, 71: 343—372.
Masuda, Y. and Aoki, K., 1979. Trace element variation in the volcanic rocks from the Nasu zone, northeast Japan. *Earth Planet. Sci. Lett.*, 44: 139—149.
Masuda, Y., Nishimura, S., Ikeda, T. and Katsui, Y., 1975. Rare-earth and trace elements in the Quaternary volcanic rocks of Hokkaido, Japan. *Chem. Geol.*, 15: 251—271.
McCarthy, T.S. and Hasty, R.A., 1976. Trace element distribution patterns and their relationship to the crystallization of granitic melts. *Geochim. Cosmochim. Acta*, 40: 1351—1358.
McCarthy, T.S. and Kable, E.J.D., 1978. On the behavior of rare-earth elements during partial melting of granitic rock. *Chem. Geol.*, 22: 21—29.
McCurry, P. and Wright, J.B., 1977. Geochemistry of calc-alkaline volcanics in northwestern Nigeria, and a possible Pan-African suture zone. *Earth Planet. Sci. Lett.*, 37: 90—96.
Mertzman, S.A., 1977. The petrology and geochemistry of the Medicine Lake volcano, California. *Contrib. Mineral. Petrol.*, 62: 221—247.
Mineyev, D.A., 1963. Geochemical differentiation of the rare earths. *Geochemistry (U.S.S.R.)*, 12: 1129—1149.
Moller, P., Morteani, G., Hoefs, J. and Parekh, P.P., 1979. The origin of the ore-bearing solution in Pb-Zn veins of the western Harz, Germany, as deduced from rare-earth element and isotope distributions in calcites. *Chem. Geol.*, 26: 197—215.
Morgan, J.W. and Wandless, G.A., 1980. Rare earth element distribution in some hydrothermal minerals: evidence for crystallographic control. *Geochim. Cosmochim. Acta*, 44: 973—980.
Morgan, J.W., Ganapathy, R., Higuchi, H. and Kröhenbühl, H., 1976. Volatile and siderophile trace elements in anorthositic rocks from Fiskenaesset, West Greenland: comparison with lunar and meteoritic analogues. *Geochim. Cosmochim. Acta*, 40: 861—887.
Mysen, B.O., 1979. Trace element partitioning between garnet peridotite minerals and water-rich vapor: experimental data from 5 to 30 kb. *Am. Mineral.*, 64: 274—287.
Mysen, B.O. and Boettcher, A.L., 1975. Melting of a hydrous mantle, II. Geochemistry of crystals and liquids formed by anatexis of mantle peridotite at high pressures and high temperatures as a function of controlled activities of water, hydrogen, and carbon dioxide. *J. Petrol.*, 16: 549—590.

Nicholls, I.A. and Ringwood, A.E., 1973. Effect of water on olivine stability in tholeiites and production of SiO_2-saturated magmas in the island arc environment. *J. Geol.*, 81: 285—300.

Noble, D.C., Korringa, M.K., Church, S.E., Bowman, H.R., Silberman, M.L. and Heropoulos, C.E., 1976. Elemental and isotopic geochemistry of non-hydrated quartz latite glasses from Eureka Valley Tuff, east-central California. *Geol. Soc. Am. Bull.*, 87: 754—762.

Noble, D.C., Rigot, W.L. and Bowman, H.R., 1979. Rare-earth-element content of some highly differentiated ash-flow tuffs and lavas. In: C.E. Chapin and W.E., Elston (Editors), Ash Flow Tuffs, *Geol. Soc. Am.*, *Spec. Paper*, 180: 77—85.

O'Nions, R.K. and Pankhurst, R.J., 1974. Rare-earth element distribution in Archean gneisses and anorthosites, Godthab area, West Greenland. *Earth Planet. Sci. Lett.*, 22: 328—338.

O'Nions, R.K. and Pankhurst, R.J., 1978. Early Archean rocks and geochemical evolution of the earth's crust. *Earth Planet. Sci. Lett.*, 38: 211—236.

Peccerillo, A., Poli, G., Sassi, F.P., Zirpoli, G. and Mezzacasa, G., 1979. New data on the upper Ordovician acid plutonism in the eastern Alps. *Neues Jahrb. Mineral., Abh.*, 137: 162—183.

Peppard, D.F., Mason, G.W. and Hucher, I., 1962. Stability constants of certain lanthanide (III) and actinide (III) chloride and nitrate complexes. *J. Inorg. Nucl. Chem.*, 24: 881—888.

Perfit, M.R., Brueckner, H., Lawrence, J.R. and Kay, R.W., 1980. Trace element and isotopic variations in a zoned pluton and associated volcanic rocks, Unalaska Island, Alaska: a model for fractionation in the Aleutian calc-alkaline suite. *Contrib. Mineral. Petrol.*, 73: 69—87.

Phelps, D., 1979. Petrology, geochemistry, and origin of the Sparta quartz diorite-trondhjemite complex, northeastern Oregon. In: F. Barker (Editor), *Trondhjemites, Dacites, and Related Rocks*. Elsevier, Amsterdam, pp. 547—579.

Philpotts, J.A., Schnetzler, C.C. and Thomas, H.H., 1966. Rare earth abundances in an anorthosite and a mangerite. *Nature*, 212: 805.

Price, R.C. and Taylor, S.R., 1977. The rare earth element geochemistry of granite, gneiss, and migmatite from the western metamorphic belt of south-eastern Australia. *Contrib. Mineral. Petrol.*, 62: 249—263.

Pride, C. and Muecke, G.K., 1980. Rare earth element geochemistry of the Scourian Complex, NW Scotland — evidence for the granite-granulite link. *Contrib. Mineral. Petrol.*, 73: 403—412.

Ringwood, A.E., 1974. The petrological evolution of island arc systems. *J. Geol. Soc. London*, 130: 183—204.

Rogers, J.J.W., Hodges, K.V. and Ghuma, M.A., 1980. Trace elements in continental-margin magmatism, II. Trace elements in Ben Ghnema batholith and nature of the Precambrian crust in central North Africa. *Geol. Soc. Am. Bull.*, 91: 1742—1788.

Rose, W.I., Grant, N.K. and Easter, J., 1979. Geochemistry of the Los Chocoyos Ash, Quezaltenango Valley, Guatemala. In: C.E. Chapin and W.E. Elston (Editors), *Ash Flow Tuffs, Geol. Soc. Am., Spec. Paper*, 180: 87—99.

Saunders, A.D., Tarney, J., Stern, C.R. and Dalziel, I.W.D., 1979. Geochemistry of Mesozoic marginal basin floor igneous rocks from southern Chile. *Geol. Soc. Am. Bull.*, 90: 237—258.

Seifert, K.E., 1978. Anorthosite-mangerite relations on Baker Mountain, N.Y. *Geol. Soc. Am. Bull.*, 89: 245—250.

Seifert, K.E., Voigt, A.F., Smith, M.F. and Stensland, W.A., 1977. Rare earths in the Marcy and Morin anorthosite complexes. *Can. J. Earth Sci.*, 14: 1033—1045.

Shaw, D.M., 1978. Trace element behavior during anatexis in the presence of a fluid phase. *Geochim. Cosmochim. Acta*, 42: 933—943.

Shul'gin, L.P. and Koz'min, Y.A., 1963. Kinetics of the europium (III)-europium (II) oxidation-reduction reaction. *Russian J. Inorg. Chem.*, 37: 1003—1004.

Simmons, E.C. and Hanson, G.N., 1978. Geochemistry and origin of massif-type anorthosites. *Contrib. Mineral. Petrol.*, 66: 119—135.

Simmons, E.C. and Hedge, C.E., 1978. Minor-element and Sr-isotope geochemistry of Tertiary stocks. Colorado mineral belt. *Contrib. Mineral. Petrol.*, 67: 379—396.

Smith, I.E.M., Taylor, S.R. and Johnson, R.W., 1979. REE-fractionated trachytes and dacites from Papua, New Guinea and their relationship to andesite petrogenesis. *Contrib. Mineral. Petrol.*, 69: 227—233.

Tarney, J., Weaver, B. and Drury, S.A., 1979. Geochemistry of Archean trondhjemitic and tonalitic gneisses from Scotland and East Greenland. In: F. Barker (Editor), *Trondhjemites, Dacites, and Related Rocks.* Elsevier, Amsterdam, pp. 275—300.

Taylor, S.R., 1968. Geochemistry of andesites. In: L.H. Ahrens (Editor), *Origin and Distribution of the Elements.* Pergamon, Oxford, pp. 559—583.

Taylor, S.R., 1969. Trace element chemistry of andesites and associated calc-alkaline rocks. In: *Proceedings of the Andesite Conference. Oreg., Dep. Geol. Miner. Ind., Bull.*, 65: 43—63.

Taylor, S.R., Ewart, A. and Capp, A.C., 1968. Leucogranites and rhyolites: trace element evidence for fractional crystallization and partial melting. *Lithos*, 1: 179—186.

Taylor, S.R., Capp, A.C., Graham, A.L. and Blake, D.H., 1969. Trace element abundances in andesites. *Contrib. Mineral. Petrol.*, 23: 1—26.

Thorpe, R.S. and Francis, P.W., 1979. Variations in Andean andesite compositions and their petrogenetic significance. *Tectonophysics*, 57: 53—70.

Thorpe, R.S., Potts, P.J. and Francis, P.W., 1976. Rare earth data and petrogenesis of andesite from North Chilean Andes. *Contrib. Mineral. Petrol.*, 54: 65—78.

Thorpe, R.S., Potts, P.J. and Sarre, M.B., 1977. Rare earth evidence concerning the origin of granites of the Isle of Skye, northwest Scotland. *Earth Planet. Sci. Lett.*, 36: 111—120.

Thorpe, R.S., Francis, P.W. and Moorbath, S., 1979. Rare earth and strontium isotope evidence concerning petrogenesis of North Chilean ignimbrites. *Earth Planet. Sci. Lett.*, 42: 359—367.

Van Leeuwen, P. and Graf, J.L., 1981. Geochemical evidence regarding the formation of manganese and iron deposits in the Urucum region, Mato Grosso do Sul, Brazil. *Geol. Soc. Am., Abstr.*, 13(5): 263—264.

Walsh, J.N., Beckinsale, R.D., Skelhorn, R.R. and Thorpe, R.S., 1979. Geochemistry and petrogenesis of Tertiary granitic rocks from the Island of Mull, northwest Scotland. *Contrib. Mineral. Petrol.*, 71: 99—116.

Weaver, B.L., 1980. Rare-earth element geochemistry of Madras granulites. *Contrib. Mineral. Petrol.*, 71: 271—279.

Wendlandt, R.F. and Harrison, W.J., 1979. Rare earth partitioning between immiscible carbonate and silicate liquids and CO_2 vapor: results and implications for the formation of light rare earth-enriched rocks. *Contrib. Mineral Petrol.*, 69: 409—419.

White, C.M. and McBirney, A.R., 1977. Some quantitative aspects of orogenic volcanism in the Oregon Cascades. *Geol. Soc. Am., Mem.*, 152: 369—388.

Whitford, D.J., Nicholls, I.A. and Taylor, S.R., 1979. Spatial variations in the geochemistry of Quaternary lavas across the Sunda arc in Java and Bali. *Contrib. Mineral. Petrol.*, 70: 341—356.

Winkler, H.G.F., 1979. *Petrogenesis of Metamorphic Rocks.* Springer, Berlin.

Yajima, T., Higuchi, H. and Nagasawa, H., 1972. Variation of rare earth concentrations in pigeonitic and hypersthenic rock series form Izu-Hakone region, Japan. *Contrib. Mineral. Petrol.*, 35: 235—244.

Zielinski, R.A. and Lipman, P.W., 1976. Trace-element variations at Summer Coon volcano, San Juan Mountains, Colorado, and the origin of continental-interior andesite. *Geol. Soc. Am. Bull.*, 87: 1477—1485.

Zielinski, R.A., Lipman, P.W. and Millard, H.T., Jr., 1977. Minor-element abundances in obsidian, perlite, and felsite of calc-alkalic rhyolites. *Am. Mineral.*, 62: 426—437.

Chapter 9

THE MOBILITY OF THE RARE EARTH ELEMENTS IN THE CRUST

SUSAN E. HUMPHRIS

9.1. Introduction

One of the main objectives in geochemical research is to understand the evolution of the wide variety of igneous rocks that make up much of the Earth's crust from the material that is believed to constitute the mantle. In achieving this goal, geochemists have used the mineralogical and chemical compositions of variably fractionated rocks to model such processes as partial melting and crystal fractionation that occur during the solidification of magmas. These methods are based on the assumption that the mineralogies and chemical compositions of volcanic rocks are unaffected by post-consolidation weathering, alteration, or metamorphic reactions, and hence they reflect the primary petrogenetic characteristics of the rock.

In recent years, various studies have demonstrated that the major, and many trace, elements are mobile during both submarine and aerial weathering and alteration processes (e.g., Thompson, 1973; Vallance, 1974; Floyd, 1976; Wood et al., 1976; Humphris and Thompson, 1978a, b) and are therefore unreliable for petrogenetic studies on rocks that have been subjected to any degree of alteration. This has led to an increase in the use of the relative abundances of selected trace elements (mainly Ti, Hf, Th, Ta, Y, Zr, and Nb) and the REE in altered and metamorphosed rocks both for modelling partial melting and crystal fractionation processes (e.g., Condie and Baragar, 1974; Sun and Nesbitt, 1978; Venturelli et al., 1979), and for classifying rocks into particular tectonic settings (Pearce and Cann, 1973; Ferrara et al., 1976; Winchester and Floyd, 1976). The usefulness of the rare earths for such analyses requires that the basic assumption of immobility of these elements during weathering and alteration processes still holds. However, there is increasing evidence that, during some alteration processes, the abundances of the REE, and their distribution patterns when considered relative to chondrite elemental concentrations, may be changed sufficiently to affect significantly any petrogenetic interpretation. It is therefore important to evaluate the degree of mobility of these elements during various types of alteration processes in order to assess their applicability to petrogenetic modelling.

9.2. Factors affecting the degree of mobility of the REE

It is becoming evident from the studies completed on both oceanic and continental rocks that there is no simple relation between the degree of mobility or immobility of the REE, metamorphic grade, or the type of rock being altered. For example, palagonitization of submarine glasses can result in LREE enrichment, and a uniform increase in the HREE, in the rims of pillow basalts relative to the interiors (Ludden and Thompson, 1978, 1979), or in depletion of all the REE (Staudigel et al., 1979). Studies of both oceanic and continental spilites have shown that this alteration process can have either no effect on the distribution of the REE (Herrmann et al., 1974) or can result in enrichment of the REE (Hellman and Henderson, 1977). Finally, the mobility of REE during alteration of granites is highly variable (e.g., Alderton et al., 1980; Exley, 1980). It is therefore apparent that similar alteration processes do not always produce consistent trends in the REE mobility, such as has been indicated by Hellman et al. (1977, 1979) and Sun and Nesbitt (1978). Hence, rather than being able to predict the behaviour of the REE from a knowledge of the type of alteration to which a rock has been subjected, the complete environment in which the weathering or alteration processes occurred must be taken into consideration.

Any system in which either weathering (used in this chapter specifically to indicate chemical reactions occurring at low (present ambient) temperatures) or alteration (reactions occurring at high (greater than present ambient) temperatures) of rocks is taking place generally involves interaction between mineral phases and a fluid phase. Additional properties, such as temperature, an open or closed system, variable fluid/rock ratios, may or may not be important in determining the fate of the REE depending on the particular situation. The overall gains and losses of the REE observed during any weathering or alteration process will be a function of the relative importance of several factors:

(1) The abundances of the REE in the unaltered rock, their distribution and sites of concentration in the mineral phases within the rock, and the relative stability of the mineral phases with respect to the fluid.

(2) The concentration of the REE in the fluid, the partitioning behaviour of the REE between the mineral phases and the fluid, and the ability of the fluid to transport the REE out of the system.

(3) The ability of the secondary minerals formed during the reactions to accommodate the REE released from the original minerals.

In this chapter, each of these factors and its effect on REE mobility will be considered. The underlying principles that determine the direction of REE transport will be emphasized, and specific examples illustrating their importance will be discussed.

In discussing the mobility of the REE and their exchange between rocks and fluids, it is usual to compare the chondrite-normalized REE patterns of fresh and altered samples. In doing this, it should be borne in mind that changes may be observed which are due only to volume changes, or to a change in the content of another constituent of the rock. For any quantitative assessment of the REE exchange, the data have to be normalized to some constant factor. Several methods have been used for this procedure in the literature, usually involving the assumption that one element is unaffected by the alteration process, and therefore its concentration is the same in both fresh and altered rock. Where such data are discussed in this chapter, the normalization procedure will be noted.

9.3. Igneous crystallization history — the distribution of REE in the unaltered rock, and their susceptibility to weathering and alteration

One of the main controls on potential lanthanide mobility is the crystallization history of the individual lavas, since this affects the sites of concentration and distribution of the REE before subsequent weathering and alteration. The solid/liquid partitioning of the REE during crystallization of a magma has been discussed in detail by Haskin (Chapter 4), and typical partition coefficients for the major phases in the common igneous rock types have been presented in Chapter 1. All the partition coefficients are very low, indicating that the REE tend to remain in the liquid phase. However, of these minerals, clinopyroxene is one of the most important major phases both in terms of removing the REE from the liquid, and also selectively enriching the liquid in the LREE. Conversely, olivine contains one or two orders of magnitude less REE than the other major constituents of most intermediate and basic rocks, and therefore tends to act as a diluent (Herrmann et al., 1974). Hence, the whole-rock REE content of these rocks is strongly influenced by the modal abundance of olivine.

Accessory minerals, such as allanite, apatite, zircon, and sphene also play important roles in the distribution of the REE, because the typical partition coefficients for these elements between accessory minerals and their host acidic rocks are one to two orders of magnitude greater than those for the common rock-forming minerals (see Chapter 1). These high partition coefficients therefore indicate that these minerals act as sites of concentration for the REE. This is particularly important for acidic volcanic rocks and granites where, not only are the REE more abundant than in basic volcanic rocks, but they also often contain a wide range of accessory minerals. In fact, Alderton et al. (1980) have shown that in some granites, only 50% of the whole-rock REE content is present in the major mineral phases, the remainder being concentrated in the accessory minerals — apatite, zircon, and sphene. Hence, if these accessory minerals are particularly susceptible to any weathering or alteration processes, then the whole-rock REE pattern may be drastically changed.

The susceptibility of the mineral phases in any rock to reaction with a fluid phase will then be important in the degree of mobility of the whole-rock REE. For example, if olivine is altered, very little or no change would be expected in the REE pattern, whereas breakdown of clinopyroxene would have a much greater effect, causing a change in the abundances of all the REE, and might result in selective mobility of the HREE.

The sequence of the susceptibility of minerals was deduced by Goldich (1938) from observations on weathering. His series, which was later systematically explained by consideration of the energies of formation of the various silicate groups (Keller, 1957), predicts that, of the common rock-forming minerals, olivine would be the most susceptible to weathering. This would be followed in sequence by pyroxene, hornblende, biotite, alkali feldspar, and quartz. This generalized order of weathering is commonly observed in both continental and oceanic crustal rocks, although variations do occur depending on the rock type and on the anion composition of the fluid. Fluids with a particular composition may, in some instances, be capable of destabilizing minerals out of sequence. For example, the presence of fluoride in a hydrothermal fluid can increase the mobility of zirconium by permitting ZrF_6^{2-} to occur. Hence, the attack of fluoride-rich fluids will destabilize zircon more readily than fluoride-poor fluids, thereby enhancing the mobility of the REE (Alderton et al., 1980).

The other phase that should be considered is glass which, in quenched submarine volcanic rocks, can often comprise most of the volume of the rock. Glass will generally be attacked first during weathering and alteration processes. Its REE content, and hence the mobility of these elements observed in rocks where only the glass phase has been altered, will depend on the crystallization history of the rock, i.e., the mineral phases that have crystallized and their effect on concentrating the REE in the liquid, the degree of fractionation, etc.

An illustration of the importance of the igneous crystallization history can be seen in a comparison of the REE patterns from zeolitized basaltic flows from eastern Iceland (Wood et al., 1976) and from Mull, northwestern Scotland (Humphris et al., 1978), both of which were probably altered under rather similar environmental conditions. Wood et al. (1976) compared the chemical variation along a vertical section through a post-glacial basaltic flow on the Reykjanes peninsula with a zeolitized Tertiary flow from Reydarfjordur, eastern Iceland. Sufficient of the pre-metamorphic mineralogy of the Reydarfjordur flow was preserved to conclude that it originally closely resembled the Reykjanes basalt petrographically. Both contained several percent olivine, augite and plagioclase phenocrysts in a groundmass of granular augite intergrown with plagioclase laths, sparse Fe-Ti oxides, and rare olivine. The abundant spaces between the groundmass grains were filled with pale brown glass. In the zeolitized flow, this interstitial glass has been converted to fine-grained assemblages of zeolites and hydrated ferromagnesian minerals, with mesolite and stilbite infilling vesicles throughout the flow. In addition, zones with maximum alteration show complete replacement of olivine and incipient conversion of plagioclase to hydrous minerals, while augite remains unaltered. Typical chondrite-normalized REE patterns for both flows are shown in Fig. 9.1a. In the unaltered Reykjanes flow, the patterns are generally flat, and all variability can be attributed entirely to varying concentrations of the three phenocryst phases. In the zeolitized Reydarfjordur flow, variability greater than could be explained by phenocryst distribution was recorded for the LREE. As most of this variation occurred near the vesicular margins of the flow, Wood et al. (1976) suggested that the LREE are variably modified during zeolite facies metamorphism.

The basaltic lava from Mull is almost holocrystalline and contains olivine, large poikilitic titanaugites enclosing small plagioclase subhedra, and titanomagnetites. The secondary alteration is concentrated mainly at the base of

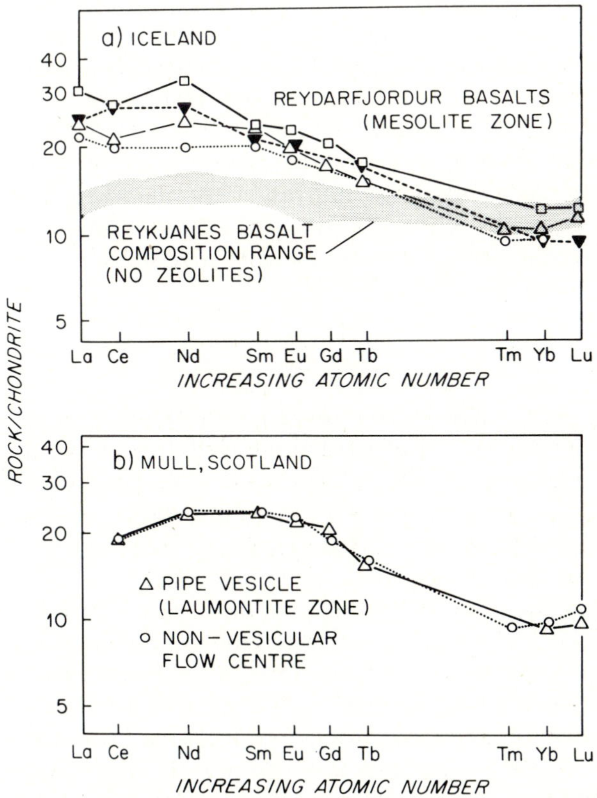

Fig. 9.1. Chondrite-normalized REE patterns for unaltered basalts and their zeolitized equivalents from (a) Iceland and (b) Mull, Scotland (data from Humphris et al., 1978).

the flow where there is an amygdaloidal layer with zeolite-filled vesicle pipes which penetrate vertically into the flow. Two chondrite-normalized REE patterns are shown in Fig. 9.1b: one from the flow centre, and the other from the pipe vesicle zone. The two patterns are virtually identical, suggesting that the REE were immobile during alteration of this flow.

Humphris et al. (1978) have attributed these two different effects of zeolitization on the REE patterns primarily to the sites of concentrations of these elements within the mineral assemblages. In the case of the Iceland basalts, the phase most readily attacked by the circulating fluids is the interstitial glass. This glass will be rich in total REE because all the early crystallizing phases in this basalt have partition coefficients of less than one. In addition, the glass will be characterized by a high LREE/HREE ratio. Of these three phases — olivine, plagioclase, and pyroxene — augite has the highest partition coefficients, averaging about 1.0 for the HREE and about 0.3 for the LREE (see Chapter 1), and so the LREE will tend to remain in the melt. Hence, by altering the interstitial glass, there will be mobilization of the REE, and preferential mobilization of the LREE.

In contrast to this, the Mull basalts are almost entirely holocrystalline. In the absence of glass, the hydrothermal fluids will react first with olivine and then plagioclase. Both of these have very low partition coefficients for REE, and hence the whole-rock REE patterns do not change during alteration.

The varying degrees of mobility of the REE during zeolitization of these basaltic flows can therefore be attributed to an interplay between the different sites of concentration of the REE in the original mineral assemblages, and the susceptibility of those phases to alteration by the fluid.

9.4. The fluid phase

The fluid phase plays an important role during weathering and alteration reactions in that it provides a means of transporting elements into and out of the system. In addition, the chemistry of the fluid can determine not only the types of reactions that occur, but changes in its chemistry can also influence the eventual fate of the elements, e.g., reprecipitation in a secondary phase, or complete removal from the system. The composition of the fluid will evolve as it migrates through the rocks; hence, similar rocks along the path of the fluid can be exposed to fluids of very different compositions. Any treatment of these systems must therefore take into account that they are open systems with continuous flow-through of a fluid phase during the alteration process. In addition, in most cases, interactions between rocks and seawater will not reach equilibrium, and hence cannot be considered as a series of equilibrium reactions. The extent of equilibration will be governed by the fluid/rock ratio and, to a lesser extent, the length of time that a rock is exposed to a migrating fluid, the composition of which is the same. Conditions that meet the requirements for equilibrium to be approached are met only rarely in natural systems. One example of such a case is low-temperature weathering of sea-floor basalts; an example of this, and its effect on the REE will be discussed later in this section.

In order to assess the ability of a fluid to transport REE out of a system, it is necessary to consider the concentrations of the REE in the fluid, the solubility of the REE in that fluid, and the partitioning of the REE between the mineral phases and the fluid. There are several types of fluids that participate in alteration and weathering processes — for the following discussion, they are divided into three categories:

(1) *Seawater* — this can be involved in both low-temperature weathering and high-temperature alteration of oceanic rocks.

(2) *Groundwater* — this refers to simple low-temperature weathering of continental rocks by percolating groundwater.

(3) *Metasomatic and hydrothermal fluids* — this includes fluids of widely varying compositions, but refers to all those fluids (except seawater, which is dealt with separately) involved in alteration processes at temperatures that are above the present ambient temperature. Examples of such fluids are post-magmatic fluids involved in metasomatism of continental rocks, or those participating in reactions resulting in the formation of hydrothermal ore deposits.

The important factors in the chemical compositions of these three groups for the transport of the REE will be discussed, and examples given of their effects on the mobility and redistribution of the REE.

Seawater

Seawater is an important fluid phase in alteration processes. Hydrothermal circulation near mid-ocean ridges exposes rocks to interactions with seawater at high temperatures, resulting in hydrous metamorphism of the basalts to grades ranging from zeolite to amphibolite. Considerable major and trace element mobility during these reactions has been observed in studies of both naturally occurring samples (e.g., Humphris and Thompson, 1978a, b) and in experiments (e.g., Hajash, 1975; Mottl and Holland, 1978). In addition, exposure of rocks to seawater at ambient bottom-water temperatures results in a weathering process during which exchange of major and trace elements occurs between the solid and fluid phases (e.g., Hart, 1970; Thompson, 1973).

The concentrations of the REE in seawater are extremely low, ranging from 1 to 100 pmol/kg seawater (Goldberg et al., 1963; Høgdahl, 1967; Høgdahl et al., 1968). The solubility products of the trivalent REE hydroxides, phosphates and carbonates are shown in Fig. 9.2. Although there are few data, carbonate salts are the least soluble, and the HREE compounds are generally less soluble than those of the LREE. The

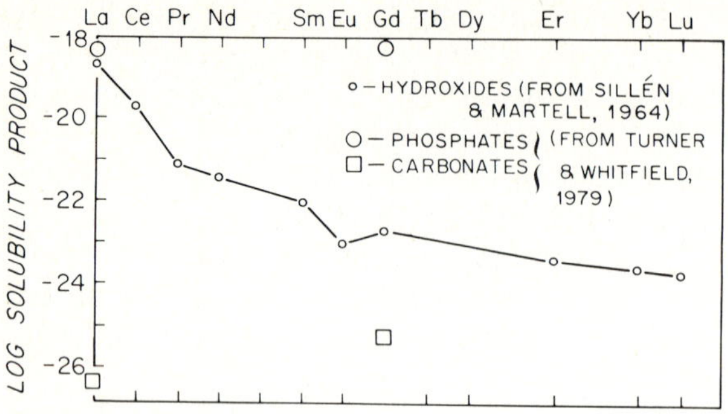

Fig. 9.2. Solubility products for the trivalent REE hydroxides, carbonates, and phosphates.

concentrations observed in seawater are several orders of magnitude below the limits set by the most likely insoluble salts (Turner and Whitfield, 1979); hence, seawater is undersaturated with respect to the REE. In the dissolved phase, the lanthanides are complexed quite effectively, in particular with carbonate and, to a lesser extent, with sulphate and chloride, as shown in Table 9.1. However, a significant proportion of each element exists as the free ion, with the percentage decreasing with increasing atomic number up to Lu, where only 7% of the element is present as the free ion.

The chondrite-normalized REE pattern for seawater (Fig. 9.3) indicates that seawater shows a large depletion of Ce compared with the concentration of the other REE. Among the mostly trivalent REE, Ce is the only element which can be oxidized to a tetravalent state. It is most likely removed from seawater by particulate scavenging (Buat-Menard and Chesselet, 1979) and is also fractionated into ferrmanganese deposits (Glasby, 1973; Piper, 1974). Other features of the REE in seawater are the decreasing enrichment with increasing atomic number of the LREE, and the relatively constant, although somewhat enriched, pattern of the HREE. The latter feature can be explained by the formation of more stable inorganic and organic complexes by the HREE than by the LREE (Goldberg et al., 1963; Sillén and Martell, 1964).

TABLE 9.1

Percent speciation of selected REE in seawater (from Turner and Whitfield, 1979)

	Free	OH^-	F^-	Cl^-	SO_4^{2-}	CO_3^{2-}
La	31	4	1	16	16	33
Ce	22	4	1	11	21	41
Eu	11	7	1	5	8	68
Gd	11	4	1	6	13	65
Lu	7	7	1	2	3	81

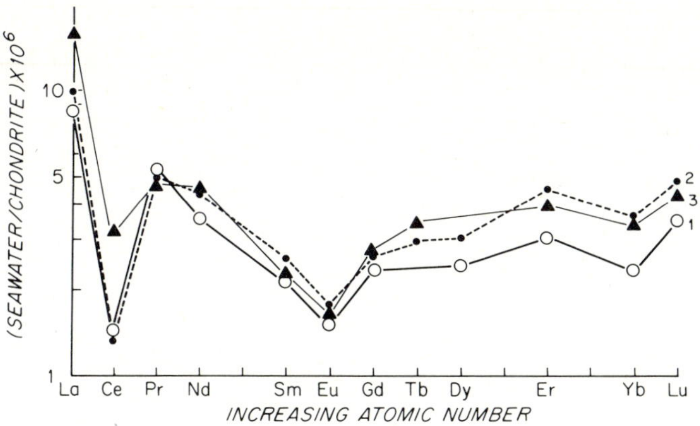

Fig. 9.3. Chondrite-normalized REE patterns for seawater. *1* = one sample from the Pacific Ocean (Goldberg et al., 1963); *2* = mean of sixteen samples from the equatorial Atlantic Ocean (Høgdahl et al., 1968); *3* = mean of two samples from the North Atlantic (Høgdahl, 1967).

An additional factor which needs to be taken into consideration is the water/rock ratio, which will determine the extent of equilibration of the REE between the rock and the fluid phase. For example, recent studies of the REE mobility during palagonitization of the glassy rims of oceanic basalts have produced conflicting results. Ludden and Thompson (1978, 1979) studied the REE patterns in fresh and altered tholeiitic basalts that were dredged from the floor of the Atlantic Ocean at 23°N (Fig. 9.4a). After normalizing to a constant composition of the HREE, they observed a significant enrichment of the LREE in the palagonitized rinds relative to the partially altered crystalline interiors, with some elements being enriched by as much as four orders of magnitude. They suggested that, for these samples where weathering has occurred under conditions of a high water/rock ratio, the LREE enrichment could be explained by equilibration with the abundances of REE found in seawater, the glassy rinds equilibrating at a faster rate.

A completely different variation of the REE patterns of palagonitized rinds was seen in a study of basalts drilled from the oceanic basement (Fig. 9.4b). Staudigel et al. (1979) observed loss of all the REE from the glassy rinds during the formation of palagonite, with no preferential mobilization of any of the REE. Both Staudigel et al. (1979) and Ludden and Thompson (1979) attribute these conflicting results to the differences in the alteration environments — in particular, the water/rock ratios and the exposure times. The dredged samples are exposed to seawater for a considerable period of time, and are therefore weathered under conditions of a high water/rock ratio. The drilled samples were recovered from deep within the oceanic basement, where circulation of water is limited both temporally

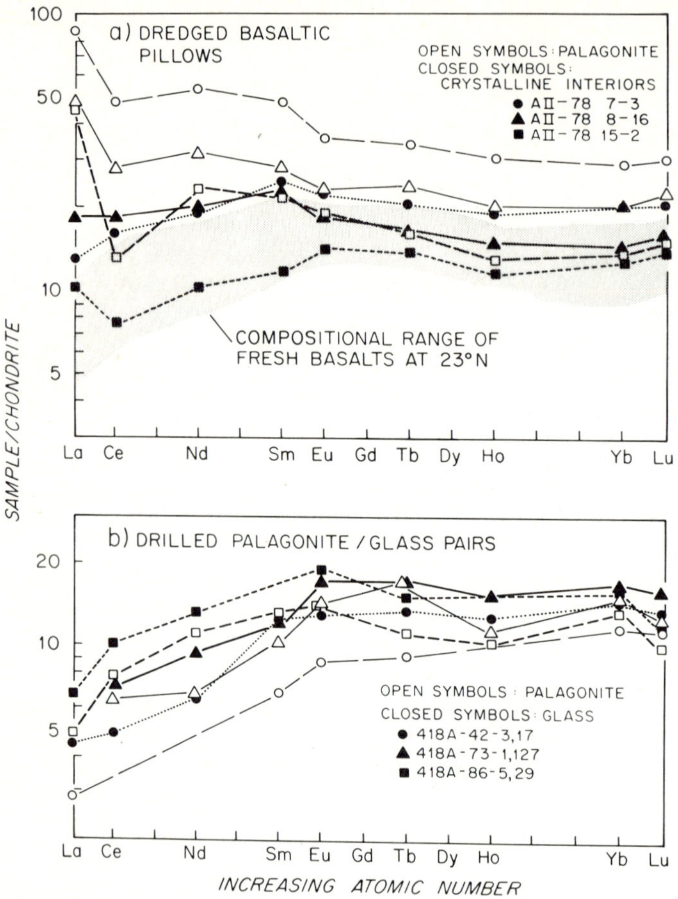

Fig. 9.4. Chondrite-normalized REE patterns for (a) the palagonitized rims and interiors of dredged pillow basalts and (b) palagonite/glass pairs from drilled basalts (data from Ludden and Thompson, 1979, and Staudigel et al., 1979).

and spatially. In addition, the composition of the water may have changed due to previous reactions with basalt along its path through the oceanic crust. Hence, these samples were palagonitized in an environment characterized by a low water/rock ratio, and therefore equilibration with the REE abundances in seawater did not take place. Instead, breakdown of glass caused mobilization of a significant fraction of the REE, which were then transported in the fluid phase.

The importance of exposure time and the water/rock ratio is also illustrated during alteration to zeolite facies. In a study of the Point Sal ophiolite in California (Menzies et al., 1977) fairly conclusive evidence suggests that the alteration fluid was seawater. The upper lavas, which are zeolitized and also contain smectite, were probably exposed to seawater for about 15 Ma before being isolated. These lavas show mobilization of the LREE and a marked negative Ce anomaly (Fig. 9.5a), similar to that observed in seawater (Goldberg et al., 1963), marine precipitates (e.g. Robertson and Fleet, 1976), and in weathered submarine basalts (Frey et al., 1974; Robertson and Fleet, 1976). Menzies et al. (1977) attributed this to a long exposure time of the lavas to seawater, and hence a high water/rock ratio which, since many of the secondary minerals are hydrated, imparted the REE characteristics of seawater to the zeolitized lavas. In contrast, at a stratigraphically lower level in the ophiolite sequence, the rocks had been completely altered to an assemblage consisting dominantly of epidote and quartz. These greenstones showed little or no change in the REE patterns, shown in Fig. 9.5b. Menzies et al. (1977) suggested that the apparent immobility of the REE at these lower levels resulted from the limited exposure of the rocks to seawater at low temperatures, which did not allow for equilibration between the REE contents of the seawater and rock, and the ability of the secondary minerals to retain the REE within their structures.

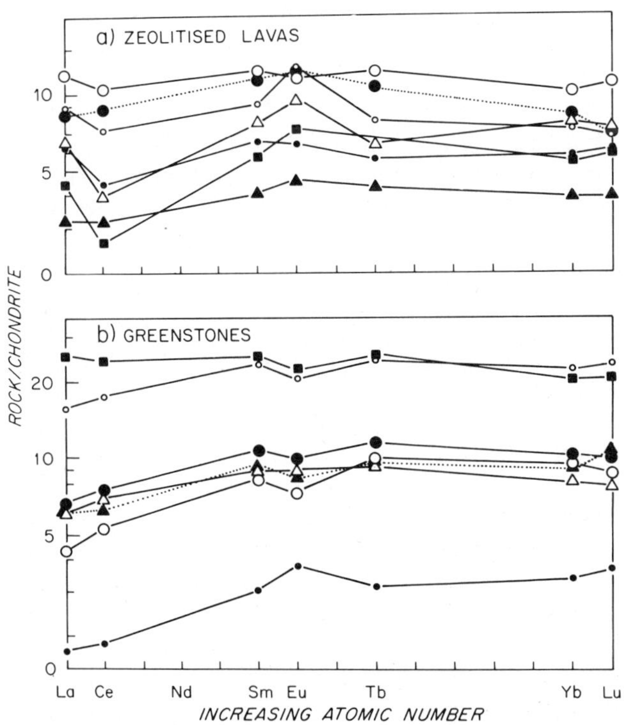

Fig. 9.5. Chondrite-normalized REE patterns of zeolitised lavas (a) and greenstones (b) from the Point Sal ophiolite, California (data from Menzies et al., 1977).

Groundwater

Considerably less work has been done on the REE mobility during chemical weathering of continental rocks than on submarine rocks. In continental systems, percolation of rain water through the rocks will result in low-temperature chemical weathering reactions that will slowly break down the primary minerals, possibly resulting in mobility of the REE. The chemistry of groundwaters is clearly very dependent on the physicochemical environments through which it has passed. Few reliable data are available on the REE contents of groundwater. The only rivers for which there are adequate data are the Garonne-Dordogne river system (with a salinity of 0.1°/$_{oo}$) and the upper reaches of the Gironde Estuary (with a salinity of 0.42°/$_{oo}$). The concentrations of the individual REE are very low and vary between 6 and 480 $\times$ 10^{-10} g/l (Martin et al., 1976). The chondrite-normalized REE pattern for river water (Fig. 9.6) is similar to that of North Atlantic shale, showing enrichment in the LREE relative to the HREE. The dissolved lanthanides are most likely transported in the fluid phase in the form of carbonate, and perhaps organic complexes (Balashov et al., 1964). The stability of complexes with alkali carbonates and organic compounds increases from La to Lu. Therefore a higher content of HCO_3^- in natural waters will cause a higher solubility of the HREE compared with the LREE (Herrmann, 1978).

pH can also significantly influence the behaviour of the REE. The pH of rain water equilibrated with atmospheric CO_2 is 5.7. A decrease in pH will favour solution of the REE and their transport either as complexes or as free ions. An increase in pH can result in one or all of three processes:

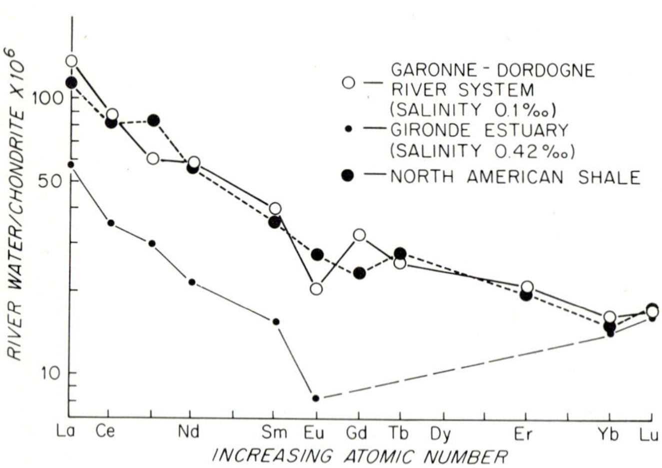

Fig. 9.6. REE profiles in river water compared with average North American shale (data from Martin et al., 1976 and Haskin et al., 1966).

precipitation of the REE as hydroxides or carbonates, exchange of the REE for H^+ on accessible mineral exchange sites, and adsorption onto the surfaces of minerals (Balashov et al., 1964; Stumm and Morgan, 1970).

A good illustration of the importance of fluid chemistry on the mobility of the REE during weathering of continental rocks was described by Nesbitt (1979). He systematically studied a granodiorite weathering profile from the Torrongo granite, east central Victoria, Australia, in which the weathering occurred as the result of water percolating through the soil zone and along fractures in the rock. Fig. 9.7 indicates that the REE are progressively enriched in the successively more altered samples. After normalization of the data assuming TiO_2 is immobile during weathering, calculations show that the HREE are enriched between 100 and 200% (Fig. 9.8). This indicates that the REE have been added to the weathered rock and, from the nature of the enrichment, they were introduced into the system along the fractures. The REE pattern of the residual weathering products (*E* in Fig. 9.7), which consist primarily of kaolinite and illite, generally show depletion of these elements compared with the material in the weathering profile. However,

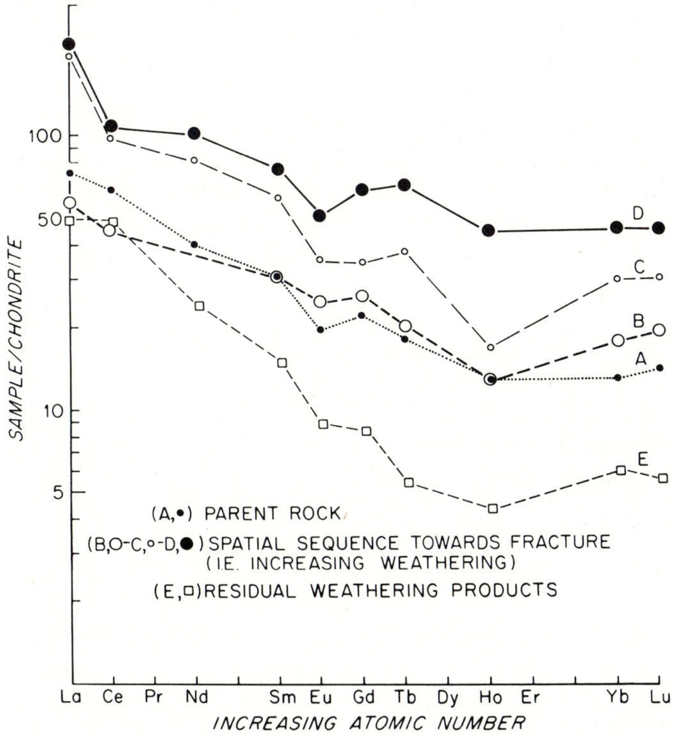

Fig. 9.7. Chondrite-normalized REE from a weathering sequence through granodiorite (data from Nesbitt, 1979).

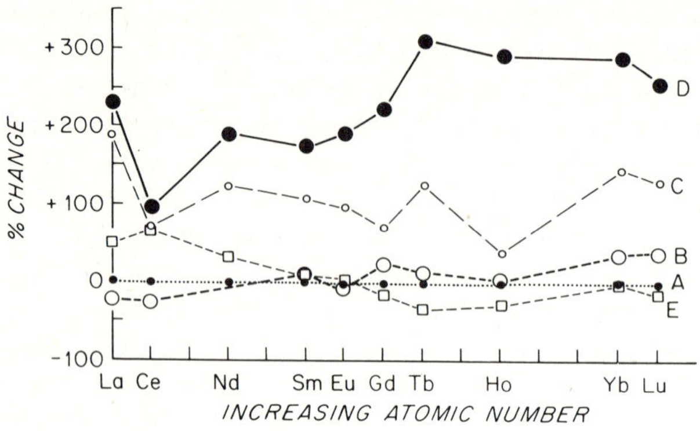

Fig. 9.8. Percent changes in the REE content of altered rocks (*B—D*) and residual weathering product (*E*), compared with parent rock (*A*), after normalizing assuming constant TiO_2 (data from Nesbitt, 1979).

after normalization of the data, it appears that only the HREE show depletion, while the LREE are enriched relative to the parent rock.

Nesbitt (1979) suggested a simple weathering scheme, based primarily on changes in fluid chemistry, to explain these observations. Uptake of CO_2 by rain water percolating through the organic-rich soil zone results in a decrease in pH to values of 4—5. In addition, dissolution of organic acids may also occur. The residual clay minerals have a limited capacity to neutralize the water; consequently, the acidic water percolates through the soil zone removing soluble major, minor, and rare earth elements. These solutions then migrate down the fractures into the granodiorite, where weathering of the mineral assemblage to clays occurs, with a concomitant increase in the pH by reactions of the type:

$$2NaAlSi_3O_8 + 3H_2O = Al_2Si_2O_5(OH)_4 + 2Na^+ + 2OH^- + 4SiO_2(aq)$$

As the solution becomes more basic, the relatively soluble REE are precipitated as hydroxides or carbonates, incorporated into secondary minerals, or adsorbed on to mineral surfaces. Hence, some of the REE are effectively being removed by acidic fluids from regions where the parent rock has previously been weathered to clays, and transported to sites where the rocks are actively being weathered, and where the REE are redeposited in response to a change in pH.

Such a process could explain the observed depletion of the HREE in the residual deposits and their enrichment in the weathering profile. Nesbitt (1979) has suggested that the selective HREE enrichment may be due to mineralogical controls, such as those discussed in section 9.3, and possibly also to preferential formation of complexes by the HREE (relative to the LREE), with carbonate ligands, thereby facilitating their transport to the weathering profile.

This weathering scheme cannot, however, explain the LREE enrichment relative to the parent rock seen in both the residual deposits and the weathering profile (Fig. 9.7). This observation indicates that LREE are being introduced into the system and taken up by the secondary minerals. The best information available on the likely REE pattern of the fluid phase involved in these reactions is that of river water (Fig. 9.6). If the river water REE patterns are assumed to be indicative of the distribution of the REE in groundwater, then the similarity between these patterns is striking. Hence, the patterns observed in the weathering profile, particularly the LREE enrichment, are probably a reflection of the REE abundance in the water, and the extent of equilibration will depend on the exposure time and, more importantly, the water/rock ratio.

Metasomatic and hydrothermal fluids

Fluids that are included in this category are all assumed to be involved in alteration reactions occurring at temperatures above present ambient. The chemical compositions of the fluids vary widely depending on the alteration process but include, for example, magmatic and post-magmatic fluids involved in metasomatism of continental rocks (e.g., Mineyev, 1963) or formation of ore deposits (e.g. Graf, 1977), and per-alkaline fluids responsible for fenitization reactions (e.g., Martin et al., 1978). The characteristic REE patterns in these fluids will be highly variable depending on their origin; however, a few general observations can be made regarding the transport of these elements in such fluids.

Transport of REE in hydrothermal fluids for any great distance is probably by the formation of carbonates, fluoride, or sulphate complexes. Experimental studies on the REE stability in acidic solutions rich in Cl^- or F^- have identified the most stable complexes as $RE(Cl,F)^+_2$ for most REE, and $RE(Cl,F)_3$ for La (Sillén and Martell, 1964). Consideration of the $\Delta G°$ values for solution of the REE oxide phases suggest that, when fluoride complexing is taken into account, $\Delta G°$ values become more strongly negative for solution reactions yielding $(RE)F^{2+}$, $(RE)F^+_2$, and for La, $(RE)F_3$ (Alderton et al., 1980). Although their calculations were based on thermodynamic properties at 25°C, extrapolation to higher temperatures suggests that similar reactions will still hold. Hence, the presence of fluoride significantly increases the mobility of the REE.

In a study of the differentiation of the REE by migration in the form of complexes, Mineyev (1963) summarized evidence for the importance of complexing in geochemical processes as follows:

(1) Migration of the REE in a simple ionic form cannot account for the extreme fractionation observed in natural systems. These elements must be transported in the form of complexes, because the complexes of the REE differ more from each other than ionic compounds.

(2) High concentrations of the REE, especially the HREE, which are particularly susceptible to complexing, are associated with high concentrations of alkalis and volatiles.

(3) Experimental studies have demonstrated the stability under hydrothermal conditions of complexes of the $NaYF_4$ and $Na_5Ce_3F_{14}$ type.

(4) Paragenetic association of the REE with complexing elements, such as Al, Be, Fe, Zr, and the ligands F^-, Cl^-, CO_3^{2-}, PO_4^{3-} indicates their simultaneous presence in mineral-forming solutions.

In a study of the Kazakhstan massif, composed of biotite, microcline, and reibeckite-albite granite, Mineyev was able to correlate the transport of the REE and the REE patterns in the metasomatized rocks with the fluoride content and pH of the fluid phase. The massif had undergone a series of alteration processes with fluids showing considerable variations in pH. The REE patterns showed a sequence of evolution from LREE-enriched near the centre of the massif to a Yb maximum at the periphery, with a concomitant increase in the fluoride and alkali contents of the rocks (Fig. 9.9).

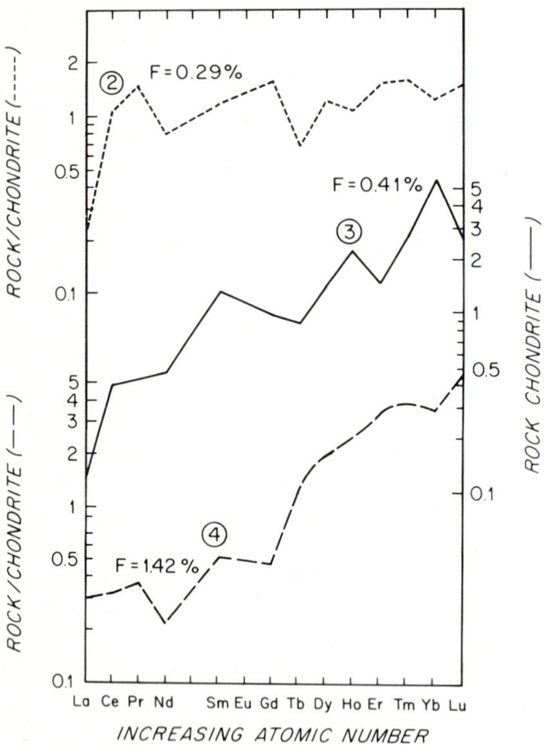

Fig. 9.9. Chondrite-normalized REE patterns for the metasomatized Kazakhstan granite massif (data from Mineyev, 1963). Samples *2*, *3*, and *4* are in sequence away from the centre of the massif.

Mineyev then explained the sequence by assuming that the REE were carried in solution as alkali fluoride complexes. A decrease in the Na^+ content due to albitization would cause precipitation of fluorides. Since the LREE complexes are the least stable, these elements would precipitate first, hence enriching the residual solutions in the HREE. Eventually, these would precipitate when the alkali content of the fluid became too low. Hence, this illustrates the importance of complexing, particularly with fluorides, in transporting the REE over great distances, and also points out the sensitivity of the complexes to changes in pH.

Another important process involving metasomatic fluids that results in REE mobility, is fenitization. In a study of fenitization of pure Cambrian quartzites from the Borralan complex, northwestern Scotland, by Martin et al. (1978), enrichment in REE, in particular the LREE, was observed. Textural, mineralogical, and compositional evidence indicated that reactions occurred between the rock and a peralkaline fluid phase that had evolved during the cooling of a pseudoleucite syenite mass. During fenitization, Si was removed from the system, while Al, and lesser amounts of Na, Mg, K, Fe, and Ca were added. Mineralogically, these changes were seen as the formation of alkali amphibole, aegirine-augite, biotite, albite, and K-feldspar, with minor sphene, apatite, pyrite, and magnetite. However, two distinct trends were seen; a sodic trend, in which the proportion of alkali amphibole and alkali pyroxene increased with the degree of fenitization, and a potassic trend, in which the fenites were much richer in biotite, and free of amphibole and pyroxenes.

Chondrite-normalized REE patterns for these fenitized quartzites (Fig. 9.10) are all LREE-enriched, and suggest increasing overall REE abundances with increasing fenitization. After a normalization procedure to account for volume changes, Martin et al. (1978) established that there is uptake of the REE by the fenites, with the LREE abundances showing the greatest increases. The lack of differences between the REE patterns showing sodic and potassic fenitization trends suggests that the specific mineralogy developed in each case had little effect on the whole-rock REE pattern. In addition, the original quartzite probably contained very low concentrations of the REE, if any at all. Hence, the REE must have been introduced into the system by the peralkaline fluid phase, and the overall REE pattern developed by the whole rock should reflect the REE distribution in the metasomatic fluid. This fluid was presumably in equilibrium with the cooling pseudoleucite syenite mass. Since there are no REE data for the syenite, it is not possible to determine whether the REE pattern of the fluid reflects a pattern of LREE enrichment in the source rock, or whether dissolution preferentially affected the LREE. However, transport of the REE in the fluid will be as complexes, those of the HREE being the most stable. As fenitization proceeds, and the alkalis are removed from solution during the formation of the secondary minerals, the LREE complexes will break down

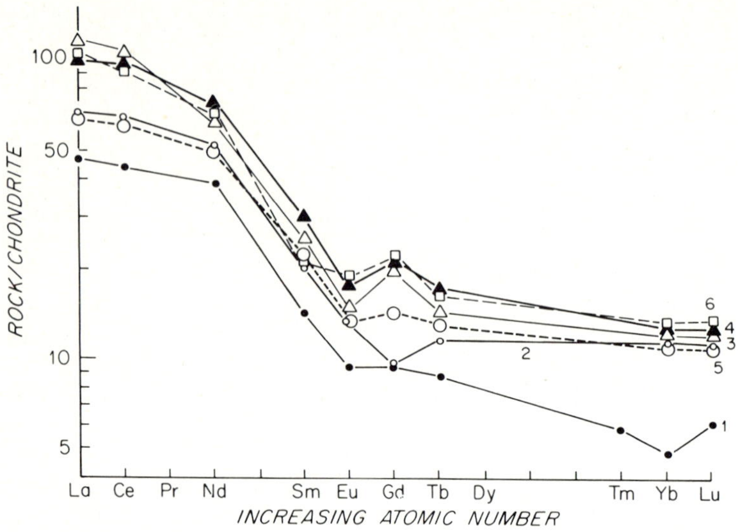

Fig. 9.10. Chondrite-normalized REE patterns of fenitized quartzites, Borralan complex, northwestern Scotland (data from Martin et al., 1978). *1—4*: order of increasing sodic fenitization; *5, 6*: order of increasing potassic fenitization.

first and precipitate or be incorporated into the secondary minerals. Further studies are clearly required to gain a complete understanding of the mobility of the REE during fenitization. However, it can be concluded that complexing is the major way in which REE are transported in hydrothermal and metasomatic fluids.

9.5. Uptake of the REE by secondary minerals

During any weathering or alteration reaction, an assemblage of secondary minerals will be formed that is characteristic of the local environment, i.e., temperature, pressure, fluid chemistry, etc. The ability of these newly-formed phases to accommodate the REE, either within their crystal structures or possibly by adsorption, will be an important factor in determining the whole-rock REE pattern of the weathered or metamorphosed sample.

The ability of a mineral to accommodate the REE will depend strongly on the ionic radius of the REE compared with that of the cation to be interchanged. Other factors that influence this substitution are the coordination number and the REE content, the importance of which was originally postulated by Khomyakov (1963). He observed that in minerals where exchange of REE does not affect the volume of the unit cell, individual REE are more readily substituted, and hence the mineral can exhibit wide variations in its REE pattern. In contrast, in those minerals in which the mole fraction of the REE is high, and thus its size significantly affects the dimensions of the unit cell, the individual REE can be less easily exchanged, and the range of possible REE substitutions will be more restricted. Khomyakov also pointed out that the ionic radius of a cation depends on its crystal environment and so, in different compounds, the best substitution for a particular cation may be different lanthanides. This was further emphasized by Haskin et al. (1966) who demonstrated that the change in preference by Ca for different REE is related to its change in coordination number in various compounds which arise from the effect of changing structural environment on cation ionic radius.

The substitution of various assemblages of REE for divalent and monovalent elements has been discussed in Chapters 1 and 2. However, very few data are available on the REE distribution in secondary minerals. Wood et al. (1976) presented chemical data for thirteen minerals common in the amygdales of zeolite facies metamorphosed eastern Iceland lavas. They comprised nine zeolites, plus apophyllite, aragonite, celadonite, and chalcedony. Only celadonite contained detectable La (>1.0 ppm) and even in this phase the REE were in concentrations too low to be measured. Chlorites exhibit different REE patterns depending on the alteration process. That formed during Fe metasomatism of granite is characterized by a REE pattern depleted in LREE and enriched in Yb and Lu relative to its precursor, and with a marked Eu anomaly (Alderton et al., 1980). In contrast, chlorites that are believed to have originated during greenschist metamorphism of ocean floor basalts show some mobility of the LREE, but overall maintain the REE pattern of their precursor (Copeland et al., 1971).

Alderton et al. (1980) have studied a series of alteration processes that have affected granitic rocks from southwest England. The approach used was to compare the REE patterns of the fresh and altered rock in order to determine the ability of the alteration products to incorporate the REE, released during breakdown of the original minerals, into their structures. Boron metasomatism — a common alteration process around zones of mineralization — resulted in depletion of all the REE and disappearance of the Eu anomaly in the altered rock (containing quartz, tourmaline and cassiterite) compared with the original granite (Fig. 9.11). The REE distribution in a tourmaline mineral separate shows a basin-shaped pattern with a slight positive anomaly. Alderton et al. concluded that, although the overall depletion could be partially explained by the dilution effect of quartz, all the LREE released from the breakdown of feldspar and mica could not be accommodated in the tourmaline and quartz. However, tourmaline has sites which could accommodate Eu^{2+} and hence Eu suffered less depletion than the other REE. This general approach of attempting to account for the REE released during alteration could be useful in that it provides a method of determining the amount of REE transported away in the fluid. However, more modal data both on the fresh and altered rocks, as

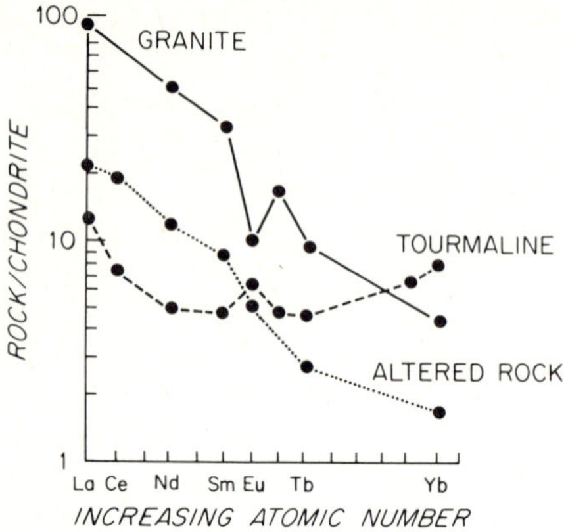

Fig. 9.11. Chondrite-normalized REE patterns for original granite, the altered rock, containing quartz, tourmaline, and cassiterite, and a tourmaline separate (data from Alderton et al., 1980).

well as the information concerning the REE distributions in secondary minerals are necessary before the relation between REE patterns of the original rock and its altered equivalent can be quantitatively analysed.

The lack of mobility of the REE during greenschist metamorphism of submarine rocks has been tentatively ascribed to stabilization of the REE by the secondary minerals, in particular, chlorite (Ludden and Thompson, 1979). This is substantiated by experiments that have demonstrated that, when a glassy tholeiite is entirely transformed to a mixed layer chlorite-smectite phase at temperatures of 150—350°C and seawater/rock ratios of 10—125, the product has a REE profile identical to that of the unaltered basalt (Menzies et al., 1979). Table 9.2 is a compilation of data concerning the REE abundances in greenstones from the Mid-Atlantic Ridge. In addition, an analysis of a chlorite separate from a greenstone (sample CH-44 DR3) is shown. The REE profiles of samples taken across a single metamorphosed pillow (sample AII-42, 1-96, Fig. 9.12a) indicate that there is virtually no change in the chondrite-normalized REE pattern during alteration. Only La shows any variation, being slightly enriched in the altered portions of the pillow. This may be an indication of retrograde reactions, causing some mobility in the LREE (Menzies et al., 1979), or could reflect uptake of La by adsorption from seawater by the chlorites and other clay minerals (Humphris et al., 1978). Such adsorption has been described in Quaternary clay-chlorite mixtures of Norway (Roaldset and Rosenquist, 1971). Fig. 9.12b and c show data from two adjacent areas of the Mid-Atlantic Ridge. Unfortunately, no samples exist in which alteration has affected only the outer rim, leaving a fresh interior. It is therefore necessary to compare the altered basalts with the field depicting the range of REE values displayed by fresh basalts from the same dredge haul. Although this technique does not

TABLE 9.2

REE contents (ppm) of greenstones and a chlorite separate from the Mid-Atlantic Ridge

	AII-42 (4°S)			AII-78 (23°N)	CH-44 (22°N)							
	1-96[a]			1-3[b]	3-2[c]	3-3[d]	3-6[d]	3-7[d]	3-13[e]	3-17[e]	3-126[e]	DR3[f]
	A	B	C									
	altered rim		fresh interior	meta-basalt	greenstones							green-stone
La	3.52	3.93	2.32	2.44	2.58	3.9	3.0	2.7	3.00	2.63	2.02	5.2
Ce	8.77	8.02	8.57	8.33	7.50	12.0	9.2	7.5	6.69	9.86	11.18	—
Nd	9.07	8.82	9.37	7.38	7.00	11.7	9.4	5.2	5.62	10.92	11.86	—
Sm	3.35	3.04	3.21	2.97	2.47	3.6	3.1	2.1	2.32	3.82	3.87	—
Eu	1.49	1.52	1.26	0.95	1.10	1.42	1.35	0.96	0.90	1.40	1.41	1.2
Gd	4.57	4.17	4.19	—	3.52	4.5	4.5	2.5	2.88	4.88	4.95	—
Tb	0.84	0.76	0.78	0.68	0.67	0.9	0.8	0.5	0.53	0.94	0.96	—
Dy	—	—	—	—	—	—	—	—	—	—	—	4.0
Ho	—	—	—	1.02	—	1.4	1.2	0.72	—	—	—	—
Yb	3.45	2.80	2.87	2.72	2.27	2.7	2.8	1.8	1.93	3.38	3.71	—
Lu	—	—	—	0.46	0.39	0.43	0.41	0.29	0.41	0.48	0.58	0.45

[a]Humphris et al. (1978).
[b]Ludden and Thompson (1979).
[c]Frey et al. (1968).
[d]Herrmann et al. (1974).
[e]This paper.
[f]Chlorite separate from DR3 greenstone (Copeland et al., 1971).

allow minor changes to be detected, the conclusion remains the same — that submarine greenschist facies metamorphism does not significantly affect the REE pattern of the rock, and suggests that the secondary minerals can accommodate the REE released during breakdown of the primary mineral assemblage of the basalt.

Immobility of the REE has also been observed in greenstones from ophiolites (e.g., Menzies et al., 1977). However, this conclusion cannot be extended to continental rocks, or other grades of metamorphism. For example, Sun and Nesbitt (1978) have indicated significantly more mobility of the LREE in Archean greenstones, and Wood et al. (1976) and Hellman et al. (1977) have demonstrated mobility of the REE during prehnite-pumpellyite facies metamorphism.

9.6. Conclusions

From the preceding discussions concerning the effects of weathering and alteration on the distribution of the REE in oceanic and continental rocks, it is clear that the chemistry of this group of elements is a complex interaction between primary and secondary processes. No simple rules referring to the direction of REE transport during different grades of metamorphism can be compiled. Evidence exists for both mobility and immobility of the REE

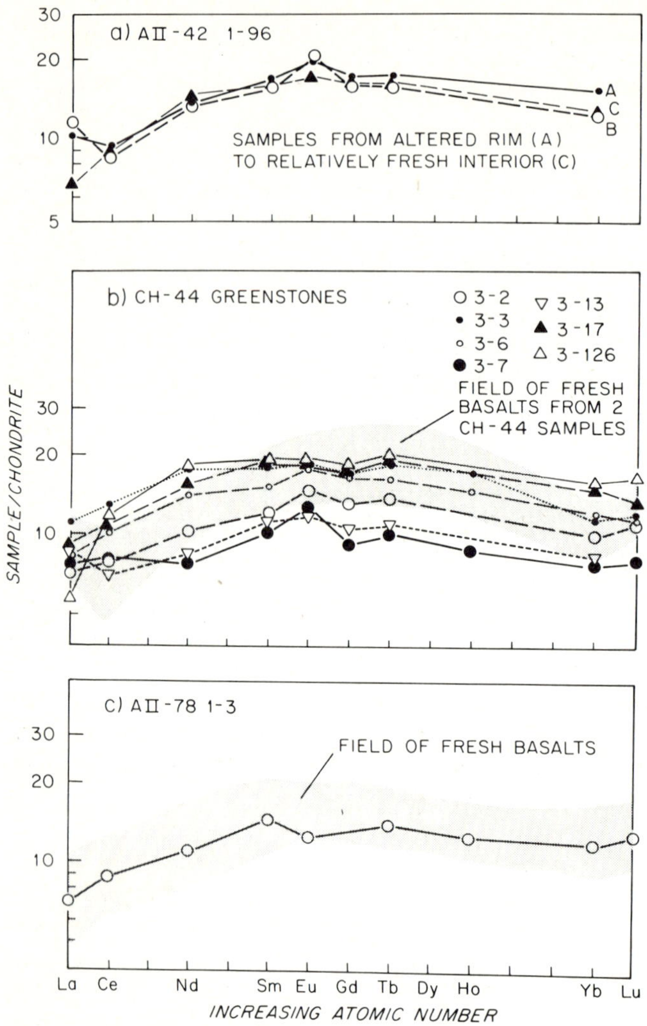

Fig. 9.12. Chondrite-normalized REE patterns of greenstones from the Mid-Atlantic Ridge (references as in Table 9.2 and in text).

in low-temperature weathering as well as all metamorphic grades. Classification of REE transport according to the fluid phase involved is not practical as, for example, variable REE trends have been observed in the studies of rock-seawater interactions at similar temperatures.

These variations are partly a function of the relative importance of the REE chemistry of the three major phases, i.e., the original rock, the fluid, and the secondary products, but also depends on the physicochemical conditions in which the weathering or alteration takes place.

The environment plays a crucial role in determining not only the type, but also the extent, of alteration. Its importance has been pointed out in reference to specific examples of alteration systems discussed in this chapter. Temperature will clearly influence the secondary minerals that are produced, as well as affecting the rates of reaction. Pressure may partly control the concentrations of volatiles in the alteration fluid, in addition to influencing the formation of secondary minerals at higher metamorphic grades. Of particular interest are the fluid/rock ratios in the system, and the exposure time of the rock to water of a similar composition, as illustrated by the differences between dredged and drilled weathered submarine basalts. These factors determine the extent of equilibration of the REE between the rock and the fluid phase, as well as the degree of alteration that can occur. The location of a rock in relation to the circulation path of the fluid through the system will partly determine the composition of the fluid with which it reacts. Hence, within a single flow unit, the REE content of the fluid will vary, and the alteration reactions occurring will be different.

It can therefore be concluded that the REE patterns of weathered and altered rocks should not be used indiscriminately for petrogenetic modelling. This does not imply that they can never be used, but that secondary processes to which the rock has been subjected, and their possible effects on the REE, should be carefully assessed. In order to do this effectively, it is necessary to be able to make some predictions concerning the release of the REE from the primary minerals that are broken down, and the likelihood of their incorporation into the secondary products, under the environmental conditions that prevailed during alteration. This would necessitate detailed mineralogical studies of the samples to identify the mineral phases involved and their modal abundances, as well as information on the environment of alteration. In addition, a better understanding of the uptake of REE by secondary minerals is required. Experimental studies are useful, but presently are limited by kinetic factors and by the oversimplifications necessary, such as the composition of the fluid, a closed system, etc. Most of the studies on natural rocks so far have concentrated on whole-rock REE patterns with only cursory information on the abundance and types of minerals involved, making any assessment of the reactions occurring impossible. Further research is required on the partitioning behaviour of the REE between fluids and secondary minerals, with detailed study of all phases in a single system subjected to a weathering or alteration process. This information can then be used as a basis for assessing the possible mobility of the REE in a particular system, and their applicability to petrogenetic modelling. Until such time, the REE patterns of weathered and altered rocks should be used for modelling partial melting and crystal fractionation processes, and for classifying rocks into particular tectonic settings, with extreme caution.

Acknowledgements

I should like to thank Geoffrey Thompson for critically reviewing and improving this manuscript. Thanks also go to Phyllis Hartley for drafting the figures. I gratefully acknowledge financial support from NSF grant No. OCE80-24629.

This is Contribution No. 4989 from the Woods Hole Oceanographic Institution.

References

Alderton, D.H.M., Pearce, J.A. and Potts, P.J., 1980. Rare earth element mobility during granite alteration; evidence from southwest England. *Earth Planet. Sci. Lett.*, 49: 149—165.

Balashov, Yu. A., Ronov, A.B., Migdisov, A.A. and Turanskaya, N.V., 1964. The effect of climate and facies environment on the fractionation of the rare earths during sedimentation. *Geochemistry (U.S.S.R.)*, 5: 951—969.

Buat-Menard, P. and Chesselet, R., 1979. Variable influence of the atmospheric flux on the trace metal chemistry of oceanic suspended matter. *Earth Planet. Sci. Lett.*, 42: 399—411.

Condie, K.C. and Baragar, W.R.A., 1974. Rare earth element distribution in volcanic rocks from Archean greenstone belts. *Contrib. Mineral. Petrol.*, 45: 237—246.

Copeland, R.A., Frey, F.A. and Wones, D.R., 1971. Origin of clay minerals in a Mid-Atlantic Ridge sediment. *Earth Planet. Sci. Lett.*, 10: 186—192.

Exley, R.A., 1980. Microprobe studies of REE-rich accessory minerals: implications for Skye granite petrogenesis and REE mobility in hydrothermal systems. *Earth Planet. Sci. Lett.*, 48: 97—110.

Ferrara, G., Innocenti, F., Ricci, C.A. and Serri, G., 1976. Ocean-floor affinity of basalts from North Apennine ophiolites: geochemical evidence. *Chem. Geol.*, 17: 101—111.

Floyd, P.A., 1976. Geochemical variation in the greenstones of S.W. England. *J. Petrol.*, 17: 522—545.

Frey, F.A., Haskin, M.A., Poetz, J.A. and Haskin, L.A., 1968. Rare earth abundances in some basic rocks. *J. Geophys. Res.*, 73: 6085—6098.

Frey, F.A., Bryan, W.B. and Thompson, G., 1974. Atlantic ocean floor: geochemistry and petrology of basalts from Legs 2 and 3 of the Deep Sea Drilling Project. *J. Geophys. Res.*, 79: 5507—5527.

Glasby, G.B., 1973. Mechanisms of enrichment of the rarer elements in marine manganese nodules. *Mar. Chem.*, 1: 105—125.

Goldberg, E.D., Koide, M., Schmitt, R.A. and Smith, R.H., 1963. Rare-earth distributions in the marine environment. *J. Geophys. Res.*, 68: 4209—4217.

Goldich, S.S., 1938. A study in rock weathering. *J. Geol.*, 46: 7—58.

Graf, J.L., 1977. Rare earth elements as hydrothermal tracers during the formation of massive sulfide deposits in volcanic rocks. *Econ. Geol.*, 72: 527—548.

Hajash, A., 1975. Hydrothermal processes along mid-ocean ridges: an experimental investigation. *Contrib. Mineral. Petrol.*, 53: 205—226.

Hart, R.A., 1970. Chemical exchange between seawater and deep ocean basalt. *Earth Planet. Sci. Lett.*, 9: 269—279.

Haskin, L.A. and Schmitt, R.A., 1976. Rare-earth distributions. In: P.H. Abelson (Editor), *Researches in Geochemistry, 2.* J. Wiley and Sons, New York, N.Y., pp. 234—258.

Haskin, L.A., Frey, F.A., Schmitt, R.A. and Smith, R.H., 1966. Meteoritic, solar, and terrestrial rare-earth distributions. *Phys. Chem. Earth*, 7: 67—321.

Hellman, P.L. and Henderson, P., 1977. Are rare earths mobile during spilitisation? *Nature*, 267: 38—40.

Hellman, P.L., Smith, R.E. and Henderson, P., 1977. Rare earth element investigation of the Cliefden outcrop, N.S.W., Australia. *Contrib. Mineral. Petrol.*, 65: 155—164.

Hellman, P.L., Smith, R.E. and Henderson, P., 1979. The mobility of the rare earth elements: evidence and implications from selected terrains affected by burial metamorphism. *Contrib. Mineral. Petrol.*, 71: 23—44.

Herrmann, A.G., 1978. Yttrium and lanthanides. In: K.W. Wedepohl (Editor), *Handbook of Geochemistry, Vol. II-5*. Springer-Verlag, Berlin, pp. 57-71-A to 57-71-O.

Herrmann, A.G., Potts, M.J. and Knake, D., 1974. Geochemistry of the rare earth elements in spilites from the oceanic and continental crust. *Contrib. Mineral. Petrol.*, 44: 1—16.

Høgdahl, O.T., 1967. Distribution of the rare earth elements in seawater. NATO Research Grant No. 203. *Central Inst. Ind. Res., Blindern, Semi-Annu. Progr. Rep.*, 4 (unpublished).

Høgdahl, O.T., Melson, S. and Bowen, V.T., 1968. Neutron activation analysis of lanthanide elements in seawater. In: R.A. Baker (Editor), *Trace Inorganics in Water. Adv. Chem. Ser.*, 73: 308—325.

Humphris, S.E. and Thompson, G., 1978a. Hydrothermal alteration of oceanic basalts by seawater. *Geochim. Cosmochim. Acta*, 42: 107—125.

Humphris, S.E. and Thompson, G., 1978b. Trace element mobility during hydrothermal alteration of oceanic basalts. *Geochim. Cosmochim. Acta*, 42: 127—136.

Humphris, S.E., Morrison, M.A. and Thompson, R.N., 1978. Influence of rock crystallization history upon subsequent lanthanide mobility during hydrothermal alteration of basalts. *Chem. Geol.*, 23: 125—137.

Keller, W.D., 1957. *Principles of Chemical Weathering.* Lucas Brothers Publishers, Columbia, Mo., 111 pp.

Khomyakov, A.P., 1963. Relation between content and composition of the rare earths in minerals. *Geochemistry (U.S.S.R.)*, 2: 125—132.

Ludden, J.N. and Thompson, G., 1978. Behaviour of rare earth elements during submarine weathering of tholeiitic basalts. *Nature*, 274: 147—149.

Ludden, J.N. and Thompson, G., 1979. An evaluation of the behavior of the rare earth elements during weathering of sea floor basalt. *Earth Planet. Sci. Lett.*, 43: 85—92.

Martin, J.-M., Høgdahl, O.T. and Phillippot, J.C., 1976. Rare earth element supply to the ocean. *J. Geophys. Res.*, 81: 3119—3124.

Martin, R.F., Whitley, J.E. and Woolley, A.R., 1978. An investigation of rare earth mobility: fenitised quartzites, Borralan complex, N.W. Scotland. *Contrib. Mineral. Petrol.*, 66: 69—73.

Menzies, M., Blanchard, D. and Jacobs, J., 1977. Rare earth and trace element geochemistry of metabasalts from the Point Sal ophiolite, California. *Earth Planet. Sci. Lett.*, 37: 203—215.

Menzies, M., Seyfried, W. and Blanchard, D., 1979. Experimental evidence of rare earth element immobility in greenstones. *Nature*, 282: 398—399.

Mineyev, D.A., 1963. Geochemical differentiation of the rare-earths. *Geochemistry (U.S.S.R.)*, 12: 1129—1149.

Mottl, M.J. and Holland, H.D., 1978. Chemical exchange during hydrothermal alteration of basalts by seawater, I. Experimental results for major and minor components of seawater. *Geochim. Cosmochim. Acta*, 42: 1103—1115.

Nagasawa, H., 1970. Rare earth concentrations in zircons and apatites and their host dacites and granites. *Earth Planet. Sci. Lett.*, 9: 359—364.

Nesbitt, H.W., 1979. Mobility and fractionation of rare earth elements during weathering of a granodiorite. *Nature*, 279: 206—210.

Pearce, J.A. and Cann, J.R., 1973. Tectonic setting of basic volcanic rocks determined using trace element analyses. *Earth Planet. Sci. Lett.*, 19: 290—300.

Piper, D.Z., 1974. Rare earth elements in the sedimentary cycle: a summary. *Chem. Geol.*, 14: 285—304.

Roaldset, E. and Rosenquist, I.Th., 1971. Unusual lanthanide distribution. *Nature*, 231: 153—154.

Robertson, A.H.F. and Fleet, A.J., 1976. The origin of rare earths in metalliferous sediments of the Troodos Massif, Cyprus. *Earth Planet. Sci. Lett.*, 28: 385—394.

Schnetzler, C.C. and Philpotts, J.A., 1968. Partition coefficients in rare-earth elements and barium between igneous matrix material and rock-forming mineral phenocrysts, I. In: L.H. Ahrens (Editor), *Origin and Distribution of Elements.* Pergamon Press, Oxford, pp. 929—938.

Schnetzler, C.C. and Philpotts, J.A., 1970. Partition coefficients of rare-earth elements between igneous matrix and rock-forming phenocrysts, II. *Geochim. Cosmochim. Acta*, 34: 331—340.

Sillén, L.G. and Martell, A.E., 1964. Stability constants of metal-ion complexes. *Chem. Soc. London, Spec. Publ.*, 17: 754 pp.

Staudigel, H., Frey, F.A. and Hart, S.A., 1979. Incompatible trace element geochemistry and $^{87}Sr/^{86}Sr$ in basalts and corresponding glasses and palagonite. In: T. Donnelly, J. Francheteau et al., *Initial Reports of the Deep Sea Drilling Project, 51, 52, 53, Part 2.* U.S. Government Printing Office, Washington, D.C., pp. 1137—1144.

Stumm, W. and Morgan, J.J., 1970. *Aquatic Chemistry*, John Wiley and Sons, New York, N.Y., 583 pp.

Sun, S.-S. and Nesbitt, R.W., 1978. Petrogenesis of Archean ultrabasic volcanics: evidence from rare-earth elements. *Contrib. Mineral. Petrol.*, 65: 301—325.

Thompson, G., 1973. A geochemical study of the low temperature interaction of seawater and oceanic igneous rock. *EOS, Trans. Am. Geophys. Union*, 54: 1015—1019.

Turner, D.R. and Whitfield, M., 1979. Control of seawater composition. *Nature*, 281: 468—469.

Vallance, T.G., 1974. Spilitic degradation of a tholeiitic basalt. *J. Petrol.*, 15: 79—96.

Venturelli, G., Capedri, S., Thorpe, R.S. and Potts, P.J., 1979. Rare-earth and other element distributions in some ophiolitic metabasalts of Corsica, western Mediterranean. *Chem. Geol.*, 24: 339—353.

Winchester, J.A. and Floyd, P.A., 1976. Geochemical magma type discrimination: application to altered and metamorphosed basic igneous rocks. *Earth Planet. Sci. Lett.*, 28: 459—469.

Wood, D.A., Gibson, I.L. and Thompson, R.N., 1976. Elemental mobility during zeolite facies metamorphism of the Tertiary basalts of eastern Iceland. *Contrib. Mineral. Petrol.*, 55: 241—254.

Chapter 10

AQUEOUS AND SEDIMENTARY GEOCHEMISTRY OF THE RARE EARTH ELEMENTS

A.J. FLEET

10.1. Introduction

The earliest work on REE in the sedimentary environment was the analysis of Palaeozoic and Mesozoic European and Japanese "shales"* by Minami (1935). Little other work was undertaken until the 1960's when a number of fundamental studies were carried out (Haskin and Gehl, 1962; Balashov et al., 1964; Wildeman and Haskin, 1965; Spirn, 1965; Haskin et al., 1966). Since then analytical and instrumental developments, especially of neutron activation analysis and mass spectrometry, and currently of ICP spectrometry (P. Henderson and R.J. Pankhurst, Chapter 13), have led to great improvements in the accuracy and precision of REE data. A burgeoning of interest in the aqueous and sedimentary geochemistry of the REE has also taken place in parallel with the growth in interest in other branches of REE geochemistry. Beside striving to define more precisely the REE contents of different natural waters, sediments and sedimentary rocks, researchers have been working to understand and quantify the processes in which the REE are involved in the sedimentary environment, and to use the REE in elucidating the formation of sedimentary deposits. The processes of continental and submarine weathering, the supply of the REE to and the behaviour of the REE in the oceans, and the formation of marine authigenic minerals, as regards both the genesis of the minerals and the removal of the REE from the oceans, have received particular attention. In this chapter an attempt is made to draw these and all the other various lines of enquiry together by considering the behaviour of the REE throughout the full spectrum of sedimentary processes.

The chapter first deals with the ways in which the REE behave during weathering and transport and in the oceans. Next the variations in the REE contents of sediments and sedimentary rocks which result from weathering, transport and deposition in the marine environment, are briefly reviewed. Finally, to complete the account, the behaviour of the REE during diagene-

*The term "shales" is used throughout this chapter loosely, as synonym for aluminosilicate mudrocks.

sis is discussed. As outlined this progression of topics sounds balanced and comprehensive. In reality the availability of data depends very much on what aspect of REE sedimentary geochemistry is under consideration: for instance REE data on manganese nodules are relatively abundant — twenty-two papers containing such data are cited in this review — in contrast the REE contents of waters of only five rivers have been studied in any kind of detail. The "state of the art" is, therefore, highlighted in this chapter and tentative suggestions are made of the sorts of new data which need to be collected in order to move towards a more balanced understanding of REE aqueous and sedimentary geochemistry.

A new approach to the behaviour of the REE in the marine environment has come through isotopic studies (e.g., O'Nions et al., 1978; Goldstein and O'Nions, 1981). A few references are made to these investigations in this chapter, but no systematic attempt can be made to incorporate the findings of these studies because our understanding of them is currently in a state of flux and has yet to be consolidated. Isotopic studies of natural waters and sediments are considered briefly, though, in Chapter 11 by C.J. Hawkesworth and P.W.C. van Calsteren as an aspect of REE isotopic studies in general.

The pioneering works of Minami (1935) and the fundamental studies of Haskin and Gehl (1962), Balashov et al. (1964), Spirn (1965), Wildeman and Haskin (1965), and Haskin et al. (1966), which are referred to above, established that the REE contents of most "shales" are very similar, being enriched in the LREE relative to the HREE when normalized to chondrites (Fig. 10.1). They also showed that the REE contents of most sediments and sedimentary

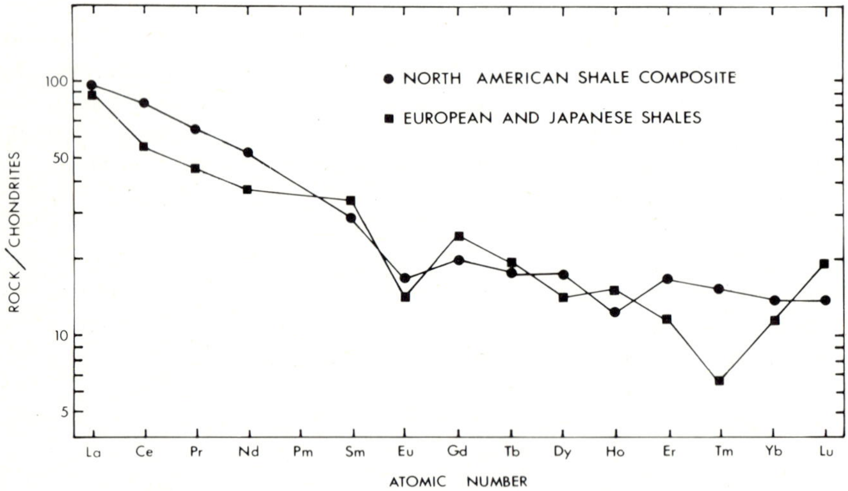

Fig. 10.1. Average REE abundances in European and Japanese shales (Minami, 1935) and in a composite of 40 North American shales (Haskin et al., 1968) compared to chondritic meteorites (Wakita et al., 1971).

rocks are similar in the relative abundance of the individual elements although they differ in absolute concentrations. A composite of 40 North American shales (Fig. 10.1) is often taken to typify sediments. The REE contents of this composite are also used for normalizing the REE contents of sediments in the same way as chondrites are typically used for normalizing those of igneous rocks (P. Henderson, Chapter 1). Shale—normalized REE contents of sediments and sedimentary minerals, when plotted against atomic number, indicate whether or not the sedimentary deposit has typical sedimentary REE contents, or identify subtle enrichments or deficiencies of single elements or group of elements.

The recognition of the similarity in REE contents of sediments in general, and of "shales" in particular, led to the idea that the average abundance of REE in sedimentary rocks could be taken to represent the average REE content of the crust, or at least the upper crust. In other words sedimentary processes could be looked on as causing homogenization of the REE fractionation which occurs during the formation of igneous rocks. Indeed, very early in the study of the sedimentary geochemistry of the REE Goldschmidt (1954, p. 317), in his classic book, summarised the concept thus:

"During the processes of weathering, soil formation, and sedimentation, all the rare earth elements are again assembled nearly completely in the hydrolysate minerals, e.g. the various clay and shale components, the bauxite minerals and the oxidates of iron and manganese".

This idea raises the question of which minerals the REE enter in the sedimentary environment. An outline answer to this question completes this introductory section and provides a very brief overview of the processes considered in this chapter.

Overall, weathering and erosion must result in only very minor amounts of REE going into solution, for only a few percent of the REE entering the oceans are dissolved (Martin et al., 1976). The bulk of the REE in eroded material is contained in clays. A substantial proportion of the REE contents of clays can be loosely held on the clays and, therefore, is available to take part in exchange reactions (e.g., Roaldset and Rosenqvist, 1971a), although such reactions prove very difficult to quantify (e.g., Martin et al., 1976). The low concentrations of REE associated with quartz, the other major silicate present in eroded material, are reflected by the fact that sandstones have lower REE contents than "shales" (e.g., Haskin et al., 1966). Indeed the REE in sandstones are probably largely contained in the clays present. Of course, if detrital grains of minerals such as apatite and zircon, in which the REE have been concentrated by igneous processes (P. Henderson, Chapter 1), are present in a sandstone in abundance they may cause the sandstone to be atypically rich in REE. The overwhelming concentration of the REE in the detrital fraction of transported material, rather than the dissolved portion, must mean that the REE pass from weathering to deposition almost exclusively without taking part, en route, in any significant chemical processes.

In contrast the REE supplied in a dissolved state to the oceans, although quantitatively insignificant in terms of the overall sedimentary mass balance, take part in various halmyrolytic processes and enter a number of authigenic minerals, notably ferromanganoan oxyhydroxides and marine apatite. The REE can substitute for Ca in marine apatite (Altschuler, 1980) just as they do in igneous apatites. Ferromanganoan oxyhydroxides, occurring as deep-sea nodules or metalliferous ridge sediments, readily accommodate the REE (e.g., Piper, 1974a). In the deep-sea nodules the REE are probably contained in both Fe-rich and P-rich phases (Elderfield et al., 1981a). Marine barites also concentrate the REE (Guichard et al., 1979), as, possibly, do deep-sea zeolites (Piper, 1974b). Amorphous phases may also significantly concentrate REE but the role of these phases are very poorly understood (Piper, 1974a). Apart from phosphatic fish debris other biogenic materials probably do not take up dissolved REE from seawater to any great extent (Elderfield et al., 1981a). A final group of minerals which the REE that are dissolved in seawater enter are the halmyrolytic minerals formed by the submarine weathering and hydrothermal alteration of oceanic igneous rocks (S.E. Humphris, Chapter 9).

10.2. Weathering and transport

Acceptance of the idea that sedimentary processes homogenize the REE fractionation which occurs during the formation of igneous rocks (see section 10.1) tends to hide the fact that fractionation of the REE can occur during weathering, and to obscure the changes which take place during sediment transport and often, though not invariably, wipe out the evidence of the weathering process. Over the last twenty years or so, detailed studies of the behaviour of the REE during continental weathering have begun to elucidate these processes. During the same period evidence has been building up that the REE are also affected, at least in some instances, by submarine weathering.

Continental weathering and transport

Exactly how the REE behave under different conditions of weathering is not clearly established. The evidence suggests that the REE can be mobilized during both warm, humid weathering and temperate weathering — Balashov et al. (1964) were the first to show that the REE are labile during the intense chemical weathering which is typical of warm, humid conditions. They studied the REE contents of two suites of sediments from the Russian Platform, one of which had formed under arid conditions and the other of which had been produced by humid weathering. They found that the argillaceous and arenaceous rocks formed under arid conditions contained similar abundances of the REE, and that their associated carbonates, although having lower REE contents, had similar LREE/HREE ratios. In contrast the REE contents of the suite of sediments formed under humid conditions varied markedly. The clays contain over one and a half times more REE than the sandstones, and the LREE/HREE ratios of the sediments decreased in the order clays, sandstones, carbonates. Balashov and his co-workers attributed the difference in REE contents between the two suites of sediments to mobilization of the REE under the humid conditions, which would, of course, cause intense chemical weathering. They also suggested that the HREE are preferentially transported in solution because they form more soluble bicarbonate and organic complexes than the LREE.

The general conclusion that the REE can be mobile during weathering has been supported by subsequent, more specific studies. Burkov and Podporina (1967) found that the REE, and in particular the LREE, were mobilized under acidic conditions when granite and granodiorite were weathered to kaolinite. Rankin and Childs (1976) showed that Fe-Mn concretions in some New Zealand soils often concentrate Ce and can concentrate all the REE. Maksimović and Roaldset (1976) reported that the REE are concentrated towards the base of some Greek and Yugoslavian bauxite deposits. They explained this by a variety of weathering processes. One of the most detailed studies carried out was undertaken by Nesbitt (1979), who also found evidence of REE mobilization during warm, humid weathering (S.E. Humphris, Chapter 9).

There is also evidence that less intense chemical weathering under cooler conditions can mobilize the REE. Roaldset and Rosenqvist (1971a) reported REE concentrations in micas from a weathered Norwegian gneiss which are 3—7 times the REE abundances in the whole weathered gneiss (Fig. 10.2). The fact that EDTA strips most of the REE contents from these micas (Fig. 10.2) indicates that the REE are susceptible to chemical changes during further weathering, erosion and transport. Interestingly Roaldset and Rosenqvist found a marked predominance in the micas from the weathered gneiss of the HREE with even atomic numbers over those with odd atomic numbers, even after normalization (Fig. 10.2). In several papers (e.g., Roaldset, 1973, 1974, 1978; Christie and Roaldset, 1979) Roaldset has argued that such behaviour is explicable in terms of the chemistry of the REE. She has emphasised that individual REE do not necessarily behave similarly in solution during weathering and so can become differentiated from one another.

The precise conditions under which the REE are mobilized during weathering merit further investigation. Of the many factors controlling this REE mobility (S.E. Humphris, Chapter 9) pH must be a major one. The work of Brown et al. (1955), who found that under acidic conditions REE are easily removed from clays but under neutral or alkaline conditions they become fixed, supports this idea; while Nesbitt (1979) stressed the

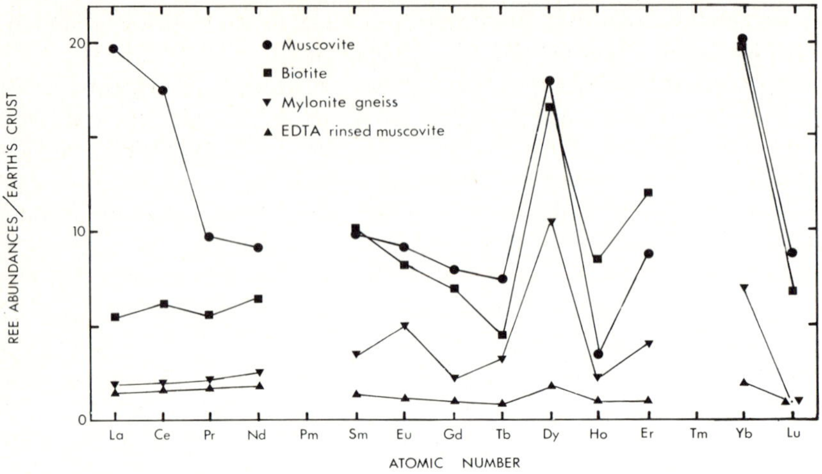

Fig. 10.2. REE abundances in weathered mylonite gneiss from Numedalen, Norway, its constituent muscovite and biotite, and EDTA-rinsed muscovite compared to the Earth's crust (from Roaldset and Rosenqvist, 1971a).

importance of pH when explaining his field observations. In general, therefore, there must be a tendency for the REE to become fixed during erosion and transport in river and surface waters: in other words mobilization of the REE may occur during weathering but is unlikely to be significant during erosion and transport.

In passing it is worth noting that the physical process of sediment sorting can determine the REE contents of sedimentary deposits if the REE are concentrated in one or more size fractions. For instance, Roaldset (1979) found that the REE contents of a Quaternary till from the Numedal area of Norway occurred mainly adsorbed on the <2-μm fraction of the sediment, and she suggested that the winnowing of the fine fraction from this material during erosion and transport could result in the relatively low REE contents of the fluvioglacial sediments of the area. The association of the REE with the 64 to 6 μm fractionation of a marine clay from the same area also led her to point out that subsand-sized detrital grains of REE-rich minerals (e.g., monozite, xeonotime) can be significant in determining the REE contents of continental sediments which have undergone relatively little sorting.

Submarine weathering

Early studies of ocean ridge basalts and their metamorphic equivalents, greenstones, indicated that the REE were not radically affected by metamorphism (Frey et al., 1968). Copeland et al. (1971) assumed that if the REE were not mobilized by metamorphism they would not be by submarine weathering and, in an elegant study, demonstrated by studying the REE contents of individual sedimentary minerals the dual provenance of minerals from a Mid-Atlantic Ridge sediment. For instance, they showed that the REE contents of the coarse-grained smectites of the sediment were similar to those of the ridge basalts and so were derived locally, but that the fine-grained smectites present contain REE abundances similar to sediments in general and therefore probably have a continental origin (Fig. 10.3).

Although the work of Copeland et al. (1971) gave support to the idea that the REE contents of ridge basalts change little during submarine weathering, later studies have shown this not to be the case. Frey et al. (1974) were the first to report data which suggested that submarine weathering affect the REE. They showed that La could be lost from basaltic glass and Ce gained. The results of such mobilization, though, can be even more striking than this. Ludden and Thompson (1978), among others, demonstrated that significant uptake of the LREE occurs, particularly in the palagonitized rinds of pillow-basalts (S.E. Humphris, Chapter 9). Studies by Bonnot-Courtois and her co-workers have confirmed this work and indicated that Ce is especially prone to enrichment in palagonites (e.g., Bonnot-Courtois, 1980). Other halmyrolytic minerals formed from basalts can contain relative REE abundances somewhat similar to those of seawater (e.g., beidellite, Desprairies and Bonnot-Courtois, 1980; celadonite, Fleet et al., 1980). Desprairies and Bonnot-Courtois (1980) have suggested that this is because these minerals form in equilibrium with seawater.

All these studies obviously underline why care must be taken in selecting material for analysis when the REE contents of submarine rocks are to be

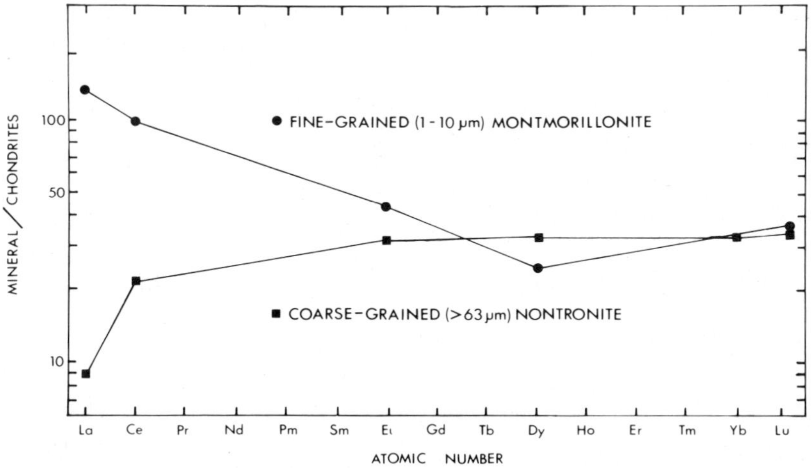

Fig. 10.3. REE abundances in fine-grained (1—10 μm) montmorillonite and coarse-grained (> 63 μm) nontronite from a Mid-Atlantic Ridge sediment (Copeland et al., 1971) compared to chondritic meteorites (Haskin et al., 1968).

used to contribute to petrogenetic studies. But above all they indicate that, as with continental weathering, much more work is needed if we are to understand the actual processes which occur. It is also important that we should work towards a quantitative understanding of the extent of REE uptake during submarine weathering, so that we can evaluate its significance in controlling the REE contents of the oceans.

10.3. The marine environment

Because of its predominant role in aqueous and sedimentary processes the marine environment has naturally been the subject of more studies and reviews than any other sedimentary environment. Goldberg et al. (1963) were the first workers to attempt to consider the marine environment as a whole. Spirn (1965) presented many useful data concerning it, and Piper (1974a) devoted most of his important summary of "REE in the sedimentary cycle" to the marine environment. These publications are discussed and others are mentioned below.

Since Høgdahl et al. (1968) reported small but significant variations in the REE contents of seawater samples from the Atlantic little other work on seawater has been published at the time of writing. In comparison there are many data on marine sediments, and in particular on manganese nodules.

REE supply to the oceans

The few data on REE supply to the oceans have been reviewed by Martin et al. (1976). They showed that the REE mainly enter the oceans incorporated in particulate material, only a few percent of the supply are dissolved. The REE contents of this detrital material are similar to those of sediments in general, being enriched in the LREE relative to the HREE and having no marked depletion or enrichment of any particular REE (Fig. 10.4a) (Høgdahl, 1970; Martin et al., 1976). This similarity indicates that any fractionation of the REE which may occur during weathering and erosion is obliterated during transport. It also suggests that detrital material, once introduced into the marine environment, accumulates there to form sedimentary deposits without undergoing significant changes in its REE contents.

The dissolved REE entering the oceans are slightly enriched in both the LREE and the HREE relative to the intermediate REE (Fig. 10.4b) (Høgdahl, 1970; Martin et al., 1976). The enrichment of the LREE reflects the greater abundances of these REE in the continental crust, but the enrichment of the HREE probably reflects the ability of these elements to form soluble complexes (e.g., Goldberg et al., 1963). How dissolved REE behave as they are transported through estuaries into the marine environment is difficult to evaluate as they are present in such low concentrations. Martin et al.

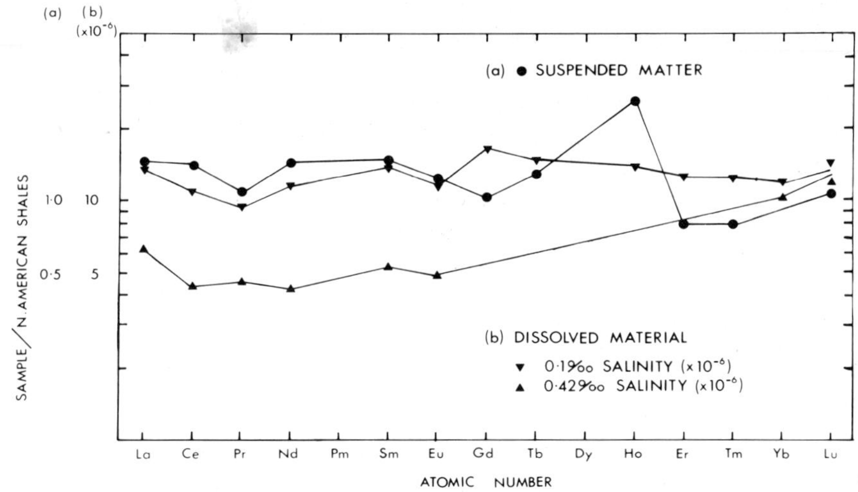

Fig. 10.4. (a) Average REE abundances in the suspended matter of rivers and (b) dissolved REE abundances in the Garonne and Dordogne river water of 0.1°/$_{oo}$ salinity and Gironde estuary water of 0.42°/$_{oo}$ salinity (Martin et al., 1976), compared to a composite of North American shales (Haskin et al., 1968).

(1976) have suggested that up to 50% of them may be removed from solution. This may be due to processes such as absorption by plankton (Turekian et al., 1973) and co-precipitation with oxyhydroxides (Aston and Chester, 1973). The LREE tend to be removed from solution before the HREE, probably because the latter are more strongly chelated, but no single REE is preferentially removed (Martin et al., 1976).

Seawater

The REE are minor constituents of seawater, having concentrations of only a few nanograms per liter (Table 10.1), which are some of the lowest for the stable elements (Turner and Whitfield, 1979). Both the LREE and HREE are enriched in seawater relative to the REE with intermediate atomic numbers (Fig. 10.5) (Goldberg et al., 1963; Høgdahl et al., 1968). In this respect seawater is similar to filtered river water (see above), and, like river water, presumably reflects the relatively high abundance of the LREE in continental crust and the ability of the HREE to form soluble complexes. This comparison hides the fact that the REE are very efficiently removed from solution as soon as they enter the oceans (Turner and Whitfield, 1979). Three observations underline this fact. Firstly the residence times of the REE are very short (Table 10.1): although it should be pointed out that these times may be minimal values, e.g., studies of Nd isotopes (Goldstein and O'Nions, 1981) suggest that the residence time of Nd given in the table is too low by a factor of about ten; possibly the residence times of the other REE are similarly too low. Secondly, the concentrations of the REE in seawater are low relative to their crustal abundances, despite their ability to form complexes in seawater especially with carbonate ions (Table 10.1). And thirdly, the REE are very undersaturated in seawater. The removal of the REE from seawater is discussed below.

Unlike river water, seawater is markedly depleted in Ce (Figs. 10.4b and 10.5). This fractionation of Ce relative to the other REE is due to Ce being removed rapidly, relative to the other REE, from the oceans, as the residence times of the REE indicate (Table 10.1). Goldberg (1961) was the first to suggest an explanation for this, which has not been superseded. He proposed that Ce^{3+} in the oceans is oxidized to Ce^{4+} and is precipitated from solution as CeO_2, while the other REE remain in the 3+ state and are lost from solution without discernible fractionation of other individual REE. An experimental study by Carpenter and Grant (1967) supports this idea. These workers found that Ce^{3+} rapidly forms colloidal ceric hydroxide in seawater with a pH of 8 or more.

A number of observations suggest that Ce removal from seawater probably occurs in the open ocean rather than in estuarine or shelf waters. Martin et al. (1976) found that throughout the Gironde Estuary the Ce/La ratio remains constant, indicating that Ce is not removed from solution in preference to La. A sample of seawater from the Barent Sea analysed by Høgdahl et al. (1968) exhibits no Ce anomaly, and nearshore waters sampled from the east coast of the United States (Carpenter and Grant, 1967) are 10—100 times

TABLE 10.1

The concentrations (Brewer, 1975), residence times (Goldberg et al., 1963) and speciation (Turner and Whitfield, 1979) of the REE in seawater.

Element	Concentration (ng l^{-1})	Residence time (years)	Seawater speciation (%) for trivalent REE					
			free	OH	F	Cl	SO_4	CO_3
La	3	440	31	4	1	16	16	33
Ce	1	80	22	4	1	11	21	41
Pr	0.6	320						
Nd	3	270						
Sm	0.05	180						
Eu	0.01	300	11	7	1	5	8	68
Gd	0.7	260	11	4	1	6	13	65
Dy	0.9	460						
Ho	0.2	530						
Er	0.8	690						
Tm	0.2	1800						
Yb	0.8	530						
Lu	0.2	450	7	7	1	2	3	81

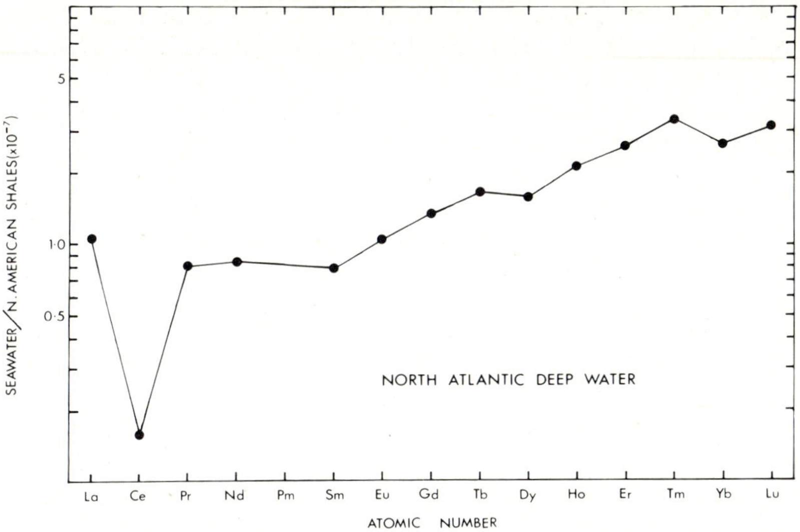

Fig. 10.5. Average REE abundances in eleven samples of North Atlantic Deep Water (Høgdahl et al., 1968) compared to a composite of North American shales (Haskin et al., 1968).

richer in Ce than samples from the adjacent Atlantic Ocean. Indirect evidence for the idea comes from minerals. The fact that they found that deep-sea cherts have negative Ce anomalies but ones formed in extensive shelves do not, led Shimizu and Masuda (1977) to suggest that Ce is depleted in the waters of open ocean but not those of shallow seas. Fleet et al. (1980) speculated along similar lines with respect to glauconites they had analysed.

The relatively few available data suggest that different water masses have distinct REE contents. Goldberg et al. (1963) like Balashov and Khitrov (1961), found that the concentrations of REE in deep water were markedly greater than those of the surface waters. Høgdahl et al. (1968) considered these data, together with their own data for samples from the Atlantic Ocean, and ones for the Indian Ocean (Balashov and Khitrov, 1961), and concluded that there is a strong relationship between REE patterns and water masses. Differences in the isotopic compositions of Nd in different ocean masses reinforces this conclusion (Piepgras et al., 1979).

Piper (1974a) considered the data of Balashov and Khitrov (1961), Goldberg et al. (1963), and Høgdahl et al. (1968). Acknowledging the paucity of the information he suggested that the deep waters of the Pacific contain higher concentrations of the REE than Atlantic deep waters. On the basis of these observations he reiterated the idea first put forward by Goldberg et al. (1963) that the REE are released into solution at depth in the oceans. This may be due to the release of the REE from lithogenous material or to the solution of biogenic material such as calcite or organic tissue. Piper favoured the latter explanation being the predominant cause, because nitrate and phosphate, two of the most important biolimiting species in ocean waters, are also more abundant in Pacific than Atlantic deep waters. Many new data are needed to test this conclusion.*

REE removal from seawater

Various processes must remove the REE from seawater and so control the REE concentrations in seawater. The possible processes could be simple inorganic precipitation resulting from the concentration limits of the insoluble salts of the REE being exceeded, the incorporation of the REE in biogenic material or hydrogenous minerals, halmyrolytic reactions between seawater and lithogenous material, and, lastly, interaction between seawater in hydrothermal solutions and the igneous oceanic lithosphere at ocean ridges. The first of these processes can be ruled out as being significant, because the concentrations of the REE in seawater are several orders of magnitude less than the concentration limits of their insoluble salts (Turner and Whitfield, 1979).

Theoretical considerations suggest that it is the ability of the REE to enter biogenic and hydrogenous phases which maintains their low concentrations in seawater and their rapid removal from solution. They do not point, however, to the REE entering preferentially either biogenic phases or hydrogenous minerals, and so do not resolve the cause of Piper's (1974a) observations which were presented at the end of the previous section. Turner and Whitfield (1979) have considered the behaviour of the REE in seawater and succinctly argued that it is the solid-state chemistry of the REE, not their aqueous chemistry, which controls their concentrations in seawater. They proposed that the preferential incorporation of the REE into biogenic and hydrogenous minerals can be explained in terms of the ion capture concepts of Goldschmidt (1954). Also, according to these authors, the LREE are incorporated more readily than the HREE into biogenic and hydrogenous minerals. This is because the effect of the coordination number, which is highest for the larger LREE, predominates over the effect of the polarizing power, which increases with atomic number. In the rest of this section the roles in removing the REE from seawater of, firstly, biogenic phases, and, then, halmyrolysates and other hydrogenous minerals, including metalliferous ridge sediments are discussed. Finally the problem of the preferential removal of Ce from seawater is reviewed.

**Note added in proof*: Since this review was written an important paper by Elderfield and Greaves (1982) on the REE in seawater has appeared which adds detail to the above discussion and highlights the needs for further research. The distribution of the REE through a water column 4.5 km deep in the eastern Central Atlantic is reported and the processes governing this distribution are discussed. Elderfield and Greaves also consider the oceanic REE cycle and conclude that the REE are rapidly cycled in the marine environment.

Although the REE contents of oceanic oozes have been determined it is impossible to use the analytical data when considering biogenic minerals as "sinks" for the REE in seawater because the oozes invariably contain lithogenous and, very probably, hydrogenous materials. Even the REE contents of calcareous oozes containing more than 90% $CaCO_3$ increase with increasing Al content, suggesting that the REE are mainly contained in any lithogenous phases (Fleet, 1977). Spirn (1965) tried to avoid this obvious "contamination" by carefully separating and cleaning *Globigerina* tests from samples of a core. She also analysed some shells and other biogenic materials from shallow-water shelf environments. Her analyses of the *Globigerina* indicate two conclusions. Firstly the *Globigerina* tests have higher REE contents than seawater (cf. Figs. 10.5 and 10.6), therefore they must concentrate the REE from seawater. And secondly the REE in the *Globigerina* are fractionated relative to one another like sediments in general (Fig. 10.1), not like seawater (Fig. 10.5). This suggests that the LREE are preferentially incorporated into these plankton relative to the HREE. Spirn's analyses of other biogenic materials, which included invertebrate shells, a coral and a swordfish bones, show the REE to be present at concentrations ten to a hundred times less than those of the *Globigerina*. These few data indicate that the growth of the calcareous plankton tests constitute the main way in which organisms remove REE from seawater.

How effective plankton are in doing this must remain open to question. Elderfield et al. (1981a) reported an analysis of a foraminiferid which contrasts with Spirn's data in two respects. Firstly they found that their sample contains at least one third to one quarter of the REE abundances which Spirn reported. Secondly, and more importantly, they showed that their sample gives a normalized plot (Fig. 10.6) which exactly parallels that of seawater (Fig. 10.5). This suggests that in being incorporated in calcareous planktonic tests from seawater the REE are not, as Spirn's work suggests, fractionated relative to one another, and therefore the carbonate of the plankton does not incorporate preferentially any REE, including Ce. Siliceous plankton must also play a role in the oceanic budget of the REE but as they only contain REE abundances similar to those of calcareous tests (Piper, 1974a; Elderfield et al., 1981a), and as they are significantly less abundant in pelagic sediments on a global scale than their calcareous equivalents, this role must be a relatively minor one.

Although the work carried out by Spirn (1965) showed that marine organisms rather than plankton have very low REE concentrations, one type of biogenic material, fish debris, is important in scavenging the REE from seawater (Bernat, 1975; Fig. 10.6). Arrenhius et al. (1957) first suggested that this is so. They proposed that biogenic apatite takes up the REE from seawater, and, where the apatite is buried it acts as a "sink" for the REE, but where the sedimentation rate is low it dissolves. Elderfield et al. (1981a), however, have concluded that the relationship between sedimentation

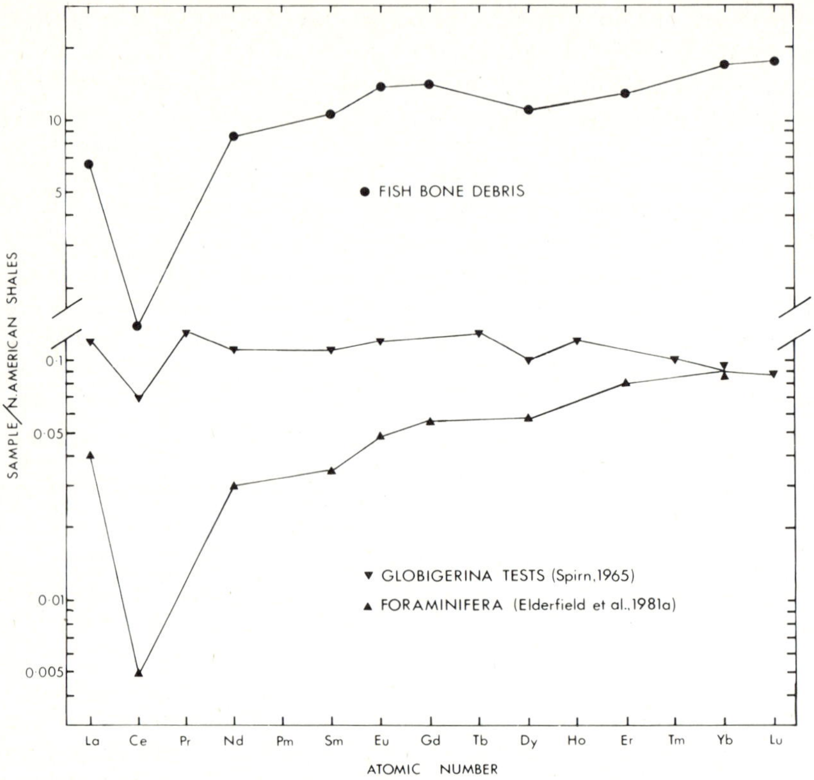

Fig. 10.6. REE abundances in fish bone debris (biogenic apatite) (Bernat, 1975), *Globigerina* tests (Spirn, 1965), and Foraminifera (Elderfield et al., 1981a) compared to a composite of North American shales (Haskin et al., 1981).

rate and REE uptake is in fact the reverse: where sedimentation is slow biogenic apatite is abundant and effectively concentrates the REE.

Assessment of halmyrolytic reactions involving lithogenous material derived from the continents is problematical because the evidence is disparate and few data are available. Estuarine halmyrolysis may lead to the removal of REE from solution, but as discussed above, this is unlikely to be significant in controlling the REE contents of seawater. Evidence of halmyrolysis in the open ocean involving sediments derived from the continents is equivocal. Martin et al. (1976), on the basis of mass balance considerations, have even argued that the REE may, in fact, be released into seawater rather than be removed from it (see below), but no definitive data exist.

Only three slender lines of evidence concerning the halmyrolysis of continental sediment are available at the time of writing. The first concerns analyses of sediments collected by sediment traps. Brewer et al. (1980)

reported results from traps deployed in the North Atlantic, which showed that the behaviour of La correlates with that of Al. This suggests that at least La is unaffected by halmyrolysis during deposition. The second tentative line of evidence comes from studies of glauconite. The available evidence indicates that, during the formation of glauconite, its precursor, which is presumably continental lithogenous material, incorporates REE from seawater. This is either a significant process, especially for Ce (Balashov and Kazakov, 1968), or it may just involve minor REE halmyrolysis (Fleet et al., 1980). The third line of evidence relates to adsorbed REE and REE in amorphous phases. In the case of amorphous material it is, of course, impossible, especially in a general review of this kind, to distinguish between the hydrogenous amorphous phases formed in the oceans and the lithogenous ones transported through estuaries. Spirn (1965) concluded that more than half the REE contents of marine clays is adsorbed on clays. O.T. Høgdahl is reported in Piper (1974a) as finding that about 50% of the REE in pelagic sediments are leachable by HCl, and that the REE in this soluble fraction give the mirror image of seawater on a normalized plot, except that Ce is not enriched. These data lead to the tentative suggestion that REE adsorption onto clays in the oceans might be an important process of REE removal from seawater. Piper (1974a) suggested that the amorphous phases of marine sediments may also be important in controlling the REE contents of seawater. He estimated that about 10% of the REE in pelagic sediments are contained in these phases, and that they are enriched in Ce. Careful studies are required if we are to learn what parts adsorption from seawater and the formation of amorphous phases in the oceans play in REE geochemistry. These studies could involve repeated selective leaching measurements (e.g., Chester and Hughes, 1967). The use of these techniques might allow the REE contents of particulate material in rivers to be compared with those of suspended matter in seawater.

The evidence for halmyrolytic reactions involving the REE and volcanic lithogenous material which is largely derived from within ocean basins is clearer. Such volcanic material undergoes halmyrolysis to yield a variety of clays, zeolites and other minerals. Smectites from the South Pacific formed by this process are depleted in Ce and enriched in the HREE relative to the other REE (Fig. 10.7) (Piper, 1974a; Courtois and Hoffert, 1977). Phillipsite, which is also believed to form by the halmyrolysis of volcanic material, contains REE which are fractionated relative to one another in a manner similar to that shown by the smectites (Fig. 10.7) (Piper, 1974b; Bernat, 1975). The REE contents of the South Pacific smectites and of phillipsite, especially their deficiency in Ce, suggest that the minerals have taken up REE from seawater during their formation. It is difficult, though, to assess the relative importance of the formation of these minerals to the oceanic REE budget, because one cannot be certain about the REE contents of their precursors. If they form mainly from volcanic material derived from volcanic

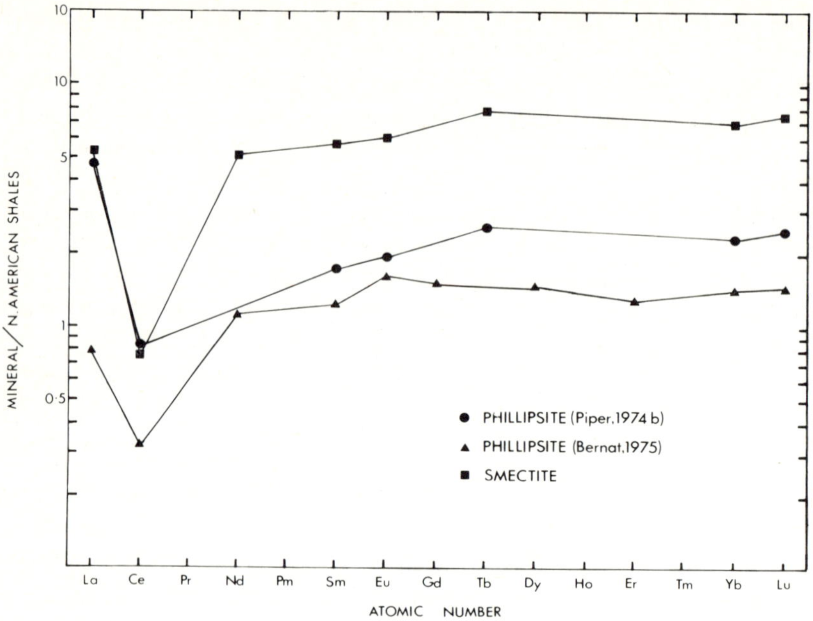

Fig. 10.7. REE abundances in phillipsites (Piper, 1974b; Bernat, 1975) and smectite from the South Pacific (Piper, 1974a) compared to a composite of North American shales (Haskin et al., 1968).

islands the REE contents of their precursors would be broadly similar to those of sediments in general. Therefore, as the REE abundances in the smectites are not dissimilar to those of sediments in general (Fig. 10.7), the uptake of REE per unit weight of smectite is probably small. The evidence relating to phillipsite, however, is equivocal. Piper's (1974b) work suggests that the REE are concentrated in phillipsite relative to sediments in general (Fig. 10.7) but Bernat's (1975) data indicate that little, if any concentration occurs. The relative abundances of halmyrolytic smectites and phillipsite in the oceans has also to be taken into account. Piper (1974a), working on the basis that the rate of sedimentation of the smectite is ten to hundred times greater than that of phillipsite, proposed that the former removed two and a half times as much of the REE from seawater as the latter (Table 10.2). Submarine weathering of basalts to give smectite and zeolites may have a similar effect but, as discussed above, the quantitative role of this process in removing the REE from seawater is unknown at present.

Other hydrogenous minerals, which may be pure precipitates or at least partly halmyrolysates (Elderfield, 1976), must help in the removal of the REE from seawater. Barite, and the minerals of phosphorites, ocean ridge hydrothermal sediments and manganese nodules are the most important of these hydrogenous minerals. Deep-sea barites contain higher abundances of

TABLE 10.2

REE sedimentation rates in the oceans (Martin et al., 1976; Piper, 1974a)

Phase in which REE removed	Annual REE sedimentation rate (10^{-9} g cm^{-2})								
	La[a]	La[b]	Ce[b]	Nd[b]	Sm[b]	Eu[b]	Tb[b]	Yb[b]	Lu[b]
Biogenic $CaCO_3$	0.06	0.49	0.36	0.46	0.08	0.02	0.01	0.04	0.01
Biogenic SiO_2		0.10	0.14	0.09	0.03	0.01	0.00	0.02	
Ferromanganese nodules	0.05	0.14	0.44	0.16	0.03	0.01	0.01	0.01	
Fish bones	0.30								
Montmorillonite	1.60	1.26	1.79	1.27	0.28	0.07	0.06	0.22	0.04
Phillipsite	0.25	0.50	0.18	0.50	0.10	0.02	0.02	0.07	0.01
Amorphous phases		0.42	1.64	0.45	0.08	0.02	0.02	0.03	0.01
Total	2.26	2.91	4.55	2.90	0.59	0.15	0.12	0.38	0.07

[a]Martin et al. (1976).
[b]Piper (1974a) for Pacific Ocean.

the REE than sediments in general (Fig. 10.8) (Piper, 1974a; Guichard et al., 1979), suggesting that the REE must be concentrated in them from sea-water. This concentration process may involve simple precipitation (Piper, 1974a), or it may involve an intermediate step during which the REE are adsorped onto and then desorped from sediments (Guichard et al., 1979).

The REE contents of marine phosphorites vary considerably but on average are enriched relative to sediments in general (Fig. 10.9) (Goldberg

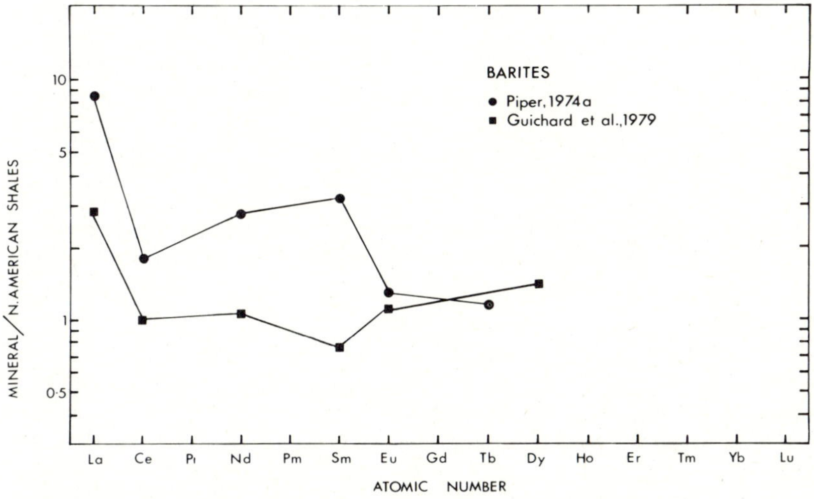

Fig. 10.8. REE abundances in barites (Piper, 1974a; Guichard et al., 1979) compared to a composite of North American shales (Haskin et al., 1968).

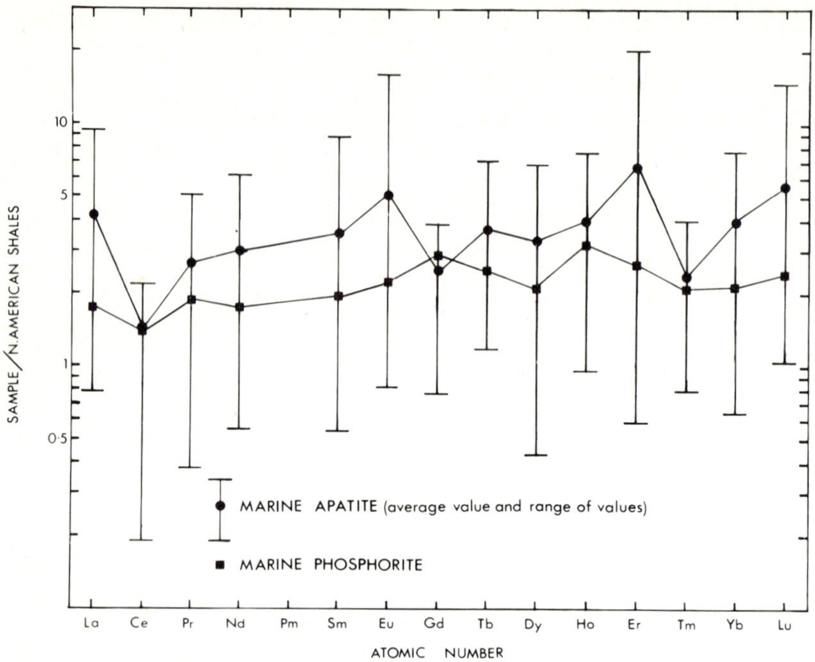

Fig. 10.9. REE abundances in marine phosphorite (Goldberg et al., 1963), and the average and range of REE abundances in marine apatite (Altschuler, 1980) compared to a composite of North American shales (Haskin et al., 1968).

et al., 1963; Altschuler et al., 1967; Altschuler, 1980). This variability seems to be due to the dilution of phosphatic minerals by lithogenous, biogenic and other hydrogenous phases: the REE are generally more concentrated in separated marine apatites than in bulk phosphorites (Fig. 10.9) (Altschuler, 1980). Phosphorites can be depleted in Ce (Altschuler et al., 1967). Altschuler (1980) attributed this deficiency and the enrichment of REE in phosphorites to precipitation or fixation of the REE from a seawater source. Not all phosphorites, however, are depleted in Ce. Those that are not may be the phosphorites which form beneath coastal upwelling systems, though more data are needed to test this correlation (H.H. Veeh, personal communication, 1981).

The REE contents of ocean ridge metalliferous sediments were first used as contributory evidence to show that these sediments form as precipitates from hydrothermal solutions of seawater which have circulated through the oceanic crust; in particular, the negative Ce anomalies exhibited by the normalized plots for these sediments were taken as indicating a seawater source at least for the REE of the sediments (e.g., Bender et al., 1971; Dymond et al., 1973; Piper and Graef, 1974). The authenticity of this process is now well established from various lines of evidence (e.g., Elderfield, 1976).

The high abundances of the REE in ocean ridge metalliferous sediments relative to those in seawater (Fig. 10.10) indicate that these sediments must scavenge significant quantities of REE from seawater. Evidence from ophiolites has shown that this scavenging can go on within the crust where oxyhydroxides are formed from hydrothermal seawater (e.g., Robertson and Fleet, 1976). The precipitation of sulphides formed from these solutions plays virtually no part in this process (Robertson and Fleet, 1977), but those "ochres" which formed from the oxidation of these sulphide scavenge REE like the other oxyhydroxides.

The chemical conditions prevailing in hydrothermal systems must control the nature and extent of REE uptake by ridge precipitates. Not all these sediments are as rich in the REE or have REE fractionated relative to another as those which were first described and are referred to above. Metalliferous

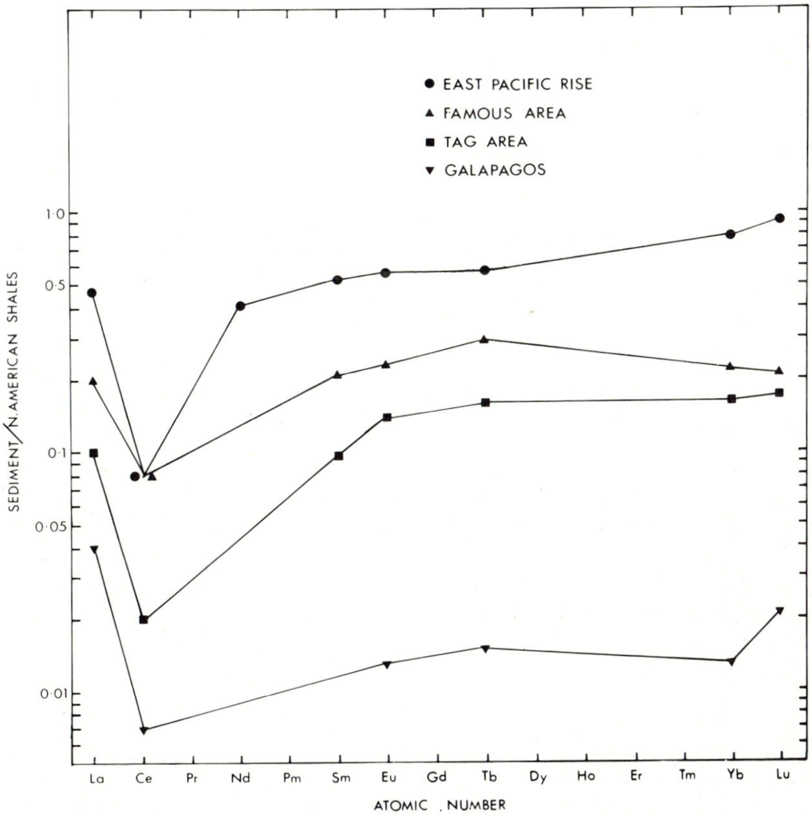

Fig. 10.10. REE abundances in East Pacific Rise crestal sediments from 39°S (average value, Piper and Graef, 1974) and hydrothermal crusts from the FAMOUS and TAG areas and the Galapagos spreading centre (samples MN 1776, MN 1749 and MN 1750 of Toth, 1980) compared to a composite of North American shales (Haskin et al., 1968).

sediments from the FAMOUS and TAG areas on the northern Mid-Atlantic Ridge and some from around the Galapagos spreading centre all have relatively low REE contents but with the REE similarly fractionated relative to one another as in seawater (Fig. 10.10) (Toth, 1980; Bonnot-Courtois, 1981; Barrett et al., 1983). Those from the Atlantis II and Chain Deeps in the Red Sea (Courtois and Treuil, 1977), from the Galapagos mounds area (Corliss et al., 1978), and from Santorini (A.J. Fleet, unpublished data) are, however, distinctly different (Fig. 10.11). Obviously care must be exercised in considering the quantitative role of metalliferous sediments in removing the REE from seawater.

Manganese nodules have been the subject of far more REE studies and reviews than any other material in the aqueous and sedimentary environments. A number of extensive and thorough reviews exist (Volkov and Fomina,

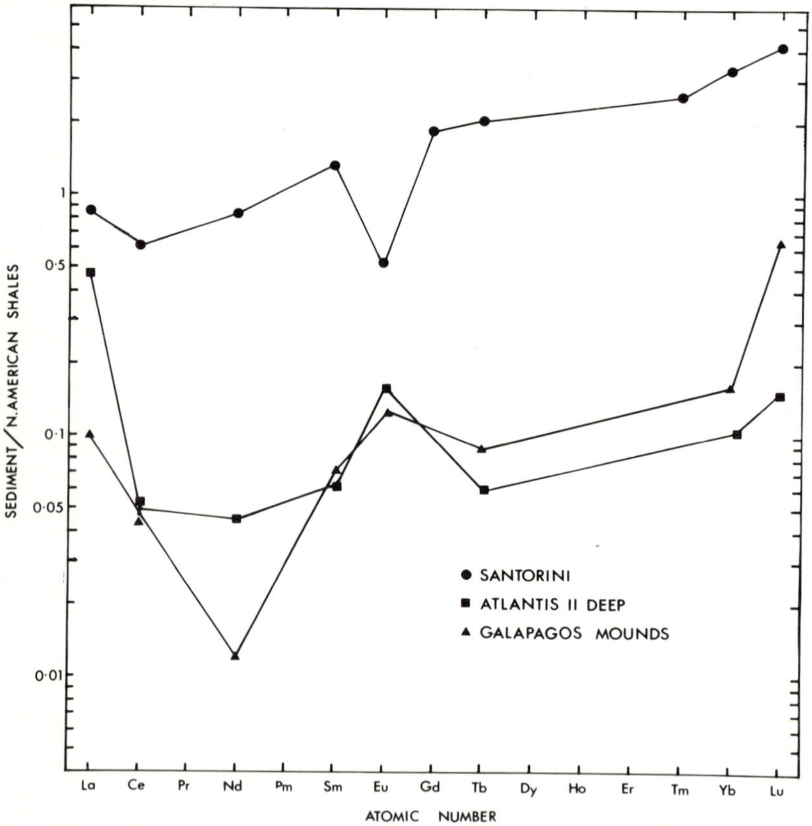

Fig. 10.11. REE abundances in metalliferous sediments from the Galapagos mounds (sample TB1 of Corliss et al., 1978), the Atlantis II Deep (280 cm depth from core 128P of Courtois and Treuil, 1977) and Santorini (A.J. Fleet, unpublished data) compared to a composite of North American shales (Haskin et al., 1968).

1967; Ehrlich, 1968; Glasby, 1973; Piper, 1974b), and various detailed regional studies have been carried out (e.g., Addy, 1979; Rankin and Glasby, 1979; Courtois and Clauer, 1980; Elderfield et al., 1981b). The major relevant features of these studies are now outlined.

Oceanic manganese nodules are enriched in the REE relative to sediments in general. In particular they are enriched in Ce and their normalized plots exhibit positive Ce anomalies (Fig. 10.12), except in a few cases which contain a significant hydrothermal component and exhibit negative Ce anomalies (Elderfield and Greaves, 1981). Otherwise, nodules have similar REE normalized plots to sediments (Fig. 10.12), as was first shown by Goldberg et al. (1963) and Spirn (1965). Being largely precipitates, rather than halmyrolysates, and widespread in occurrence, manganese nodules must act as major "sinks" for the REE, and especially for Ce. Various processes have been put forward to explain how the REE become incorporated into nodules. These have been clearly summarised by Elderfield et al. (1981a) thus:

"(1) The REE have a direct seawater source, being incorporated into the nodular oxyhydroxides by direct precipitation of the trivalent RE hydroxides (Glasby, 1973), by coprecipitation with particulate iron colloids or other nodular phases (Goldberg, 1954; Goldberg et al., 1963; Glasby, 1973; Piper, 1974b), or by an unspecified mechanism (Addy, 1979).

(2) The REE have a seawater source but are incorporated by coprecipitation after their release from biogenic carrier phases in abyssal depths, either calcite (Piper, 1974b) or fishbone apatite (Arrhenius et al., 1957; Arrhenius and Bonatti, 1965) or indeed from inorganic particles (Fomina, 1966).

(3) The REE are derived from the underlying sediments, the host phases being inorganic detrital minerals, and are incorporated into the nodules by surface exchange processes (Ehrlich, 1968; Bender, 1972).

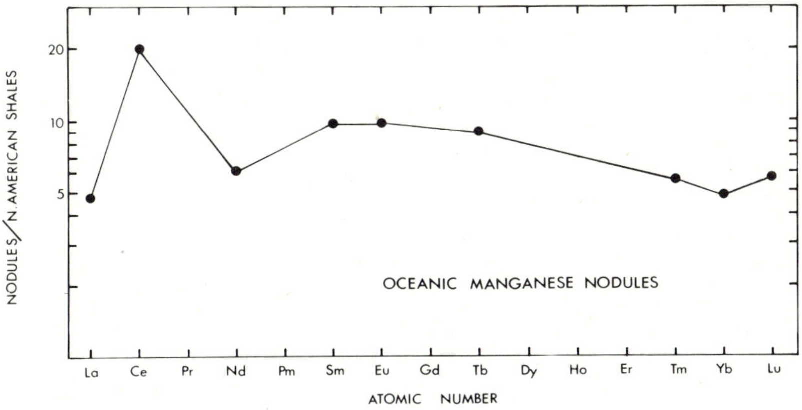

Fig. 10.12. Average REE abundances in oceanic manganese nodules (Ehrlich, 1968) compared to a composite of North American shales (Haskin et al., 1968).

Piper (1974b) also proposed a polygenetic origin whereby Pacific nodules shallower than 3000—3500 m (the approximate depth of the lysocline) gained their REE directly from seawater whereas deeper nodules gained their REE from dissolving biogenous calcite".

Elderfield et al. (1981a, b), themselves, studied a suite of Pacific nodules. They suggested that the REE contents of this suite, which is mainly composed of todorokite, reflect the proportions of a Fe-rich phase and a P-rich phase in the nodules. They believe that the Fe-rich phase has REE contents, which give a normalized plot mirroring that of seawater, while a plot of the REE contents of the P-rich phase parallels that of biogenic apatite (Fig. 10.6).

The removal of Ce from seawater has been alluded to here but not discussed. Obviously manganese nodules act as a major repository for Ce. Elderfield et al. (1981a) suggest that Ce might be associated mainly with Fe oxyhydroxide flocs and so be incorporated directly into nodules, unlike the other trivalent REE which Elderfield et al. consider probably undergo some cycling through surface sediments. Other minerals beside those of manganese nodules may preferentially incorporate Ce but this effect is difficult to assess. If such minerals take up the REE from seawater without preferential incorporation of Ce their REE contents should just reflect those of seawater. The data of Elderfield et al. (1981a) suggest this is so for biogenic calcite, but earlier work by Spirn (1965) indicate that Ce is preferentially taken up (Fleet et al., 1980). The problem of interpreting data relating to this dilemma arises from the fact that any lithogenous material in biogenic or hydrogenous phases will tend to raise preferentially the Ce content of these phases. It is impossible, therefore, to ascertain whether or not preferential Ce uptake by a phase has occurred unless the degree of "lithogenous contamination" is first assessed. In the case of biogenic minerals it is just possible to eliminate this "contamination" by exhaustive cleaning, but in the case of hydrogenous minerals this will be impossible as most are, at least in part, halmyrolysates. In the latter case, therefore, the contribution to the total Ce content may be any lithogenous material present. This aspect will have to be assessed in future studies if the chemistry of Ce in the oceans is to be evaluated in a truly quantitative fashion.

REE mass balances of the oceans

The lack of data makes it difficult to ascertain by mass balance calculations whether or not the known supply of REE to the oceans is sufficient to account for recognized REE deposition in the marine environment. Martin et al. (1976) acknowledged this but produced a mass balance for La (Table 10.2). Piper (1974a) had previously attempted a more ambitious balance for the REE as a whole (Table 10.2). He considered that the igneous oceanic lithosphere, as well as the continents, contribute REE to the marine environment, and calculated that the rate of accumulation of the REE in oceanic sediments is approximately twelve times the rate of input from both sources. Martin et al. (1976) found a deficit of a similar magnitude in the supply of La for rivers alone. They hesitated, though, to attribute any significance to this since so few data were available, and they, like Piper (1974a), only had data relating to the Garonne and Dordogne rivers and had to take these as being typical. Piper (1974a) proposed that desorption of the REE in estuaries makes up the deficit, but Martin et al. (1976) concluded that the opposite is true and the REE are, in fact, removed from solution in estuaries. As an alternative explanation they suggested that minor solution of, or desorption from, detrital material in the oceans would easily supply the deficit, if indeed there is one.

Obviously far more information is needed if a more accurate mass balance for the REE in the oceans is to be estimated. Apart from the need for further data concerning the supply of REE from rivers and estuarine processes, there seem to be three major areas where our knowledge is significantly lacking. Firstly, we need to know more of the behaviour of particulate matter in the oceans. Work in this field is already in hand, for instance through the GEOSECS programme. Secondly, we must define the ranges of REE contents of marine biogenic and hydrogenous material more rigorously. Thirdly, a quantitative assessment of the role of the igneous oceanic lithosphere needs to be made. On the one hand we need to know to what extent submarine weathering of igneous rocks act as a sink for the REE in seawater. And on the other we must assess how much, if at all, the igneous oceanic lithosphere supplies REE to the oceans. Overall does seawater gain or lose REE while undergoing hydrothermal circulation at ocean ridges and, if it does gain REE, does the gain have a significant effect on the composition of seawater or are the REE lost at or near the ridges by incorporation in metalliferous sediments?

10.4. REE contents of sediments and sedimentary rocks

The REE contents of sediments and sedimentary rocks naturally reflect the mineral contents of these deposits, and hence the processes by which the minerals formed and were incorporated into the deposits. "Shales", as mentioned at the beginning of this review, generally contain similar absolute REE abundances and have their REE fractionated relative to another similarly (Fig. 10.1) (e.g., Haskin and Gehl, 1962; Wildeman and Haskin, 1965; Spirn, 1965; Haskin and Haskin, 1966; Haskin et al., 1966). This is true for Precambrian sediments as well as for Phanerozoic ones, although the former seem to be enriched in Eu when post-Archean in age and neither enriched nor depleted in Eu when Archean in age (Fig. 10.13) (Wildeman and Condie, 1973; Wildeman and Haskin, 1973; Jenner et al., 1981). This characteristic of "shales" is generally taken to indicate that fine-grained sedimentary material is efficiently mixed during weathering, erosion and transport. The

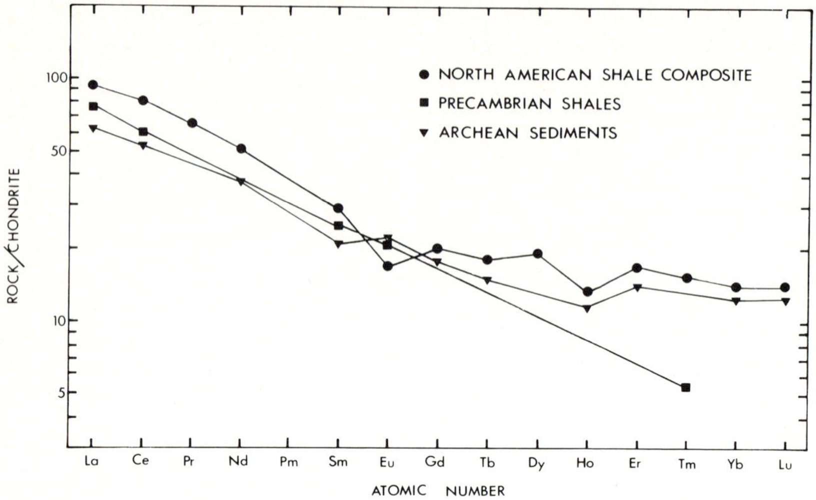

Fig. 10.13. Average REE abundances in five Precambrian shales (Wildeman and Condie, 1973), a composite of North American shales (Haskin et al., 1968), and Archean sediments from the Yellow Knife Supergroup (Slave Structural Province, Canada) (Jenner et al., 1981) compared to chondritic meteorites (Wakita et al., 1971).

REE abundances of "shales" are, therefore, taken to represent the REE contents of the upper continental crust, as Goldschmidt (1954) originally suggested. Hence the differences between the REE of Phanerozoic and Precambrian sediments presumably reflects differences in the REE compositions of the crust with time (e.g., Wildeman and Condie, 1973; Jenner et al., 1981). The absolute values are not beyond dispute (e.g., McLennan et al., 1980), but the concept is widely accepted.

A few detailed studies have shown that variations in the REE contents of "shales" can sometimes occur. The Lower Permian Havensville and Eskridge shales of Kansas and Oklahoma have REE contents which seem to vary in response to differing source areas but not to differing clay mineral contents (Cullers et al., 1975). In contrast sediments derived from very different source areas, like those deposited in the Kimmeridgian epicontinental sea of North Europe (Dypvik and Brunfelt, 1979), can have similar REE contents. Localized inputs of volcanic material often lead to clays having distinct REE contents (e.g., Dypvik and Brunfelt, 1976; Fleet et al., 1976). Lastly pelagic clays may contain anomalous REE abundances. The REE contents of pelagic clays are often, though not invariably (e.g., Courtois and Chamley, 1978), higher than those of "shales", etc. (Chester and Aston, 1976). This is due to the presence of a significant hydrogenous component in the pelagic clays, which consists of, for instance, Fe-Mn oxyhydroxide micronodules (e.g., Elderfield, 1977), which are themselves enriched in the REE (Fig. 10.14).

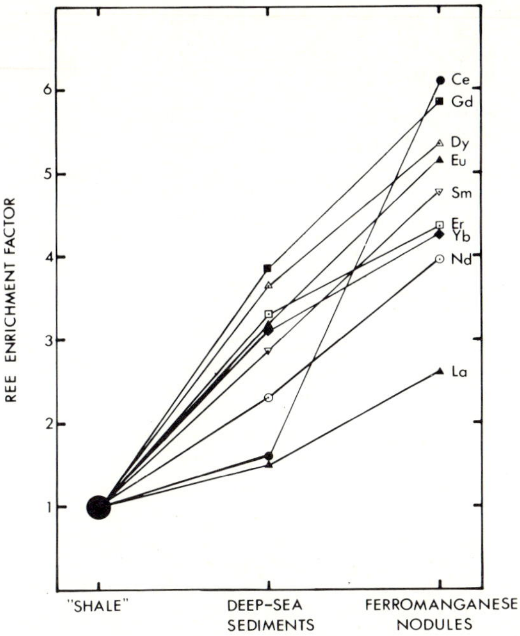

Fig. 10.14. Average REE abundances in deep-sea sediments and nodules compared to concentrations in shales (Haskin and Haskin, 1966) (after Elderfield et al., 1981a).

The individual REE contents in sediments and sedimentary rocks, other than those of aluminosilicate mudrocks, are generally fractionated relative to one another as those of "shales". The absolute REE abundances in these sediments, though, are less than those in "shales" and decrease in the order of greywackes, other sandstones and limestones (Fig. 10.15) (Haskin et al., 1966). This decrease in REE concentration presumably reflects the decreasing clay mineral and rock fragment contents and the increasing quartz and biogenic carbonate contents of these rocks, since the former constituents will contain higher REE contents than the latter ones.

10.5. Diagenesis

In general, the available data, which are few, suggest that the REE are unaffected by diagenesis. Sediments from the coast of the Gulf of Mexico, which have been buried to depths of up to about 5 km, contain REE abundances which seem to reflect their provenances not any diagenetic changes (Chaudhuri and Cullers, 1979). Precambrian sediments, some of which have undergone a degree of metamorphism, as well as diagenesis, have REE which are essentially the same as younger and present-day sediments (Wildeman and Condie, 1973; Wildeman and Haskin, 1973), suggesting that only minimal

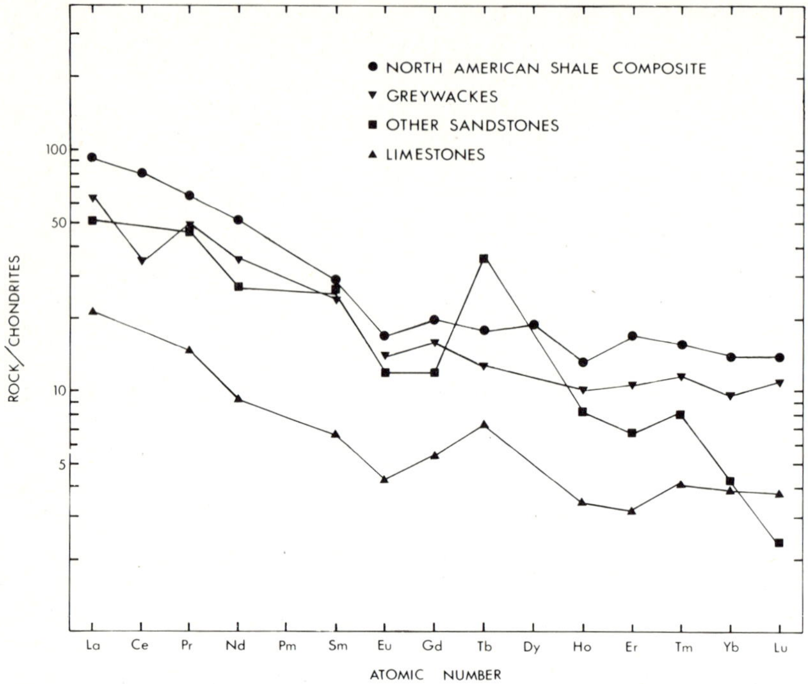

Fig. 10.15. Average REE abundances in composite of North American shales (Haskin et al., 1968), greywackes, sandstones and limestones (Haskin et al., 1966) compared to chondritic meteorites (Wakita et al., 1971).

post-depositional changes occur to the REE contents of sediments. This generally holds for biogenic as well as lithogenous sediments. For instance the suite of Precambrian sedimentary rocks studied by Wildeman and Haskin (1973) included biogenic sediments, and deep-sea cherts analysed by Shimizu and Masuda, (1977) appear to have formed from siliceous oozes without any change in their REE contents. REE mobilization, however, may take place during the metamorphic or hydrothermal alteration of limestones (Jarvis et al., 1975; Bellanca et al., 1981).

Decreases in pH to give acidic conditions might be expected to cause mobilization of the REE. There is no direct evidence, though, for this occurring during diagenesis, but the behaviour of the REE during weathering and erosion and the experimental data discussed above suggest that this is a plausible extrapolation. Furthermore, experimental work by Balashov and Girin (1969) showed that 20—95% of the REE of clays are readily leachable and, therefore, may be available for migration during diagenesis: the intermediate REE being most susceptible and the LREE least. Roaldset and Rosenqvist (1971b) obtained similar results which showed that about 80% of the REE on some lacustrine clays were adsorbed rather than held in the structures of the clays.

Diagenetic changes in Eh are most likely to effect Ce and Eu because of their ability to exist in different oxidation states under geological conditions. Courtois and Clauer (1980) found that the Ce contents of the layers in some nodules which they studied varied and speculated that this was a result of diagenetic Eh changes. No equivalent speculations concerning the diagenetic mobilization of Eu have been published, although Jarvis et al. (1975) suggested that metamorphism or metasomatism may cause Eu to be reduced from the 3+ to the 2+ oxidation state.

Acknowledgements

A number of people kindly responded to my requests for reprints, preprints and other relevant information. In particular Chantal Bonnot-Courtois, Hein de Baar and Elen Roaldset provided a wealth of invaluable information, most of which I have only touched on here without doing it fully justice. I would also especially thank Harry Elderfield for supplying an early reprint of one of his papers, and for making useful, pertinent comments on a draft of the paper. Lastly I thank Doreen Norman for her patient typing of this paper, and the management of the British Petroleum Company Ltd. for permission to publish it.

References

Addy, S.K., 1979. Rare earth element patterns in manganese nodules and micronodules from northwest Atlantic. *Geochim. Cosmochim. Acta*, 43: 1105—1115.

Altschuler, Z.S., 1980. The geochemistry of trace elements in marine phosphorites, 1. Characteristic abundances and enrichment. *Soc. Econ. Paleontol. Mineral., Spec. Publ.* 29: 19—30.

Altschuler, Z.S., Berman, S. and Cuttitta, F., 1967. Rare earths in phosphorites — geochemistry and potential recovery. *U.S. Geol. Surv. Prof. Paper*, 575-B: 1—9.

Arrhenius, G. and Bonatti, E., 1965. Neptunism and volcanism in the oceans. *Progr. Oceanogr.*, 3: 7—22.

Arrhenius, G., Bramlette, M.N. and Piciolto, E., 1957. Localization of radioactive and stable heavy nuclides. *Nature*, 180: 85—86.

Aston, S.R. and Chester, R., 1973. The influence of suspended particles on the precipitation of iron in natural waters. *Estuarine Coastal Mar. Sci.*, 1: 225—231.

Balashov, Yu.A. and Girin, Yu.P, 1969. On the reserve of mobile rare earth elements in sedimentary rocks. *Geochem. Int.*, 7: 649—659 (translation).

Balashov, Yu.A. and Kazakov, G.A., 1968. Source of the rare earths in Pacific Ocean glauconite. *Dokl. Acad. Sci. USSR*, 179: 440—442.

Balashov, Yu.A. and Khitrov, L.M., 1961. Distribution of the rare earths in the waters of the Indian Ocean. *Geochem. Int.* 9: 877—890 (translation).

Balashov, Yu.A., Ronov, A.B., Migdisov, A.A. and Turanskaya, N.V., 1964. The effect of climate and facies environment on the fractionation of the rare earths during sedimentation. *Geochem. Int.*, 10: 995—1014 (translation).

Barrett, T.J., Fleet, A.J. and Friedrichsen, 1983. Major element, rare earth element, and

O- and H-isotopic composition of metalliferous and pelagic sediments from the Galapagos mounds area, Leg 70. In: *Initial Reports of the Deep Sea Drilling Project, 70.* U.S. Government Printing Office, Washington, D.C. (in press).

Bellanca, A., Di Salvo, P., Moller, P., Neri, R. and Schley, F., 1981. Rare-earth and minor element distribution and petrographic features of fluorites and associated Mesozoic limestones of northwestern Sicily. *Chem. Geol.*, 32: 255—269.

Bender, M.L., 1972. Mechanisms of trace metal removal from the oceans. In: D.R. Horn (Editor), *Ferromanganese Deposits on the Ocean Floor.* National Science Foundation, Washington, D.C., pp. 73—80.

Bender, M.L., Broecker, W., Gornitz, V., Middel, U., Kay, R., Sun, S.S. and Biscaye P., 1971. Geochemistry of three cores from the East Pacific Rise. *Earth Planet. Sci. Lett.*, 12: 425—433.

Bernat, M., 1975. Les isotopes de l'uranium et du thorium et les terres rares dans l'environnement marin. *Cah. ORSTOM Sér. Geol.*, 7: 65—83.

Bonnot-Courtois, C., 1980. Le comportement des terres rares au cours de l'altération sous-marine et ses conséquences. *Chem. Geol.*, 30: 119—131.

Bonnot-Courtois, C., 1981. Distribution des terres rares dans le dépôts hydrothermaux de la zone FAMOUS et des Galapagos — comparaison avec les sédiments métallifères. *Mar. Geol.* 39: 1—14.

Brewer, P.G., 1975. Minor elements in seawater. In: J.P. Riley and G. Skirrow (Editors), *Chemical Oceanography, 1.* Academic Press, New York, N.Y., 2nd ed., 451—489.

Brewer, P.G., Nozaki, Y., Spencer, D.W. and Fleer, A.P., 1980. Sediment trap experiments in the deep North Atlantic: isotopic and elemental fluxes. *J. Mar. Res.*, 38: 703—728.

Brown, R.E., Parker, H.M. and Smith, J.M., 1955. Disposal of liquid wastes to the ground. In: *U.N. Internal Conference on the Peaceful Uses of Atomic Energy, 9.* United Nations, New York, N.Y., pp. 669—675.

Burkov, V.V. and Podporina, Ye.K., 1967. Rare earths in granitoid residuum. *Dokl. Acad. Sci. USSR*, 177: 691—694.

Carpenter, J.H. and Grant, V.E., 1967. Concentration and state of cerium in coastal waters. *J. Mar. Res.*, 25: 228—238.

Chaudhuri, S. and Cullers, R.L., 1979. The distribution of rare-earth elements in deeply buried Gulf coast sediments. *Chem. Geol.*, 24: 327—338.

Chester, R. and Aston, S.R., 1976. The geochemistry of deep-sea sediments. In: J.P. Riley and R. Chester (Editors), *Chemical Oceanography, 6.* Academic Press, New York, N.Y., 2nd ed., pp. 281—390.

Chester, R. and Hughes, M.J., 1967. A chemical technique for the separation of ferromanganese minerals, carbonate minerals and adsorbed trace elements from pelagic sediments. *Chem. Geol.*, 2: 249—262.

Christie, O.H.J. and Roaldset, E., 1979. Geochemical behaviour of lanthanoid elements in some clays and bauxite. *Geochem. J.*, 13: 11—14.

Copeland, R.A., Frey, F.A. and Wones, D.R., 1971. Origin of clay minerals in a Mid-Atlantic Ridge sediment. *Earth Planet. Sci. Lett.*, 10: 186—192.

Corliss, J.B., Lyle, M., Dymond, J. and Crane, K., 1978. The chemistry of hydrothermal mounds near the Galapagos Rift. *Earth Planet. Sci. Lett.*, 40: 12—24.

Courtois, C. and Chamley, H., 1978. Terres rares et minéraux argileux dans le Crétacé et le Cénozoique de la marge Atlantique orientale. *C.R. Acad. Sci. Paris*, 286: 671—674.

Courtois, C. and Clauer, N., 1980. Rare earth elements and strontium isotopes of polymetallic nodules from southeastern Pacific Ocean. *Sedimentology*, 27: 687—695.

Courtois, C. and Hoffert, M., 1977. Distribution des terres rares dans les sédiments superficiels du Pacifique sud-est. *Bull. Soc. Géol. Fr.*, 19: 1245—1251.

Courtois, C. and Treuil, M., 1977. Distribution des terres rares et de quelques éléments en trace dans les sédiments récents des fosses de la Mer rouge. *Chem. Geol.*, 20: 57—72.

Cullers, R.L., Chaudhuri, S., Arnold, B., Lee, M. and Wolf, C.W. Jr., 1975. Rare earth distributions in clay minerals and in the clay-sized fraction of the Lower Permian Havensville and Eskridge shales of Kansas and Oklahoma. *Geochim. Cosmochim. Acta*, 39: 1691—1703.

Desprairies, A. and Bonnot-Courtois, C., 1980. Relation entre la composition des smectites d'altération sous-marine et leur cortège de terre rares. *Earth Planet. Sci. Lett.*, 48: 124—130.

Dymond, J., Corliss, J.B., Heath, G.R., Field, C.W., Dasch, E.J. and Veeh, H.H., 1973. Origin of metalliferous sediments from the Pacific Ocean. *Geol. Soc. Am. Bull.*, 84: 3355—3372.

Dypvik, H. and Brunfelt, A.O., 1976. Rare-earth elements in Lower Palaeozoic epicontinental and eugeosynclinal sediments from the Oslo and Tronheim regions. *Sedimentology*, 23: 363—378.

Dypvik, H. and Brunfelt, A.O., 1979. Distribution of rare earth elements in some North Atlantic Kimmeridgian black shales. *Nature*, 278: 339—341.

Ehrlich, A.M., 1968. *Rare earth abundances in manganese nodules.* Ph.D. Thesis. Massachusetts Institute of Technology, Cambridge, Mass., 225 pp. (unpublished).

Elderfield, H., 1976. Hydrogenous material in marine sediments; excluding manganese nodules. In: J.P. Riley and R. Chester (Editors), *Chemical Oceanography, 5.* Academic Press, New York, N.Y., 2nd ed., pp. 137—215.

Elderfield, H., 1977. The form of manganese and iron in marine sediments. In: G.P. Glasby (Editor), *Marine Manganese Deposits.* Elsevier, Amsterdam, pp. 269—290.

Elderfield, H. and Greaves, M.J., 1981. Negative cerium anomalies in the rare earth element patterns of oceanic ferromanganese nodules. *Earth Planet. Sci. Lett.*, 55: 163—170.

Elderfield, H. and Greaves, M.J., 1982. The rare earth elements in sea water. *Nature*, 296: 214—219.

Elderfield, H., Hawkesworth, C.J., Greaves, M.J. and Calvert, S.E., 1981a. Rare earth element geochemistry of oceanic ferromanganese nodules and associated sediments. *Geochim. Cosmochim. Acta*, 45: 513—528.

Elderfield, H., Hawkesworth, C.J., Greaves, M.J. and Calvert, S.E., 1981b. Rare earth element zonation in Pacific ferromanganese nodules. *Geochim. Cosmochim. Acta*, 45: 1231—1234.

Fleet, A.J., 1977. *Geochemistry of drilled (DSDP) sediments from the southern Indian Ocean.* Ph.D. Thesis. University of London, London, 333 pp. (unpublished).

Fleet, A.J., Henderson, P. and Kempe, D.R.C., 1976. Rare earth element and related chemistry of some drilled southern Indian Ocean basalts and volcanogenic sediments. *J. Geophys. Res.*, 81: 4257—4268.

Fleet, A.J., Buckley, H.A. and Johnson, L.R., 1980. The rare earth element geochemistry of glauconites and celadonites. *J. Geol. Soc. London*, 137: 683—688.

Fomina, L.S., 1966. Accumulation and redistribution of rare-earth elements during formation of iron-manganese concretions in the ocean. *Dokl. Acad. Nauk USSR*, 170: 1181—1184.

Frey, F.A., Haskin, M.A., Poetz, J.A. and Haskin, L.A., 1968. Rare earth abundances in some basic rocks. *J. Geophys. Res.*, 73: 6085—6098.

Frey, F.A., Bryan, W.B. and Thompson, G., 1974. Atlantic Ocean floor: geochemistry and petrology of basalts from Legs 2 and 3 of the Deep-Sea Drilling Project. *J. Geophys. Res.*, 79: 5507—5527.

Glasby, G.P., 1973. Mechanisms of enrichment of the rarer elements in marine manganese nodules. *Mar. Chem.*, 1: 105—125.

Goldberg, E.D., 1954. Marine geochemistry, 1. Chemical scavengers of the sea. *J. Geol.*, 62: 249—265.

Goldberg, E.D., 1961. Chemistry in the oceans. In: M. Sears (Editor), *Oceanography. Am. Assoc. Adv. Sci. Publ.*, 67: 583—597.

Goldberg, E.D., Koide, M., Schmitt, R.A. and Smith, R.H., 1963. Rare earth distributions in the marine environment. *J. Geophys. Res.*, 68: 4209—4217.

Goldschmidt, V.M., 1954. *Geochemistry*. Oxford University Press, Oxford, 730 pp.

Goldstein, S.L. and O'Nions, R.K., 1981. Nd and Sr isotopic relationships in pelagic clays and ferromanganese deposits. *Nature*, 292: 324—327.

Guichard, F., Church, T.M., Treuil, M. and Jaffrezic, H., 1979. Rare earths in barites: distribution and effects on aqueous partitioning. *Geochim. Cosmochim. Acta*, 43: 983—997.

Haskin, L. and Gehl, M.A., 1962. The rare-earth distribution in sediments. *J. Geophys. Res.*, 67: 2537—2541.

Haskin, M.A. and Haskin, L.A., 1966. Rare earths in European shales: a redetermination. *Science*, 154: 507—509.

Haskin, L.A., Wildeman, T.R., Frey, F.A., Collins, K.A., Keedy, C.R. and Haskin, M.A., 1966. Rare earths in sediments. *J. Geophys. Res.*, 71: 6091—6105.

Haskin, L.A., Haskin, M.A., Frey, F.A. and Wildeman, T.R., 1968. Relative and absolute terrestrial abundances of the rare earths. In: L.H. Ahrens (Editor), *Origin and Distribution of the Elements.* Pergamon, Oxford, pp. 889—912.

Høgdahl, O.T., 1970. Distribution of the lanthanides in the waters and sediments of the River Gironde in France: a summary. In: *Marine Radioactive Studies.* IAEA Research Agreement Coordinated Programme, Morocco (unpublished).

Høgdahl, O.T., Melsom, S. and Bowen, V.T., 1968. Neutron activation of lanthanide elements in seawater. *Adv. Chem. Ser.*, 73: 308—325.

Jarvis, J.C., Wildeman, T.R. and Banks, N.G., 1975. Rare earths in the Leadville limestone and its marble derivatives. *Chem. Geol.*, 16: 27—37.

Jenner, G.A., Fryer, B.J. and McLennan, S.M., 1981. Geochemistry of the Archean Yellow Knife Supergroup. *Geochim. Cosmochim. Acta*, 45: 1111—1129.

Ludden, J.N. and Thompson, G., 1978. Behaviour of rare earth elements during submarine weathering of tholeiitic basalts. *Nature*, 274: 147—149.

Martin, J.M., Høgdahl, O. and Philipott, S.C., 1976. Rare earth element supply to the ocean. *J. Geophys. Res.*, 81: 3119—3124.

Maksimović, Z. and Roaldset, E., 1976. Lanthanide elements in some Mediterranean karstic bauxite deposits. *Trav. Com. Int. Etude Bauxites, Oxydes, Hydroxydes Alum.*, 13: 199—220.

McLennan, S.M., Nance, W.B. and Taylor, S.R., 1980. Rare earth element-thorium correlations in sedimentary rocks, and the composition of the continental crust. *Geochim. Cosmochim. Acta*, 44: 1833—1839.

Minami, E., 1935. Gehalte Seltener Erden in Europaischen and japanischen Tonschiefern. *Nachr. Ges. Wiss. Goettingen, Math.-Phys. Kl., Fachgruppe 4*, 1: 155—170.

Nesbitt, H.W., 1979. Mobility and fractionation of rare earth elements during weathering of a granodiorite. *Nature*, 279: 206—210.

O'Nions, R.K., Carter, S.R., Cohen, R.S., Evensen, N.M. and Hamilton, P.J., 1978. Pb, Nd and Sr isotopes in oceanic ferromanganese deposits and ocean floor basalts. *Nature*, 273: 435—438.

Piepgras, D.J., Wasserburg, G.J. and Dasch, E.J., 1979. The isotopic composition of Nd in different ocean masses. *Earth Planet. Sci. Lett.*, 45: 223—236.

Piper, D.Z., 1974a. Rare earth elements in the sedimentary cycle: a summary. *Chem. Geol.*, 14: 285—304.

Piper, D.Z., 1974b. Rare earth elements in ferromanganese nodules and other marine phases. *Geochim. Cosmochim. Acta*, 38: 1007—1022.

Piper, D.Z. and Graef, P.A., 1974. Gold and rare-earth elements in sediments from the East Pacific Rise. *Mar. Geol.*, 17: 287—297.

Rankin, P.C. and Childs, C.W., 1976. Rare-earth elements in iron-manganese concretions from New Zealand soils. *Chem. Geol.*, 18: 55—64.

Rankin, P.C. and Glasby, G.P., 1979. Regional distribution of rare earth and minor elements in manganese nodules and associated sediments in the southwest Pacific and other localities. In: J.L. Bishoff and D.Z. Piper (Editors), *Marine Science, 9. Marine Geology and Oceanography of the Pacific Manganese Nodule Province*. Plenum, New York, N.Y., pp. 681—697.

Roaldset, E., 1973. Rare earth elements in Quaternary clays of the Numedal area, southern Norway. *Lithos*, 6: 349—372.

Roaldset, E., 1974. Lanthanide distributions in clays. *Bull. Groupe Fr. Argiles*, 26: 201—209.

Roaldset, E., 1978. *Mineralogical and chemical changes during weathering, transport and sedimentation in different environments, with particular reference to the distribution of yttrium and the lanthanoide elements.* Ph.D. Thesis, University of Oslo, Oslo (unpublished).

Roaldset, E., 1979. Rare earth elements in different size fractions of a marine quick clay from Ullensaker, and a till from Upper Numedal, Norway. *Clay Miner.*, 14: 229—240.

Roaldset, E. and Rosenqvist, I.Th., 1971a. Unusual lanthanide distribution. *Nature Phys. Sci.*, 231: 153—154.

Roaldset, E. and Rosenqvist, I.Th., 1971b. Rare earth elements in vivianite from Lake Åsrum. *Lithos*, 4: 417—422.

Robertson, A.H.F. and Fleet, A.J., 1976. The origins of rare earths in metalliferous sediments of the Troodos Massif, Cyprus. *Earth Planet. Sci. Lett.*, 28: 285—394.

Robertson, A.H.F. and Fleet, A.J., 1977. REE evidence for the genesis of the metalliferous sediments of Troodos, Cyprus. *Spec. Publ. Geol. Soc. London*, 7: 78—79.

Shimizu, H. and Masuda, A., 1977. Cerium in chert as an indication of marine environment of its formation. Nature, 266: 346—348.

Spirn, R.V., 1965. *Rare earth distributions in the marine environment.* Ph.D. Thesis, Massachusetts Institute of Technology, Cambridge, Mass., 165 pp. (unpublished).

Toth, J.R., 1980. Deposition of submarine crusts rich in manganese and iron. *Geol. Soc. Am. Bull.*, 91: 44—54.

Turekian, K.K., Katz, A. and Chan, L., 1973. Trace element trapping in pteropod tests. *Limnol. Oceanogr.*, 18: 240—249.

Turner, D.R. and Whitfield, M., 1979. Control of seawater composition. *Nature*, 281: 468—469.

Volkov, I.I. and Fomina, L.S., 1967. Rare-earth elements in sediments and manganese concentrations of the ocean. *Lithol. Miner. Resour. (USSR)*, 5: 579—595.

Wakita, H., Rey, P. and Schmitt, R.A., 1971. Abundances of the 14 rare earth elements and 12 other rare elements in Apollo 12 samples: five igneous and one breccia rocks and four soils. *Proc. 2nd Lunar Sci. Conf., Geochim. Cosmochim. Acta, Suppl. 2*, 2: 1319.

Wildeman, T.R. and Condie, K.C., 1973. Rare earths in Archean graywackes from Wyoming and from the Fig Tree Group, South Africa. *Geochim. Cosmochim. Acta*, 37: 439—453.

Wildeman, T.R. and Haskin, L.A., 1965. Rare-earth elements in ocean sediments. *J. Geophys. Res.*, 70: 2905—2910.

Wildeman, T.R. and Haskin, L.A., 1973. Rare earths in Precambrian sediments. *Geochim. Cosmochim. Acta*, 37: 419—438.

Chapter 11

RADIOGENIC ISOTOPES — SOME GEOLOGICAL APPLICATIONS

C.J. HAWKESWORTH and P.W.C. VAN CALSTEREN

11.1. Introduction

The REE contain two radioactive decay schemes of great potential to earth scientists; ^{147}Sm decays to ^{143}Nd, and ^{176}Lu to ^{176}Hf. However, both decay extremely slowly with the result that the present range of the relevant neodymium ($^{143}Nd/^{144}Nd$) and hafnium ($^{176}Hf/^{177}Hf$) isotope ratios is very small (~0.5% compared with over 200% for $^{87}Sr/^{86}Sr$), and routine analyses have only recently been made possible by dramatic improvements in mass spectrometric techniques. As with the well known Rb-Sr and U-Pb systems the process of decay may be harnessed both as a geological "time-piece" to determine the ages of rocks and minerals, but also to evaluate the nature and evolution of their source regions. On a larger scale an understanding of how radiogenic isotopes have changed through time remains central to most models for the chemical evolution of the Earth and the other planetary bodies from which material is available.

The particular advantage of the pairs Sm-Nd and Lu-Hf is that although they each share similar chemical and physical characteristics these differ markedly from those of elements such as Rb, Sr, U, Th and Pb. Thus studies of Nd and Hf isotopes tend to complement work on Sr and Pb. Low Rb/Sr ratios are often accompanied by high Sm/Nd ratios so that the Sm-Nd decay scheme may be used to date rocks and minerals which are unsuitable for Rb-Sr. Similarly, since the REE are less mobile during metamorphism, Sm and Nd isotopes are more likely to yield reliable estimates of primary ages from metamorphosed rocks. Equally exciting, however, is the potential of *combined* isotope studies on samples from a wide range of geochemical environments. The relative fractionation of Rb/Sr and Sm/Nd reflects both the prevailing physical conditions and the compositions of the solid and liquid phases, and will with time generate characteristic differences in $^{87}Sr/^{86}Sr$ and $^{143}Nd/^{144}Nd$. Although still in its infancy this approach has provoked considerable interest and reappraisal in subjects as diverse as chemical oceanography, the movement of trace elements in the upper mantle and the effects of crustal contamination on the chemistry of magmatic rocks.

This chapter is devoted largely to neodymium isotopes and some of the problems to which they have been applied. A brief resumé of the law of radioactive decay and the derivation of the isochron equation is followed by a discussion of the geochronological applications of the Sm-Nd decay scheme and the use of model Nd, or T^{Nd}_{CHUR}, ages. Subsequent sections consider more geochemical topics and in particular the use of combined Nd and Sr isotopes in studies of the distribution of trace elements in mantle xenoliths and magmatic rocks, and on the behaviour of Nd and Sr in estuarine and oceanic environments. Finally, a general review of large-scale models for the evolution of the Earth's crust and mantle is followed by a short section on Lu and Hf isotopes.

11.2. Radioactive decay

Isotopes which are unstable decay and in so doing they emit radiation. This process of radioactive decay is *exponential* — the number of atoms which decay in a particular period of time depends on the number which are present but is independent of the chemical form or the physical condition of the element. More formally, the *law of radioactive decay* states that the number of atoms disintegrating per unit time ($-dN/dt$) is proportional to the total number of radioactive atoms (N) which are present. Thus:

$$-dN/dt = \lambda N \tag{11.1}$$

where λ is called the *decay constant* and has a characteristic value for each radioactive isotope: it represents the probability that an atom will decay in unit time.

On integrating equation (11.1):

$$N = N_0\, e^{-\lambda t}$$

or:

$$N_0 = N\, e^{\lambda t} \tag{11.2}$$

where N_0 is the number of radioactive isotopes at the formation of the sample, N is the number present today, and e is the base of the natural logarithms. t is clearly the length of time since the formation of the sample to the present day, that is, its age. However, confusion can arise because unlike those in historical time, geological dates are expressed relative to the present day. Thus the age of a rock is numerically identical to the time at which it formed — a granite which crystallized *at* 550 Ma is 550 Ma old!

The problem posed by equation (11.2) is to devise a way of determining the number of radioactive isotopes present when the sample formed (N_0).

However, since each radioactive (or parent) isotope that decays produces *one* radiogenic (daughter) isotope, N_0 equals the sum of the parent isotopes remaining at the present time (now called N_P) plus the number of daughter isotopes produced during the period of time t (N_{D*}). Substituting for N_0 in equation (11.2):

$$N_{D*} = N_P\,(e^{\lambda t} - 1) \qquad (11.3)$$

It should be noted in passing that the rate of decay of an isotope may also be expressed in terms of its *half-life* ($T_{1/2}$) — the length of time needed for *half* of a given number of radioactive isotopes to decay. Moreover, since, as stated above, the decay of one parent isotope produces one daughter, the half life is also the time at which the number of daughter isotopes produced (N_{D*}) equals the number of parent isotopes remaining (N_P). Putting both equal to one, substituting into equation (11.3) and taking logarithms demonstrates that the half-life is simply related to the decay constant λ by $T_{1/2} = \ln 2/\lambda$. For ^{147}Sm, $\lambda = 6.54 \times 10^{-12}$ a^{-1} and $T_{1/2} = 1.06 \times 10^{11}$ a.

Returning to equation (11.3) the number of daughter isotopes produced in the time t (N_{D*}) also equals the number of daughter isotopes present today (N_D) minus those present at the formation of the sample (N_{D_0}). Substituting for N_{D*} and rearranging:

$$N_D = N_P\,(e^{\lambda t} - 1) + N_{D_0} \qquad (11.4)$$

For the decay of ^{147}Sm to ^{143}Nd, ^{147}Sm is the parent isotope and ^{143}Nd is the daughter, so that:

$$^{143}\mathrm{Nd} = {}^{147}\mathrm{Sm}\,(e^{\lambda t} - 1) + {}^{143}\mathrm{Nd}_0 \qquad (11.5)$$

In practice, however, mass spectrometers are most simply used to measure relative abundances, or isotope *ratios*, and it is therefore convenient to divide all terms in equation (11.5) by an isotope not involved in the decay scheme, in this case ^{144}Nd. Thus:

$$\frac{^{143}\mathrm{Nd}}{^{144}\mathrm{Nd}} = \frac{^{147}\mathrm{Sm}}{^{144}\mathrm{Nd}}\,(e^{\lambda t} - 1) + \left(\frac{^{143}\mathrm{Nd}}{^{144}\mathrm{Nd}}\right)_0 \qquad (11.6)$$

and this is the practical equation used for calculating t and $(^{143}\mathrm{Nd}/^{144}\mathrm{Nd})_0$ from measurements of present-day ^{143}Nd/^{144}Nd and ^{147}Sm/^{144}Nd ratios. It is known as an *isochron* equation and similar expressions are used widely for other decay schemes such as $^{87}\mathrm{Rb} \rightarrow {}^{87}\mathrm{Sr}$ and $^{176}\mathrm{Lu} \rightarrow {}^{176}\mathrm{Hf}$. $(^{143}\mathrm{Nd}/^{144}\mathrm{Nd})_0$ is the Nd isotope ratio of a sample *at the time of its formation* and it is called the *initial* Nd isotope ratio.

The significant point about equation (11.6) is that it is in the form of a straight line, $y = mx + c$. A straight line on a graph of $^{143}Nd/^{144}Nd$ versus $^{147}Sm/^{144}Nd$ has a slope of $(e^{\lambda t} - 1)$ and an intercept on the y-axis of $(^{143}Nd/^{144}Nd)_0$. Such a diagram is called an *isochron diagram* and the age "t" may be calculated from the slope of the straight line — the *isochron*. However, for that age to be correct the following assumptions must be upheld: (a) all the samples must be of the same age, (b) at the time of their formation they must have had the same $^{143}Nd/^{144}Nd$ ratio, i.e., they would have plotted on a horizontal line on an isochron diagram such as Fig. 11.1, and (c) between the formation of the samples and the present day the $^{143}Nd/^{144}Nd$ and $^{147}Sm/^{144}Nd$ ratios must have changed *only* by the process of radioactive decay; no Sm or Nd can have been added to, or lost from, the system.

In addition, a graph of $^{143}Nd/^{144}Nd$ against time "t" also results in a straight line, but in this case the slope reflects the $^{147}Sm/^{144}Nd$ ratio. Such a diagram is called an *isotope evolution* diagram (e.g., Fig. 11.3).

Finally, in this section, a note of warning. Analytical techniques for measuring Nd isotope ratios have only been developed comparatively recently, and there is still no agreement about the relative abundances of even the stable isotopes of Nd. In practice each individual $^{143}Nd/^{144}Nd$ ratio measured on a mass spectrometer is normalized to a particular ratio of two unradiogenic Nd isotopes which are assumed to be constant in nature (this corrects for mass fractionation between ^{143}Nd and ^{144}Nd during the analysis). However,

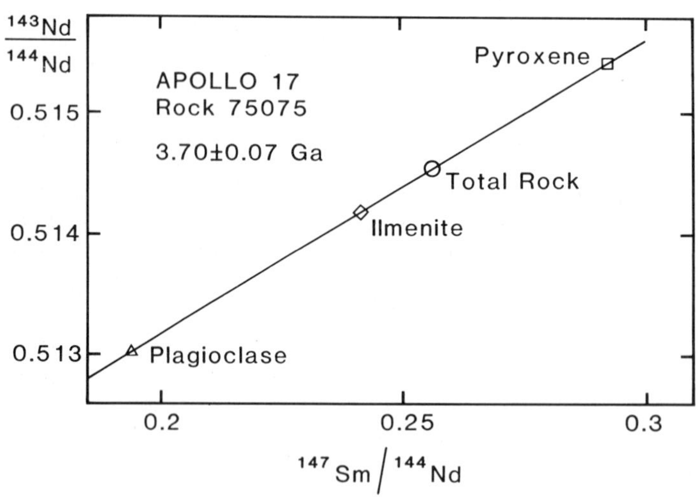

Fig. 11.1. Sm-Nd isochron diagram for basalt 75075 collected on the Apollo 17 mission (Lugmair et al., 1975b). The slope of the best-fit straight line through the four data points corresponds to an age of 3.70 ± 0.07 Ga, and its intercept on the $^{143}Nd/^{144}Nd$ axis gives the *initial* Nd isotope ratio (0.50825 ± 12).

because there are essentially two schools of thought on the overall isotope composition of Nd, different laboratories effectively normalize to different numbers. This is developed further in the Appendix ("Analytical techniques"), but the net result is that different laboratories have published what on the surface appear to be widely different $^{143}Nd/^{144}Nd$ ratios on the same samples. Great care must therefore be taken to ensure that published Nd isotope analyses are normalized to the same value for the unradiogenic Nd isotope ratios, before they are compared, plotted on the same diagram, etc. In this chapter we discuss $^{143}Nd/^{144}Nd$ ratios normalized to $^{146}Nd/^{144}Nd$ = 0.7219 after O'Nions et al. (1977), with the result that the U.S.G.S. standard rock BCR-1 gives a present-day $^{143}Nd/^{144}Nd$ ratio of ~0.51262 compared with 0.51184 reported by, for example, DePaolo and Wasserburg (1976a) (see also Appendix, "Analytical techniques").

11.3. Geochronology — the dating of rocks and minerals

The $^{147}Sm \rightarrow {}^{143}Nd$ decay scheme was first used as a geochronological tool in the dating of meteorites and lunar samples. Lugmair et al. (1975a) obtained an age of 4.56 ± 0.08 Ga with an initial $^{143}Nd/^{144}Nd$ ratio of 0.50677 ± 10 from the Juvinas achondrite. Another achondrite, Angra dos Reis, yielded a similar age, 4.55 ± 0.04 Ga and an even more precise initial Nd isotope ratio of 0.50682 ± 5 (Lugmair and Marti, 1977); whereas ages of 4.4—3.3 Ga have been obtained on samples of lunar rocks (e.g., Lugmair et al., 1975b, see Fig. 11.1 and Papanastassiou et al., 1977).

The significance of these results is fourfold:

(1) They demonstrated that the Sm-Nd dating technique was capable of precision comparable to that of Rb/Sr.

(2) The calculated ages were in accord with results obtained by U/Pb and Rb/Sr techniques, and so provided support for $\lambda(^{147}Sm) = 6.54 \times 10^{-12}$ a^{-1} (Lugmair and Marti, 1978).

(3) Achondrites are difficult to date by Rb/Sr, due to their low Rb/Sr ratios, but amenable to Sm/Nd. A new class of meteorites was therefore opened to chronologic studies (see also Unruh et al., 1977, and Nakamura et al. 1976).

(4) They provided estimates of the initial $^{143}Nd/^{144}Nd$ ratios of solar system material which are used widely in discussions of planetary and, in particular, terrestrial evolution.

Following the work of Lugmair and his co-workers on extra-terrestrial materials, the ^{147}Sm-^{143}Nd decay scheme has been successfully applied to the dating of terrestrial rocks and minerals. For whole rocks, its strength is that in contrast to the alkali and alkaline earth elements, the rare earths are relatively unaffected by secondary processes of alteration and metamorphism (Chapter 9). Its limitations lie in the long half-life of ^{147}Sm (1.06×10^{11} a) and the comparatively restricted range of Sm/Nd ratios in most crustal rocks (see Fig. 11.6), which together have ensured that up to now most precise whole-rock ages have been reported on Archaean (>2.5 Ga) examples.

Fig. 11.2 illustrates the precise Sm-Nd whole-rock isochron corresponding to an age of 3.54 ± 0.03 Ga obtained on rock units (the Onverwacht volcanics) whose Rb-Sr systems had been significantly disturbed (Allsopp et al., 1973; Jahn and Shih, 1974; Hamilton et al., 1979b). In addition, Hamilton et al. (1979a) determined an age of 2.92 ± 0.05 Ga on Lewisian gneisses from northwestern Scotland and cited it as evidence that the Sm/Nd whole-rock techniques can "see back through", or "remain undisturbed" by even granulite facies metamorphism.

Although precise ages of Archaean rocks are important to the study of geological processes early in Earth's history, arguably the most significant contribution of these and other whole-rock studies is their initial Nd isotope ratios. These are plotted on a graph of $^{143}Nd/^{144}Nd$ against time in Fig. 11.3 and they lie close to a straight line which, as indicated in equation (11.6),

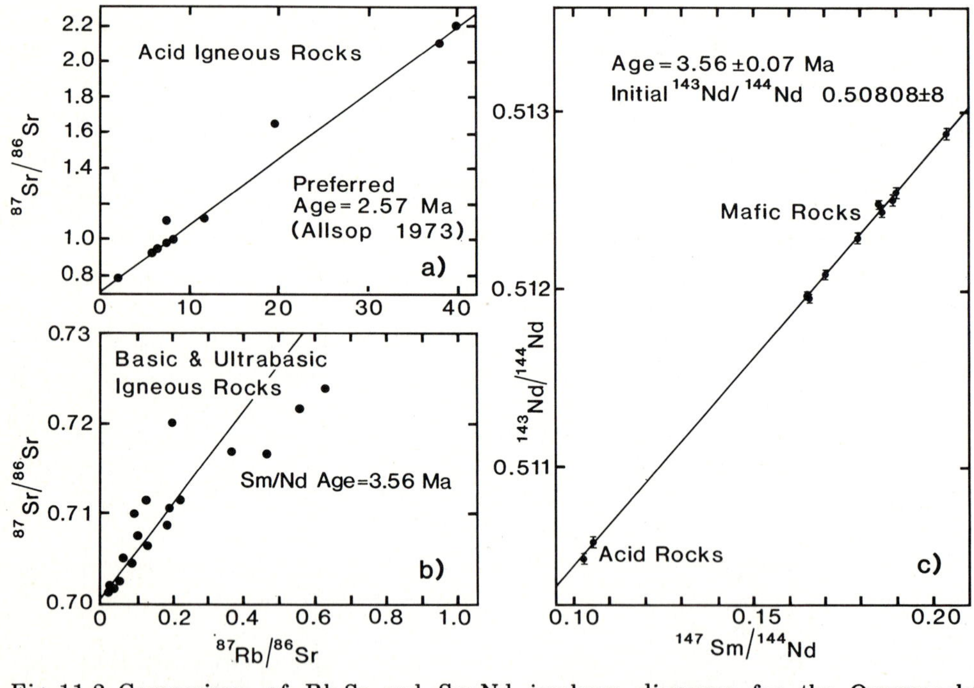

Fig. 11.2. Comparison of Rb-Sr and Sm-Nd isochron diagrams for the Onverwacht volcanics, South Africa (after O'Nions et al., 1979a; data from Allsopp et al., 1973; Jahn and Shih, 1974; Hamilton et al. 1979b). Contrast the scatter of Rb-Sr results on the basic and ultrabasic rocks and what appears to be a secondary (or metamorphic) Rb/Sr age on the acid rocks, with the precise age of 3.56 ± 0.07 Ga obtained using the Sm-Nd technique.

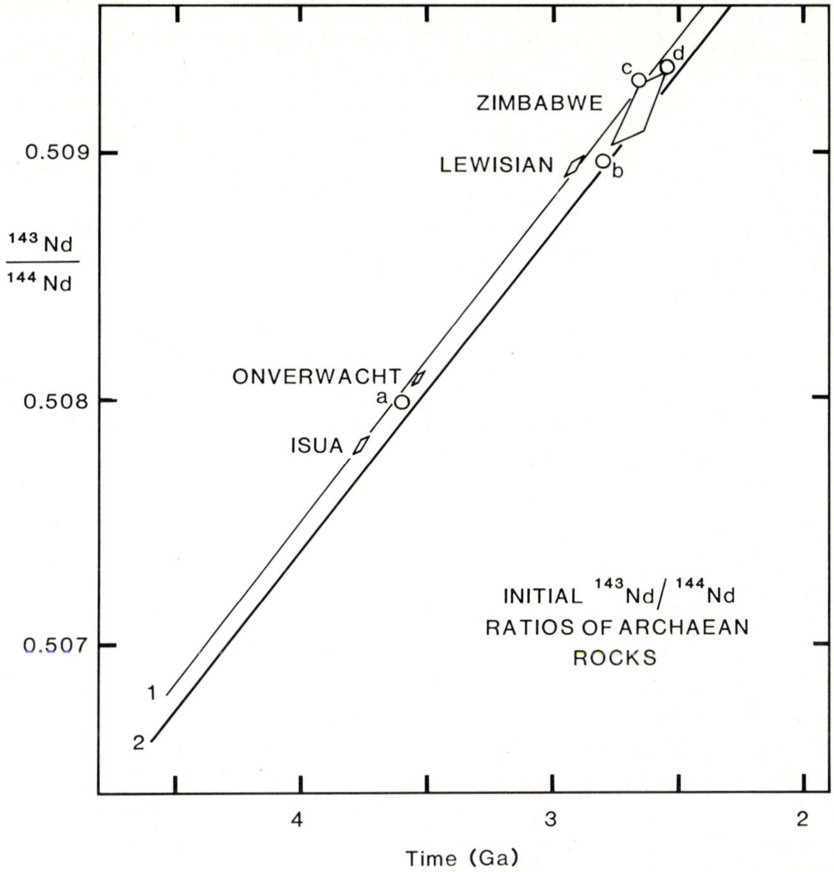

Fig. 11.3. Initial $^{143}Nd/^{144}Nd$ ratios of Archaean rocks. Those of the Isua, Onverwacht, Lewisian and Zimbabwe samples were determined from Sm-Nd whole-rock isochrons (Hamilton et al., 1977, 1979a, b). The initial ratios of the Amitsoq gneiss (*a*), Fiskenaesset anorthosite (*b*), Preissac-Lacorne Batholith (*c*), and the Great Dyke (*d*) were calculated from single whole-rock analyses (DePaolo and Wasserburg, 1976a, b). Line *1* corresponds to Sm/Nd = 0.31 and $^{143}Nd/^{144}Nd$ = 0.50682 at 4.55 Ga, whereas for line *2* Sm/Nd = 0.325 and $^{143}Nd/^{144}Nd$ = 0.50663 at 4.6 Ga ago.

section 11.2, describes the evolution of a system with constant $^{147}Sm/^{144}Nd$. For these terrestrial samples that trend corresponds to the evolution of a system or reservoir which has $^{147}Sm/^{144}Nd$ and $^{143}Nd/^{144}Nd$ ratios similar to those in average chondrites. Moreover, since magmatic rocks formed early in Earth's history are more likely to provide a reliable estimate of the trace element and isotope ratios of the bulk Earth, the trend in Fig. 11.3 is a striking indication that the bulk Earth had chondritic $^{147}Sm/^{144}Nd$ and $^{143}Nd/^{144}Nd$ ratios. It prompted DePaolo and Wasserburg (1976a) to introduce the acronym CHUR ("chondritic uniform reservoir"), and at least for the Sm-Nd decay scheme CHUR is synonomous with the bulk Earth.

In detail, however, there is still some confusion over what values should be accepted for average chondrites. Early estimates (DePaolo and Wasserburg, 1976a; O'Nions et al., 1977, 1979a) adopted an average Sm/Nd ratio of ~0.31 (Masuda et al., 1973, Evensen et al., 1978) and assumed that the bulk Earth, or CHUR, had the same initial Nd isotope ratio as Juvinas or Angra dos Reis 4.55 Ga ago (line *1*, Fig. 11.3). This allowed a present-day $^{143}Nd/^{144}Nd$ ratio of the average bulk Earth to be calculated as ~0.51262. Recently Jacobsen and Wasserburg (1980) presented more results on a number of chondrites, which were re-evaluated by Wasserburg et al. (1982). These indicate an average Sm/Nd ratio of 0.325 for chondritic material with a present-day $^{143}Nd/^{144}Nd$ ratio of 0.51264 (normalized to $^{146}Nd/^{144}Nd$ = 0.7219; Appendix 11.1). This corresponds to an initial Nd isotope ratio of ~0.50663 4.6 Ga ago (line *2*, Fig. 11.3), which is significantly lower than those reported for Angra dos Reis (Lugmair and Marti, 1977) and Juvinas (Lugmair et al., 1975a). However, the latter are basaltic achondrites — they probably represent differentiated rocks from older planetary material and as such they might well have both higher $^{143}Nd/^{144}Nd$ ratios and slightly younger ages.

In summary, the $^{147}Sm \rightarrow {}^{143}Nd$ decay scheme has been used very successfully to determine precise whole-rock ages on Archaean rocks, and their initial Nd isotope ratios have been shown to plot close to the evolution of a reservoir with $^{147}Sm/^{144}Nd$ and $^{143}Nd/^{144}Nd$ ratios of average chondrites. The exact evolution path of CHUR, or the bulk Earth, is still unresolved and, as illustrated in Fig. 11.3, this hinders the interpretation of many results particularly those on Archaean rocks. CHUR is effectively the reference to which isotope results on terrestrial samples are compared and if, for example, line *2* (Sm/Nd = 0.325) is preferred then the rocks from Isua and the Onverwacht appear to have been derived from source material which was more radiogenic (higher $^{143}Nd/^{144}Nd$) than CHUR. Alternatively if line *1* (Sm/Nd = 0.31) is chosen it effectively destroys the evidence for relatively high $^{143}Nd/^{144}Nd$ source regions in the Archaean (Fig. 11.3). Clearly at present models for the early evolution of the upper mantle and crust are uncomfortably sensitive to our choice of CHUR.

Traditionally there are two main reasons for dating individual minerals phases: to obtain a more precise age by extending the range of parent/daughter (Sm/Nd) ratios, or to unravel polymetamorphic and cooling histories in samples where the minerals are younger than their host rocks. As with the dating of achondrites, the advantage of the ^{147}Sm-^{143}Nd technique is that it makes it possible to date minerals which are very difficult to study by other techniques — e.g., several minerals with very low Rb/Sr ratios have high Sm/Nd ratios and thus with time generate favourable $^{143}Nd/^{144}Nd$.

The factors which control the distribution of REE in a mineral are many and complex, but they include the composition of both the rock and the other phases which are present (Chapter 1). For example, van Breemen and Hawkesworth (1980) reported a range of Sm/Nd ratios in garnets from 0.33

in a mica schist to 3.5 in a basic granulite and noted that there was no clear relationship between Sm/Nd and the composition of the different garnets. In general, however, garnets and pyroxenes are the common minerals which tend to be depleted in LREE and thus have high Sm/Nd ratios suitable for dating. Typical values of Sm/Nd in common minerals are: pyroxene, 0.2—0.5; plagioclase and alkali feldspars, 0.1—0.3; garnet, 0.3—3.5; biotite, ~0.2; amphibole, 0.1—0.5; apatite, 0.2—0.4; and zircon, 0.2—0.6.

Jacobsen and Wasserburg (1979a) reported ages of 501 ± 13 and 508 ± 6 Ma on clinopyroxene-plagioclase pairs separated from samples of the Bay of Islands ophiolite, Newfoundland. Whole-rock analyses revealed significant variations of initial $^{143}Nd/^{144}Nd$ within the ophiolite complex which effectively frustrated attempts to make precise age determinations by whole-rock measurements. Griffin and Brueckner (1980) used a similar approach to investigate the age of metamorphism preserved in the eclogites of the Caledonian belt of western Norway. Five out of the six garnet-clinopyroxene pairs analysed give two-point isochron ages of 407—447 Ma, and they concluded that, although the protoliths had existed as crustal rocks long before the Caledonian, these Sm-Nd ages probably record a blocking temperature close to the peak of metamorphism. However, at present the blocking conditions of the Sm-Nd system are poorly understood.

Under most metamorphic conditions Sm and Nd appear to be less mobile than Rb and Sr, suggesting that the former isotope system may remain "undisturbed" to higher temperatures. There are no published data on the rate of diffusion of Sm and Nd in garnet for example, yet petrofabric and geothermometric studies suggest that at least the major element compositional zoning is preserved up to 640—680°C before it becomes erased by higher diffusion rates at higher temperatures (Yardley, 1977; Råheim and Green, 1975). Consistent with this evidence van Breemen and Hawkesworth (1980) obtained a garnet-feldspar age on a lower crustal xenolith which simply reflected its time of emplacement (355 ± 10 Ma). Although the rocks themselves are probably of Lower Proterozoic age the minerals appear to have been reset isotopically en route to the surface: moreover, if the trivalent REE re-equilibrate under such conditions it may again revive doubts over the significance of metamorphic pressures and temperatures determined on both lower crustal and upper mantle xenoliths (e.g., Graham and Upton, 1978).

11.4. Model Nd, or $T_{\mathrm{CHUR}}^{\mathrm{Nd}}$, ages

In the previous section it was emphasized that most of the initial Nd isotope ratios of Archaean rocks were very similar to the $^{143}Nd/^{144}Nd$ ratios in a chondritic reservoir (CHUR) at the time of their formation (Fig. 11.3). Moreover, that is in sharp contrast with the results on younger rocks which exhibit significant variations in initial $^{143}Nd/^{144}Nd$ ratios (Fig. 11.4) corre-

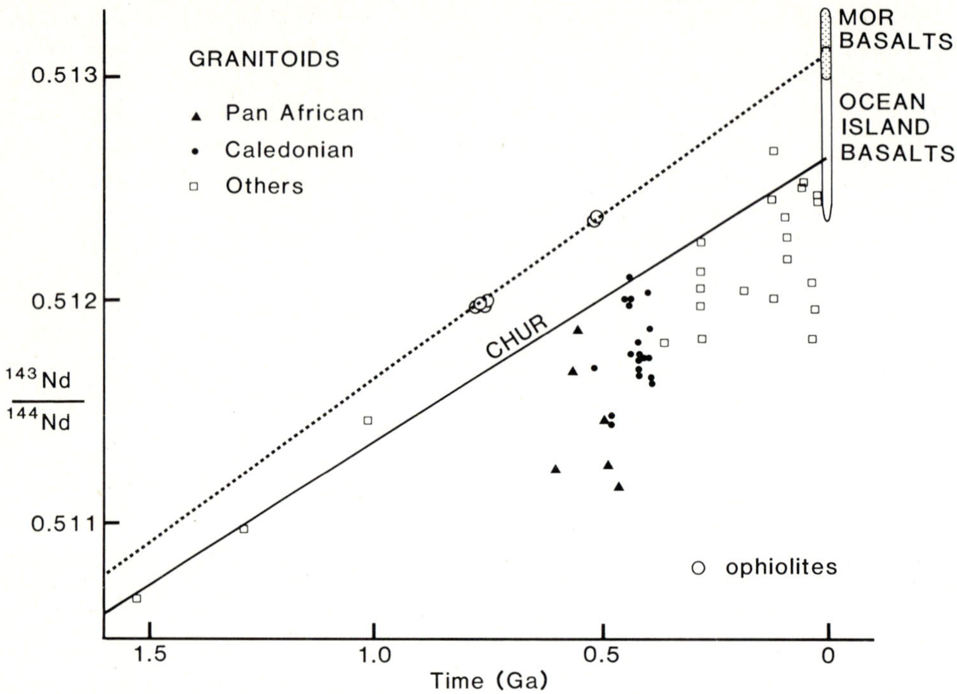

Fig. 11.4. Nd isotope evolution diagram illustrating the variation in initial $^{143}Nd/^{144}Nd$ ratios observed in granites less than 1.5 Ga old, compared with those from recent MORB and ocean island basalts (data from Allègre and Othman, 1980; Bokhari and Kramers, 1981; Hamilton et al., 1980; Hawkesworth et al., 1981; Jacobsen and Wasserburg, 1979a). The dashed line depicts the evolution of LREE-depleted mantle with Sm/Nd = 0.37 and $^{143}Nd/^{144}Nd$ = 0.5131 at the present day.

sponding to a range of $^{143}Nd/^{144}Nd$ of ~0.5133—0.5113 at the present day. In general, mantle-derived rocks tend to have higher $^{143}Nd/^{144}Nd$ ratios than CHUR, implying that their source regions had high Sm/Nd ratios (i.e., were depleted in LREE) for most of their history; whereas crustal rocks tend to be less radiogenic in Nd isotopes than CHUR — their source rocks had low Sm/Nd ratios. Model Nd, or T^{Nd}_{CHUR}, ages are simply an expression of how long their respective source regions may have had REE ratios different to those in CHUR. As with most model ages, they are readily calculated but demand cautious interpretation.

T^{Nd}_{CHUR} ages were pioneered by DePaolo and Wasserburg (1976a) and McCulloch and Wasserburg (1978) and Fig. 11.5 illustrates how they are calculated. Consider a sample of granite which was emplaced 500 Ma ago: at the time of its emplacement it plotted at point Y, and since then it has evolved along the line YZ, whose slope reflects its Sm/Nd ratio (see equation (11.6)). The $^{143}Nd/^{144}Nd$ ratio at Y is the initial Nd isotope ratio of the granite, and it was clearly *lower* than that of CHUR at that time. If we

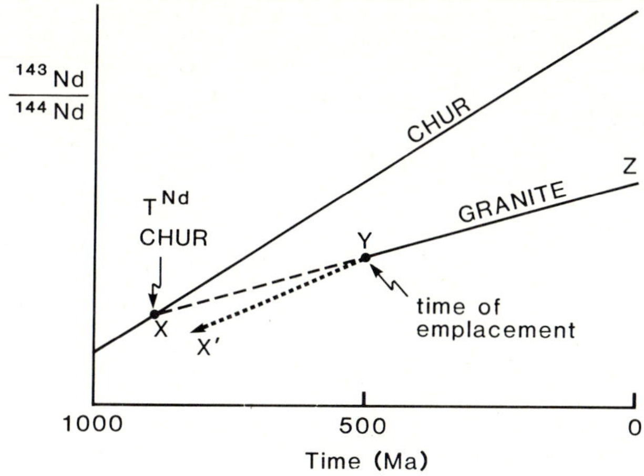

Fig. 11.5. The calculation of a model Nd, or $T_{\text{CHUR}}^{\text{Nd}}$, age for a sample of granite emplaced 500 Ma ago (see also Appendix).

assume, for the purposes of this discussion, that the Sm/Nd ratio in the granite sample was the same as that in its source region, then before 500 Ma the source region was evolving along the dashed line XY. At point X it plotted on the evolution path for CHUR — and that corresponds to its $T_{\text{CHUR}}^{\text{Nd}}$ age.

$T_{\text{CHUR}}^{\text{Nd}}$ ages are thus model Nd ages which assume that the evolution of particular samples may be usefully described in terms of a two-stage history. Stage 1 is within CHUR, and stage 2 is with an Sm/Nd ratio similar to that observed today. For each problem it is therefore necessary to evaluate (a) whether a two-stage model is applicable, and (b) whether the particular rock type is likely to have an Sm/Nd ratio similar to that in its source region.

To consider mantle-derived rocks first, basalts exhibit a wide range of Sm/Nd ratios (Fig. 11.6) and most of them probably reflect large enough degrees of partial melting for such trace element ratios to be similar to those in their source rocks (Chapter 4). However, as discussed in a later section, there are compelling reasons why the evolution of the upper mantle *cannot* be described satisfactorily by a two-stage model, and thus $T_{\text{CHUR}}^{\text{Nd}}$ ages are at best minimum estimates for how long Sm/Nd ratios different to CHUR may have existed. Crustal rocks by contrast have a limited range in Sm/Nd, and these are significantly lower than those in the bulk of the Earth's mantle — compare with the value for CHUR in Fig. 11.6. Moreover, it suggests that the formation of new continental crust is the process which causes the greatest change in Sm/Nd ratios, and that thereafter they remain comparatively unaffected by subsequent remelting or even erosion and sedimentation (McCulloch and Wasserburg, 1978). If so, then variations in Nd isotope ratios in crustal rocks will largely reflect the time they, or their precursors, were derived from the upper mantle — that is, their $T_{\text{CHUR}}^{\text{Nd}}$ age (Fig. 11.5).

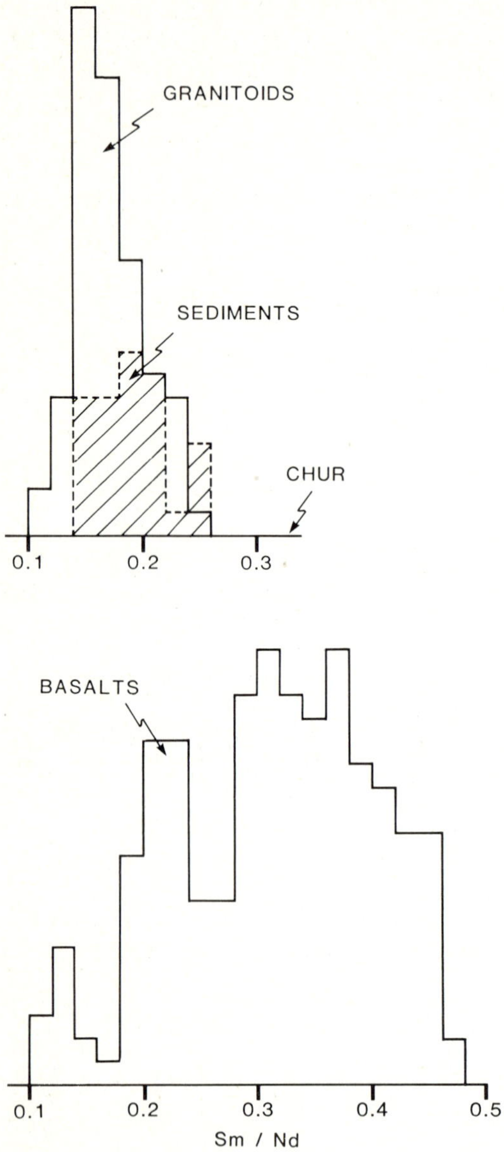

Fig. 11.6. Sm-Nd ratios determined on a range of intraplate and mid-ocean ridge basalts, granitoids, and continental sediments.

In practice the low and relatively uniform Sm/Nd ratios of crustal rocks imply that even if fractionation does occur within the crust it only results in small changes in what were already small Sm/Nd ratios, and so will have relatively little effect on the calculated T^{Nd}_{CHUR} ages. There are also several reasons why such model ages on continental rocks are often likely to be *minimum* estimates for the time they or their precursors were derived from the mantle.

(1) The mantle from which they were derived might have been depleted relative to CHUR and thus plot above CHUR on a Nd isotope evolution diagram (Fig. 11.4).

(2) If remelting in the crust does fractionate Sm/Nd ratios, those in the melt are likely to be *less* than those in the original source rock. In our example in Fig. 11.5 the source region would therefore have evolved along a steeper slope (higher Sm/Nd ratio) before 500 Ma (perhaps along the dotted line $X'Y$) and so be *older* than the calculated T^{Nd}_{CHUR} age.

(3) Any mixing of mantle and crustal material will result in a "hybrid" T^{Nd}_{CHUR} age which will be younger than the crustal component.

Notwithstanding these notes of caution much useful information has been obtained from T^{Nd}_{CHUR} ages. In the Pan-African (550—450 Ma) granites of Namibia (Fig. 11.4) they range from 0.5 to 2.2 Ga, indicating that the Damara orogenic event resulted in both the generation of new continental crust and the remobilization of old, 2.0-Ga basement material (Hawkesworth et al., 1981; Hawkesworth and Marlow, 1983). In Scotland both Caledonian granites (Hamilton et al., 1980) and lower crustal xenoliths from the Midland Valley (van Breemen and Hawkesworth, 1980; G.R. Davies, unpublished) have T^{Nd}_{CHUR} ages of 0.9—1.3 Ga, suggesting that at least some of the former are crustal remelts and providing new evidence for the existence of Proterozoic crust in these areas. Similar arguments have been put forward for depleted, mantle-derived rock types. Samples from the ~500-Ma Bay of Islands ophiolite in Newfoundland have an average T^{Nd}_{CHUR} age of 1.9 Ga (Jacobsen and Wasserburg, 1979a), whereas those from possible Pan-African ophiolites in Namibia (Hawkesworth et al., 1981) and Arabia (Bokhari and Kramers, 1981) yield values of 2.2—2.5 Ga. This is illustrated in Fig. 11.4 where the ophiolite samples are seen to lie close to an evolution path corresponding to a typical MORB Sm/Nd ratio of 0.37 (see Fig. 11.6) and a T^{Nd}_{CHUR} age of 2.4 Ga. Although a two-stage model is not strictly applicable (see section 11.7) these results do demonstrate that at least some of the Earth's mantle has been depleted in LREE since the Archaean.

T^{Nd}_{CHUR} ages have also been applied to sedimentary rocks, where once again their usefulness hinges on whether the Sm/Nd ratio of, in this case, a sediment is similar to that in its source material. If so, then T^{Nd}_{CHUR} ages of sediments provide an estimate of the age of their source terrain. McCulloch and Wasserburg (1978) introduced this approach and cited evidence that the processes of erosion, sedimentation and metamorphism do *not* fractionate the REE significantly (see also Chapters 1 and 9). They demonstrated that whereas the Archaean sediments analysed tended to have T^{Nd}_{CHUR} ages similar to their estimated geological ages, many of those reported on younger sediments were significantly older than the rocks themselves, indicating that they contain a component of "older" continental material. A similar relationship has recently been observed in the Damara sediments of Namibia (C.J. Hawkesworth, P.W.C. van Calsteren and A.R. Gledhill, unpublished) where T^{Nd}_{CHUR} ages range from 1.0 to 2.0 Ga in rocks deposited 950—750 Ma ago. Moreover, in this case other isotope data are available which testify to the reliability of the T^{Nd}_{CHUR} ages. Sediments with T^{Nd}_{CHUR} ages of ~2.0 Ga lie on basement orthogneisses of that age, and contain detrital zircons up to 2.0 Ga old (Brigueu et al., 1980). Although further work is urgently needed to evaluate the assumptions inherent in the calculation of T^{Nd}_{CHUR} ages for sediments, they appear to offer an exciting new approach to the study of the evolution of continental crust.

11.5. Isotopes as tracers in petrogenesis

Radiogenic isotopes are used widely in the study of magmatic rocks. However, in many instances their significance is still debated and it is helpful to retain a clear idea of their two essential applications: (a) to identify components from different sources (subducted ocean floor basalt, continental crust, etc.) which may have contributed to a particular magmatic suite, and (b) to constrain models for the evolution of the source regions of magmatic rocks. The time scale for the former is short and similar to that over which most geological and petrogenetic processes operate, whereas the time scale for the latter is long and likely to be measured in billions, rather than millions, of years. In this section we shall look at the processes of crustal contamination of continental magmatic rocks, enrichment of incompatible elements in the upper mantle, and magma genesis along destructive plate margins, before considering the longer time scale in a brief discussion of models for the evolution of the crust and mantle in section 11.7.

Although the Sm-Nd system clearly stands as an important geochronological technique in its own right, many geochemical studies have benefited from the combined application of both Nd and Sr isotopes. Fig. 11.7 illustrates the variation of $^{143}Nd/^{144}Nd$ and $^{87}Sr/^{86}Sr$ in recent basalts and the striking feature is that the majority of analyses appear to define a single broad negative correlation. Mid-ocean ridge basalts have low $^{87}Sr/^{86}Sr$ and high $^{143}Nd/^{144}Nd$, and intraplate volcanics which characteristically have higher concentrations of incompatible elements tend also to exhibit higher Sr isotope and lower Nd isotope ratios. This negative correlation was first recognized by DePaolo and Wasserburg (1976b) and O'Nions et al. (1977) who used it to estimate the Rb/Sr ratio of the bulk Earth. The Nd isotope results on Archaean rocks suggest that the bulk Earth has Sm/Nd and Nd isotope ratios similar to chondrites (Figs. 11.3 and 11.4) and its present-day $^{143}Nd/^{144}Nd$ ratio is therefore ~0.51264 (section 11.3). Assuming that Rb/Sr and Sm/Nd have fractionated coherently throughout the differentiation and

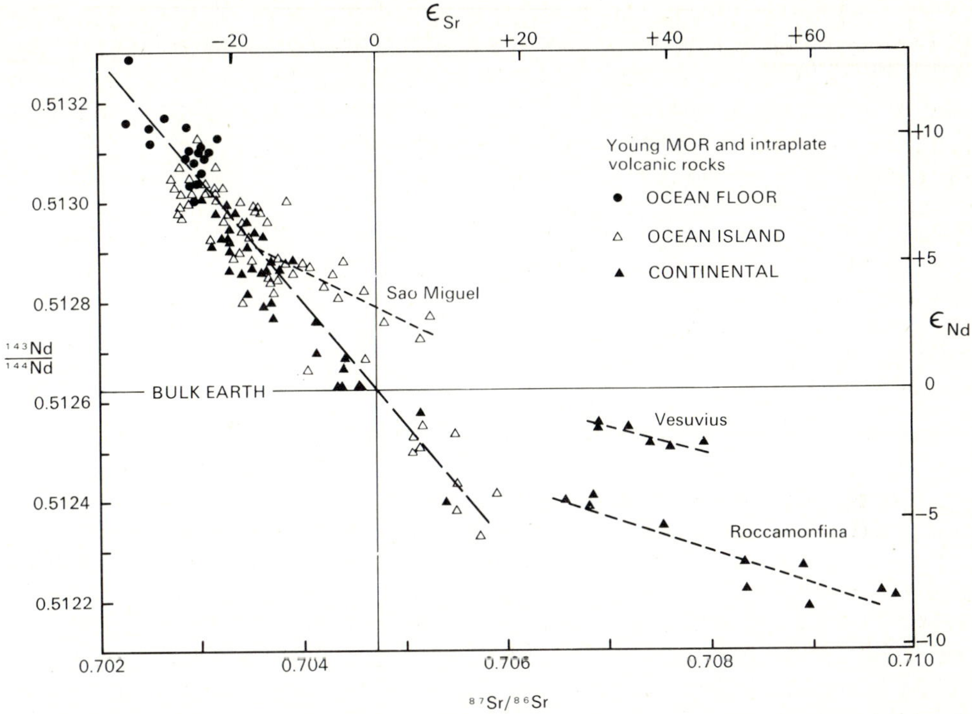

Fig. 11.7. Variation in Nd and Sr isotopes in recent volcanic rocks. The majority plot on a broad negative correlation, often referred to as the "mantle array", from which the present-day $^{87}Sr/^{86}Sr$ ratio of the bulk Earth is estimated to be 0.7047. Similar isotope variations are observed in rocks from ocean island and continental settings. Samples in the top left-hand quadrant are "depleted" isotopically relative to the bulk Earth, whereas those, in the bottom right-hand quadrant are relatively "enriched". A few rocks plot to the right of the main correlation (group II of Hawkesworth et al., 1979a). For calculation of ϵ_{Nd} and ϵ_{Sr} see text, equations (11.7) and (11.8).

evolution of the mantle, the present-day $^{87}Sr/^{86}Sr$ ratio of the bulk Earth may be estimated from the correlation between Nd and Sr isotopes for the majority of mantle-derived volcanic rocks. At $^{143}Nd/^{144}Nd = 0.51264$, $^{87}Sr/^{86}Sr = 0.7047$ (Fig. 11.7) — which corresponds to an Rb/Sr ratio for the bulk Earth of approximately 0.03. Moreover, since this is significantly lower than that in chondrites, the acronym CHUR cannot be used to describe the evolution of Rb-Sr in the Earth and thus for combined Nd and Sr isotope studies we shall discuss results relative to the bulk Earth.

A limitation of the diagram of $^{143}Nd/^{144}Nd$ against $^{87}Sr/^{86}Sr$ is that it can only be used to compare rocks of the same age and they must be plotted relative to the isotope ratios of the bulk Earth *at that time*. To overcome this problem and to allow greater flexibility, Nd and Sr isotope results are often reported *relative* to those of the bulk Earth and the notation adopted here is that introduced by DePaolo and Wasserburg (1976a). The initial isotope ratio of a sample formed t million years ago is presented as fractional deviations from the isotope ratio of a uniform reservoir (UR) at that same time. Thus for neodymium:

$$\epsilon_{\mathrm{Nd}}^{\mathrm{UR}} = \left[\frac{{}^{143}\mathrm{Nd}/{}^{144}\mathrm{Nd}_{\mathrm{sample}(t)}}{{}^{143}\mathrm{Nd}/{}^{144}\mathrm{Nd}_{\mathrm{UR}(t)}} - 1 \right] \times 10^4 \qquad (11.7)$$

and for strontium:

$$\epsilon_{\mathrm{Sr}}^{\mathrm{UR}} = \left[\frac{{}^{87}\mathrm{Sr}/{}^{86}\mathrm{Sr}_{\mathrm{sample}(t)}}{{}^{87}\mathrm{Sr}/{}^{86}\mathrm{Sr}_{\mathrm{UR}(t)}} - 1 \right] \times 10^4 \qquad (11.8)$$

In practice, for terrestrial samples the "uniform reservoir" is the best estimate for the bulk Earth, and the superscript UR is usually omitted. The values of ϵ_{Nd} and ϵ_{Sr} for recent basalts are illustrated in Fig. 11.7 for which it was assumed that the bulk Earth was formed 4.60 Ga ago, and that $({}^{143}\mathrm{Nd}/{}^{144}\mathrm{Nd})_{\mathrm{I}} = 0.50663$, ${}^{147}\mathrm{Sm}/{}^{144}\mathrm{Nd} = 0.1967$, $({}^{87}\mathrm{Sr}/{}^{86}\mathrm{Sr})_{\mathrm{I}} = 0.69898$, and ${}^{87}\mathrm{Rb}/{}^{86}\mathrm{Sr} = 0.0847$.

The calculation of present-day ${}^{143}\mathrm{Nd}/{}^{144}\mathrm{Nd}$ and ${}^{87}\mathrm{Sr}/{}^{86}\mathrm{Sr}$ ratios for the bulk Earth illustrates a second important feature of the results on recent basalts — to which we shall return later. Not only do the majority of mantle-derived rocks plot on a single trend in Fig. 11.7, but they also tend to have lower ${}^{87}\mathrm{Sr}/{}^{86}\mathrm{Sr}$ and higher ${}^{143}\mathrm{Nd}/{}^{144}\mathrm{Nd}$ ratios than the bulk Earth. Irrespective of whether these rocks are presently enriched or depleted in incompatible elements, they were derived from source regions which for much of their history must have had lower Rb/Sr and higher Sm/Nd ratios than the bulk Earth. Moreover, most models for the isotopic evolution of the Earth's crust and mantle assume that this depletion of the upper mantle is due primarily to the formation of continental crust (see section 11.7).

Crustal contamination of continental magmatic rocks

This is a topic which at present is both confused and controversial, but it is extremely important not least because without some understanding of the effects of crustal contamination on the chemistry of magmatic rocks we cannot hope to assess their source regions in the sub-continental mantle. In practice much of the confusion has arisen from the different perspectives of individual authors which emerges in the different implications attached to terms like "crust" and "mantle". On one hand many isotope geochemists have been concerned with chemical changes, reflected in variations in isotope ratios, which take place over very long periods of time. The evolution of the Earth has been satisfactorily modelled in terms of just the crust, depleted or residual mantle, and primitive mantle with the same composition as the

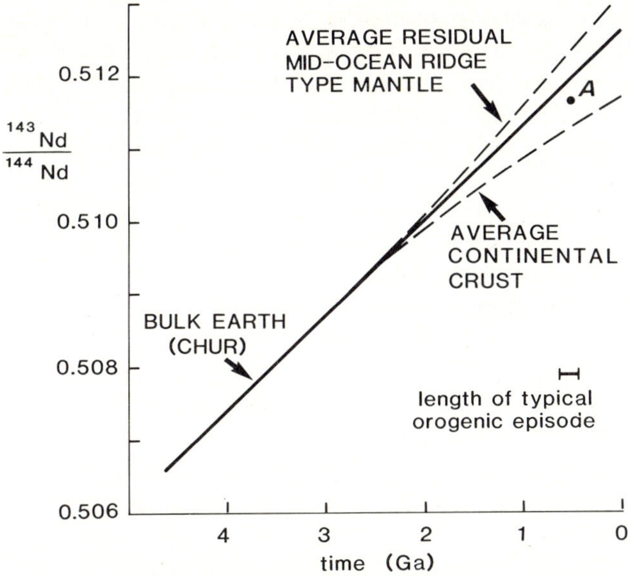

Fig. 11.8. An Nd isotope evolution diagram illustrating the estimated changes in $^{143}Nd/^{144}Nd$ in the bulk Earth, average depleted mid-ocean ridge-type mantle, and average continental crust over the 4.6 Ga since the formation of the Earth. Point *A* represents a hypothetical continental andesite erupted 500 Ma ago with an intermediate initial Nd isotope ratio (for discussion, see text).

bulk Earth (Fig. 11.8, and see also further discussion in section 11.7). In this context the crust and residual mantle are represented by average values; they have average isotope ratios and an average age of approximately 2 Ga; and these inevitably reflect the evolution of the crust and the mantle through most of the 4.6 Ga of Earth history — an immense period of time compared with that within which most geological processes operate.

Geologists, petrologists and traditional geochemists investigating, for example, a magmatic province or an orogenic belt are looking at comparatively brief episodes in the Earth's history. They are faced with suites of rocks that vary in age and composition, and which are often difficult to reconcile with those simple concepts of an average crust and mantle. Specifically, confusion arises in attempting to describe any rock or source region of intermediate composition, such as the andesite (*A*) in Fig. 11.8. In the context of the long time scale models for the Earth it is perfectly justified to describe an intermediate composition in terms of the two chosen end-members, the crust and residual mantle. *However, and this cannot be emphasized enough, that is not to suggest that such an intermediate composition was formed by physical mixing or chemical interaction between mantle and crustal material within the short time scale of whatever regional geological event is being studied.* In the case of the andesite in Fig. 11.8 its intermediate isotope ratio could be due to a number of causes including: (a) melting relatively young crust; (b) differentiation from basaltic magma derived from "enriched" upper mantle (see below); (c) melting a sediment which contained both volcanic (mantle-derived) and crustal detritus; and (d) interaction between a mantle-derived magma and the continental crust through which it passed en route to the surface. For a geologist considering such possibilities it can be positively misleading to be told by an isotope geochemist that because the particular sample has an intermediate isotope ratio it may be usefully described as a mixture of (average) crustal and mantle material — the more so when it requires just a tiny shift in innuendo to the conclusion that it is a "mantle-derived magma which has been contaminated by continental crust"!

There are many examples where confusion is at least partly due to such differences in the frames of reference set by isotope geochemists, and those set by petrologists and geologists concerned with comparatively short-term processes. "Crustal contamination" has also been used differently by different authors and so for the purposes of this discussion we shall regard it simply as contamination of mantle-derived magmas with crustal material *after they have left their source region*. It carries the clear implication that the initial isotope composition of the magmatic rock sampled on the surface was not the same as that in its source rock.

When $^{87}Sr/^{86}Sr$ ratios in continental volcanic rocks exceed those in oceanic areas, a common interpretation has been that crustal contamination has occurred (see Faure and Powell, 1972). A similar argument can be put forward using Nd isotopes since most mantle-derived rocks have higher $^{143}Nd/^{144}Nd$, and most crustal rocks lower $^{143}Nd/^{144}Nd$ than CHUR (Figs. 11.4 and 11.8). However, whereas the continental rocks do have consistently lower Sm/Nd ratios than CHUR (Fig. 11.6), basaltic, and thus presumably mantle-derived, rocks exhibit a range of Sm/Nd ratios, many of which are similar to those in crustal rocks. Assuming that these trace element ratios are similar to those in their source regions, then it clearly follows that segments of the upper mantle with low Sm/Nd ratios will generate the same relatively low $^{143}Nd/^{144}Nd$ ratios as crustal rocks *in the same period of time*. The problem is whether such enriched material can remain undisturbed in a convecting upper mantle for a sufficient length of time; ironically, the most likely place for that to happen is in the sub-continental mantle — within "old" continental lithosphere. Thus continental magmatic rocks, which are obviously the only ones that can be affected by crustal contamination as we have described it, are also those which are most likely to have unusual isotope compositions — e.g., high $^{87}Sr/^{86}Sr$ and low $^{143}Nd/^{144}Nd$.

The advent of combined Nd and Sr isotope studies has brought about a refinement in what is essentially the same argument. Many authors have suggested that since the majority of mantle-derived rocks plot on a single negative correlation between $^{143}Nd/^{144}Nd$ and $^{87}Sr/^{86}Sr$ (Fig. 11.7), then rocks which plot off that trend do so because they have been contaminated with continental crust (e.g., DePaolo and Wasserburg, 1979; Menzies and Murthy, 1980b). The most striking example is perhaps in the Tertiary rocks of Skye (Carter et al., 1978) where some of the basalts plot to the left of the main correlation and the granites plot to the right (Fig. 11.9). Analyses of Lewisian basement rocks were also available (Hamilton et al., 1979a) and Carter et al. (1978) therefore concluded that the basalts were contaminated with between 5 and 50% of Lewisian granulite material (low $^{87}Sr/^{86}Sr$, low $^{143}Nd/^{144}Nd$) and that the granites were contaminated with Lewisian amphibolite facies gneisses (high $^{87}Sr/^{86}Sr$, low $^{143}Nd/^{144}Nd$). This conclusion is consistent with that reached from Pb isotope studies (Moorbath and Welke, 1969; Dickin, 1981) although it is still unclear how such contamination takes place and what effect it has on the major and trace element geochemistry of such rocks.

On a more general level, since it has now been shown that some

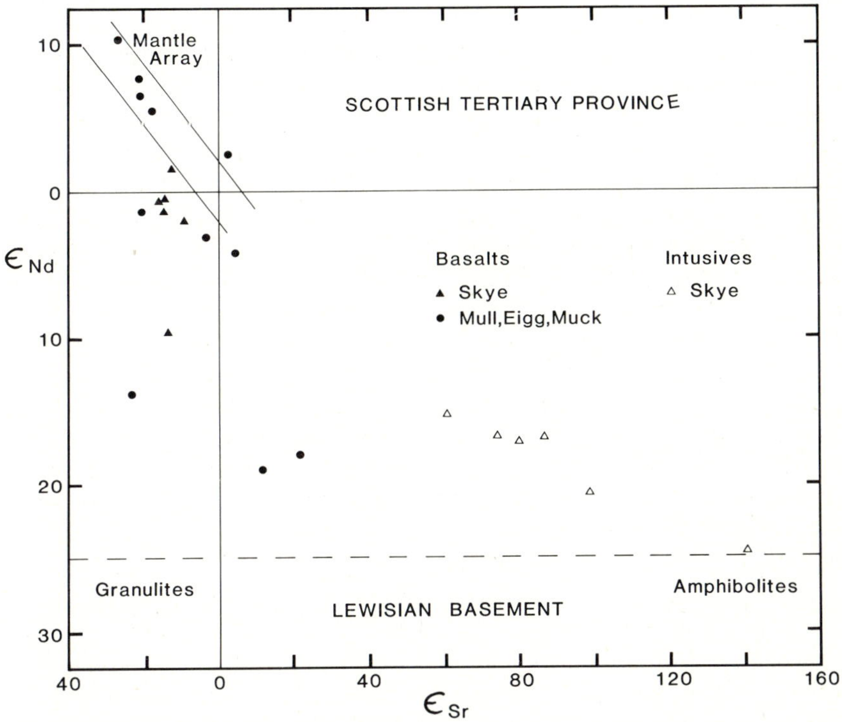

Fig. 11.9. Initial Nd and Sr isotope ratios of Tertiary igneous rocks from west Scotland compared to the "mantle array" and granulite- and amphibolite-facies gneisses from the Lewisian Basement. Data from Carter et al. (1978) who concluded that basalts in the bottom left-hand quadrant were contaminated with granulite material and that the granites reflected interaction with amphibolite facies gneisses.

"uncontaminated" mantle rocks do plot off the main trend between Nd and Sr isotopes (see Fig. 11.10), it is clearly naive to conclude that crustal contamination has occurred *simply because particular samples no longer plot on that main trend.* In practice, it is most important that the study of several isotope systems be linked closely with detailed petrology and geochemistry and probably one of the best examples of such a multi-disciplinary approach has been in the study of the volcano of Roccamonfina in central Italy.

The potassium-rich lavas of Roccamonfina have seen a number of hypotheses come and go over the past few years. They range from regional mixing between crustal and mantle end-members (Turi and Taylor, 1976; Vollmer, 1976, 1977), to anomalously enriched mantle (Appleton, 1972; Cox et al., 1976), to the present consensus that whereas primitive rock types

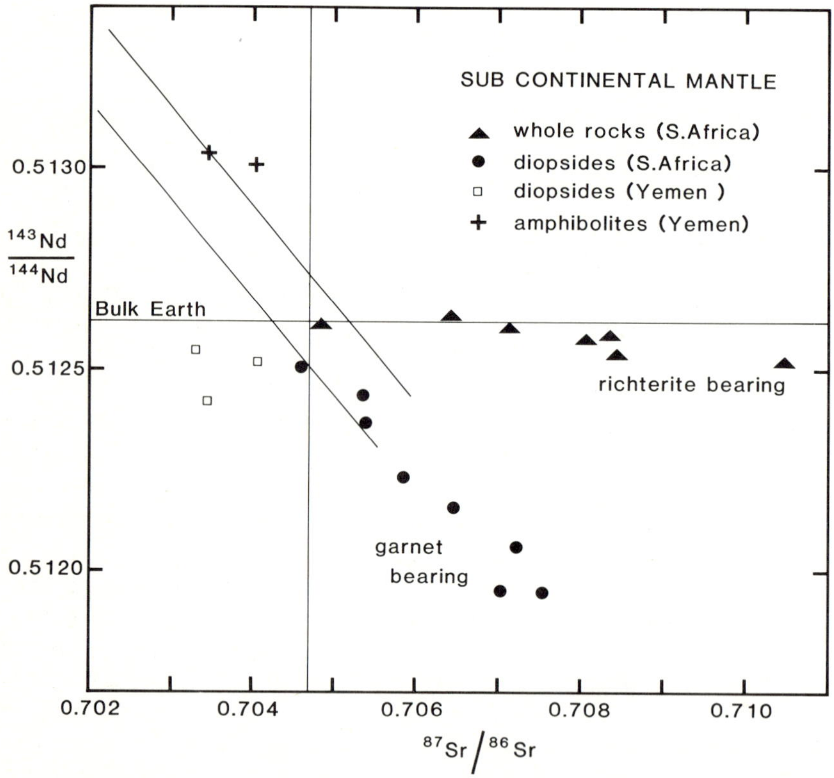

Fig. 11.10. Present-day $^{143}Nd/^{144}Nd$ and $^{87}Sr/^{86}Sr$ ratios in separated minerals and whole-rock xenoliths from the sub-continental mantle compared with the main correlation in most mantle-derived rocks (Menzies and Murthy, 1980a, b; C.J. Hawkesworth and A.J. Erlank, unpublished). Within the South African samples metasomatism appears to result in the formation of the K-rich amphibole richterite, usually at the expense of garnet (A.J. Erlank, personal communication).

reflect derivation from an enriched mantle source (Hawkesworth and Vollmer, 1979) the more evolved magmas experienced some crustal contamination during fractional crystallization in near-surface magma chambers (H.P. Taylor et al., 1979). The lessons to be learnt from these various studies are that stable isotopes, and $\delta^{18}O$ in particular, remain the best indication as to whether contamination has taken place (see H.P. Taylor, 1980) and that ^{87}Sr is often more sensitive than ^{143}Nd. (It is also worth noting that P.N. Taylor et al., 1980, have argued elsewhere that Pb isotopes are probably even more sensitive to the effects of crustal contamination than Sr.)

In the lavas of Roccamonfina $\delta^{18}O$ changes, as a function of chemical composition, from as low as +5.5 to +6.5‰ at 46—47% SiO_2 to over +10‰ at 60% SiO_2 (H.P. Taylor et al., 1979). The more primitive rock types have $\delta^{18}O$ values similar to those of many "uncontaminated" mantle-derived rocks (+5.5 to 6.5‰; H.P. Taylor, 1968), but the increase of $\delta^{18}O$ with increasing SiO_2 is greater than that which might be due to fractional crystallization (Anderson et al., 1971; Matsuhisa, 1979). This suggests that the high $^{87}Sr/^{86}Sr$ (0.7065—0.7096) and low $^{143}Nd/^{144}Nd$ (0.5124—0.5122) ratios in the primitive rock types do not reflect crustal contamination en route to the surface but were generated within the upper mantle — even though they plot well to the right of the main correlation in Fig. 11.7. In addition these rocks contain up to 1500—2000 ppm Sr so that it would require relatively large amounts of crustal strontium to cause a significant change in their $^{87}Sr/^{86}Sr$ ratios (Hawkesworth and Vollmer, 1979).

The clear demonstration that crustal contamination has contributed to the evolution of the higher-SiO_2 rocks at Roccamonfina raises the question of how such contamination takes place. Several models have been discussed including magma mixing (between crustal- and mantle-derived magmas), bulk assimilation of country rocks, and selective assimilation or isotopic exchange. Magma mixing is arguably a separate process since it is often recognizable on textural and mineralogical grounds (e.g., Eichelburger, 1975, 1978), and the equations to describe simple two-component mixing are well documented (Vollmer, 1976; Langmuir et al., 1978). For both bulk and selective assimilation, the magma must provide heat to the surrounding crustal rocks, which in turn encourages crystallization. The more evolved rock types will also be more prone to contamination because they represent smaller volumes of residual magma, and they are likely to have spent more time within the crust. Thus, as seen at Roccamonfina, the changes due to crustal contamination should be more pronounced in the more evolved rocks. Moreover, such arguments imply that for suites of related magmatic rocks, plots of $^{87}Sr/^{86}Sr$ (or $\delta^{18}O$) against a differentiation index such as SiO_2 provide reasonable method of assessing whether significant contamination with crustal material has taken place (Hawkesworth, 1982). H.P. Taylor (1980) has argued that similar tests may be applied using diagrams of $\delta^{18}O$ versus $^{87}Sr/^{86}Sr$.

In summary there is good evidence that crustal contamination of continental magmatic rocks does take place, and that where it has been documented in detail it tends to be accompanied by increasing differentiation. However, it has also been shown that combined isotope and major and trace element studies can satisfactorily distinguish the effects of contamination from those which are characteristic of a particular source region. The results of such combined studies suggest that ^{87}Sr may be more sensitive to contamination than ^{143}Nd so that it often results in a shallow trend, displaced to relative high ^{87}Sr/^{86}Sr, on the diagram of Nd against Sr isotopes (Hawkesworth, 1979). Finally it is important to retain a clear distinction between the effects of crustal contamination as defined here, and the interpretation of isotope ratios in source rocks which for the purposes of understanding longer-term processes may sometimes be described as mixtures of average crustal and mantle material.

Enrichment of incompatible elements in the upper mantle

As indicated in the introduction to this section the majority of mantle-derived rocks have lower ^{87}Sr/^{86}Sr and higher ^{143}Nd/^{144}Nd ratios than the bulk Earth, and their source rocks must therefore have been relatively depleted in Rb and Nd for much of their history. However, many of these magmatic rocks are enriched in incompatible elements and it is most unlikely that they were derived from a depleted source. We must, therefore, conclude that the enrichments of their source took place comparatively recently so that there was insufficient time for that enrichment to have affected the isotope ratios significantly. Such "decoupling" between trace element and isotope ratios is surprisingly common in mantle-derived rocks, and care must be taken to ensure that when a rock is described as "enriched" or "depleted" it is clear whether it refers to the trace element and/or the isotope composition.

A second point is that because Rb and Sr and the REE have such different chemical characteristics, different processes may result in distinctive fractionation patterns between Sr and Nd isotopes. For example, the alkali elements are more soluble in H_2O-rich fluids which should therefore cause a relative enrichment in Rb and Sr, and perhaps in ^{87}Sr.

Hawkesworth et al. (1979a) argued that the Nd and Sr isotope results available on uncontaminated basalts at that time could usefully be considered in two groups. Group I was simply those basalts which plotted on the main correlation (Fig. 11.7), whereas group II contained those samples displaced to relatively high ^{87}Sr/^{86}Sr ratios. Although the latter clearly represented a minute fraction of the upper mantle sampled by magmatism, examples were available in both oceanic (São Miguel) and continental (Roccamonfina and Vesuvius) areas and their importance lay in the fact that they must involve a different component to those in group I. Since then several studies have reported isotope results on upper mantle material transported to the surface in volcanic diatremes (Menzies and Murthy, 1980a, b; C.J. Hawkesworth and A.J. Erlank, unpublished) and as illustrated in Fig. 11.10 they plot both on, and to the left and the right of the main correlation between Nd and Sr isotopes. The picture is therefore more complex, but it is still useful to compare the variation in isotope ratios with that of Sm/Nd and Rb/Sr ratios in mantle-derived rocks (Fig. 11.11).

The most striking feature of the anti-correlation between Sm/Nd and Rb/Sr in Fig. 11.11 is that, in sharp contrast to the variation in $^{143}Nd/^{144}Nd$ and $^{87}Sr/^{86}Sr$ (Fig. 11.7) it does not include the point for the Sm/Nd and Rb/Sr ratios of the bulk Earth — which were themselves deduced from isotope studies. Also shown in Fig. 11.11 are the results of batch melting calculations on garnet lherzolite which was assumed to have the same Sm/Nd and Rb/Sr ratios as the bulk Earth (Hawkesworth et al., 1978). They demonstrate that for amounts of partial melting exceeding 5—10% little displacement occurs between melts and their source rocks when plotted in Fig. 11.11 and, since with time samples which plot in a particular quadrant in Fig. 11.11 should

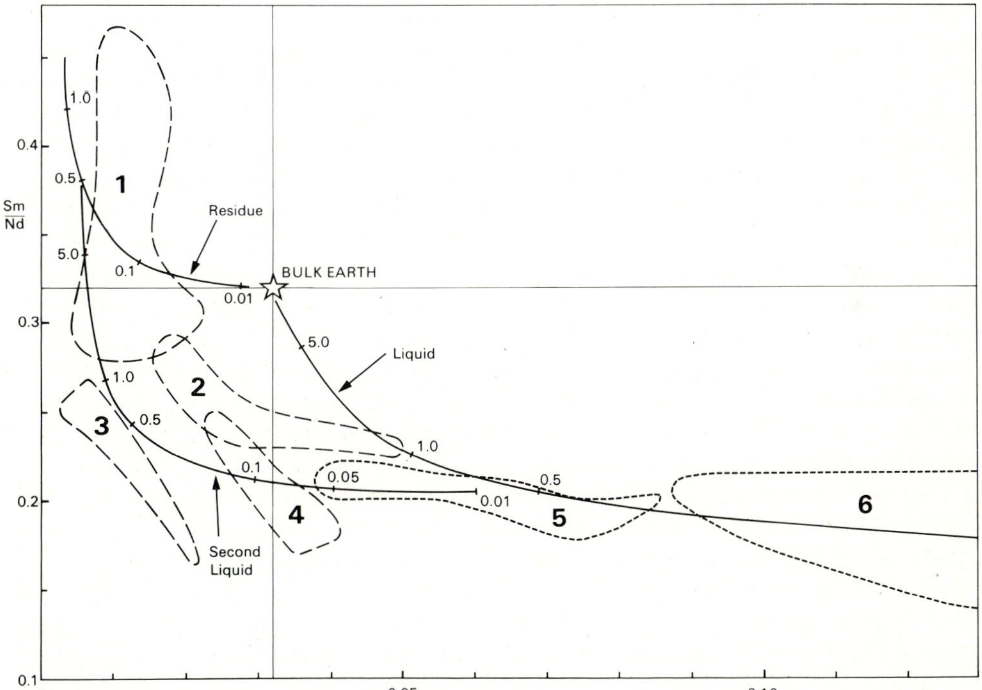

Fig. 11.11. Plot of Sm/Nd versus Rb/Sr in the more primitive rock types from Iceland (*1*), Hawaii (*2*) Etna (*3*), Patagonia (*4*), São Miguel (*5*), and Roccamonfina (*6*) (after Hawkesworth et al., 1979b). The results of batch melting calculations are from Hawkesworth et al. (1978) and have the melt fractions in percent.

evolve into the same quadrant on the equivalent isotope diagram (Figs. 11.7 and 11.10), they allow the following points to be emphasized:

(1) As indicated earlier, most mantle-derived rocks have lower $^{87}Sr/^{86}Sr$ and higher $^{143}Nd/^{144}Nd$ ratios than the bulk Earth (they plot in the top left-hand quadrant of Fig. 11.7) and yet many of them have relatively high Rb/Sr and low Sm/Nd ratios (Fig. 11.11). It was, therefore, suggested that the enrichment in incompatible elements observed in the basalts was due to comparatively recent enrichment of their source rocks.

(2) The majority of incompatible-element-enriched basalts plot in the bottom left-hand quadrant of Fig. 11.11 — they have both low Sm/Nd and Rb/Sr ratios. Thus although the enrichment process fractionated the REE, it appears to have increased the *amounts* of Rb and Sr without significantly changing their Rb/Sr ratios. Prolonged residence in this quadrant of Fig. 11.11 will result in Nd and Sr isotope compositions plotting to the left of the main isotope correlation (Fig. 11.7), and it is noticeable that such compositions have so far only been reported from continental volcanic rocks — Ataq (Menzies and Murthy, 1980a), the Oslo Rift (Jacobsen and Wasserburg, 1978), and Karoo and Tertiary volcanics from Southern Africa (C.J. Hawkesworth and J.S. Marsh, unpublished). Since "uncontaminated" mantle phases have been shown to plot to the left of the main correlation (Fig. 11.10) and the continental lithosphere is arguably the one area where material might be left undisturbed for long periods of time, the fact that the Scottish Tertiary basalts also have low Sm/Nd and Rb/Sr ratios leaves nagging doubts over the necessity of invoking up to 50% crustal contamination to explain their low Nd and Sr isotope ratios (Carter et al. 1978; Fig. 11.9).

(3) Few incompatible-element-enriched basalts have Rb/Sr ratios greater than ~0.04 (e.g., Fig. 11.11), but those that do also appear to have relatively high $^{87}Sr/^{86}Sr$ ratios and plot on comparatively flat-lying trends on diagrams of $^{143}Nd/^{144}Nd$ against $^{87}Sr/^{86}Sr$ (Fig. 11.7). They correspond to the group II rocks of Hawkesworth et al. (1979a) and show some isotope and trace element similarities with the results obtained on metasomatised K-richterite-bearing upper mantle xenoliths (Erlank, 1976). Erlank and Shimizu (1977) reported $^{87}Sr/^{86}Sr$ ratios of 0.705—0.710 and demonstrated that because of high Rb/Sr ratios (0.1—0.8), such isotopic differences could have been generated in ~150 Ma. Since they show little variation in Sm/Nd (0.12—0.16) or $^{143}Nd/^{144}Nd$ (0.5125—0.5126; C.J. Hawkesworth and A.J. Erlank, unpublished) they, like the group II basalts, are relatively enriched in both Rb and ^{87}Sr (Fig. 11.10), and it is tempting to ascribe both to similar processes. The metasomatism observed in the mantle xenoliths is characterized by the formation of the amphibole richterite: it is therefore H_2O-rich and, presumably for that reason, results in relatively high concentrations of alkali elements — which are also typical of the group II basalts (Hawkesworth and Vollmer, 1979; Hawkesworth et al., 1979a).

In summary, the majority of mantle-derived rocks which are enriched in incompatible elements have low Sm/Nd ratios but also relatively low Rb/Sr (Fig. 11.11). With time they would therefore tend to evolve to the left of the main correlation between Nd and Sr isotopes (Fig. 11.7), and the fact that few samples do not plot there suggests that the enrichment of incompatible elements in their source rocks took place comparatively recently so that there was insufficient time for their isotope ratios to be affected significantly. They contrast with a much smaller number of rocks which have both relatively high Rb/Sr and $^{87}Sr/^{86}Sr$ ratios (group II). The curves for batch melting on Fig. 11.10 indicate that melting in the range 2—8% will cause a reduction in Sm/Nd in the liquid with little increase in Rb/Sr. Although these are clearly approximate figures, they suggest that the group I type of enrichment may be due to the migration of small partial melts, whereas that of group II reflects the movement of H_2O-rich, and hence alkali-rich, fluids — "mantle metasomatism" in the strict sense.

We have discussed some specific differences in the types of enrichment of incompatible elements observed in mantle-derived rocks, but there is also the question of *when* such enrichment took place. In principle, since we are dealing with radiogenic isotopes it should be possible to estimate the age of an enrichment event, but there are obstacles. Ages are calculated using the isochron equation (equation (11.6)) and, as in Figs. 11.1 and 11.2, that requires that there is a relationship between $^{147}Sm/^{144}Nd$ and $^{143}Nd/^{144}Nd$ due to the decay of ^{147}Sm to ^{143}Nd over the period of time t. However, it has been argued above that in many enriched basalts the isotope and trace element ratios are "decoupled": there is no relationship between $^{143}Nd/^{144}Nd$ and Sm/Nd ratios, and in that case it is only possible to deduce a maximum age for the enrichment event.

In other areas positive correlations have been observed between Sm/Nd and the *initial* $^{143}Nd/^{144}Nd$ ratios within suites of related basalts (e.g., Fig. 11.12). Often referred to as "mantle isochrons" (Brooks et al., 1976) their slope only reflects the age of some mantle event if the Sm/Nd and $^{143}Nd/^{144}Nd$ ratios of each sample faithfully reflect those in its source region, and if the latter has remained undisturbed since the mantle event. In practice, however, two-component mixing can also generate straight-line relationships on an isochron diagram (e.g., Langmuir et al., 1978), and to assess the significance of a particular "age" it is necessary to enquire whether it can be reproduced using a different decay scheme. At Roccamonfina, for example, "ages" determined on the same samples range from ~2.0 Ga for Sm-Nd (Fig. 11.12), to ~500 Ma for Rb-Sr (Cox et al., 1976), to almost 0 Ma for $^{207}Pb/^{204}Pb$ versus $^{206}Pb/^{204}Pb$ (Vollmer and Hawkesworth, 1980). Hawkesworth and Vollmer (1979), therefore, argued that these "isochrons" carried no age significance and were probably generated by comparatively young (less than 90 Ma) mixing processes within the upper mantle. At this time we are not aware of any ages calculated from "mantle isochrons" which

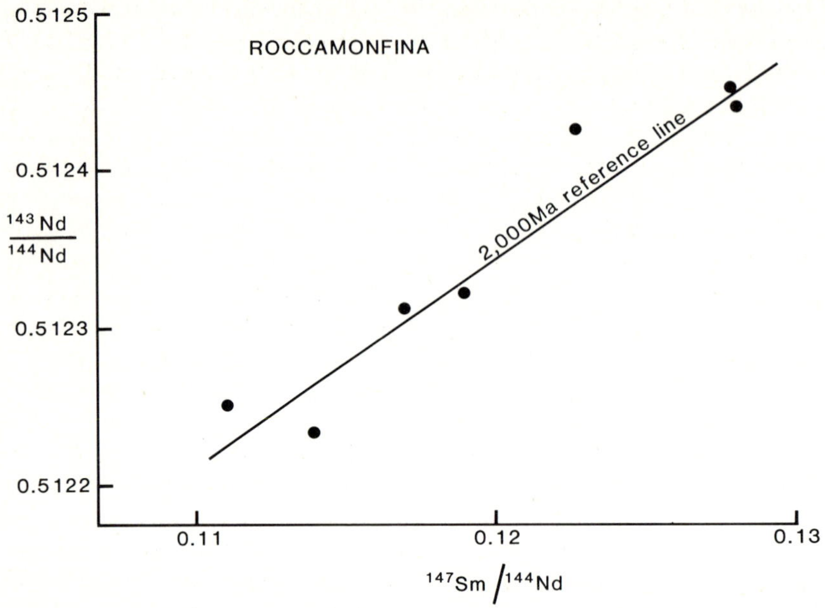

Fig. 11.12. Variation of initial $^{143}Nd/^{144}Nd$ and $^{147}Sm/^{144}Nd$ ratios in the primitive lavas of Roccamonfina (Hawkesworth and Vollmer, 1979). The slope of the straight line corresponds to an "age" of 2000 Ma, but since different ages were obtained from Rb-Sr and Pb-Pb analyses on the same samples they are thought to have no age significance: they could reflect comparatively recent two-component mixing (see also Langmuir et al., 1978).

have stood the test of being reproduced by more than one decay scheme, and conclude that the incompatible-element-enriched nature of many volcanic rocks reflects the relatively recent movement of such elements into their mantle source regions — perhaps within the last two hundred million years.

Destructive plate margins

Crustal material is both created and destroyed along destructive plate margins and the net balance effectively determines at least recent trends in the evolution of the crust — and hence the upper mantle. Radiogenic isotopes have been widely used, partly because they form the basis of most models for crustal evolution, and partly because the different materials which might contribute to magmas generated in this environment tend to have different isotope ratios (see Hawkesworth, 1979, 1982, for review). They may include material from both the subducted ocean crust (fresh and altered basalt and continental and oceanic sediment) and the over-riding mantle wedge, with the added complication that in continental areas the magmas may also suffer contamination en route to the surface. The study of at least the isotope and trace element geochemistry of destructive plate margin magmas thus embraces a range of topics from the behaviour of elements in sediments and the oceanic cycle to the effects of crustal contamination!

Our discussion of the Nd and Sr isotope ratios in recent "uncontaminated" mantle rocks emphasized that the majority plotted on a broad negative correlation between $^{143}Nd/^{144}Nd$ and $^{87}Sr/^{86}Sr$ ratios (Fig. 11.7). Thus before they are affected by any input of material from the subducted oceanic crust, the rocks of the over-riding mantle wedge are likely to plot on that main correlation between Nd and Sr isotopes. In a very small number of cases they might plot on a flat-lying trend similar to the group II rocks. Similarly, for the subducted crust the majority of unaltered MORB plot at the low $^{87}Sr/^{86}Sr$, high $^{143}Nd/^{144}Nd$ end of the main correlation, whereas the ocean island rocks, which are volumetrically much less significant, tend to have higher $^{87}Sr/^{86}Sr$ and lower $^{143}Nd/^{144}Nd$ ratios (Fig. 11.7).

Combined Nd and Sr isotope studies have been reported on island arc volcanic rocks from the South Sandwich Islands (Hawkesworth et al., 1977), the Marianas (De Paolo and Wasserburg, 1977), New Britain (DePaolo and Johnson, 1979), the Lesser Antilles (Hawkesworth et al., 1979c; Hawkesworth and Powell, 1980), and Indonesia (Whitford et al., 1980). The results are summarized in Fig. 11.13 and the majority are displaced to high $^{87}Sr/^{86}Sr$ ratios relative to the main trend of results on basalts from mid-ocean ridge and intraplate settings. For some reason, which is not understood, the rocks from the Atlantic island arcs appear to be displaced to higher $^{87}Sr/^{86}Sr$ ratios than those from the southwest Pacific although one possibility is that it is due to differences in the isotope ratios of the over-riding mantle wedge. For example, Hawkesworth (1979) pointed out that $^{87}Sr/^{86}Sr$ ratios in back-arc basalts from the Marianas are as low as 0.7026 so that ratios of 0.7030 in the arc volcanics (Fig. 11.13) might still represent a shift to higher $^{87}Sr/^{86}Sr$. Work is presently in progress to test this suggestion, but it serves to highlight problems which may arise from comparing individual arc basalts with the broad average trend between $^{143}Nd/^{144}Nd$ and $^{87}Sr/^{86}Sr$ for most mantle-derived rocks.

For an explanation of these relatively high $^{87}Sr/^{86}Sr$ ratios in island arc magmas, it is necessary to turn to the other components of the subducted oceanic crust. First, hydrothermal systems along mid-ocean ridges tend to increase the $^{87}Sr/^{86}Sr$ ratios of the altered basalts without significantly changing $^{143}Nd/^{144}Nd$. This presumably reflects the high concentration of Sr in seawater (8 μg g^{-1}) compared with that of Nd (3 pg g^{-1}), and the net effect is to displace altered MORB to the right of the main correlation between Nd and Sr isotopes (Fig. 11.14). Second, seawater and hence authigenic sediments, have essentially constant $^{87}Sr/^{86}Sr$ = 0.7091 but $^{143}Nd/^{144}Nd$ varies from 0.5125 to 0.5119 depending on the particular ocean (Piepgras and Wasserburg, 1980). Continental detritus increases their Sr and reduces their Nd isotope ratios (Fig. 11.14) but in general all sediments tend to have relatively high $^{87}Sr/^{86}Sr$. Thus the displacement to higher $^{87}Sr/^{86}Sr$

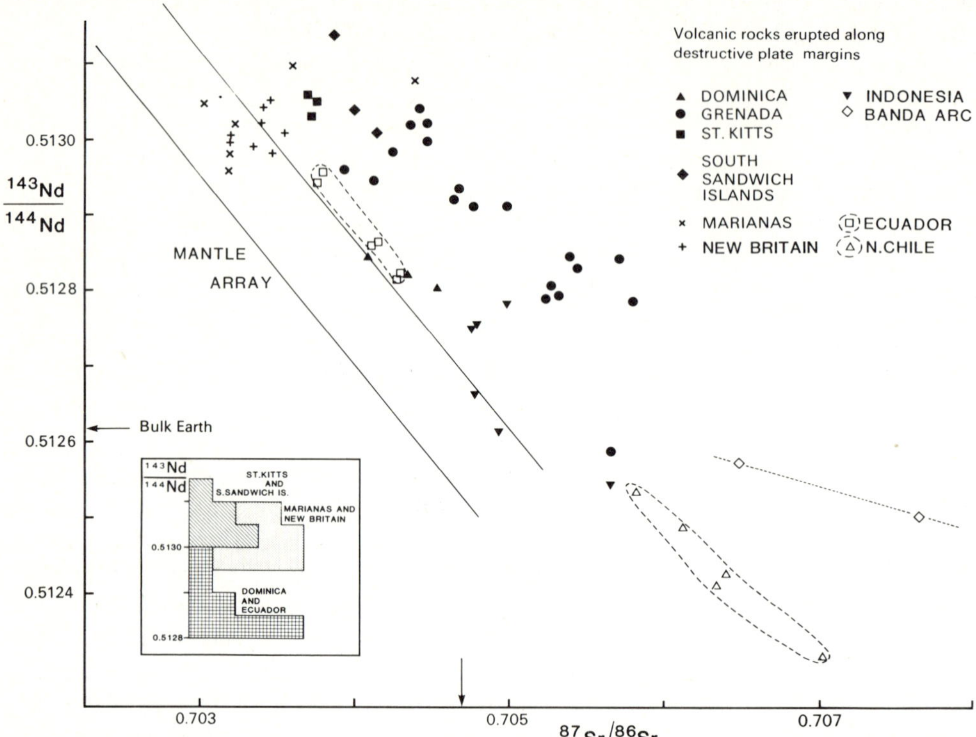

Fig. 11.13. Variation in Nd and Sr isotopes in recent volcanic rocks erupted along destructive plate margins (for data sources, see text). Two further samples from the Banda Arc plot off the diagram, but on the trend of the dotted line (Whitford et al., 1980). *Inset*: Histogram of $^{143}Nd/^{144}Nd$ ratios in island arc tholeiites (St. Kitts and South Sandwich Islands) and mature calc-alkaline provinces (Dominica and Ecuador). The rocks from New Britain and the Marianas are transitional in that although many are calc-alkaline, few are significantly enriched in LREE.

ratios observed in many island arc rocks is attributed to a significant component from either altered ocean floor basalt and/or subducted sediment.

It is well documented that compared with volcanic rocks from mid-ocean ridge and intraplate environments, those from island arcs are also preferentially enriched in elements such as Ba, Cs, Rb, K, Sr (henceforth referred to as "alkaline") and Pb relative to less mobile elements such as the REE, Zr and Y (Gill, 1974; Hawkesworth et al., 1977; Kay, 1977; Pearce, 1982). Hawkesworth and Powell (1980) emphasized this point using a diagram of Sm/Ce against Sr/Ce, on which the relatively high Sr/Ce ratios of the island arc rocks are believed to be the trace element analogue of the displacement to high $^{87}Sr/^{86}Sr$ revealed by combined Nd and Sr isotope studies (Fig. 11.13). Both are attributed to the release of material from the

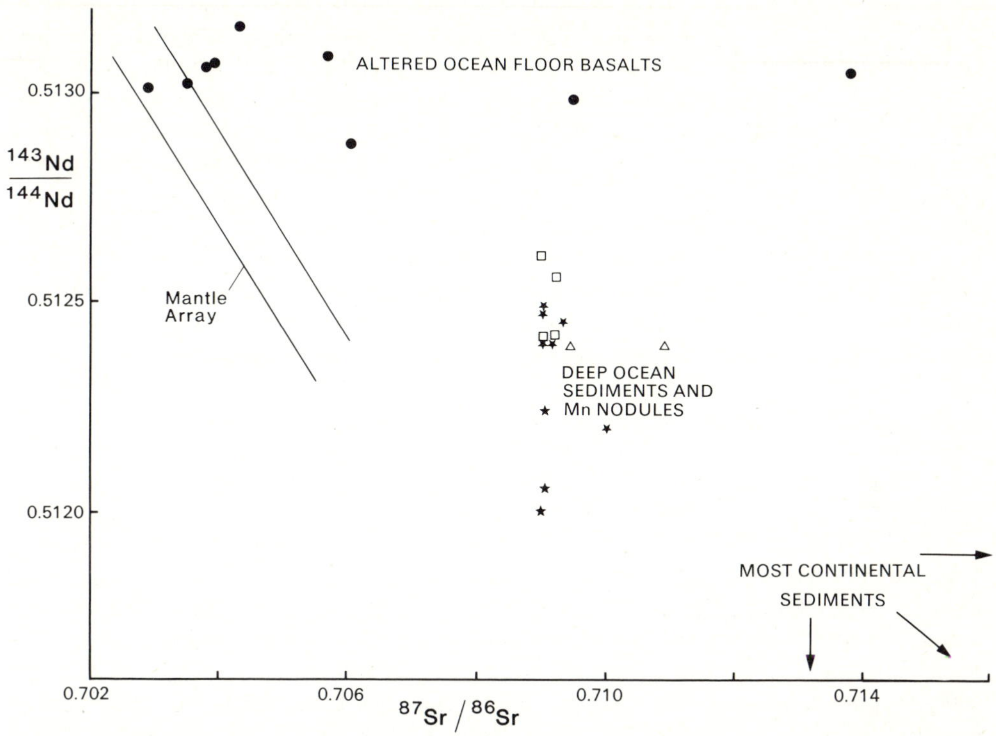

Fig. 11.14. $^{143}Nd/^{144}Nd$ and $^{87}Sr/^{86}Sr$ ratios in hydrothermally altered ocean floor basalts, deep ocean sediments and Mn-nodules (⋆) from the Atlantic and Pacific Oceans, and continental sediments (McCulloch and Wasserburg, 1978; O'Nions et al., 1978; Piepgras et al., 1979, and Elderfield et al., 1981).

subducted slab, but what about less mobile elements such as the REE — are they also remobilized?

The histogram in Fig. 11.13 demonstrates that there appears to be a small but significant shift in $^{143}Nd/^{144}Nd$ from the island arc tholeiites of the South Sandwich Islands and St. Kitts to the more enriched calc-alkaline rocks of Dominica and Ecuador. Most models for the genesis of arc tholeiites attribute them to partial melting in the upper mantle wedge above the subduction zone (Ringwood, 1974). Typically they are relatively enriched in "alkaline" elements and ^{87}Sr, and yet depleted in LREE (e.g., Hawkesworth et al., 1977; Kay, 1977; Hawkesworth and Powell, 1980). This requires the addition of material containing "alkaline" elements (and ^{87}Sr) but little or no REE, which in turn suggests that their high $^{143}Nd/^{144}Nd$ ratios (Fig. 11.13) reflect those of their source region in the over-riding mantle wedge.

The results on calc-alkaline rocks are perhaps more ambiguous, principally because, although they are still relatively enriched in "alkaline" elements and ^{87}Sr, they also contain higher concentrations of less mobile elements, such as the LREE, Zr and Y, than arc tholeiites. In addition they tend to have lower $^{143}Nd/^{144}Nd$ (Fig. 11.13) and more radiogenic Pb isotope ratios (e.g., Donnelly et al., 1971; Sun, 1979) than tholeiitic rocks. Such differences in isotope and trace element abundances might reflect either differences in the material introduced from the subducted slab and/or chemical variations which existed in the over-riding mantle wedge before subduction commenced. The latter implies that coincidently calc-alkaline rocks are generated when subduction occurs beneath more "enriched" mantle, whereas tholeiites result if it occurs beneath "depleted" mantle. That is a little unlikely and so it would appear that the tendency for calc-alkaline rocks to have lower $^{143}Nd/^{144}Nd$ ratios is due to a greater contribution of "continental" Nd introduced from the subducted ocean crust (see Kay, 1980, and Hawkesworth, 1979, 1982, for more detailed discussions). The rocks from the Banda Arc appear to be an extreme example since their high $^{87}Sr/^{86}Sr$ ratios, and the flat-lying trend in Fig. 11.13 are accompanied by very high $\delta^{18}O$ values attributed to the melting of subducted sediments (Magaritz et al., 1978; Whitford et al., 1980).

Combined Nd and Sr isotope studies have therefore demonstrated that magmas erupted along destructive plate margins contain a significant component from the subducted ocean crust. Moreover, since that component reflects the influence of seawater (either via authigenic sediments or the hydrothermal alteration of ocean floor basalts) whose composition is in turn largely governed by the influx of continental material to the oceans (section 11.6), a significant proportion of at least the more mobile trace elements being added to the crust above subduction zones is being "recycled" — those trace elements have been in the crust before! Recent estimates for the amount of recycled Sr and Nd in island arc volcanic rocks are in the range 20—40% and 10—35%, respectively (Kay, 1980).

Most of this discussion has been illustrated with examples from oceanic areas but some results are available on rocks generated along continental margins (Hawkesworth et al., 1979b; Tilton, 1980). Those from Ecuador are generally believed not to have been affected by contamination with crustal material en route to the surface (Francis et al., 1977; Harmon et al., 1981) and combined Nd and Sr isotope results show that like so many island arc volcanics, they have relatively high $^{87}Sr/^{86}Sr$ compared with the main correlation for most mantle-derived rocks (Fig. 11.13). Thus even though most authors have argued that, for example, continental andesites and island arc tholeiites are generated by different processes (e.g., Ringwood, 1974, and references therein), their $^{143}Nd/^{144}Nd$ and $^{87}Sr/^{86}Sr$ ratios suggest that similar processes have influenced at least their Sr isotope geochemistry (Hawkesworth et al., 1979b).

The results on the andesites from northern Chile are more ambiguous; they have higher Sr and lower Nd isotope ratios (Fig. 11.13) and one interpretation is that this reflects crustal contamination of magmas similar to those erupted in Ecuador (Francis et al., 1977). The crust beneath northern Chile is certainly thicker than that beneath Ecuador and this is usually cited as the reason that the northern Chilean magmas are more prone to contamination (Thorpe and Francis, 1979). However, not only do some of the northern Chilean magmatic suites show no evidence that such contamination was accompanied by increasing differentiation, they also tend to have different trace element ratios to "uncontaminated" andesites from, for example, Dominica or Ecuador (Hawkesworth, 1982; Hawkesworth et al., 1982). Thus while crustal contamination may well have contributed to the present composition of some northern Chilean andesites, an alternative model is that other more evolved suites may have been generated by remelting pre-existing crustal material. This envisages that the development of a thicker crust both provides a physical barrier with the result that fewer mantle-derived magmas reach the surface, and promotes intracrustal melting as the isotherms rise in response to radioactive decay.

Such controversy over models for the generation of the northern Chilean andesites highlights the point made in the introduction to this section. On the long time scale of models for the evolution of the crust and upper mantle, the higher $^{87}Sr/^{86}Sr$ of such andesites may be taken to reflect a "greater crustal component". Yet on the much shorter time scale over which magmas are generated their isotope ratios do not necessarily imply that their formation involved a physical mixing process between crustal- and mantle-derived end-members.

11.6. Isotope variations in seawater

Chemical oceanography is concerned with the behaviour of elements as they enter, move about in, and are removed from the waters of the ocean basins. They are added from the continents and by interaction with the rocks of the oceanic crust, and removed by the formation of authigenic material on the ocean floor. On the broader scale the oceans represent the avenue through which much of the continental crust is returned, via subduction, to the upper mantle.

The fractionation of individual REE (e.g., Goldberg et al., 1963; Høgdahl et al., 1968; Elderfield et al., 1981) is discussed in Chapter 10 and this section is restricted to the application of radiogenic isotopes. The variation in Sm/Nd ratios in the rocks of the continental crust and the upper mantle (Fig. 11.6) has resulted in different $^{143}Nd/^{144}Nd$ ratios, so that the Nd isotope composition of seawater will be sensitive to the age, and the proportions of crustal and mantle-derived REE. Similar arguments can be made for the isotopes of Pb and Sr (e.g., Chow and Patterson, 1962; Spooner, 1976).

The average length of time an element remains in seawater is called its *residence time* t_A, defined by:

$$t_A = \frac{A}{dA/dt}$$

where A is the total amount of the element dissolved in the oceans, and dA/dt is the amount of the element introduced into or removed from the ocean per unit time, assuming steady state (Goldberg et al., 1963). It is estimated that it takes about 10^3 years for the oceans "to turn over", that is, for them to have a reasonable chance of mixing. Thus elements which have residence times that are significantly *longer* than the estimated ocean turn-over time should become well mixed and isotopically homogeneous in seawater. Conversely if the residence time is much *shorter* than the turn-over time, complete mixing cannot occur and so seawater will have variable isotope ratios reflecting regional source compositions. Sr and Nd illustrate this point well. The former has a residence time of ~20 Ma, much longer than the estimated turn-over time of 10^3 years, and the present-day Sr isotope composition of seawater is very uniform ($^{87}Sr/^{86}Sr = 0.7091$; Faure and Powell, 1972). Nd by contrast has a very short residence time (~270 years; Goldberg et al., 1963, and more recent discussion by Piepgras and Wasserburg, 1980) and so we may predict that seawater will have variable $^{143}Nd/^{144}Nd$ ratios. Note also that because ^{87}Rb and ^{147}Sm have such long half-lives compared to the residence times of Sr and Nd, there is insufficient time of $^{87}Sr/^{86}Sr$ and $^{143}Nd/^{144}Nd$ to be changed significantly by radioactive decay while these elements reside in the oceans.

The Nd isotope composition of seawater can be determined indirectly by analysing samples such as Mn-nodules which are derived largely from seawater and which concentrate the REE (O'Nions et al., 1978; Piepgras et al., 1979; Elderfield et al., 1981), or directly by analysing Nd which has been coprecipitated from seawater (Piepgras and Wasserburg, 1980). Selected results are presented in Fig. 11.15; several points can be emphasized:

(1) There is a very large spread in the measured $^{143}Nd/^{144}Nd$ ratios, 0.5119—05125, although for samples from any one ocean, the range is comparatively restricted.

(2) Mn-nodules and seawater from the same ocean have very similar $^{143}Nd/^{144}Nd$ ratios, confirming that most of the Nd in the Mn-nodules was derived ultimately from seawater. Moreover, since most of the nodules have the same $^{87}Sr/^{86}Sr$ ratio as that of seawater, the resultant variation in $^{143}Nd/^{144}Nd$ with little change in $^{87}Sr/^{86}Sr$ (Fig. 11.14) simply reflects the variations in the different ocean masses.

(3) Piepgras et al. (1979) suggested that the shift to slightly higher $^{143}Nd/^{144}Nd$ ratios observed in several "hydrothermal" sediments (Fig. 11.15) reflects a small contribution from mid-ocean ridge-type mantle.

(4) All the measured $^{143}Nd/^{144}Nd$ ratios are less than that of the bulk Earth at the present day, demonstrating that the Nd isotopes composition of seawater is controlled dominantly by the input of material from

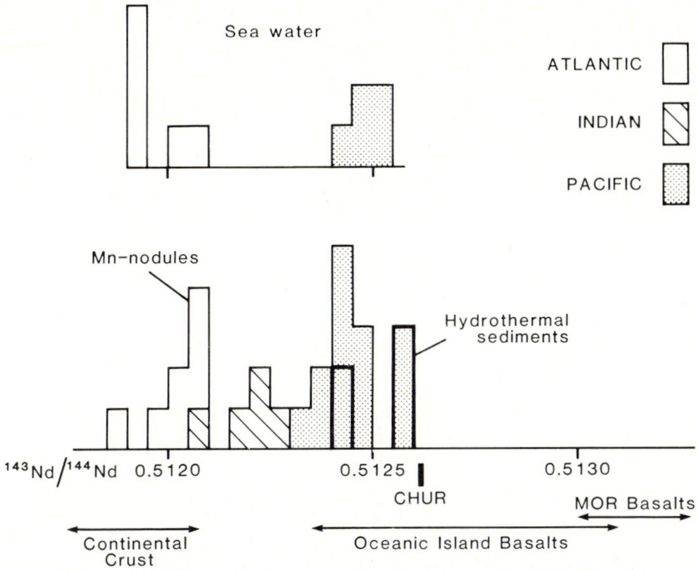

Fig. 11.15. Histogram of $^{143}Nd/^{144}Nd$ ratios in seawater, Mn-nodules, and "hydrothermal" Fe-Mn sediments (after Piepgras and Wasserburg, 1980). Data also from O'Nions et al. (1978), Piepgras et al. (1979), and Elderfield et al. (1981).

continental areas (Fig. 11.15). However, precise estimates of the proportions of continental and mantle-derived Nd are difficult because of the lack of $^{143}Nd/^{144}Nd$ measurements on samples of river water.

The problems in interpreting the actual $^{143}Nd/^{144}Nd$ ratio of a portion of ocean water may be usefully illustrated by considering the differences in the composition of Nd in the Atlantic and Pacific Oceans. Accepting the generalized values for $^{143}Nd/^{144}Nd$ in the continental crust from Fig. 11.15 suggests that it contributes at least 80% of the Nd in the Atlantic Ocean and a minimum of 50% in the Pacific (Piepgras and Wasserburg, 1980), which in turn suggests that for some reason the Pacific Ocean receives a much greater contribution from the oceanic upper mantle. Alternatively it can be argued that the continental component in the Atlantic is derived from much older crust and thus has lower $^{143}Nd/^{144}Nd$. Piepgras et al. (1979), using a typical crustal Sm/Nd ratio of ~0.19 (see Fig. 11.6), calculated average $T_{\mathrm{CHUR}}^{\mathrm{Nd}}$ ages for the Atlantic, Indian and Pacific Ocean samples of 1.2 Ga, 0.85 Ga and 0.35 Ga, respectively. Whatever the details of this greater continental component in the Atlantic Ocean water it is interesting to note that continental drainage patterns suggest that it receives much more fresh water than the Pacific (Budyko, 1974) and that lead in Atlantic Mn-nodules also tends to be more radiogenic than that in the Pacific (Chow and Patterson, 1962).

In summary, Nd has a short residence time in seawater: there is, therefore, insufficient time for complete mixing to occur and so different masses of ocean water retain different $^{143}Nd/^{144}Nd$ ratios. This is most strikingly illustrated between ocean basins (Fig. 11.15) but there is some evidence from the Atlantic Ocean that such differences may be maintained even within a single ocean. The samples of Atlantic water from below 3000 m are less radiogenic and more uniform in composition than those from shallower depths (Piepgras and Wasserburg, 1980). Such isotope variations contrast with the uniform $^{87}Sr/^{86}Sr$ ratio of present-day seawater, but they indicate that, providing the isotope composition of seawater is faithfully reproduced by fossil debris and chemical sediments, combined Nd and Sr isotope studies offer exciting new insight into the nature and evolution of ancient ocean basins.

11.7. Models for the evolution of the Earth's crust and mantle

The last two sections concentrated on the application of radiogenic isotopes to the study of processes which were instantaneous compared with the rate of decay of the parent isotope, so that isotope ratios were essentially used as tracers in the different processes. We turn now to the much longer span of time since the formation of the Earth and to attempts to assess the cumulative effect of these "short-term" geological processes in models for the evolution of the crust and at least the uppermost mantle. Such models provide a unique approach to understanding the history of the Earth and as such they have often been the long-term objective behind more localized isotope studies.

The majority of analysed Archaean igneous rocks have initial Nd isotope ratios similar to that of a chondritic reservoir (CHUR) at the time of their formation (Fig. 11.3), and yet at the present-day $^{143}Nd/^{144}Nd$ ratios measured in comparatively young rocks vary from 0.5133 to 0.5113 (e.g. Fig. 11.4). Clearly the differentiation of the Earth into segments, or domains, with different Sm/Nd ratios has in time resulted in significant variations in the isotope composition of neodymium. The questions to be addressed are how and when such differentiation occurred, and whether it happened continuously or in a few major catastrophic events?

All models assume that the Earth had initial Sm/Nd and $^{143}Nd/^{144}Nd$ ratios similar to chondrites, and that the "depleted" characteristics of much of the upper mantle sampled by recent magmatism (lower $^{87}Sr/^{86}Sr$, higher $^{143}Nd/^{144}Nd$ than the bulk Earth, Fig. 11.7) are due ultimately to the formation of the crust (relatively high Rb/Sr, low Sm/Nd). The simplest approach is to envisage that the Earth has evolved in two discrete stages: the first as a uniform reservoir similar to CHUR, and a second in which it contains both stable continental crust and depleted mid-ocean ridge-type mantle. The timing of the onset of stage 2 is given by the T^{Nd}_{CHUR} ages of average continental crust and mid-ocean ridge-type mantle, and an obvious first requirement is that both should agree. Average compositions are notoriously difficult to determine accurately but using those compiled by Jacobsen and Wasserburg (1979b), the T^{Nd}_{CHUR} ages for depleted mantle ($^{143}Nd/^{144}Nd$ = 0.51313, Sm/Nd = 0.395) and continental crust ($^{143}Nd/^{144}Nd$ = 0.51172, Sm/Nd = 0.192) are 1.8 Ga and 1.7 Ga, respectively.

One point which applies to this and more complex models should be clarified immediately. The slow rate of decay of ^{147}Sm to ^{143}Nd ensures that it takes time, often several 100 Ma, to generate significant differences in $^{143}Nd/^{144}Nd$. Thus to describe the evolution of the early history of the Earth in terms of a uniform reservoir similar to CHUR is not, paradoxically, to imply that there was no continental crust — or depleted mantle. It is simply that no such chemical differences can have been preserved for long enough to generate significant variations in $^{143}Nd/^{144}Nd$. Models for the evolution of the Earth based on radiogenic isotopes must inevitably be concerned not so much with the *generation* of chemical variations, as with their tendency to become increasingly *stabilized* later in Earth history. The implication is that as the amount of heat produced within the Earth decreases, there is less energy to dissipate and so segments of different composition are more likely to be left undisturbed for long periods of time.

In practice, simple two-stage models are not viable for a number of reasons. Initially there is the conceptual problem of understanding what it was that brought about the comparatively sudden change from stage 1 to stage 2. Secondly, such models are inconsistent with the distribution of, for example, Sm/Nd and Rb/Sr in recent basalts (Fig. 11.11), assuming that that is similar to the variations in the source regions of these basalts. The calculated curves for the partial melting of garnet-bearing peridotite in Fig. 11.11 suggest that it is not possible to derive samples which plot in the bottom left-hand quadrant from peridotite with trace element ratios similar to those of the bulk Earth in a single differentiation event. The majority of incompatible-element-enriched basalts plot in that bottom left-hand corner and probably reflect the migration of trace elements from source regions that had been depleted previously (low Rb/Sr, high Sm/Nd) (Hawkesworth et al., 1979a). Thirdly, the recent values of Sm/Nd and $^{143}Nd/^{144}Nd$ for CHUR proposed by Jacobsen and Wasserburg (1980) suggest that some portions of the upper mantle may have had relatively high $^{143}Nd/^{144}Nd$, indicative of LREE depletion, in the Archaean (Fig. 11.3). Although this interpretation is perhaps open to question in view of the controversy over the precise Sm-Nd composition of CHUR, it is at least consistent with the relatively low initial $^{87}Sr/^{86}Sr$ ratios of many Archaean volcanic rocks. Fourthly, it is not possible to generate the observed negative correlation between Nd and Sr isotopes in recent basalts (Fig. 11.7) after any single differentiation event (Hawkesworth et al., 1978; Jacobsen and Wasserburg, 1979a; O'Nions et al., 1979b).

More complex models for the evolution of the mantle and the crust have been recently described by Jacobsen and Wasserburg (1979b), O'Nions et al. (1979b), and Allègre and Othman (1980). Often they build on previous models developed to explain variations in other isotope systems and for a detailed explanation the reader is referred to the primary literature sources. This section offers only a brief resume of some of the ideas behind such models and an indication of the conclusions that result. Fig. 11.16 illustrates two models described by Jacobsen and Wasserburg (1979b). In the first, all new continental crust is derived by equilibrium partial melting of undepleted, or primitive, mantle leaving a residuum of depleted mantle. Apart from when new crust is generated there is no subsequent interaction between the three reservoirs of continental crust, depleted and undepleted mantle; and with time the masses of crust and depleted mantle increase at the expense of undepleted mantle. In the second model continental crust is derived from a reservoir of mantle which is homogenized after each crust-forming event so that the whole reservoir becomes increasingly depleted as more and more crust is generated. In both cases the size of the mantle reservoir from which the crust is derived is unknown, and the oceanic crust is presumed to be part of the depleted mantle reservoir (it is formed from that reservoir and subducted back into it within 200 Ma). Moreover, because Jacobsen and Wasserburg (1979b) were concerned with the formation of *continental* crust and the effect that had on the residual mantle, they felt it safe to assume that the transport of material is solely from the mantle to the crust and initially at least did not treat the flux of crustal material back into the mantle. In contrast, O'Nions et al. (1979b) considered the evolution of an

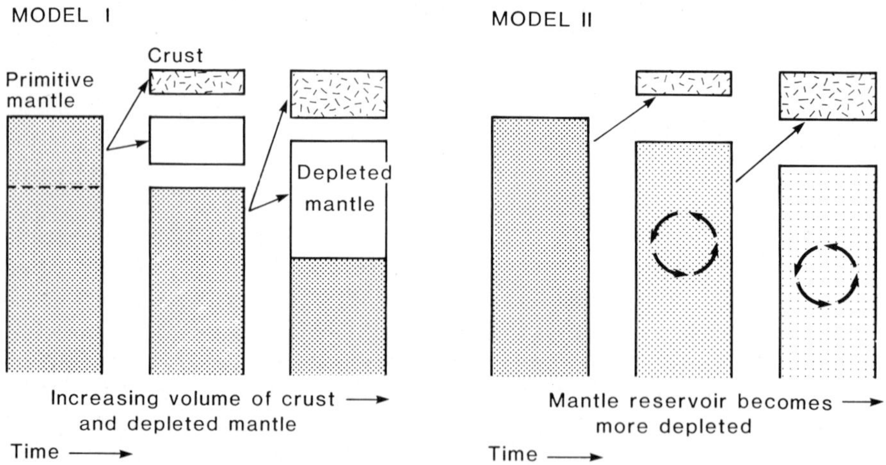

Fig. 11.16. Cartoons illustrating the major processes in models I and II of Jacobsen and Wasserburg (1979b) for the formation of continental crust and the evolution of the crust-mantle system. In model I, depleted and undepleted mantle remain separate, whereas in model II, the residue after partial melting is instantly mixed back into the remainder of the depleted mantle. In both models a sizeable mass of mantle is *not* involved in the generation of continental crust.

outer "crustal" reservoir 50 km thick and, since this includes the oceanic crust, they inevitably arrived at much larger figures for the size of the return flux into the mantle.

The models describe the transport of stable, radioactive, and radiogenic isotopes between reservoirs and the continuing changes due to radioactive decay within each reservoir. They are constrained by estimates of trace element and isotope abundances in the bulk Earth at the time of its formation and in the continental crust and depleted (mid-ocean ridge-type) mantle at the present day (see also Fig. 11.8 and discussion). Such calculations demonstrate that the formation of continental crust from the upper mantle can explain the broad present-day variations in isotope and trace element abundances (e.g., Figs. 11.4, 11.6 and 11.7). In addition, they suggest:

(1) Continental crust and depleted mantle only stabilized for long enough to generate significantly different isotope ratios comparatively late in the Earth's history. Thus the models of Jacobsen and Wasserburg (1979b) (Fig. 11.16) derived mean ages of 1.8—1.5 Ga and 1.8 Ga for the continental crust and depleted mantle, respectively, and those authors concluded that most of the continental crust formed in the time interval 3.0—0.5 Ga ago. O'Nions et al. (1979b) argued that provided that crustal growth was simply related to the rate of increase of potassium in the outer layer, the rate of formation of new crust was greatest between 3.5 and 2.5 Ga.

(2) Not all the Earth's mantle has been involved in, or has contributed to, the generation of new crust. Estimates of the amount of mantle required vary from 30% (Jacobsen and Wasserburg, 1979b) to 50% (O'Nions et al., 1979b), but both suggest that the lower mantle should be regarded as distinct geochemically from the upper mantle, and that, in turn, constrains models for convection in the mantle. It appears, for example, not to be compatible with whole mantle convection.

11.8. Lu-Hf isotopes

The isotope of ^{176}Lu decays to ^{176}Hf by β^- decay and with a half-life of 3.5×10^{10} a. However, hafnium is extremely refractory and it is only recently that analytical techniques have been developed for the precise measurement of Hf isotope ratios. Patchett and Tatsumoto (1980a, b) determined ^{176}Hf/^{177}Hf on eucrite meteorites and recent basalts; they used ^{179}Hf/^{177}Hf = 0.7325 to correct for laboratory mass fractionation and reported ^{180}Hf/^{177}Hf = 1.88651 ± 12 from 12 analyses of the Hf standard JMC 45. The isochron equation for the Lu-Hf decay scheme may be derived in the same way as that for Sm-Nd in section 11.2:

$$\frac{^{176}\mathrm{Hf}}{^{177}\mathrm{Hf}} = \frac{^{176}\mathrm{Lu}}{^{177}\mathrm{Hf}}(e^{\lambda t} - 1) + \left(\frac{^{176}\mathrm{Hf}}{^{177}\mathrm{Hf}}\right)_0 \tag{11.9}$$

where t is the age of the sample, λ is the decay constant, and $(^{176}Hf/^{177}Hf)_0$ is the initial Hf isotope ratio.

The first Lu-Hf whole-rock isochron was reported by Patchett and Tatsumoto (1980a) and it is reproduced in Fig. 11.17a. The eucrite meteorites constitute a coherent group of cumulate and non-cumulate igneous rocks which appear to have been produced during a major planetary differentiation 4.55 Ga ago. This age has been confirmed by a number of Rb-Sr, U-Pb, and Sm-Nd studies (e.g., Lugmair et al., 1975a) and thus Patchett and Tatsumoto (1980a) were able to use their analyses to estimate the half-life for the β^- decay of ^{176}Lu. Assuming that the whole-rock isochron in Fig. 11.17a corresponds to an age of 4.55 Ga, $T_{1/2}$ (^{176}Lu) = $3.53 \pm 0.14 \times 10^{10}$ a ($\lambda(^{176}Lu) = 1.96 \pm 0.08 \times 10^{-11}$ a^{-1}). Moreover, the initial $^{176}Hf/^{177}Hf$ (0.27973 ± 12) is the best available estimate of the initial Hf isotope ratio of the Earth and the other inner planets (see also the discussion of initial $^{143}Nd/^{144}Nd$ in Juvinas, section 11.3).

Hf isotope ratios have also been determined in a number of recent basalts (Patchett and Tatsumoto, 1980b). $^{176}Hf/^{177}Hf$ varies from 0.2828 to 0.2835 in those analysed, which is similar to the range of $^{143}Nd/^{144}Nd$ in the same rocks (0.5125—0.5133), and the positive correlation between Hf and Nd isotopes (Fig. 11.17b) suggests that Lu/Hf and Sm/Nd behaved coherently throughout the evolution of their upper mantle source regions. The estimated present-day $^{143}Nd/^{144}Nd$ ratio of 0.51264 for the bulk Earth (section 11.3) corresponds to a present-day $^{176}Hf/^{177}Hf$ ratio of ~0.28295, and assuming an initial Hf isotope ratio similar to that of eucrite meteorites (0.27973 ± 12 at 4.55 Ga) gives a Lu/Hf ratio of 0.25 for the bulk Earth

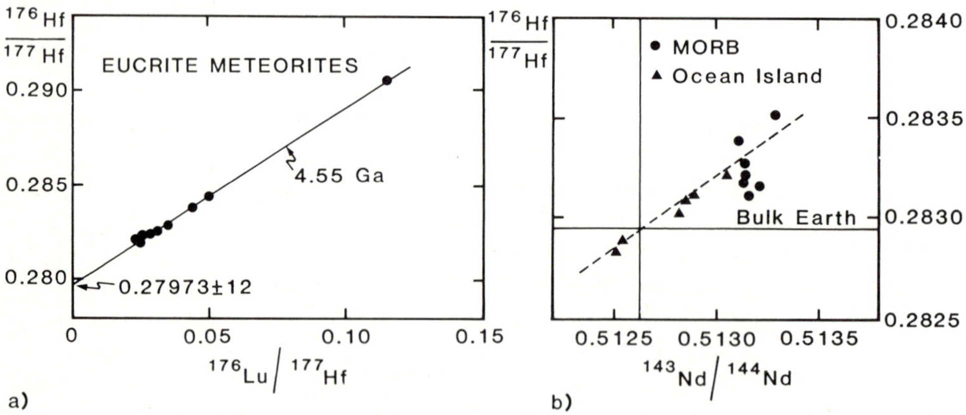

Fig. 11.17. (a) Lu-Hf whole-rock isochron for eucrite meteorites (Patchett and Tatsumoto, 1980a). The initial $^{176}Hf/^{177}Hf$ ratio is 0.27973 ± 12. (b) $^{176}Hf/^{177}Hf$ and $^{143}Nd/^{144}Nd$ ratios in recent MORB and ocean island basalts (after Patchett and Tatsumoto, 1980b). Estimated ratios for the bulk Earth at the present day are $^{143}Nd/^{144}Nd$ = 0.51264 and $^{176}Hf/^{177}Hf$ = 0.28295.

(Patchett and Tatsumoto, 1980b). Although there are still surprisingly few analyses of both Lu and Hf in chondrites, the estimate for Lu/Hf in the bulk Earth is comparable with, for example, Lu/Hf = 0.244 in the Juvinas achondrite (Patchett and Tatsumoto, 1980b).

Finally, most mantle-derived rocks appear to have more radiogenic Hf isotope ratios than the bulk Earth (Fig. 11.17b) indicating that their mantle source regions had relatively high Lu/Hf ratios for much of their history. Since these ratios are accompanied by high $^{143}Nd/^{144}Nd$ and low $^{87}Sr/^{86}Sr$, high $^{176}Hf/^{177}Hf$ (and hence high Lu/Hf ratios) must be typical of trace-element-depleted mantle. Hafnium is partitioned strongly into the liquid phase during partial melting so that high and variable Lu/Hf ratios only occur in residual or cumulate rock types. Conversely, enriched basalts and continental rocks have low and relatively uniform Lu/Hf ratios.

11.9. Concluding remarks

This chapter has illustrated some of the ways in which the Sm-Nd and Lu-Hf decay schemes have been applied to a wide range of geological and cosmological problems, and we conclude with just two general points. The first is the spectacular manner in which the study of Nd isotopes complements that of the more widely used Sr isotopes, and how considerable progress has been made through *combined* isotope studies. Low Rb/Sr ratios tend to be accompanied by high Sm-Nd so that rocks and minerals which are difficult to date by Rb-Sr may be dated by Sm-Nd. Similarly, rocks with low, uniform $^{87}Sr/^{86}Sr$ ratios often have relatively high and variable $^{143}Nd/^{144}Nd$. The different chemical characteristics of Rb and Sr compared with the rare earths cause differences in the relative fractionation of Rb/Sr and Sm/Nd, and hence with time of $^{87}Sr/^{86}Sr$ and $^{143}Nd/^{144}Nd$, depending on, for example, whether they are being transported in silicate liquids or H_2O-rich fluids. More specifically, combined Nd and Sr isotope studies have demonstrated how a present-day $^{87}Sr/^{86}Sr$ ratio of 0.704 can be generated by widely different processes in ocean island and island arc settings. Finally, even in the oceans, Sr and Nd conveniently have residence times which are respectively longer and shorter than the estimated ocean turn-over time. Consequently, ocean waters tend to have uniform $^{87}Sr/^{86}Sr$ and variable $^{143}Nd/^{144}Nd$ which has interesting implications for palaeo-oceanography.

The second point is a plea to the non-specialist to become more familiar with the possible applications of radiogenic isotopes. Traditionally, students of geology have shied away from isotopes as either being "too chemical", or in the mistaken, but occasionally cultivated belief that because the analytical techniques are sophisticated the concepts in isotope geology must also be difficult. We hope that this chapter goes some way to dispelling that myth, and more especially that it illustrates the confusion that can arise from the different frames of reference of isotope geochemists, interested in the main with planetary evolution, and of geologists and petrologists concerned with comparatively short-term process.

Acknowledgements

A. Gledhill, G. Davies, G. Rogers and N. Rogers contributed to various aspects of this review. R.K. O'Nions and G.N. Hanson kindly acted as referees, John Taylor prepared the diagrams, and Janet Dryden cheerfully survived the problems of typing a continually changing manuscript.

Appendix

Analytical techniques

Sm and Nd may be separated for isotope analysis by ion exchange techniques. They usually involve two or maybe three columns and several methods have been published (Lugmair et al., 1975b; O'Nions et al., 1977; DePaolo and Wasserburg, 1976a). Within the mass spectrometer Nd may be analysed as simple Nd, or as the oxide species NdO^+ — in which case it is necessary to correct for oxygen as well as neodymium mass fractiona-

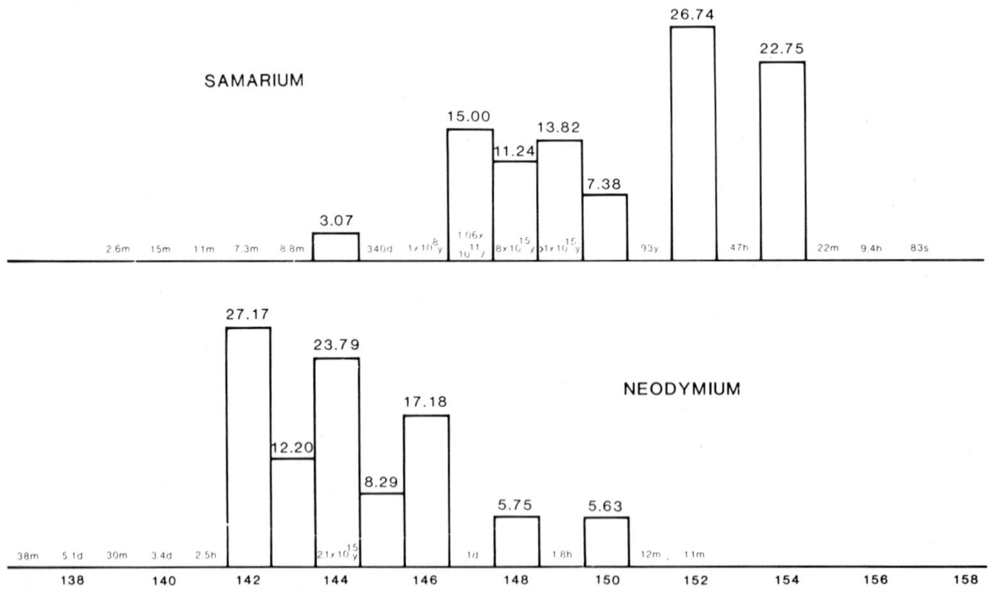

Fig. 11.18. Isotopes of Neodymium and Samarium, with half-lives where applicable (isotope abundances after Russ et al., 1971, and Hamilton et al., 1977; see Table 11.1).

tion (e.g., Wasserburg et al., 1982). However, as intimated in the text, it is the correction for the latter which at present causes most confusion in the literature.

Fig. 11.18 illustrates the different isotopes of Sm and Nd. Mass numbers 150, 148, and 144 are common to both elements, although ^{144}Sm only represents ~3% of Sm. Nonetheless there is also overlap with cerium (^{142}Ce) and so it is important that a good separation of Nd from Sm and Ce is achieved in the chemical preparation.

Mass fractionation

Light isotopes are more readily ionized than heavy isotopes and thus the ratio of two isotopes emitted from the same filament will change during the analysis of a single sample. Correction for such mass fractionation assumes that one of the isotope ratios measured is constant in nature and that the amount of fractionation is proportional to the difference between the mass numbers. A ^{143}Nd/^{144}Nd ratio normalized to one value of, for example, ^{146}Nd/^{144}Nd can be recalculated to another value using the following approximation:

$$\left(\frac{^{143}\mathrm{Nd}}{^{144}\mathrm{Nd}}\right)_N = \left(\frac{^{143}\mathrm{Nd}}{^{144}\mathrm{Nd}}\right)_M \left[1 + \Delta M\left(\frac{(^{146}\mathrm{Nd}/^{144}\mathrm{Nd})_N}{(^{146}\mathrm{Nd}/^{144}\mathrm{Nd})_M} - 1\right)\right]$$

where ΔM = (143 − 144)/(146 − 144) = − 1/2, N = converted ratios, and M = presented ratios.

TABLE 11.1

Isotope ratios and abundances of Sm and Nd

Ratios	Abundances (%)
Samarium (atomic weight = 150.35)	
^{144}Sm/^{154}Sm = 0.13516	^{144}Sm = 3.075
^{147}Sm/^{154}Sm = 0.65918	^{147}Sm = 15.00
^{148}Sm/^{154}Sm = 0.49419	^{148}Sm = 11.24
^{149}Sm/^{154}Sm = 0.60750	^{149}Sm = 13.82
^{150}Sm/^{154}Sm = 0.32440	^{150}Sm = 7.38
^{152}Sm/^{154}Sm = 1.17537	^{152}Sm = 26.74
	^{154}Sm = 22.75
Neodymium (atomic weight = 144.24)	
^{142}Nd/^{144}Nd = 1.141827	^{142}Nd = 27.17
^{145}Nd/^{144}Nd = 0.348417	^{145}Nd = 8.290
^{146}Nd/^{144}Nd = 0.7219[a]	^{146}Nd = 17.18
^{148}Nd/^{144}Nd = 0.241578	^{148}Nd = 5.748
^{150}Nd/^{144}Nd = 0.236418	^{150}Nd = 5.625
^{143}Nd/^{144}Nd = 0.512638[b]	^{143}Nd = 12.20
	^{144}Nd = 23.79

[a] Value used for normalization.
[b] Value in bulk Earth reservoir, indistinguishable from BCR-1.

Conversion of Sm/Nd to $^{147}Sm/^{144}Nd$

The conversion factor f is given by:

$$f = \frac{\text{abundance } ^{147}\text{Sm}}{\text{abundance } ^{144}\text{Nd}} \times \frac{\text{atomic weight Nd}}{\text{atomic weight Sm}}$$

Thus using the values in Table 11.1:

$$f = \frac{0.1500}{0.2379} \times \frac{144.24}{150.35} = 0.6049 \quad \text{and:} \quad ^{147}\text{Sm}/^{144}\text{Nd} = f \times \text{Sm/Nd}.$$

Note that this conversion factor was calculated for BCR-1 which has a present-day $^{143}Nd/^{144}Nd$ ratio of ~0.51264. Since f depends on the abundance of ^{143}Nd it will be different for samples with different $^{143}Nd/^{144}Nd$ ratios. The decay of ^{147}Sm reduces its abundance, but since its half-life is long compared with the age of the solar system the abundance of ^{147}Sm is regarded as constant.

Calculation of T^{Nd}_{CHUR} *ages*

The T^{Nd}_{CHUR} age is the time at which a sample of particular present-day Sm/Nd and $^{143}Nd/^{144}Nd$ ratios had the same $^{143}Nd/^{144}Nd$ ratio as CHUR (Fig. 11.5). From that time, T^{Nd}_{CHUR}, to the present day the evolution of both the sample and CHUR may be described using equation (11.6):

$$\left(\frac{^{143}\text{Nd}}{^{144}\text{Nd}}\right)_{\text{CHUR}} = \left(\frac{^{143}\text{Nd}}{^{144}\text{Nd}}\right)_{T^{\text{Nd}}_{\text{CHUR}}} + \left(\frac{^{147}\text{Sm}}{^{144}\text{Nd}}\right)_{\text{CHUR}} (e^{\lambda T^{\text{Nd}}_{\text{CHUR}}} - 1)$$

and for the sample (S):

$$\left(\frac{^{143}\text{Nd}}{^{144}\text{Nd}}\right)_{\text{S}} = \left(\frac{^{143}\text{Nd}}{^{144}\text{Nd}}\right)_{T^{\text{Nd}}_{\text{CHUR}}} + \left(\frac{^{147}\text{Sm}}{^{144}\text{Nd}}\right)_{\text{S}} (e^{\lambda T^{\text{Nd}}_{\text{CHUR}}} - 1)$$

Since ($^{143}Nd/^{144}Nd$) at time T^{Nd}_{CHUR} is the same for both CHUR and the sample, we may substitute one equation in the other and solve for T^{Nd}_{CHUR}:

$$T^{\text{Nd}}_{\text{CHUR}} = \frac{1}{\lambda} \ln\left(\frac{(^{143}\text{Nd}/^{144}\text{Nd})_{\text{S}} - (^{143}\text{Nd}/^{144}\text{Nd})_{\text{CHUR}}}{(^{147}\text{Sm}/^{144}\text{Nd})_{\text{S}} - (^{147}\text{Sm}/^{144}\text{Nd})_{\text{CHUR}}} + 1\right)$$

where all the isotope ratios are present-day ratios.

References

Allègre, C.J. and Othman, D.B., 1980. Nd-Sr isotopic relationship in granitoid rocks and continental crust development: a chemical approach to orogenesis. *Nature*, 286: 335—342.

Allsopp, H.L., Viljoen, M.J. and Viljoen, R.P., 1973. Strontium isotope studies of the mafic and felsic rocks of the Onverwacht Group of the Swaziland sequence. *Geol. Rundsch.*, 62: 902—917.

Anderson, A.T., Clayton, R.N. and Mayeda, T.K., 1971. Oxygen isotope thermometry of mafic igneous rocks. *J. Geol.*, 79: 715—729.

Appleton, J.D., 1972. Petrogenesis of potassium-rich lavas from the Roccamonfina Volcano, Roman region, Italy. *J. Petrol.*, 13: 425—456.

Bokhari, F.Y. and Kramers, J.D., 1981. Island arc character and late Precambrian age of volcanics at Wadi Shwas, Hijas, Saudi Arabia: geochemical and Sr and Nd isotopic evidence. *Earth Planet. Sci. Lett.*, 54: 409—422.

Brigueu, L., Lancelot, J.R., Valois, J.-P. and Walgenwitz, F., 1980. Géochronologie U-Pb et genèse d'un type de minéralization uranifère: les alaskites de Goanikontès (Namibie) et leur encaissant. *Bull. Cent. Rech. Explor.-Prod. Elf-Aquitaine*, 4: 759—811.

Brooks, C., James, D.E. and Hart, S.R., 1976. Ancient lithosphere: its role in young continental lithosphere. *Science*, 193: 1086—1094.

Budyko, M.I., 1974. *Climate and Life*. Academic Press, New York, N.Y., 508 pp.

Carter, S.R., Evensen, N.M., Hamilton, P.J. and O'Nions, R.K., 1978. Neodymium and strontium isotope evidence for crustal contamination of continental rocks. *Science*, 202: 743—747.

Chow, J.J. and Patterson, C.C., 1962. The occurrence and significance of lead isotopes in pelagic sediments. *Geochim. Cosmochim. Acta*, 26: 263—306.

Cox, K.G., Hawkesworth, C.J., O'Nions, R.K. and Appleton, J.D., 1976. Isotopic evidence for the derivation of some Roman Region volcanics from anomalously enriched mantle. *Contrib. Mineral. Petrol.*, 56: 173—180.

DePaolo, D.J. and Johnson, R.W., 1979. Magma genesis in the New Britain island arc: constraints from Nd and Sr isotopes and trace element patterns. *Contrib. Mineral. Petrol.*, 70: 367—379.

DePaolo, D.J. and Wasserburg, G.J., 1976a. Nd isotopic variations and petrogenetic models. *Geophys. Res. Lett.*, 3: 249—252.

DePaolo, D.J. and Wasserburg, G.J., 1976b. Inferences about magma sources and mantle structure from variations of $^{143}Nd/^{144}Nd$. *Geophys. Res. Lett.*, 3: 743—746.

DePaolo, D.J. and Wasserburg, G.J., 1977. The sources of island arcs as indicated by Nd and Sr isotopic studies. *Geophys. Res. Lett.*, 4, 10: 465—468.

DePaolo, D.J. and Wasserburg, G.J., 1979. Petrogenetic mixing models and Nd-Sr isotopic patterns. *Geochim. Cosmochim. Acta*, 43: 615—627.

Dickin, A.P., 1981. Isotope geochemistry of Tertiary igneous rocks from the Isle of Skye, N.W. Scotland. *J. Petrol.*, 22: 155—189.

Donnelly, T., Rogers, J.J.W., Pushker, P. and Armstrong, R.L., 1971. Chemical evolution of igneous rocks of the eastern West Indies: an investigation of thorium, uranium and potassium distributions and lead and strontium isotope ratios. *Geol. Soc. Am., Mem.*, 130: 181—224.

Eichelburger, J.C., 1975. Origin of andesite and dacite: evidence of mixing at Glass Mountain in California and at other circum-Pacific volcanoes. *Geol. Soc. Am. Bull.*, 86: 1381—1391.

Eichelburger, J.C., 1978. Andesites in island arcs and continental margins: relationships to crustal evolution. *Bull. Volcanol.*, 41: 480—500.

Elderfield, H., Hawkesworth, C.J., Greaves, M.J. and Calvert, S.E., 1981. Rare earth element geochemistry of oceanic ferromanganese nodules and associated sediments. *Geochim. Cosmochim. Acta*, 45: 513—528.

Erlank, A.J., 1976. Upper mantle metasomatism as revealed by potassic richterite-bearing xenoliths from kimberlites. *EOS, Trans. Am. Geophys. Union*, 57: 597.

Erlank, A.J. and Shimizu, N., 1977. Strontium and Sr-isotope distributions in some kimberlite nodules and minerals. *Abstr., 2nd Int. Kimberlite Conf., Santa Fe, N.M.*

Evensen, N.M., Hamilton, P.J. and O'Nions, R.K., 1978. Rare-earth abundances in chondritic meteorites. *Geochim. Cosmochim. Acta*, 42: 1199—1212.

Faure, G. and Powell, J.L., 1972. *Strontium Isotope Geology*. Springer-Verlag, Berlin, 188 pp.

Francis, P.W., Moorbath, S. and Thorpe, R.S., 1977. Strontium isotope data for the Recent andesites in Ecuador and North Chile. *Earth Planet. Sci. Lett.*, 36: 197—202.

Gill, J., 1974. Role of underthrust oceanic crust in the genesis of a Fijian calc-alkaline suite. *Contrib. Mineral. Petrol.*, 43: 29—45.

Goldberg, E.D., Koide, M., Schmitt, R.A. and Smith, R.H., 1963. Rare-earth distributions in marine environment. *J. Geophys. Res.*, 68: 4209—4217.

Graham, A.M. and Upton, B.G.J., 1978. Gneisses in diatremes, Scottish Midland Valley: petrology and tectonic implications. *J. Geol. Soc. London*, 135: 219—228.

Griffin, W.L. and Brueckner, H.K., 1980. Caledonian Sm-Nd ages and a crustal origin for Norwegian eclogites. *Nature*, 285: 319—321.

Hamilton, P.J., O'Nions, R.K. and Evensen, N.M., 1977. Sm-Nd dating of Archaean basic and ultrabasic nodes. *Earth Planet. Sci. Lett.*, 36: 263—268.

Hamilton, P.J., Evensen, N.M., O'Nions, R.K. and Tarney, J., 1979a. Sm-Nd systematics of Lewisian gneisses: implications for the origin of granulites. *Nature*, 1977: 25—28.

Hamilton, P.J., Evensen, N.M., O'Nions, R.K., Smith, H.S. and Erlank, A.J., 1979b. Sm-Nd dating of the Onverwacht Group volcanics, Southern Africa. *Nature*, 279: 298.

Hamilton, P.J., O'Nions, R.K. and Pankhurst, R.J., 1980. Isotopic evidence for the provenance of some Caledonian granites. *Nature*, 287: 279—284.

Harmon, R.S., Thorpe, R.S. and Francis, P.W., 1981. Petrogenesis of Andean andesites from combined O-Sr relationships. *Nature*, 290: 396—399.

Hawkesworth, C.J., 1979. $^{143}Nd/^{144}Nd$, $^{87}Sr/^{86}Sr$ and trace element characteristics of magmas along destructive plate margins. In: M.P. Atherton and J. Tarney (Editors), *Origin of Granite Batholiths: Geochemical Evidence.* Shiva, pp. 76—89.

Hawkesworth, C.J., 1982. Isotopic characteristics of magmas erupted along destructive plate margins. In: R.S. Thorpe (Editor), *Orogenic Andesites.* J. Wiley and Sons, New York, N.Y., pp. 549—571.

Hawkesworth, C.J. and Marlow, A., 1983. Isotope evolution of the Damara orogenic belt. In: R. McG. Miller (Editor), *Evolution of the Damara Orogen, South West Africa/Namibia. Geol. Soc. S. Afr., Spec. Publ.* (in press).

Hawkesworth, C.J. and Powell, B.M., 1980. Magma genesis in the Lesser Antilles island arc. *Earth Planet. Sci. Lett.*, 51: 297—308.

Hawkesworth, C.J. and Vollmer, R., 1979. Crustal contamination versus enriched mantle: $^{143}Nd/^{144}Nd$ and $^{87}Sr/^{86}Sr$ evidence from the Italian volcanics. *Contrib. Mineral. Petrol.*, 69: 151—169.

Hawkesworth, C.J., O'Nions, R.K., Pankhurst, R.J., Hamilton, P.J. and Evensen, N.M., 1977. A geochemical study of island-arc and back-arc tholeiites from the Scotia Sea. *Earth Planet. Sci. Lett.*, 36: 253—263.

Hawkesworth, C.J., Norry, M.J., Roddick, J.C. and Vollmer, R., 1978. The significance of trace element modelling calculations for the evolution of Sr and Nd isotopes in the mantle. *U.S. Geol. Surv., Open-file Rep.*, 78—701: 162—164.

Hawkesworth, C.J., Norry, M.J., Roddick, J.C. and Vollmer, R., 1979a. $^{143}Nd/^{144}Nd$ and $^{87}Sr/^{86}Sr$ ratios from the Azores and their significance in LIL-element enriched mantle. *Nature*, 280: 28—31.

Hawkesworth, C.J., Norry, M.J., Roddick, J.C., Baker, P.E., Francis, P.W. and Thorpe, R.S., 1979b. $^{143}Nd/^{144}Nd$, $^{87}Sr/^{86}Sr$ and incompatible element variations in calc-alkaline andesites and plateau lavas from South America. *Earth Planet. Sci. Lett.*, 43: 45—57.

Hawkesworth, C.J., O'Nions, R.K. and Arculus, R.J., 1979c. Nd and Sr isotope geochemistry of island arc volcanics, Grenada, Lesser Antilles. *Earth Planet. Sci. Lett.*, 45: 237—248.

Hawkesworth, C.J., Kramers, J.D. and Miller, R. McG., 1981. Old model Nd ages in Namibian Pan-African Rocks. *Nature*, 289: 278—282.

Hawkesworth, C.J. Hammill, M., Gledhill, A.R., van Calsteren, P.W.C. and Rogers, G., 1982. Isotope and trace element evidence for late-stage intra-crustal melting in the high Andes. *Earth Planet. Sci. Lett.*, 58: 240—254.

Høgdahl, O.T., Bowen, B.T. and Melson, S., 1968. Neutron activation analysis of lanthanide elements in seawater. *Adv. Chem. Ser.*, 73: 308—325.

Jacobsen, S.B. and Wasserburg, G.J., 1978. Nd and Sr isotopic study of the Permian Oslo rift. *U.S. Geol. Surv., Open-file Rep.*, 78—701: 194—196.

Jacobsen, S.B. and Wasserburg, G.J., 1979a. Nd and Sr isotopic study of the Bay of Islands ophiolite complex and the evolution of the source of mid-ocean ridge basalt. *J. Geophys. Res.*, 84: 7429—7445.

Jacobsen, S.B. and Wasserburg, G.J., 1979b. The mean age of mantle and crustal reservoirs. *J. Geophys. Res.*, 84: 7411—7427.

Jacobsen, S.B. and Wasserburg, G.J., 1980. Sm-Nd isotopic evolution of chondrites. *Earth Planet. Sci. Lett.*, 50: 139—155.

Jahn, B.J. and Shih, C.Y., 1974. On the age of the Onverwacht Group, Swaziland sequence, South Africa. *Geochim. Cosmochim. Acta*, 38: 873—875.

Kay, R.W., 1977. Geochemical constraints on the origin of Aleutian magmas. In: M. Talwani and W. Pitman, (Editors), *Island Arcs, Deep Sea Trenches and Back-Arc Basins. Am. Geophys. Union, Maurice Ewing Ser.*, 1: 229—242.

Kay, R.W., 1980. Volcanic arc magmas: implications of a melting-mixing model for element recycling in the upper-crust mantle system. *J. Geol.*, 8: 497—552.

Langmuir, C.H., Vocke, R.D., Hanson, G.N. and Hart, S.R., 1978. A general mixing equation with applications to Icelandic basalts. *Earth Planet. Sci. Lett.*, 37: 380—392.

Lugmair, G.W. and Marti, K., 1977. Sm-Nd-Pu timepieces in the Angra dos Reis meteorite. *Earth Planet. Sci. Lett.*, 35: 273—284.

Lugmair, G.W. and Marti, K., 1978. Lunar initial $^{143}Nd/^{144}Nd$; differential evolution of the Lunar crust and mantle. *Earth Planet. Sci. Lett.*, 39: 349—357.

Lugmair, G.W., Scheinin, N.B. and Marti, K., 1975a. Search for extinct ^{146}Sm, 1. The isotopic abundance of ^{142}Nd in the Juvinas meteorite. *Earth Planet. Sci. Lett.*, 27: 79—84.

Lugmair, G.W., Scheinin, N.B. and Marti, K., 1975b. Sm-Nd age and history of Apollo 17 basalt 75075: evidence for early differentiation of the Lunar exterior. *Proc. 6th Lunar Sci. Conf., Geochim. Cosmochim. Acta, Suppl.*, 6, 2: 1419—1429.

Margaritz, M., Whitford, D.J. and James, D.E., 1978. Oxygen isotopes and the origin of high $^{87}Sr/^{86}Sr$ andesites. *Earth Planet. Sci. Lett.*, 40: 220—230.

Masuda, A., Nakamura, N. and Tanaka, T., 1973. Fine structures of mutually normalized rare earth patterns of chondrites. *Geochim. Cosmochim. Acta*, 37: 239—248.

Matsuhisa, Y., 1979. Oxygen isotopic composition of volcanic rocks from the East Japan island arc and their bearing on petrogenesis. *J. Volcanol. Geotherm. Res.*, 5: 271—296.

McCulloch, M.T. and Wasserburg, G.J., 1978. Sm-Nd and Rb-Sr chronology of continental crust formation. *Science*, 200: 1002—1011.

Menzies, M.A. and Murthy, V.R., 1980a. Nd and Sr isotope geochemistry of hydrous mantle nodules and their host alkali basalts: implications for local heterogeneities in metasomatically veined mantle. *Earth Planet. Sci. Lett.*, 46: 323—334.

Menzies, M.A. and Murthy, V.R., 1980b. Enriched subcontinental mantle: Nd and Sr isotopes in diopsides from kimberlite nodules. *Nature*, 283: 634—636.

Moorbath, S. and Welke, H., 1969. Lead isotope studies on igneous rocks from the Isle of Skye, north-west Scotland. *Earth Planet. Sci. Lett.*, 5: 217—230.

Nakamura, N. Tatsumoto, M., Nunes, P.D., Unruh, D.M., Schwab, A.P. and Wildeman, T.R., 1976. 4.4 b.y. old clast in Boulder 7, Apollo 17: a comprehensive geological study by U-Pb, Rb-Sr and Sm-Nd methods. *Proc. 7th Lunar Sci. Conf., Geochim. Cosmochim. Acta, Suppl.*, 7: 2309—2333.

O'Nions, R.K., Hamilton, P.J. and Evensen, N.M., 1977. Variations in $^{143}Nd/^{144}Nd$ and $^{87}Sr/^{86}Sr$ ratios in oceanic basalts. *Earth Planet. Sci. Lett.*, 34: 13—22.

O'Nions, R.K., Carter, S.R., Cohen, R.S., Evensen, N.M. and Hamilton, P.J., 1978. Pb, Nd and Sr isotopes in oceanic ferromanganese deposits and ocean floor basalts. *Nature*, 273: 435—438.

O'Nions, R.K., Carter, S.R., Evensen, N.M. and Hamilton, P.J., 1979a. Geochemical and cosmochemical applications of Nd-isotope analysis. *Annu. Rev. Earth Planet. Sci.*, 7: 11—38.

O'Nions, R.K., Evensen, N.M. and Hamilton, P.J., 1979b. Geochemical modelling of mantle differentiation and crustal growth. *J. Geophys. Res.*, 84: 6091—6101.

Papanastassiou, D.A., DePaolo, D.J., Tera, F. and Wasserburg, G.J., 1977. An isotopic triptych on mare basalts: Rb-Sr, Sm-Nd, U-Pb. In: *Lunar Science VIII*. Lunar Science Institute, Houston, Texas, pp. 750—752.

Patchett, P.J. and Tatsumoto, M., 1980a. Lu-Hf total-rock isochron for eucrite meteorites. *Nature*, 288: 571—574.

Patchett, P.J. and Tatsumoto, M., 1980b. Hafnium isotope variations, in oceanic basalts. *Geophys. Res. Lett.*, 7: 1077—1080.

Pearce, J.A., 1982. Trace element characteristics of lavas from destructive plate boundaries. In: R.S. Thorpe (Editor), *Orogenic Andesites.* J. Wiley and Sons, New York, N.Y., pp. 525—548.

Piepgras, D.J. and Wasserburg, G.J., 1980. Neodymium isotopic variations in seawater. *Earth Planet. Sci. Lett.*, 50: 128—138.

Piepgras, D.J., Wasserburg, G.J. and Dasch, E.J., 1979. The isotopic composition of Nd in different ocean masses. *Earth Planet. Sci. Lett.*, 45: 223—236.

Råheim, A. and Green, D.H., 1975. *P-T* paths of natural ecologites during metamorphism — a record of subduction. *Lithos*, 8: 317—326.

Ringwood, A.E., 1974. The petrological evolution of island arc systems. *J. Geol. Soc. London*, 130: 183—204.

Russ, G.P., Burnett, D.S., Lingenfelter, R.E. and Wasserburg, G.J., 1971. Neutron capture on ^{149}Sm in lunar samples. *Earth Planet. Sci. Lett.*, 13: 53—60.

Spooner, E.T.C., 1976. The strontium isotopic composition of seawater, and seawater—oceanic crust interaction. *Earth Planet. Sci. Lett.*, 31: 167—174.

Sun, S.S., 1980. Lead isotopic study of young volcanic rocks from mid-ocean ridges, ocean islands and island arcs. *Philos. Trans. R. Soc. London, Ser. A*, 297: 406—446.

Taylor, H.P., 1968. The oxygen isotope geochemistry of igneous rocks. *Contrib. Mineral. Petrol.*, 19: 1—17.

Taylor, H.P., 1980. The effects of assimilation of country rocks by magmas on $^{18}O/^{16}O$ and $^{87}Sr/^{86}Sr$ systematics in igneous rocks. *Earth Planet. Sci. Lett.*, 47: 243—254.

Taylor, H.P., Giannetti, B. and Turi, B., 1979. Oxygen isotope geochemistry of potassic igneous rocks from the Roccamonfina volcano, Roman comagmatic region, Italy. *Earth Planet. Sci. Lett.*, 46: 81—106.

Taylor, P.N., Moorbath, S., Goodwin, R. and Petrykowski, A., 1980. Crustal contamination as an indicator for the extent of early Archean continental crust: Pb isotopic evidence from the late Archean gneisses of west Greenland. *Geochim. Cosmochim. Acta*, 44: 1437—1453.

Thorpe, R.S. and Francis, P.W., 1979. Petrogenetic relationships of volcanic and intrusive rocks of the Andes. In: M.P. Atherton and J. Tarney (Editors), *Origin of Granite Batholiths: Geochemical Evidence*. Shiva, pp. 65—75.

Tilton, G.R., 1980. Isotopic studies of Cenozoic Andean calc-alkaline rocks. *Carnegie Inst. Washington Yearb.*, 78: 298—304.

Turi, B. and Taylor, H.P., 1976. Oxygen isotope studies of potassic volcanic rocks of the Roman province, Central Italy. *Contrib. Mineral. Petrol.*, 55: 1—31.

Unruh, D.M., Nakamura, N. and Tatsumoto, M., 1977. History of the Pasamonte achondrite: relative suspectibility of the Sm-Nd, Rb-Sr and U-Pb systems in metamorphic events. *Earth Planet. Sci. Lett.*, 37: 1—12.

van Breemen, O. and Hawkesworth, C.J., 1980. Sm-Nd isotopic study of garnets and their metamorphic host rocks. *Trans. R. Soc. Edinburgh*, 71: 97—102.

Vollmer, R., 1976. Rb-Sr and U-Th-Pb systematics of alkaline rocks: the alkaline rocks from Italy. *Geochim. Cosmochim. Acta*, 40: 283—295.

Vollmer, R., 1977. Isotopic evidence for genetic relations between acid and alkaline rocks in Italy. *Contrib. Mineral. Petrol.*, 60: 109—110.

Vollmer, R. and Hawkesworth, C.J., 1980. Lead isotopic composition of the potassic rocks from Roccamonfina (South Italy). *Earth Planet. Sci. Lett.*, 47: 91—101.

Wasserburg, G.J., Jacobsen, S.B., DePaolo, D.J., McCulloch, M.T. and Wen, J., 1982. Precise determinations of Sm/Nd ratios, Sm and Nd isotopic abundances in standard solutions. *Geochim. Cosmochim. Acta*, 46: 2311—2323.

Whitford, D.J., White, W.M., Jezek, P.A. and Micholls, I.A., 1980. Nd isotope composition of recent andesites from Indonesia. *Carnegie Inst. Washington Yearb.*, 78: 304—308.

Yardley, B.W.D., 1977. An empirical study of diffusion in garnet. *Am. Mineral.*, 62: 793—780.

Chapter 12

THE ECONOMIC IMPORTANCE OF THE RARE EARTH ELEMENTS

C.R. NEARY and D.E. HIGHLEY

12.1. Introduction

The REE are relatively recent additions to the inventory of useful raw materials and since their first commercial exploitation in the latter part of the 19th century, there has been a dramatic change in their industrial uses and the types of deposit worked.

Interest in rare earth minerals originated in 1883 with the development of incandescent gas mantles containing rare earth and zirconium oxides. These were replaced by the improved thoriated gas mantles which were commercially introduced in 1891. Until the early 20th century commercial interest in rare earth minerals was confined to monazite, principally as a source of thorium, although the gas mantles also contained a small percentage of cerium. Initially thorium was derived from residues already accumulated from earlier mantle production and this was the start of the efforts, still evident today, to match the ratio of rare earths produced from the ore minerals to the ratio of demand for the individual elements. A wide range of new uses have been developed for the REE and currently Ce, La, Nd, Pr, Sm, Gd and Eu are of commercial importance, in addition to mischmetal, a mixture of the REE in metallic form. Yttrium (atomic number 39) has similar chemical properties and a close mineralogical association to the REE. It is also of considerable economic importance and is sometimes included in discussions of the REE, as it is in this chapter.

Historically most commercial sources of REE have been monazite-bearing placer deposits in which liberation and concentration has been undertaken by the natural agencies of weathering, erosion, transportation and deposition. Although monazite-rich sands occur locally, in most commercial operations monazite is present only as a minor constituent of the heavy mineral sand and is usually produced as a by-product.

Small quantities of monazite were recovered from fluviatile deposits in North Carolina, U.S.A., from 1887, but initially the major world source of supply was the rich beach sand deposits of Brazil where large-scale production began in 1895. Similar deposits were developed in India and elsewhere but since 1967 Australian heavy mineral sand deposits, particularly those in

Western Australia, have become the major source of monazite, currently accounting for some 60—70% of world output, equivalent to about 30% of total world REE production.

Primary concentrations of rare earth minerals of a grade sufficiently high to justify exploitation are rare. However, the rare earth fluocarbonate, bastnäsite, superseded monazite as the major source in 1965. The carbonatite deposit at Mountain Pass in California is both the only significant source of bastnäsite and provides the world's main supply of REE, accounting for about 50% of total production. During the 1970's the U.S.A. and Australia dominated world REE production. More recently China has emerged as a major producer of bastnäsite and REE.

12.2. Abundance and ore mineralogy

The crustal abundance of REE varies markedly from element to element (see Chapter 1) and with the exception of yttrium, the light sub-group members are far more abundant than the HREE (Table 12.1). The REE have an average combined crustal abundance of 183 ppm, including yttrium. Rare earth elements occur in many mineral species but only a very few are sufficiently rich in them and occur in such concentrations to be of economic interest. By far the most important are monazite ($(RE,Th,Y)PO_4$) and bastnäsite ($RE(CO_3)F$), which together have accounted for almost all of historical world production. Total world production of monazite and bastnäsite is estimated to be 520,000 and 340,000 t respectively to the end of 1980. Commercial monazite concentrates commonly contain 55—60% REO (rare earth oxide), 3—10% thoria, some yttrium and small amounts of uranium. Bastnäsite has a slightly higher REE content but contains little or no thorium and yttrium. Both bastnäsite and monazite contain a predominance of LREE whilst xenotime ($(Y,RE)PO_4$), which is produced in minor quantities (with a cumulative production of perhaps about 1000 t to 1980) is the major source of yttrium and HREE (see Table 12.1). REE have also been recovered from brannerite, euxenite and apatite (see Chapter 2).

12.3. Geological characteristics of the deposits

Introduction

Rare earth minerals occur in a wide range of geological environments of both a primary and secondary nature. The limited number of mined deposits reflects not only the comparative rarity of significant concentrations but also the still relatively small demand for the elements themselves. Deposits of potential economic interest are both widely distributed (Fig. 12.1) and

TABLE 12.1

Distribution of rare earths in the main ore minerals

	Monazite[a] REO (%) (RE,Th,Y)PO_4	Bastnäsite[b] REO (%) RE(CO_3)F	Xenotime[c] REO (%) (Y,RE)PO_4	Apatite[d] REO (%) $(Ca,RE)_5[(P,Si)O_4]_3(O,F)$
La_2O_3	23.7	33.2	0.5	25.1
CeO_2	45.6[1]	49.1	5.0	45.0[4]
Pr_6O_{11}	5.0	4.3	0.7	3.9
Nd_2O_3	17.2	12.0	2.2	14.0
Sm_2O_3	2.5	0.78	1.9	1.6
Eu_2O_3	0.05	0.11	0.2	0.5
Light REO total	94.05	99.49	10.5	90.1
Gd_2O_3	1.5	0.17	4.0	1.5
Tb_4O_7	0.04	0.016	1.0	0.1
Dy_2O_3	0.68	0.031	8.7	1.0
Ho_2O_3	0.05	50 ppm	2.1	0.1
Er_2O_3	0.21	35 ppm	5.4	0.15
Tm_2O_3	0.02	8 ppm	0.9	0.02
Yb_2O_3	0.12	13 ppm	6.2	0.08
Lu_2O_3	0.04	1 ppm	0.4	—
Y_2O_3	2.4	0.09	60.8	4.3
Heavy REO total	5.06	0.318	89.5	7.25
Total REO in marketable concentrates	55—60%	60—70%	42—51%	<1%

Sources:
[a]Yoganup monazite (recalculated), Capel, Western Australia, Westralian Sands Ltd. Cerium as Ce_2O_3.
[b]Bastnäsite concentrate, Mountain Pass (typical of current production). Courtesy of Molycorp Inc., 1982.
[c]Xenotime, Malaysia (Moore, 1979).
[d]Apatite, Kola Peninsula (Lounamaa et al., 1980). Cerium as CeO_2 or Ce_2O_3 unclear from original source.

varied in their geological association so that simple classification of deposits is difficult. However, REE are currently almost entirely produced from (a) alkaline rocks and carbonatites, and (b) placer deposits. A number of other occurrences are of potential economic interest: for example, apatite may be enriched in REE, particularly where associated with alkaline igneous rocks, and commercial recovery is potentially feasible during the production of phosphoric acid for fertilizer manufacture.

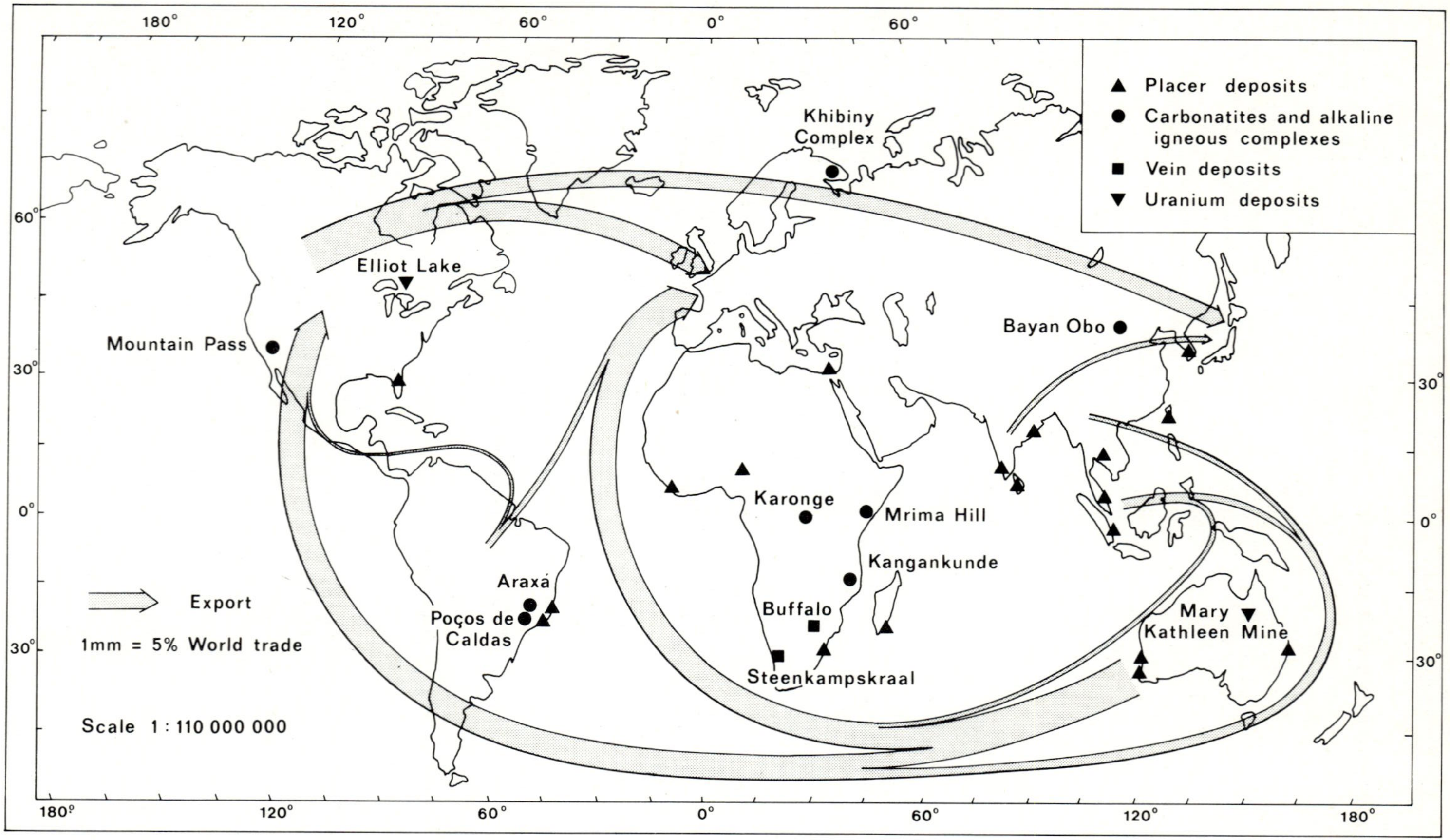

Fig. 12.1. Location of the major REE deposits with approximate late 1970's exports of "bulk" rare earth materials (ores, crude chloride or mischmetal) from the major producing countries (destinations within Europe are not differentiated).

Alkaline rocks and carbonatites

Many alkaline igneous rocks, and particularly associated carbonatites, are enriched in REE, both in rare earth minerals and by virtue of substitution within such other minerals as apatite and pyrochlore ($(Na,Ca,Ce)_2Nb_2O_6F$). Enrichment may also occur in secondary deposits developed as residual accumulations on carbonatites as a result of deep chemical weathering (Deans, 1966, 1978; Semenov, 1974).

REE-bearing carbonatites have been recognized in many parts of the world but by far the most important economically occurs in the Mountain Pass district of San Bernardino County, California. The deposit, now the world's leading source, was discovered in 1949 and until recently was the world's largest known concentration of rare earth minerals. REE mineralization, chiefly as bastnäsite, occurs in carbonatites intruded into a Precambrian basement complex consisting mainly of granite gneiss (Olson et al., 1954). The carbonatite occurs in thin veins mostly less than 2 m in width, but the large Sulphide Queen carbonatite orebody, on which mining is currently based, is some 730 m in length and up to 210 m in width. The ore contains on average 12% bastnäsite, 20% baryte, 10% quartz and silicate minerals and 60% carbonate minerals, chiefly calcite. The REO content of the ore is usually in the range 5—15%, averaging about 7%, but locally concentrations of up to 40% REO are present. In January 1981 proved and probable ore reserves at Mountain Pass were reported to be about 40 Mt assaying 7.68% REO (Annual Report, Union Oil Co. America, 1980). A number of other occurrences associated with alkaline and carbonatite complexes have been reported elsewhere in North America (Adams and Staatz, 1973; Rose, 1979) although none are currently producing.

In Africa several REE-bearing carbonatites have been recognized and some have attracted commercial interest. Production has been confined to the bastnäsite-bearing veins at Karonge southeast of Bujumbura in the Republic of Burundi. Here the REE mineralization, consisting mainly of bastnäsite and monazite, occurs in veins and veinlets often forming a dense stockwork. Quartz, baryte and goethite are the common gangue minerals. The mineralization is believed to be related to the hydrothermal stage of a carbonatitic differentiation (Van Wambeke, 1977). The bastnäsite contains 31.1% Ce_2O_3, 39.4% $(La,Pr,Nd)_2O_3$, 0.25% Y_2O_3 and 0.10% ThO_2 (De Kun, 1965). Small amounts have been produced, although the deposit was reported to have been abandoned in 1979.

At Mrima Hill, 80 km southwest of Mombassa, Kenya, rare earth minerals are concentrated in the thick residual soils overlying a deeply weathered carbonatite. Although much of the ferruginous residuum contains only 4% REO, locally average values of up to 14% have been obtained from exploratory shafts sunk to bedrock. The REE are mainly concentrated in earthy cryptocrystalline monazite associated with baryte, gorceixite ($BaAl_3(OH)_7P_2O_7$)

and also pyrochlore. Reserves grading about 5% REO have been estimated at 6 Mt (Deans, 1966) and there are proposals to develop a mine. The deposit is also one of the richest concentrations of niobium in the world, but the fine-grained and friable nature of the pyrochlore has prevented the development of a viable processing method for its recovery (Harris and Jackson, 1966). Monazite also occurs in strontianite-rich carbonatitic dykes and veins at Kangankunde Hill in southern Malawi. Grades are very variable but average 5—10% monazite with about 14% strontianite and minor baryte. Monazite also occurs in the thin residual soil overlying the deposits. Resources have been estimated at 350,000 t of "ore" grading 5.6% monazite to a depth of 30 m, although mineralization continues in depth (Holt, 1965). Other REE-bearing carbonatites in Africa include Wigu Hill in Tanzania and Nkombwa Hill in Zambia. Concentrations of REE, principally in the mineral britholite (a complex silicate and phosphate of the cerium metals and calcium, see Chapter 2) occur in the alkaline rocks of the Pilansberg complex in South Africa. Some 50,000 t of high-grade rare earth ore have been outlined (Coetzee, 1976; von Backstrom, 1976).

In Brazil, rare earth minerals, pyrochlore and apatite occur in three separate deposits in residual soils overlying the deeply weathered Barrairo do Araxá carbonatite complex near Araxá in Minas Gerais. The complex is the world's principal source of niobium and apatite is also exploited on an increasing scale. The REE occur in an earthy monazite and goyazite ($SrAl_3P_2O_7(OH_7)$) and total resources have been estimated at 120,000 t REO (McNeil, 1979). There is no production, however, because the ore is difficult to upgrade. REE and thorium are also closely associated with a large magnetite stockwork cutting alkaline igneous rocks at Morro do Ferro, on the Poços de Caldas plateau between Minas Gerais and São Paulo states (Wedow, 1967). The alkaline rocks (syenite and phonolite) have been deeply weathered to depths of 100 m and are locally enriched in REE and thorium in the vicinity of magnetite veins. The REE occur in allanite, particularly in its weathering products, and also in bastnäsite. Although easily worked, the ore is difficult to upgrade and there is no production. In the centrally enriched zone the ore grades 1% ThO_2 and 4% REO and resources are estimated at over 1 Mt to depths not exceeding 15 m (Wedow, 1967).

REE have been reported from a number of areas in China but the largest resources are associated with the Bayan Obo (Pai-yun-o-po) iron ore deposits situated on the plateau of the Ulanqab district of Nei Monggol in northern China. The iron ore deposits occur as tabular bodies and lenses in steeply dipping dolomites, the principal iron ore minerals being magnetite, specularite and hematite with which rare earth minerals are intercalated (K.Y. Lee, 1970). The rare earth mineralogy of the deposits is complex. Some of the more important minerals include the nioboaeschynite-titanaeschynite series ($(Ce,Th)(Nb,Ti)_2(O,OH)_6$), bastnäsite and monazite. The deposits extend 18 km from east to west with an average width of 2—3 km and consist of three major orebodies; the East and Main, which were being worked in 1980 and the West, where exploration is being carried out. The Main orebody contains 35% Fe and 6.19% REO and the East orebody 33% Fe and 5.71% REO. High-grade areas may contain up to 10% REO, however. The REE content of the Western orebody is significantly lower being of the order of 1% REO (Argall, 1980). Reserves of rare earth oxides in the Main and East orebodies alone have been placed at 20 and 15 Mt, respectively, making these deposits by far the largest identified in the world to date. Emplacement of the alkaline carbonate and iron-rich solutions from which iron and rare earth-niobium minerals were precipitated are believed to be genetically related to Hercynian alkaline granite, syenite and gabbroic rocks which outcrop to the south of the deposits (K.Y. Lee, 1970). The geochemistry of the deposits suggests a possible carbonatite association. Iron ore from Bayan Obo is transported 150 km south to Baotou (Pao-t'ou) for beneficiation and smelting and since 1974 a rare earth concentrate has been recovered. A range of rare earth products of the cerium sub-group are produced, the 60% REO concentrate having a relatively high europium content (0.2% Eu_2O_3).

REE mineralization similar to the Mountain Pass deposit occurs in carbonatite veins, small stocks and stockwork zones associated with the pseudoleucite syenite of the Lugin Gol pluton in the southern part of the Gobi Desert in the Mongolian People's Republic (Kovalenko, et al., 1976). The veins, which are up to 1 m in thickness have high REE contents, due to bastnäsite, with a maximum of 16% REO and Y combined.

Vein deposits

Only rarely are non-carbonatitic vein deposits sufficiently enriched in REE to make them economically attractive. However, the monazite-bearing vein at Steenkampskraal, halfway between Cape Town and the Orange River in the Republic of South Africa was the world's major source of monazite between 1952—1959 and 1962-63. Some 57,000 t of concentrate were produced up to 1965 when the mine finally closed, although the equivalent of 20,000 t of concentrate remain unworked (von Backstrom, 1976). The deposit is emplaced in an elongated, sub-vertical shear zone some 300 m in length, with a maximum width of 4 m, which traverses ancient granite-gneiss. Mineralization has been proved to a depth of 150 m and monazite and apatite account for about 80% of the ore. The monazite concentrate contained 45% REO and 6% thoria.

Monazite, together with some bastnäsite, comprise up to 2% of the ore at the Buffalo fluorspar mine near Naboomspruit in South Africa, the fluorspar mineralization occurring mainly as simple or composite veins in leptite, which is present as roof pendants in the Bushveld Granite (Watson and Snyman, 1975). The veins have a variable thickness but average about 10 mm, the fine-grained monazite occurring preferentially along the vein-leptite contact. By-product recovery of a monazite concentrate from the flotation tailings is being considered using size classification, gravity and high intensity magnetic separation (J.M. Jacobs, written communication) and there is a potential output of 2500—3000 t/year.

A rare earth concentrate was produced as a by-product of galena from the Korsnas mine on the Gulf of Bothnia in Finland. Total production between 1961 and 1972, when mining ceased, was 860,000 t of ore assaying 3.57% Pb and 0.91% REO. The deposit occupies a vein cutting Precambrian migmatitic mica gneiss and in addition to the ore minerals (galena, apatite and monazite), the gangue consists of calcite, feldspar (mainly microcline), and diopside (Isokangas, 1978).

Primary monazite has also been reported to occur at Bou Naga some 104 km southwest of Akjoujt in Mauritania. The deposit, containing 3.66% REO, was worked for 18 months between 1968 and 1970.

Placer deposits

Monazite has a high specific gravity (5), and is resistant during weathering and transportation. Placer concentrations, therefore, occur under favourable conditions usually in association with other heavy minerals such as ilmenite, zircon and rutile. Monazite-bearing placers have been widely reported in fluviatile, lacustrine and deltaic deposits but by far the most important economically are beach and dune sand deposits. These are formed by a combination of tide, long-shore currents, wave and wind action, the detrital monazite probably having passed through several cycles of erosion before the final deposition of the ore (Overstreet, 1967). The formation of heavy mineral beach sand deposits takes place over relatively short periods of geological time. With the exception of the Precambrian Blind River Formation in Ontario, consolidated placer deposits have not been of economic importance as sources of REE. Production is confined primarily to relatively unconsolidated beach deposits of Holocene to Late Tertiary age.

Monazite is by far the most common rare earth mineral occurring in placer deposits although minor quantities of xenotime are recovered and the multi-oxide euxenite has been worked, primarily for its niobium content. Monazite is usually concentrated in thin layers but in many commercial operations it often comprises less than 0.1% of the ore mined which may contain as little as 5—10% total heavy minerals. However, much richer deposits are worked in India. Monazite is normally produced, therefore, as a by-product of other heavy minerals, the most important being ilmenite, zircon and rutile, although cassiterite and gold are important in some countries.

Placers were the major source of REE until 1965 when large-scale mining of primary bastnäsite commenced. With the exception of the brief production at Steenkampskraal in South Africa, placer deposits continue to be the principal source of monazite. Beach deposits in Brazil and India were formerly the most important sources but Australia has now become the leading producer.

In Australia monazite is produced (Table 12.2) chiefly in Western Australia and to a much more limited extent on the east coast in Queensland and New South Wales.

In Western Australia the heavy mineral sands are worked principally for ilmenite, although rutile, zircon, leucoxene and monazite are important by-products. Total production of heavy mineral concentrates amounted to 1,614,778 t in 1980, of which about 0.8% was monazite. Mining has been undertaken since 1956 in the Geographe Bay area between Bunbury and Busselton in the southwest and since 1974 at Eneabba, 250 km north of Perth, where about 40 Mt of recoverable heavy mineral sands have been identified (Phillips and Bassett, 1979). The Eneabba deposits are currently the world's major source of monazite and have accounted for the major increase in production shown in Table 12.2. In the southwest, the deposits are located on the Swan Coastal Plain on or close to shorelines (strandlines)

TABLE 12.2

Australia: monazite concentrate production (in tonnes)

	1974	1975	1976	1977	1978	1979	1980
Queensland	25	14	832	622	—	—	499
New South Wales[a]	1003	988	457	910	233	1798	290
Western Australia	2549	3505	4021	7847	14,759	14,542	13,290

[a]Shipments.
Sources: Ward, 1979, 1980, 1981; Bureau of Mineral Resources, Geology and Geophysics, Canberra, A.C.T. (written communication, 1982).

TABLE 12.3

Australian monazite resources (in 10^3 tonnes)

	1975	1976	1977	1978	1979	1980
Economic resources						
demonstrated	307*	432	361	351.3	328.7	332.6
inferred	—	5	3.5	3.5	3.5	3.5
Sub-economic resources						
undifferentiated	78	54	183.1	185.01	163.1	158.7
Total	385	491	547.6	539.81	495.3	494.8

*includes inferred.
Sources: Ward, 1979, 1980, 1981; Bureau of Mineral Resources, Geology and Geophysics, Canberra, A.C.T. (written communication, 1982).

of Pleistocene to Holocene age which now stand variously between sea level and 76 m above it (Welch et al., 1975). Mineralization occurs to a depth of 25 m and over a width of up to 1200 m. Monazite is typically concentrated towards the base of a deposit but comprises less than 1%, and usually less than 0.5%, of the heavy mineral concentrate, itself averaging some 12—15% of the ore (Baxter, 1977). At the Capel processing plant very limited quantities of xenotime concentrate (27 t in 1980 containing 8158 kg Y_2O_3) have been recovered from some Geographe Bay sands.

The Eneabba heavy mineral sands were deposited in Late Tertiary or Early Pleistocene times (Lissiman and Oxenford, 1975), the fossil shorelines occurring up to 50 km inland and at higher elevations than those worked in the Geographe Bay area. The Eneabba deposit consists of an association of beach, estuarine and dune sands about 12 km long and up to 1200 m wide. Their monazite content varies throughout the deposit from 0 to 35% of the heavy mineral fraction (Baxter, 1977) but generally constitutes less than 1% of the heavy mineral concentrates produced. Modern beach deposits are small in size and of no economic significance, although offshore exploration has been carried out.

The east coast of Australia, for about 2000 km between the Hawkesbury River north of Sydney in New South Wales and north of Rockhampton in Queensland, is fringed with Holocene to late Pleistocene sand beaches and dunes which extend up to 13 km inland. Natural concentrations of heavy minerals are principally worked for their rutile and zircon contents and monazite constitutes less than 0.8% of the heavy mineral fraction (McKellar, 1975).

Australian monazite resources are mainly on the west coast with the major proportion occurring in the Midlands area to the north of Perth.

Monazite was discovered in Brazil in 1880 in beach sand deposits at Alcobaça, Bahia State. The major deposits are situated intermittently along the coast from Rio de Janiero to Canavieriras in Bahia State. However, the mineral has most consistently been worked along the coast of the states of Espirito Santo, from which the major production has been derived, and Rio de Janiero (Maciel and Cruz, 1973). The deposits, which are of low grade and occur in raised terraces, elevated beaches and bars, are mainly derived from Tertiary sandstones (Leonardos, 1974). Initially monazite-rich sands were worked from storm beaches but monazite is now obtained as a co-product with ilmenite and zircon. Reserves of monazite in Espirito Santo have been estimated at 33,000 t principally in the vicinity of Cachoeiro de Itapemirim (Anonymous, 1979a).

The principal beach placer deposits in southern India are in the Manavalakurichi area, Kanniyakumari district, Tamil Nadu, and between Neendakara and Kayankulam in the Chavara region of Kerala State. At Manavalakurichi they contain about 5% monazite out of a total heavy mineral content of some 80% (Karve et al., 1966) and are the principal source of monazite in India. In Kerala State they contain 1—2% monazite (Poulose, 1972) out of a total heavy mineral content of some 75%. In both cases ilmenite is the major product together with smaller amounts of zircon, rutile, garnet and sillimanite. Similar deposits occur elsewhere in India and a new project to develop beach sands at Gopalpur in Orissa State is being undertaken.

Monazite production in the U.S.A. has been somewhat sporadic but started in the late 19th century from fluviatile deposits in Cleveland County, North Carolina, to provide thorium for the manufacture of gas lamp mantles (Garnar, 1981). However, monazite has also been recovered from beach sand deposits in Florida and Georgia as a by-product of ilmenite, leucoxene and zircon recovery. In recent years production has been based on the Green Cove Springs and Boulougne deposits in northern Florida (Garnar, 1981). At Green Cove Springs the deposit is from 16 to 21 km in length, up to 1200 m wide and approximately 6 m thick; the Boulougne deposit is 4.5 km in length, 800—1200 m wide and up to 7.5 m thick. Both represent elevated Pleistocene beach sands containing 3—4% heavy minerals which accumulated as beach ridges during periods of standstill in a general regressive sequence (Pirkle et al., 1974). Mining of the deposits ceased in 1978, the Boulougne deposit being exhausted, but production at Green Cove Springs resumed in 1980.

In Malaysia significant quantities of monazite and small amounts of xenotime are currently produced from the heavy mineral residues ("amang") of alluvial tin mining that consist mainly of ilmenite. The major tin-producing areas are in Perak, where the monazite concentrates have a composition of 54.5% REO, 5.5% ThO_2 (Alexander et al., 1964), and Selangor. Similarly in Thailand, monazite together with ilmenite, zircon and tantalite, occurs in small concentrations associated with alluvial tin, and small quantities are recovered and sold periodically. Minor amounts of xenotime are also recovered. Monazite also occurs as an accessory in tin placer deposits in Indonesia and beach placer deposits of potential economic interest occur on the west coast of Kalimantan and the southwestern coast of Sumatra. Heavy mineral placer deposits of fluvial origin, in which monazite is often the most abundant heavy mineral, occur at numerous localities in South Korea, and littoral deposits are present along the east and west coasts. The source rocks are granite gneiss and granitic rocks which cover about one half of the country (Kim, 1968). Commercial recovery of monazite has been confined to stream placers but beach sand deposits also have potential — grades of 0.2—0.4% monazite having been reported from Taewha on the east coast (Noakes, 1968a). Monazite of both black and yellow varieties together with zircon, ilmenite and rutile occur in beach placer deposits associated with offshore bars on the southwestern coast of Taiwan. Heavy mineral contents range between 0.88 and 5.56%, of which monazite (mostly black) accounts for between 9.6 and 13.5% (Noakes, 1968b). The black pellet-like monazite is depleted in thorium (1.3% ThO_2) and enriched in europium (0.35% Eu_2O_3) compared to other monazites but is reported to contain an unacceptably high acid-insoluble residue (31%) for normal processing. The pellets consist of cemented aggregates of monazite and other minerals and are the result of both weathering and authigenic growth (Matzko and Overstreet, 1977). Monazite-bearing beach placer deposits which occur on the west coast of Sri Lanka south of Colombo contain 15% monazite on average. Minor quantities of monazite, typically with a high thoria content (10%), are produced by the Geological Survey Department. The largest heavy mineral sand deposit in Sri Lanka occurring on the north-east coast at Pulmoddai contains 0.3% monazite but is worked only for ilmenite, zircon and rutile (Herath, 1977). Monazite is also believed to be recovered as a by-product of alluvial gold mining in the U.S.S.R., and monazite is produced from placer deposits in Guangdong and Guangxi provinces in China.

Placer deposits containing monazite have been recorded from numerous other localities (Overstreet, 1967). Minor quantities of monazite are recovered from tin placers in the Plateau Province of northern Nigeria, and monazite is also obtained as a by-product of working alluvial and eluvial tin in Kivu region in the eastern part of Zaire. Monazite was formerly produced from the Nile delta (De Kun, 1965) and from ancient beach and dune deposits on the south-southeastern coast of Madagascar near Antete and Vohibarika, where proved reserves of monazite have been estimated at 32,000 t. Production ceased in the late 1960's. Proposals have been made to develop heavy mineral beach sand deposits containing monazite in Liberia (Rosenblum, 1974). Quaternary dune sands containing heavy minerals occur on both the east and west coasts of South Africa, but monazite is present in only small quantities ranging from trace amounts to 0.3% of the heavy mineral concentrate (Hammerbeck, 1976). Ilmenite, zircon and rutile are currently exploited on a large scale at Richards Bay on South Africa's east coast. The sand as mined contains 5—15% heavy minerals including 0.04% monazite, which is believed to be recovered.

Fossil placer deposits older than the late Tertiary have not been utilized as commercial sources of REE. The origin of the one possible exception is in doubt; the pyritic matrix of a quartz-pebble conglomerate of Precambrian (Huronian) age in the Elliot Lake area of Ontario, Canada, contains minute grains of uraninite, brannerite and monazite and an yttrium concentrate has been recovered intermittently as a by-product of uranium production. The ore contains 0.11% U_3O_8, 0.028% ThO_2 and 0.057% REO (Wood, 1979). The distribution of the REE is variable but is approximately 14.3% LREE and 85.7% HREE including some 51.4% Y_2O_3 (Rose, 1979). Shipments of yttrium concentrate in 1977 amounted to 30.4 t (Wood, 1979), but recovery has become uneconomic and production ceased in 1978 (Cranstone, 1981). Higher concentrations of REE have been reported from similar deposits at

Agnew Lake, 65 km east of Elliot Lake, but REE are not recovered and uranium mining is to cease.

Apatite deposits

Apatite in phosphate rock is often enriched in REE, which could be recovered during phosphoric acid manufacture — the major end-use for the world production of 135 Mt of phosphate rock in 1980. There appears to be little likelihood of utilizing this resource, however, while less costly sources are available. Apatites of marine origin, on which world phosphoric acid production is predominantly based, contain between 110 and 1550 ppm REO (Robinson, 1948). Characteristically in sedimentary phosphate rock there is an enrichment in yttrium and a deficiency in cerium, reflecting the composition of seawater (Altschuler et al., 1967; Altschuler, 1980). Up to 1000 ppm Y has been reported from certain phosphatic horizons within the Permian Phosphoria Formation of western U.S.A. (Gulbrandsen, 1966) and 1500 ppm from Middle Cambrian phosphate rock in the Georgina Basin of northwest Queensland (De Keyser and Cook, 1972).

Apatite deposits with sufficiently high REE contents to be considered as potential commercial sources are found associated with alkaline igneous complexes. Selected specimens of apatite from the Khibiny alkaline igneous complex in the Kola Peninsula, U.S.S.R., for example, contain up to 4.90% REO (Volkova and Melentiev, 1939), although the marketable apatite concentrate produced contains only 0.85% REO (Golovanov et al., 1969). On this basis resources within the complex amount to some 9 Mt REO. By-product recovery of REE from Kola apatite during phosphate fertilizer manufacture is now believed to be confined to the U.S.S.R. Apatite concentrates produced from the phoscorite and pyroxenite zones of the Palabora complex in South Africa contain 0.6 and 0.8% REO, respectively, and the feasibility of their recovery has been considered (Russell, 1977). Apatite from the Siilinjarvi carbonatite deposit in Finland averages 0.4% REO (Puustinen, 1971), but this is regarded as too low for commercial recovery (N. Lounamaa, written communication).

Other potential sources of REE-bearing apatite include magnetite deposits in which apatite occurs as a deleterious impurity and must be removed prior to iron- and steelmaking, for example at Kiruna in northern Sweden.

Other deposits

Pegmatites may be enriched in REE and have been mined locally on a very small scale. For example, a few hundred kilograms of yttrotantalite and gadolinite were recovered from pegmatites at Cooglegong in Western Australia (Hill, 1975), and bastnäsite has been recovered from pegmatites in the Ambatofinandrahana region of Madagascar.

REE mainly contained in the uranium-bearing minerals allanite and stillwellite were formerly produced from the Mary Kathleen uranium-bearing skarn deposit in Queensland (Hawkins, 1975). Uranium mining at Mary Kathleen ceased in 1982.

A potentially large source of REE has recently been identified in South Australia. The large Olympic Dam copper deposit, situated in the Roxby Downs area some 650 km north-northwest of Adelaide, contains significant amounts of REE in the ore zone, mainly present as bastnäsite and florencite. The Cu-U-Au-REE mineralization is both stratabound and transgressive and occurs within a thick sequence of monomict and polymict, iron-rich, sedimentary breccias of Proterozoic age, over a vertical thickness of up to 350 m (Roberts and Hudson, 1983). The deposit is estimated to contain 2000 Mt of ore grading 1.6% Cu, 0.64 kg/t U_3O_8 and 0.6 g/t Au. REE are associated with copper, and particularly uranium, typical combined La and Ce values for composite bulk samples falling in the range 0.24—0.45% (Roberts and Hudson, 1983).

Easily worked deposits of REE have recently been developed in China at Nan-lu in southern Jiangxi province; the precise nature of these deposits is as yet unclear, but reserves have been placed at 1 Mt REO.

12.4. World reserves and resources

An assessment of world reserves and resources by the U.S. Bureau of Mines and U.S. Geological Survey is shown in Table 12.4, with the major ore types also indicated. At current rates of production these reserves represent some 200 years supply. They do not include, however, either the very large reserves of REE associated with the Bayan Obo iron ore deposits in China, which alone have been estimated at 35 Mt of rare earth oxides in the orebodies currently being worked, or the recently discovered South Australian deposit (Olympic Dam).

The grade of rare earth ore reserves varies markedly depending on the type of deposit. For example, the high-grade primary bastnäsite deposit at Mountain Pass averages 7% REO but, in contrast, the overall grade of the ore mined in some heavy mineral placer deposits may be as low as 300 ppm REO, little above the crustal abundance levels (150 ppm). However, the constituent minerals have already been liberated by natural processes and, more importantly, the by-product nature of monazite recovery, allows the extremely low grades to be economically worked. A disadvantage is that monazite production and reserves are dependent on the viability of the primary product recovery. Hence the classification of heavy mineral sands as rare earth reserves depends on the economic recovery of other minerals.

It seems likely that further REE deposits will continue to be discovered in a number of geological environments, and in this context carbonatites form excellent exploration targets. In addition the wider use of by-product

TABLE 12.4

World resources of REE (in 10^3 tonnes contained rare earth oxides)

	Major ore type	REE		Yttrium	
		reserves	other resources	reserves	other resources
North America					
U.S.A.	Bastnäsite-bearing carbonatite	4540	23,590	3.2	38.1
Canada	Uranium ores	225	1000	2.2	23.6
South America					
Brazil	Beach placers, carbonatite and alkaline rocks	320	165	2.3	7.7
Europe					
U.S.S.R.	By-product of apatite processing	455	8620	1.5	11.8
Finland, Norway, Sweden	Carbonatite, alkaline rocks	55	110	0.2	0.7
Africa					
Madagascar	Beach placers (monazite)	20	40	n.a.	n.a.
South Africa	Monazite vein	5	15	n.a.	n a.
Egypt	Fluviatile placers (monazite)	10	90	n.a.	n.a.
Malawi	Carbonatite	15	45	n.a.	n.a
Other	—	20	30	n.a.	n.a.
Asia					
Malaysia	Alluvial tin placers (monazite and xenotime)	25	45	0.5	1.7
India	Beach placers (monazite)	910	1815	18.1	36.3
Korea	Placers (monazite)	45	90	0.9	2.7
Sri Lanka	Beach placers (monazite)	10	10	0.2	0.3
Other	—	10	20	0.2	0.7
Australia	Heavy mineral placers (monazite)	365	225	5.4	18.1
World total[a]		7010	36,290	34.7	145.1

The tonnages are based on Moore (1979) and converted from short tons. Figures have been rounded to the nearest 5000 tonnes for the REE and 100 tonnes for Y.
n.a. = not available.

[a]Data may not add to totals because of rounding.

production from phosphoric acid manufacture and, possibly uranium recovery, is conceivable in the future. No foreseeable shortage of REE through inadequate resources seems likely, therefore.

12.5. Mining, mineral processing, extraction and separation

The mining and processing techniques used in the extraction of rare earth minerals depends on the physical and chemical characteristics of the ore deposit. Thus the exploitation and beneficiation of loosely consolidated heavy-mineral-bearing sands is very different from that used for primary bastnäsite deposits.

A variety of methods are used to work unconsolidated placer deposits. In the larger operations, floating dredges with suction or bucket wheel excavators are used under wet conditions whereas front end loaders and scrapers are used for dry mining. The processing of the ore initially depends on the recovery of a high-grade heavy mineral concentrate by wet gravity methods. Initially the oversize material is rejected by dry and wet vibratory screening and the product is then deslimed. Recovery of the heavy mineral concentrate is undertaken by Humphreys spiral or Reichert cone concentrators, which make use of the density differential between the heavy minerals and quartz-rich gangue. Tailings are usually pumped back to worked-out areas, which are subsequently restored. The heavy mineral concentrate is then dewatered and dried before being separated into its individual component minerals by a combination of gravity, magnetic and electrical methods.

The beneficiation techniques adopted by Westralian Sands Ltd. at their Capel plant in Western Australia illustrate the principles involved (Anonymous, 1979b). A simplified flow diagram is shown in Fig. 12.2. The screened and dried before being separated into its individual component minerals by a (*1* in Fig. 12.2) for the initial removal of an ilmenite concentrate. The non-magnetic fraction is scavenged on roll magnets (*2*) from which a magnetic fraction is fed to crossbelt magnetic separators (*3*). Here a secondary magnetic ilmenite concentrate is produced as well as, following electrostatic separation (*4*), a high-titanium altered-ilmenite concentrate from the non-magnetic and conductor fraction. The non-conductor fraction together with the non-magnetics from the roll magnets are further treated by wet gravity methods using Reichert spirals (*5*), to remove quartz and other gangue minerals such as kyanite, staurolite and spinels. This concentrate is then fed into attrition cells (*6*) where hydrochloric acid is added to remove clay minerals and prepare the surface for subsequent high-tension and electrostatic separation. The concentrate is washed, dewatered and dried (*7*), the drier discharge temperature being critical for later separations.

After screening, the —0.5-mm fraction is fed to high-tension roll separators (*8*). The resultant conducting mineral stream feeds the leucoxene plant and

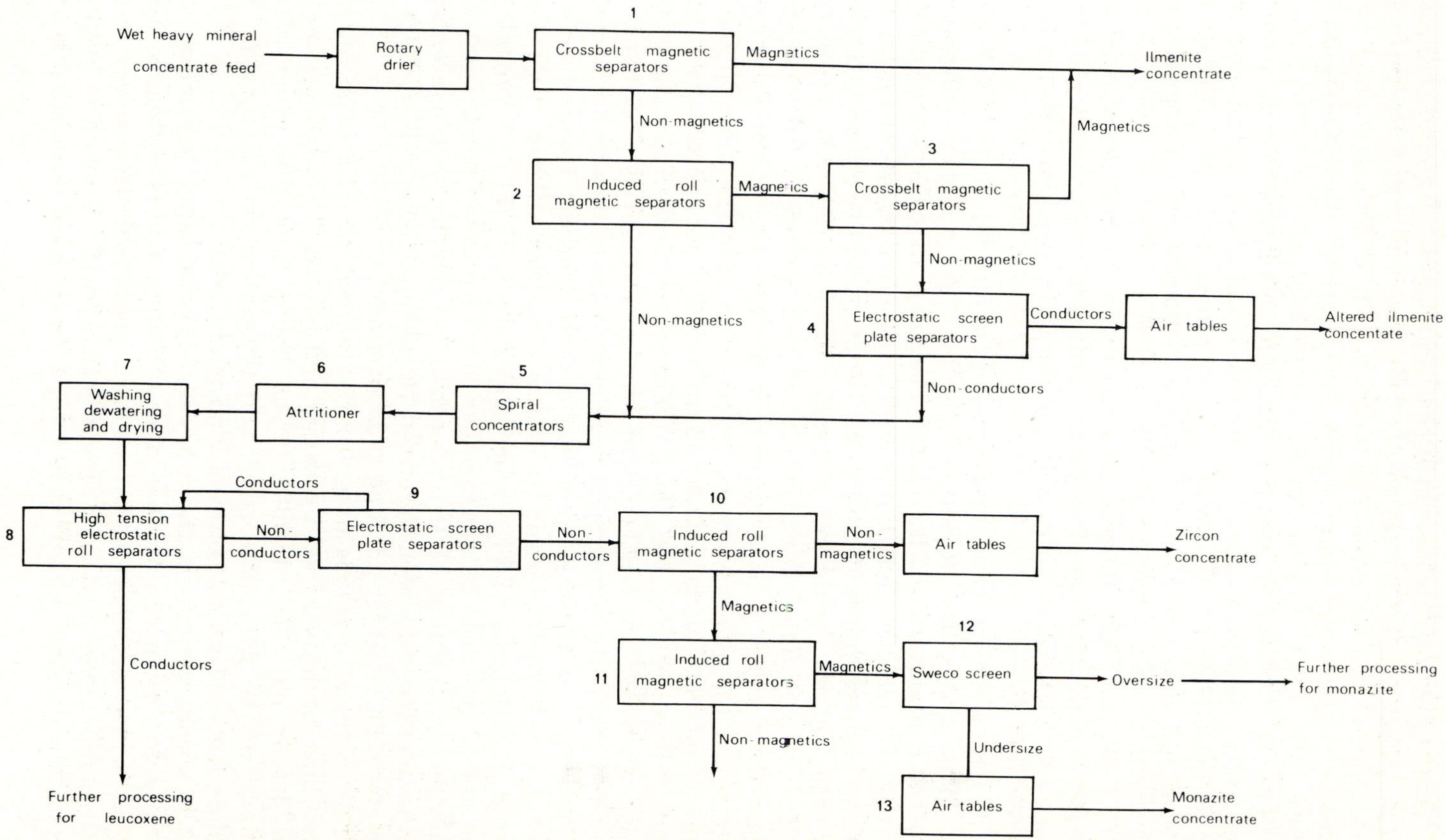

Fig. 12.2. Production flow sheet for the recovery of heavy minerals by Westralian Sands Ltd. (based on Anonymous 1979a).

the non-conducting stream, consisting principally of zircon, monazite and quartz, passes to the zircon plant. The non-conducting stream is cleaned using electrostatic screenplate separators (*9*) to remove minor rutile and leucoxene (altered ilmenite) which are returned, via high-tension rolls to remove any zircon and monazite, to the leucoxene plant as a conductor fraction. The stream is then fed through high-intensity roll magnets (*10*) from which the non-magnetic fraction is mainly zircon. The magnetic fraction is mainly monazite which is then passed through a further stage of induced roll magnets (*11*) to clean out remaining non-magnetic minerals such as zircon and some aluminosilicates. The magnetic product is screened (*12*) at 210 μm and the oversize product is stockpiled for later monazite recovery by wet tabling. The screen undersize is air-tabled (*13*) to reduce zircon and TiO_2 levels to give a final monazite product assaying 55% REO and 6% ThO_2.

The bastnäsite ore at Mountain Pass in California occurs in close association with baryte, celestite, strontianite, calcite and a quartz gangue. Run of mine ore contains some 7.25% REO. The ore is mined by conventional open-pit methods and beneficiation is by flotation and chemical treatment. After primary and secondary crushing, the ore is ground to a size such that 96% is less than 150 μm, firstly in a rod mill and then a ball mill. A four-stage pre-flotation conditioning is carried out in heat agitators. The temperature is progressively raised to boiling point in the presence of conditioning agents, after which the slurry is cooled to 60°C for froth flotation. In the rougher flotation circuit, baryte and strontium minerals are depressed and bastnäsite floats. After cleaning, the final flotation concentrate averages 60% REO (range 57—65%). The REO content may be increased by a further 5—10% by leaching with hydrochloric acid to remove calcium and strontium carbonates. Additionally, calcination removes carbon dioxide from the bastnäsite, increasing the REO content to 90% as a mixture of rare earth oxides and fluorides (Johnson, 1966; Parker and Baroch, 1971). A process for the recovery of rare earth oxides together with by-product barium and strontium sulphates, by reaction of the bastnäsite ore directly with concentrated sulphuric acid has also been described (Eisele and Bauer, 1974). This method eliminates the need for grinding and flotation.

The chemical methods used for the removal of REE from their concentrates have been described by Parker and Baroch (1971), Kaczmarek (1980) and Murthy and Gupta (1980). In the case of monazite, which requires the removal of the phosphate anion and the separation of thorium, either an acid or alkaline (caustic) process is normally used.

The caustic process is the more widely used since it removes the phosphate anion more readily than does sulphuric acid (Kaczmarek, 1980) and has the added advantage of producing a marketable by-product in the form of tri-sodium phosphate. Finely-ground monazite is digested in hot caustic soda and maintained at 140—150°C for a few hours. A water soluble tri-sodium phosphate is recovered from the insoluble rare earth, yttrium and thorium hydroxides. A number of processes may be used to remove thorium from this mixture, for example, it is possible to selectively dissolve rare earth hydroxides in dilute hydrochloric acid, leaving insoluble thorium hydroxide which is recovered by filtration. Similarly, xenotime may also be digested by either sulphuric acid or caustic soda.

At Mountain Pass the bastnäsite flotation concentrate is calcined and leached with 30% hydrochloric acid to convert trivalent REE to soluble chlorides. The tetravalent cerium remains undissolved and the slurry is thickened and filtered, resulting in a filter cake grading 65—70% REO and containing 55—60% CeO_2. This precipitate is repulped, filtered, dried and marketed as a cerium concentrate. The leach solution is subsequently treated by solvent extraction methods to produce europium oxide with a purity of 99.99% and Sm, Gd, Pr and Nd compounds as by-products. A lanthanum-rich mixture of REE is also recovered (Harrah, 1967).

At Elliot Lake in Canada, an yttrium concentrate was, until recently, obtained from the residual ion-exchange solutions after leaching uranium ores with sulphuric acid. The residual solutions were treated with air and lime to oxidize iron and adjust the pH, and the resultant slurry was thickened and filtered. The filtrate, clarified and acidified with sulphuric acid, was fed to the solvent extraction circuit, where an organic solvent removed yttrium and REE in a countercurrent centrifugal extractor. This loaded solvent was stripped with nitric acid from which a REE-rich product was precipitated with lime and ammonia. The filtered and dried product graded 60—70% REO including 30—35% Y_2O_3 (Lucas and Ritcey, 1975).

The recovery of REE from apatite during phosphoric acid production based on nitric acid and sulphuric acid dissolution is technically feasible. In the production of phosphoric acid by the reaction of phosphate rock with sulphuric acid most of the REE are lost into the phosphogypsum (Ionescu et al., 1980). However, REE from apatite are completely dissolved in nitric acid and recovery has been carried out on a commercial scale in the U.S.S.R., Poland (Bril, 1964) and Finland (Lounamaa et al., 1980). Production was based on Kola apatite, which contains about 0.85% REO. REE were recovered in Finland from Kola apatite and a low-grade indigenous rare earth concentrate, during the period 1965—1972, by nitric acid dissolution and selective precipitation with ammonia. Commercial recovery is now believed to be confined to the U.S.S.R. but there is still considerable interest in the process elsewhere. Recent research suggests that a combined precipitation and solvent extraction method based on tributyl phosphate is technically feasible and potentially economic when the apatite concentrates contain more than about 0.5—1% REO. The exact lower limit is dependent on the distribution of individual REE (Lounamaa et al., 1980).

For some applications the REE can be used without separation and in others a broad separation into light and heavy sub-groups is sufficient.

However, to gain advantage of their individual properties, high-purity metals or compounds must be obtained. Because of their similar chemical properties, separation and purification has presented a major problem. A number of different techniques have been utilized. These include valency change reactions by either selective oxidation or reduction, fractional crystallization, which depends on small differences in solubilities of various REE salts, and fractional precipitation, involving the addition of a precipitant to a mixture in an amount insufficient for complete precipitation (Pings, 1969). For both fractional crystallization and precipitation a large number of individual separations are required to achieve the desired purity. Ion exchange, introduced on a commercial scale in the mid-1950's, and more recently solvent extraction, have now mainly replaced these older methods allowing high-purity separations (99.99%) to be achieved. In the ion-exchange process trivalent REE ions are adsorbed by resins. The REE ion is then removed (eluted) differentially from the resin, by a complexing solution which has an affinity for smaller ions, leaving larger ions on the resin. After repeated separations a high-purity product can be produced. However, the process requires large capital investment, is slow, discontinuous and is no longer used for large-scale separations.

Solvent extraction, also referred to as liquid-liquid extraction and introduced in the early 1960's, is today the most important process used for separating the individual REE (Kaczmarek, 1980). The process utilizes the relative affinities of various REE between a liquid solvent, generally a mixture of organic compounds, which is immiscible or only partially miscible in an aqueous feed solution containing the REE phases. Selected REE ions migrate from the aqueous feed solution which is usually the rare earth chloride hexahydrate ($RECl_3 \cdot 6H_2O$), into the organic solvent which forms the extract. After scrubbing to remove impurities the organic solvent is separated from the aqueous feed and the REE taken from the loaded solvent into a new aqueous phase. Adequate mixing of the aqueous feed solution and the organic solvent is essential in obtaining good separations and this is achieved by multi-stage countercurrent contacting of the fluids. Continuous processing is feasible with this technique and large-scale extractions of exceptional purity are possible.

Certain rare earth metals are produced by the electrolysis of the fused anhydrous rare earth chloride or oxide (Morrice and Wong, 1979). These include mischmetal, the only rare earth metal to be produced in quantity, Ce, La, Pr and Nd. Metallothermic reduction is used to produce the heavier metals (Murthy and Gupta, 1980). Bastnäsite is also smelted with ferrosilicon in an electric furnace to obtain rare earth silicide.

12.6. Historical production and demand

Monazite was the major source of the REE until 1965 when bastnäsite from the Mountain Pass mine in California became the leading source of supply. Total cumulative production in the Western World of rare earth oxides up to and including 1980 is estimated to be nearly 0.5 Mt and although production in the U.S.A. has only become very important during the last twenty years its cumulative output is clearly dominant (Fig. 12.3). World mine production of REE is currently dominated by the U.S.A. and Australia, although a number of other countries produce significant quantities (Table 12.5). China, with the world's largest reserves, has recently started production and offers considerable potential for the future.

Interest in monazite originated in 1883 with the development of incandescent gas mantles. The first non-commercial lanthania-zirconia gas mantles were brittle and emitted a cold blue-green light. These were improved to the still-used Auer incandescent 99% thoria, 1% ceria mantles, commercially introduced in 1891 (Overstreet, 1967). As the REE occur together in the same minerals and have very similar chemical behaviour, they are produced together. The production of the thorium mantle created large quantities of surplus REE and this mismatch of production and demand is still a

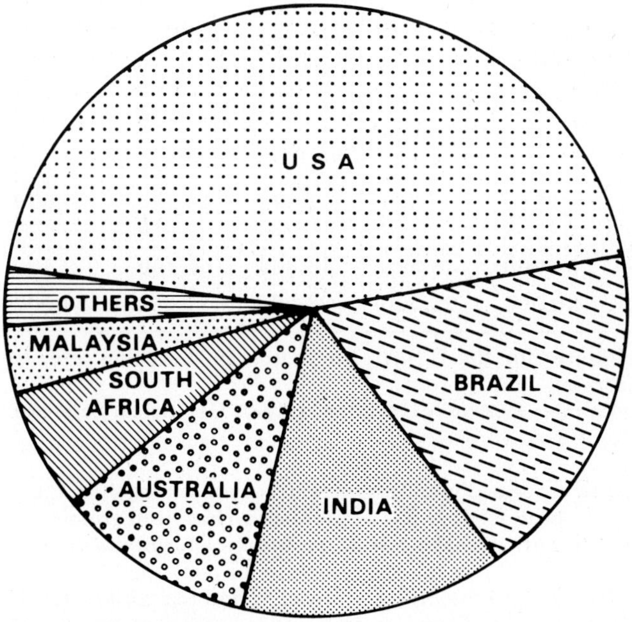

Fig. 12.3. Pie diagram showing the distribution (by country) of the total cumulative REE production 1885—1980.

TABLE 12.5

Production of rare earth minerals (in tonnes)

	1975	1976	1977	1978	1979	1980
Burundi						
Bastnäsite	82	139	165	40	30[e]	n.a.
Nigeria						
Monazite	20[e]	20[e]	20[e]	20[e]	20[e]	20[e]
Zaire						
Monazite	298	125	96	77	92	51
Canada						
Yttrium concentrates	34.9	26.3	30.4	—	—	—
U.S.A.						
Bastnäsite[a]	14,940	13,039	15,359	14,148	16,515	15,986
Monazite[a]	750[e]	600[e]	600[e]	1250[e]	1200[e]	700[e]
Brazil						
Monazite	1403	1610	2441	2541	1900	2532
India						
Monazite	3000	2994	2734	3303	3254	4210[e]
Malaysia						
Monazite[b]	3285	1879	1976	1607	466	347
Xenotime[b]	25	79	35	—	—	—
Sri Lanka						
Monazite	6	1	—	18	4	63
Thailand						
Monazite	367	—	—	—	32	152
Xenotime	—	—	50	—	6	52
Australia						
Monazite	4507	5310	9379	14,992	16,340	14,079
Xenotime	20	7	12	23	19	27
World total[c,d]	39,200	34,900	43,500	48,300	51,700	49,400
Estimated REO content	23,500	20,900	26,100	28,900	31,000	29,500

Note: With the exception of the figures for the U.S.A., the figures in this table refer to gross tonnage of rare earth concentrates.
Source: Minerals Strategy and Economics Research Unit, Institute of Geological Sciences.
n.a. = not available
[a] REO content.
[b] Exports. Xenotime now exported as processed yttrium concentrate containing 60% Y_2O_3.
[c] Expressed as weight of concentrates produced; U.S.A.'s production has been grossed up to concentrate equivalent on the assumption that both its bastnäsite and monazite concentrates contain 60% REO.
[d] Rare earth minerals are also produced in China and the Soviet Union (ca. 2300 t REO) and possibly, on a small scale, in Indonesia, Democratic Republic of Korea and South Africa.
[e] Estimated.

characteristic of the industry. Currently only a third of the REE are in regular demand to match their availability (Cannon, 1980) which is in proportion to their natural ratios in the ore minerals. Consequently the elements, such as Dy, Ho, Lu, Th, and Yb, which have few industrial uses, are being stockpiled for possible future use. Efforts to use the excess production of some REE have persisted since the turn of the century when the "waste" from thorium mantle production, combined with the necessity for a simple gas ignition system, resulted in the discovery of a pyrophoric metal consisting of 70% mischmetal and 30% iron.

Initially the major world source of supply of monazite was the rich beach sand deposits of Brazil where large-scale production began in 1895 (Fig. 12.4). Peak production of 6460 t was achieved in 1909 but output subsequently declined markedly (Parker and Baroch, 1971). Similar deposits were developed in India in 1911, which soon became the major producer, but the introduction of electric lighting after World War I caused a

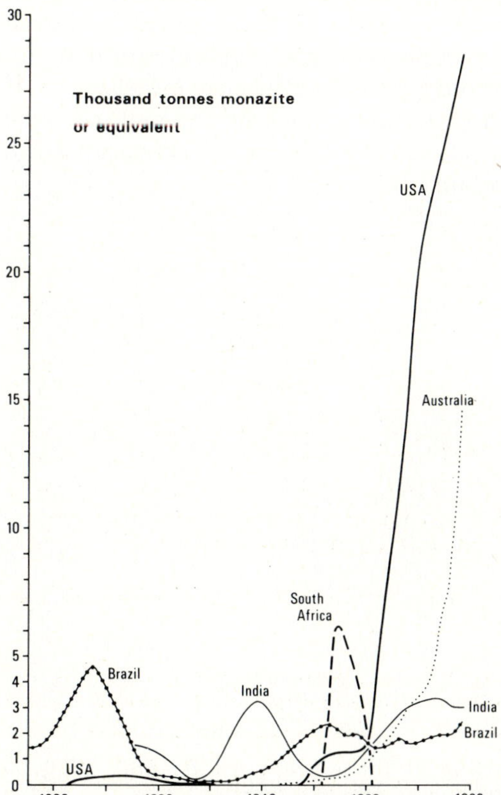

Fig. 12.4. Major producing countries of REE since 1895. As monazite and bastnäsite recalculated to a monazite equivalent. Trends smoothed from a ten-year running average.

dramatic decline in demand for thorium, and thus monazite. World monazite production fell to below 700 t during the period 1922—1931 (Parker and Baroch, 1971; Fig. 12.5).

Before 1930 industrial usage was dominated by the mixed REE or those that were easy to separate. Until World War II REE had only limited applications as mischmetal for lighter flints and as an additive to cast iron, and as mixed compounds for glass colouring and carbon arc lightning. The possible use of thorium as a nuclear fuel triggered renewed exploration for monazite during the period 1949—1959 and also led to Brazil and India prohibiting or controlling the export of monazite in 1946 and 1951, respectively. Monazite production from placer deposits began elsewhere, in Australia, South Korea, Malaysia, the U.S.A., Sri Lanka, Indonesia, Nigeria, Thailand and Madagascar. The monazite-bearing vein deposit at Steenkampskraal in the Republic of South Africa was the major world source for most of the period 1953—1963 with a mine capacity of 9000 t of monazite concentrates a year. A large demand for thorium as a fissionable source did not materialize, however.

The properties and applications of the individual REE were intensively investigated between 1930 and 1960, especially because of their concentration in the wastes from thorium production. Markets developed for lanthanum in optical glass, cerium in polishing media and didymium* in the glass industry.

Since 1967, Australian beach sand deposits have become the major source of monazite, currently accounting for about 60—70% of world production with Western Australia dominating production.

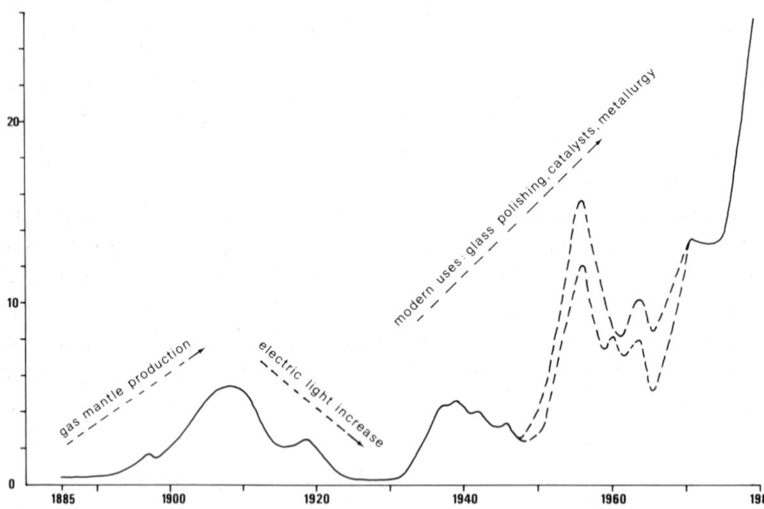

Fig. 12.5. Production of monazite concentrates in the western world 1885—1979 (in 10^3 tonnes). Dashed lines represent maximum and minimum estimates.

*Mixture of REE, enriched in Pr and Nd, left after the removal of Ce and La.

The bastnäsite deposit at Mountain Pass, was discovered in 1949. Production started in 1952 and continued on a limited scale until 1965 when development of a new red phosphor for colour television tubes based on europium and yttrium oxides resulted in a dramatic increase in demand for bastnäsite, since it is relatively rich in Eu (0.1% Eu_2O_3). Mountain Pass has retained its position as the largest source of REE in the world. China has emerged as a significant producer of bastnäsite and minor quantities of bastnäsite have been produced in Burundi and Madagascar.

The main source of yttrium until 1966 was xenotime, derived principally from placer deposits in Malaysia, Thailand and Australia. Yttrium recovery from the acid-leach residues of uranium ores in the Elliot Lake area of Ontario started in 1966 and this was the major source until 1970. Oversupply resulting from increased stocks and lower requirements in the production of phosphors caused shipments of yttrium concentrate to cease in 1971. Production was resumed by one company in 1973 and continued until the end of 1977 but there was no production in either 1978 or 1979 (Cranstone, 1981).

The historical variation in the supply diversification for REE has three main stages. From the start of commercial production until the late 1940's there were two major suppliers with, for most of the time, one country very dominant (Fig. 12.6). For the next twenty years there were four or five substantial producers with no single dominant country. This period of high supply diversity has been followed (since the late 1960's) by the rapid rise in importance of Australian monazite, and bastnäsite from the U.S.A. With the current production from politically stable developed countries satisfying the large demand from the industrialized nations (Fig. 12.1) and with other major deposits already discovered, the vulnerability of REE supply to political interference is low. However, the strategic nature of the thorium and uranium contents (~0.3% U_3O_8) in monazite has restricted the export of the mineral from certain countries.

There is a significant price differential between monazite and bastnäsite entering world trade. Quoted prices in 1981 for 70% REO bastnäsite concentrate was U.S. \$2.20 per kg contained REO and U.S. \$0.78 per kg contained REO for Australian monazite (minimum 55% REO). This difference reflects a number of factors; bastnäsite concentrates have a higher REO content, no thorium and uranium, and have processing advantages as well as a more useful distribution of REE.

12.7. Current uses

The total annual consumption in the western world is about 30,000 t, expressed as rare earth oxide content. Growth in demand, as approximately represented by increase in production, has been substantial over the last four

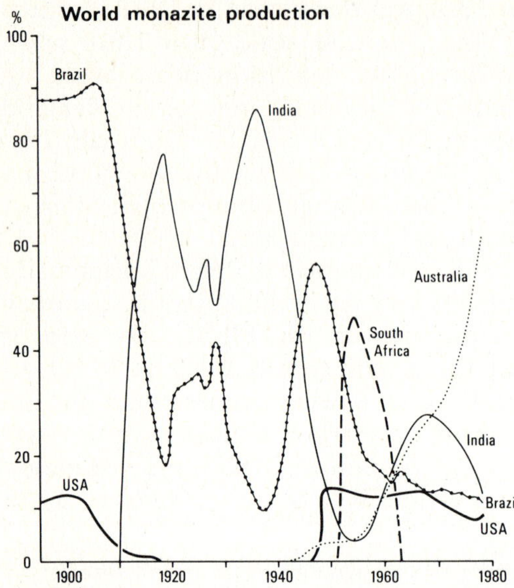

Fig. 12.6. Percentage production of monazite concentrates in the six major producing countries in the western world (1895—1979). Trends smoothed from a ten-year running average.

decades. The growth in production averaged about 10% over the period 1945—1979 but with a slower 7% average growth rate between 1965 and 1979. Demand for REE is normally considered to be from four market sectors. For the major consuming country, the U.S.A., the consumption pattern in each year since 1973 has been remarkably stable (according to the U.S. Bureau of Mines). This comprises: petroleum and cracking catalysts, 41—45%; metallurgical uses, 33—37%; ceramics and glass, 16—19%; and miscellaneous uses, including magnets, electrical and research applications, 4—8%. No matching data are available for world consumption although a broadly comparable estimate for 1980 was: metallurgy, 34%; catalysts and chemicals, 35%; glass and ceramics, 31%; and phosphors and electronic applications, <1%.

For particular applications, the REE are required in various degrees of purity according to whether the general properties of the group or specific properties of individual elements are needed. To generalize, the REE are produced in three grades of purity (Greinacher, 1980): Firstly, they may either be entirely unseparated and in the proportions that they occur naturally in the ore ("natural" mischmetal is about 53% Ce, 25% La, 16% Nd, 5% Pr, 2% other) or be the residue after the extraction of a specific rare earth (e.g., cerium-free mischmetal is about 42% La, 42% Nd, 14% Pr, 2% other). The use of large quantities of unseparated REE in the steel industry, in catalysts and in glass polishing are examples of these end-uses.

Secondly, a simple chemical separation may yield products that can contain 60—90% of the individual element desired. This type of product is used where the associated elements are merely diluents that are not detrimental to product effectiveness and the cost is therefore kept to a minimum. An example is the production of permanent magnets. Thirdly, the very expensive preparation of extremely pure individual rare earth oxides (98—99.999%) is very costly and is reserved for small uses where purity is essential, such as in phosphors and electronic applications. There is a major contrast between the use of mixed or pure rare earth compounds. About 98% by weight of the REE produced are used in the mixed form in metallurgy, chemicals and catalysts, glass manufacture and polishing compounds. By value, however, the use of individual REE in phosphors is the most important single use.

Catalysts

Between a quarter and a third of total REE production is used in catalytic cracking in the petroleum industry. This and other catalytic functions use mixed REE rather than specific members of the group. Amorphous silica-alumina catalysts were used in catalytic cracking of petroleum feedstocks until 1964 when zeolite promoted catalysts were introduced. These contain a proportion of crystalline rare earth aluminosilicate (zeolite) and are substantially superior to amorphous catalysts in their activity and selectivity for gasoline production.

The contained REE improve the thermal and hydrothermal stability of the zeolite. The rare earth zeolite contains on average about 4% REO and the catalysts currently contain between approximately 5% and 25% of rare earth zeolite depending on the level of activity required. The balance is a matrix generally consisting of silica and/or alumina, and clay. The major outlet for these catalysts is in the fluid catalytic cracking process, where they are used in the form of a microspheroidal powder. A typical fluid catalytic cracking unit has a catalyst inventory of approximately 100 t, to which fresh catalyst is regularly added at a rate of about 1% a day to counterbalance catalyst losses and deactivation. Rare earth recovery from deactivated catalysts is not currently practised.

Rare earth chlorides are also used as catalysts in the recovery of chlorine from by-product hydrochloric acid. Rare earth phosphate, containing either lanthanum or cerium, is used as a catalyst to produce creosol and xylenol (Moore, 1979). The removal of sulphur from hydrogen sulphide can be assisted by a Claus catalyst: an alumina carrier stabilized by REE which impart an improved resistance to aging. The selectivity of the ferric oxide catalyst used in the dehydrogenation of ethylbenzene to styrene is improved by the addition of CeO_2.

A potentially large use of rare earth catalysts is in the pollution control of exhaust gas emissions. The acceptable levels of carbon monoxide (CO), unburnt hydrocarbons and oxides of nitrogen (NO_x) in motor exhaust gases have been continually reduced by successive U.S. legislation since 1968 and the first use of catalytic convertors was after the 1975-76 legislation. The general requirements are that the NO_x should be reduced to nitrogen, the CO and hydrocarbons oxidized to carbon dioxide and water vapour, without the catalyst being poisoned easily, for example, by lead. Most emission catalysts consist of a fine dispersion of platinum and palladium coated on to alumina pellets or on to a single honeycomb structure made of alumina and silica. This does not affect the NO_x and the addition of another active agent, e.g. rhodium, is required.

Rare earth additions have two distinct functions. Firstly, any REE or REE mixture is used to stabilize the gamma-alumina carrier, decreasing its tendency to revert to the alpha form and loose its large surface area. In addition to this, catalytic action itself requires three way catalysis. The oxidation of CO and hydrocarbons requires an oxidizing atmosphere while the reduction of NO_x requires a reducing atmosphere. These reactions can only proceed simultaneously in a limited redox domain, but the use of cerium enlarges the domain owing to the oxygen buffering effect of the equilibrium between Ce^{3+} and Ce^{4+} (Ph. Poirier, written communication, 1981). The most promising REE alternative to the present catalysts is a lanthanum-cobalt oxide which approaches the platinum types for efficiency and would be considerably cheaper. If REE prove to be suitable this would be the most significant development in the REE market this century, offering a high-volume demand with replacement requirements on a considerable scale. From 1981 onwards the U.S.A. motor industry is expected to become a significant consumer of REE (Poirier, 1981).

Cerium is also an effective catalyst component in hydrocarbon-oxygenation reactions, such as in self-cleaning ovens, although there is strong competition from other compounds of large surface area, such as the oxides of the alkaline earths and the transition metals (Ross, 1975).

Ferrous metallurgy

The most important uses of the REE in ferrous metallurgy are in the control of the content and shape of sulphide inclusions in steel and in the production of nodular cast iron (also known as ductile iron or spheroidal graphite iron) (Raman, 1976). Other functions are to improve the hot workability of the exotic super alloys and in the production of special highly oxidation resistant alloys such as Fecralloy. This is a ferritic stainless steel containing chromium, aluminium and yttrium which, as it is less prone to mechanical and thermal shock than ceramics, might be used to replace these as a substrate for automobile emission control catalysts.

The presence of sulphur causes steel to develop cracks when rolled or forged at high temperatures. This "hot shortness" is due to the melting of inclusions of an iron sulphide compound in the steel, particularly if these are present as a semi-continuous low-strength "chain" along the grain boundaries. A conventional solution to this is to add manganese which combines with the sulphur and almost neutralizes its undesirable effects. However, due to the relative plasticity of the manganese sulphide and the steel matrix at hot rolling temperatures, the sulphide particles deform in a manner related to the nature of the shaping operation. Thus, in plate products, the sulphide particles are flattened and elongated in the rolling direction. The presence of these elongated particles causes anisotropy of the mechanical properties, particularly of ductility. For example, there is a lower ductility in the transverse direction compared with the longitudinal direction in conventionally rolled plate. However, ductility is lowest overall in the short transverse direction; that through the plate thickness.

Major industrial applications where this effect is important include pipes for oil and gas transmission and tubular members fabricated from thick plate for large structures, such as oil field platform supporting towers, or jackets. A further important sector where anisotropy of ductility must be minimized is in light-section plate and strip made from high-strength low-alloy (HSLA), or microalloyed steels. These are used as pressings for structural components in the automobile industry and for applications such as truck side frames and telescoping crane-booms.

For these specialist applications, the effects of the manganese sulphide inclusions can be suppressed by a combination of desulphurization of the steel and the addition of elements which will form more stable sulphides than manganese. Suitable REE additions form rare earth sulphide and oxysulphide phases which are globular in morphology and do not deform significantly on subsequent hot working of the steel (Bennett and Sandell, 1974; Wilson et al., 1974; Wells, 1978).

Although significant improvements in the anisotropy of properties is obtained, (see, for example, Hirschorn, 1980), the demand for mischmetal in Europe for steelmaking has fallen since 1976. For the bulk steelmaking applications, this has been the result of intensive development of cheaper processes for the desulphurization of steel, principally by the use of calcium compounds. These may be added as special slag mixtures during secondary steelmaking, or by injection into the ladle (D. Dulieu, written communication, 1981). By these techniques, sulphur contents as low as 0.002% can be obtained in tonnage production. It is possible also to inject calcium and magnesium deep into the liquid steel, in the form of alloys, which minimize losses of the elements due to their high vapour pressure (Luyckx, 1980).

As a result of these developments in both desulphurization and modification of inclusion morphology by the use of calcium treatments, there is less scope for the use of the REE. Additionally there are certain drawbacks to their use in steels, such as the possibility of arc sputter problems in welding. Where used in the U.K., additions of REE to the steel are made as mischmetal or perhaps in the form of a steel-clad mischmetal wire, or rare earth silicide, in concentrations of 0.5 kg/t per 0.01% S (Ross, 1975). Additions are made as the last inoculant after deoxidation and desulphurization (Luyckx and Jackman 1978), in order to utilize cheaper ferrosilicon or aluminium powder for deoxidation which would otherwise consume the REE.

The use of high-duty cast iron instead of more expensive forgings, for example, as crank shafts in the automobile industry and centrifugally cast pipes for water, relies on the development of high strength and ductility, i.e., a toughness more resembling steel than cast iron. Ductile cast iron, containing all of the graphite in nodular form, exhibits better mechanical properties than iron with partly nodular graphite or iron which contains all the graphite in flake form. REE have added benefits over the cheaper nodularizing agent, magnesium (see, e.g., Wilson, 1979). For example, the REE, especially cerium, counteract the deleterious effects of nodularizing inhibitors such as Ti, Sb, Al, Cu, Te and Bi and increase the level of nucleation in the treated metal. Also, unlike the more conventional magnesium additions, the REE do not boil from the molten iron. There is thus no flaring and fuming and less air pollution in the local working environment (Neumeier and Betts, 1978). At present there is no commercially acceptable process for producing nodular iron which does not use magnesium as the primary nodularizing element, and it appears that the combined additions of magnesium and cerium/mischmetal will be used for the foreseeable future. Most nodular iron currently produced has a cerium content of 0.003—0.006%. The cerium is usually added in the form of mischmetal which is alloyed with the magnesium alloy or made as a separate addition with the magnesium alloy. The average yield of weight, of casting to weight of metal poured, is about 60%, so for every 1 Mt of nodular iron castings produced the equivalent consumption of mischmetal is about 2000 t.

Non-ferrous metallurgy

REE are used as minor additions to alloys of magnesium, the most important, aluminium, bronze and brass. A range of magnesium alloys with added mischmetal (up to 3%), Zn (up to 6%) and Zr (1%) have been developed for the aerospace industry. Due to the precipitation of intermetallic compounds they have high thermal stability (Greinacher, 1980). As well as low density, some alloys possess high tensile strength, good vibrational damping and good creep resistance in the range 200—250°C and are used in helicopter gearboxes, transmission housings, aircraft landing wheels and satellites. Some alloys developed in the U.S.S.R. contain specific individual REE (Raman, 1977). Die-cast magnesium-alloy automobile engine parts have REE added to reduce microporosity and increase the "castability" of the alloy. Various modifications to the properties of welded constructions are also possible using REE additions to welding wire. Magnesium alloys containing up to 3% Nd or Y have also been developed for applications in bone surgery.

Micro-additions of LREE to aluminium and its alloys variously increase hardness, tensile strength and ductility; improve heat, vibration and oxidation corrosion resistance; and modify the porosity of cast metal and the non-metallic inclusion shape (Raman, 1977). The additions are particularly interesting for use in high-tension transmission lines where beneficial physical properties are obtained without an increase in electrical resistance (Ross, 1975). The aluminium-silicon die-casting alloys have a potential use in die-cast engine blocks. The required low thermal expansion, adequate hardness and wear resistance are all improved by the addition of mischmetal whilst maintaining a satisfactory machinability.

The presence of small amounts of REE increases the oxidation and sulphidation resistance of nickel-based, or cobalt-based, superalloys subjected to reaction with hot exhaust fumes in gas turbines. Other uses with similar demands such as nichrome heating elements also require the improved oxidation resistance imparted by REE. Niobium alloys that are used to carry molten lithium in nuclear reactors, contain up to 3.6% Y to reduce oxygen embrittlement at welds.

REE additions to copper alloys improves their strength at low temperature and ductility at high temperature. Particularly in deep-drawing leaded brasses and bronzes the addition of mischmetal improves hot workability, eliminating grain-boundary tears and edge cracks (Raman, 1977), normally associated with lead melting, by the formation of a high-temperature mischmetal-lead compound. In the U.S.A., alloys such as 72% Cu, 5% Zn and 23% Pb are routinely treated with mischmetal (Ross, 1975).

Pyrophoric alloys

Due to its highly exothermic reaction with oxygen, mischmetal ignites readily if finely divided. This property is retained when the metal is alloyed with iron to give a mechanically stronger alloy called ferrocerium. This is either cast or extruded to form the familiar lighter flints. The composition of the pyrophoric alloy having excellent sparking properties and used in the cast method is 74% mischmetal, 23% iron, 2% copper and 1% magnesium (Ross, 1975). In recent years most manufacturers have produced flints by the extrusion method using a composition such as 78.3% mischmetal, 19.4% iron, 0.3% ferrosilicon, 0.5% zinc and 1.5% magnesium (K. Clarke, written communication, 1981).

Permanent magnets

During the last decade commercial use has been made of the demagnetization resistance of rare earth/cobalt inter-metallic mixtures of the $RECo_5$

type. The REE may be Y, La, Ce, Pr, Nd or mischmetal, although the most widely manufactured is the Sm-Co type. The resulting magnet is 2—5 times stronger than alnico (Al, Ni, Co) magnets (Becker, 1970).

Fabrication can be by casting or powder metallurgy. In the casting method the alloy ($RECo_5$; about 70% Co) is induction melted under an argon atmosphere, quenched and heat treated at 500°C with the resulting coercivity dependent upon the precise conditions of manufacture. The powder metallurgical technique is the most common, if not the only method in commercial use. The initial alloy is prepared by reduction of samarium oxide by calcium in the presence of cobalt powder (C.S. Brown, written communication). This is milled, magnetically orientated, pressed, sintered in argon at about 1150°C and heat treated for several hours at 875°C (Ross, 1975; R.W. Lee, 1979). This method allows the easy fabrication of intricate shapes as well as dense packing giving a remnant flux density close to the saturation of the massive material. However, compositional control is difficult making the product expensive. Higher production rates can be achieved by polymer-bonding the REE-Co magnetic materials but the resulting magnets have less magnetic strength.

The application of these magnets is governed by the need for higher field strengths or lower-weight products but is limited by the relatively high cost of materials and fabrication. Uses include electronic watches, travelling wave tube amplifiers used in communication satellites, and aircraft tachogenerators, where non-REE substitutes would be inferior. Applications for the more expensive sintered types are, therefore, in the more critical end-uses, i.e., military and aerospace motors where weight saving is important; in the electronics industry for magnetron valves, microwave isolators; in difficult atmospheres or environments; and in motors for the machine tool industry that require good torque at high and low speeds. An increase in demand is expected from the machine tool industry. The uses of the polymer-bonded varieties include semi-domestic and less demanding applications. Development of Sm_2Co_{17}-sintered or -bonded compositions is currently in progress. These are 50% stronger again than the $SmCo_5$ type because they do not have a hexagonal structure and so are not highly anisotropic with regard to magnetization. In Japan, more complex formulae are under research and development, where Fe, Cu, Zr, Mn, Cr and Ni may replace the cobalt to some extent to give the molecular proportions samarium 2, transition metal 17. These products are produced commercially only in Japan and on a small scale. Processing is more expensive but the content of samarium is lower and some cobalt is replaced by less expensive materials so that raw material costs are lower. The next generation of magnets of the sintered and bonded type will probably have these compositions.

Sintered magnets are now very stable. Problems in the early days arose because low-temperature sintering gave a density less than 93% of the theoretical value. The resulting continuous porosity allowed the REE to react slowly with the atmosphere. The present high-temperature sintering gives a porosity of only 4% and so no continuous porosity (permeability). In contrast, polymer-bonded material suffers slow degradation due to the inherent permeability of the polymer. This causes no problems at room temperature but above 60°C there is definite degradation and oxidation of the REE powder.

In the future, increasing use could be justified on the grounds of weight reduction in portable appliances, such as electric sewing machines, and in the automobile industry due to the smaller magnetic parts in, e.g., fuel gauges, electronic ignition, motors for windscreen wipers, heater fans and alternators. REE-Co earth magnets also eliminate the need for bearings in aircraft tachometers (Anonymous, 1973), resulting in weight saving and an increase in service life. Another application has been as an attachment mechanism for jewellery such as in necklace clasps and ear-rings with the jewel mounted on one magnet which is held in place by another magnet held behind the ear. Demand for these goods fluctuates markedly from year to year with the current fashions.

Glass polishing compounds

Over the last 20 years, and particularly the last 10 years, cerium-based polishing compounds have almost completely replaced zirconia and red iron oxide rouges in glass polishing although, in the U.S.A., zirconia is still used for certain applications and is only half the cost. The nature of the polishing process is 75% chemical and 25% physical. Although mixed rare earth oxide powders are utilized, cerium oxide has been found to be the most effective, but for reasons that are not fully understood. The technical grade rare earth oxide powders that are normally used, contain about 40—45% CeO_2, although high-Ce polishing powders containing up to 90% CeO_2 are produced. These are diluted with other materials during use. The main demand for glass polishing is in the specialized field of optical glass, with the ophthalmic industry probably being the largest market. Other markets include television face plates, mirrors and plate glass; float glass does not require polishing.

Glass and ceramic manufacture

The REE are also used in the glass industry for decolourizing, as colouring agents and in the production of special glasses. Decolourization of glass is required because of the presence of iron. As the blue colour produced by iron in the divalent state is much more intense than the yellow produced by trivalent iron, decolourizing depends on both the oxidation of ferrous to ferric iron and the addition of a complementary colour to neutralize any residual colour. Selenium is the usual decolourizing agent used in lime-soda-silica glasses but as ceric oxide is a particularly effective oxidizing agent, considerable interest has been shown in its use in recent years, mainly for glass used in containers but also in tableware. Ceric oxide is usually introduced as a cerium concentrate with a minimum CeO_2 content of 60%. There are two ways of using it in the glass batch. At low levels (a few hundred grams per tonne of sand) ceric oxide controls the oxidation state of selenium, maintaining it in its pink-coloured zero valence state, which complements the iron-green colour. In this way high-cost selenium is efficiently used. Alternatively, at higher concentrations (1—2 kg/t of sand) cerium oxidizes the iron to the ferric state. Small additions of cobalt oxide are also introduced to complement any residual colour, but no selenium is required. Ceric oxide has been found to work most efficiently as a decolourizer in the presence of sulphates, which are used as glass refining agents. Glass with iron contents of up to 0.1% Fe_2O_3 have been successfully decolourized in this way. In lead crystal glassware cerium concentrate (3—7 kg/t of sand) also acts as a refining agent replacing the very poisonous arsenious oxide.

The absorption spectrum of a number of REE is characterized by narrow peaks which produce attractive brightly coloured glasses (Poirier, 1981). As colouring agents they are, however, limited by price to use in optical glass, ophthalmic tints and high-quality decorativeware. Neodymium oxide, which produces a blue to purple red colour, is the most widely used, but others include praseodymium oxide (green), erbium oxide (pink) and holmium oxide (pale yellow). Ceric oxide is also used in special glass such as ophthalmic glasses for eye protection against ultraviolet light. REE may also be used as colouring agents in the new field of laser glass which is coloured as a consequence of the visible optical absorption required to "pump" the lasing action (Bamford, 1977).

Lanthanum oxide is used in amounts up to 40% in certain special optical glasses, such as camera lenses, to produce glass with a low dispersion and high refractive index. It is also being used in glass for fibre optics (Anonymous, 1979c). Additions of rare earth oxides and thoria are also reported to confer enhanced resistance to alkali attack of glass fibres used in fibre-reinforced cement products.

In the refractories industry yttrium oxide (melting point 2415°C) is probably most widely used in stabilizing the cubic modification of zirconia refractories. This prevents the formation of other phases which are accompanied by large volume changes. Yttria is also used as a refractory coating and in crucibles.

At high temperature gadolinia and yttria become conductive. Cerium oxide ceramics doped with these find application as solid electrodes, for instance, in the high-temperature electrolysis of water. Other new applications include the so-called lambda sensors for oxygen determination of, for example, automobile exhaust gases (Greinacher, 1980) in order to optimize the air-fuel ratio. The most common type used to date has been yttria-stabilized zirconia (Kennard, 1980).

In the ceramics industry small quantities of certain rare earth oxides are used as pigments in coloured glazes. The most important is praseodymium oxide which with zirconium silicate produces a bright yellow glaze (praseodymium yellow). Due to the high affinity of the REE for oxygen, the pigment has the advantage of being stable at high temperatures and has mainly replaced tin/vanadium stain.

Phosphors

Rare earth phosphors for colour television tubes, X-ray intensifying screens and fluorescent lighting tubes is one of the most important uses developed for the REE in recent years.

Luminescent materials can absorb energy during bombardment by photons (commonly ultraviolet light), electrons or positive ions. This energy is released as visible or near-visible light, either immediately after absorption (within 10^{-8} s), termed fluorescence, or after more than 10^{-8} s, called phosphorescence. The emission can be aided by doping with an activator which increases the intensity of phosphorescence.

Rare earth luminescence has been known and studied since before the turn of the century. High-purity Eu_2O_3 in combination with Y_2O_3, yttrium oxysulphide (Y_2O_3S) or yttrium orthovanadate (YVO_4) has been used in colour-television picture tubes since the mid-1960's (Pings, 1969). Europium-doped oxides and oxysulphides, particularly Y_2O_3(Eu), Gd_2O_3(Eu) and Y_2O_3S(Eu), give higher intensity and have superseded the use of the vanadate as the red phosphor. Of these, Y_2O_3S(Eu) is most commonly used (Tecotzky, 1973). (Silver-activated zinc cadmium sulphide was used previous to the vanadate as the red phosphor but it does not have sufficiently high luminosity to match the emission available from green and blue phosphors and these had to be deliberately deintensified to obtain a good colour balance. The new phosphors allow full use of the intensity of the green and blue phosphors, and result in a 40% gain in overall tube brightness.) This application requires high-purity (99.99%) rare earth oxides.

In medical X-ray radiography an intensifying screen is used with matching film to increase film exposure and hence decrease the X-ray dosage to patients (Tecotzky, 1973), and permits the use of slower films of lower silver content. The double screen or foil is placed on either side of a double emulsion film and is excited, by the X-rays, to emit light in the visible range which exposes the film. The "amplification" allows image production with decrease in patient exposure time by a factor of 2—4 (Rabatin, 1980). Calcium tungstate has been used for decades as an X-ray phosphor but barium sulphate doped with europium, as well as a number of other compounds, are now being used. The market for new X-ray screen phosphors is based on terbium-activated gadolinium or lanthanum oxysulphide in the U.S.A. and Europe. Phosphors such as terbium-lanthanum oxybromide or terbium-gadolinium oxysulphide have a higher efficiency than calcium tungstate when used with a matching film. The rare earth oxides have to be of very high purity. Phosphors containing Eu, Tb, and Y are also used in improved fluorescent lamps, and industrial mercury vapour lamps are improved by an addition of dysprosium (Pings, 1969). High-pressure mercury vapour lamps are good emitters of blue, green, yellow and ultraviolet radiation but are poor emitters of red light (Mathers, 1973). Consequently, unless corrected for this deficiency, they distort the true colour of many objects and are not suitable for applications in which good colour retention is necessary. Colour-correcting phosphors, placed on the inner surface of a transparent quartz envelope surrounding the arc tube discharge source, emit red light in response to excitation by the ultraviolet radiation. The use of europium-yttrium vanadate is important to improve both the colour emission and yield of the high-vapour-pressure lamps. Europium is now being used in low-pressure mercury discharge tubes (fluorescent lamps) to provide a "warmer" light. The conventional calcium halo-phosphate-coated tubes have been partly replaced by more expensive tubes utilizing Y, Eu, Ce and Tb, in Europe, the U.S.A. and Japan. Special fluorescent lamps are being produced containing rare earth phosphors for photocopying machines and medical treatments. A detailed review of fluorescent-lamp phosphors is given by Butler (1980).

Lasers and microwaves

Laser action involves stimulation of an active material by an external source to raise a proportion of the medium (which is usually atomic or molecular) to a higher-energy level (excited state) than the normal ground level. The excited atoms may then return spontaneously to a lower energy state, emitting energy, or if stimulated whilst in the excited state will emit energy which is in phase with, and in addition to, the stimulating radiation.

Although the earliest research work on optical masers was done with pink ruby (aluminium oxide containing about 0.05% Cr) the potential of material doped with at least one REE ion (Eu^{3+}) was recognized also in the late 1950's (Schawlow, 1963). Practically all the REE have been evaluated in lasers, as crystals, in glasses, in chelates, and in liquids. The electrons responsible for the optical properties of the REE ions are situated deep within the electron cloud and are usually well shielded from external perturbing influences. It is this property that accounts for their characteristically sharp line emission and success in laser application (Lempicki and Samelson, 1967). In newer approaches to "lasing" substances the host materials, including REE, can aid in the excitation of the active dopant ions, i.e., reduce the energy consumption of the laser. Such substances are, for example, the Nd-doped yttrium-aluminium-garnet (YAG; $Y_3Al_5O_{12}$) whose short-wavelength beam is useful in cutting, scribing semi-conductors, drilling and welding (Roskill Information Services Ltd., 1980).

Laser application can utilize either the coherence of the light (e.g., for very accurate measurement of distance) or the very high brightness (energy per unit area) as a local intense heat source; for example, in surgery (Er- and Ho-doped YAG), in welding and to rough cut wire drawing dies from diamonds (Herriott, 1968). Neodymium-doped glass lasers are being evaluated as energy sources in laser-fusion research (Roskill Information Services Ltd., 1980).

Single crystals of iron oxide-bearing yttrium and gadolinium oxides with the garnet structure have been used in microwave electronics for several years. These garnets can be operated in low magnetic fields in the lower microwave frequencies. Yttrium-iron-garnets (YIG) and gadolinium-iron-garnets (GIG) are widely used as ferrite materials in microwave applications (Moore, 1979).

Bubble magnetic memories

Conventional memory systems for electronic computers can be built from tiny ring-shaped ferrite cores strung on a mesh of fine wires or from transistor circuits laid down on tiny silicon chips (Bobeck and Scovil, 1971). Data storage can also exploit the domains of magnetization that exist in all magnetic materials. Magnetic disc stores use a thin layer of γ-Fe_2O_3 or an evaporated cobalt film as the magnetic material, and information is stored on circular tracks as reversals of magnetization (Brown, 1973).

When a thin film of a low-coercivity anisotropic magnetic material, with the "easy" axis of magnetization parallel to the film, is placed in a magnetic field parallel to this easy axis, irregular magnetic domains appear. As the field is increased these domains form into cylinders of uniform size standing across the film. These appear circular in plan view, hence are called "bubbles", have a reverse magnetic polarity to the surrounding material, and are stable until the field is increased to an extent that they collapse and the film becomes uniformly magnetized. Magnetic "bubbles" can be stable over a considerable range of conditions, moved from point to point at high velocity and store binary information at high density. The presence of a bubble can designate a binary digit "1" and its absence "0" and the stored information can be read by a sensitive magnetic detector. Each domain is only a few micrometres across. Since the bubble is stable in the presence of a small magnetic field which can be supplied by a permanent magnet, the bubble domain memory is non-volatile, that is, the information is not lost when the power to the memory is off (Nielsen, 1979). The domain mobility and bubble size vary with the nature of the magnetic material.

A simple geometric bubble shape and unimpaired mobility is only possible in a boundary-free continuum, e.g., a single crystal with very few inclusions or defects that modify the regular crystal field. The thickness of the thin film of magnetic materials must be approximately the same as the bubble radius and so the memory element must be a film supported on a substrate rather than a free plate. Possible substances are the orthoferrites ($REFeO_3$) although they have rather large bubble diameters (15—399 μm). The smaller bubbles tend to be temperature-sensitive, they are difficult to grow as single crystals, and there are no good substrate materials. Using hexagonal ferrites the mobility of the bubble domains was too low.

The synthetic garnets ($RE_3Fe_5O_{12}$) with their large compositional flexibility and small bubble diameter of perhaps 3 μm are rather more appropriate. Uniaxial anisotropy is induced during the crystal growth which is normally by liquid-phase epitaxy (LPE) on a garnet substrate. The most commonly used substrate is gadolinium-gallium-garnet (GGG or $Gd_3Ga_5O_{12}$). The currently favoured garnet composition is based on the prototype yttrium-iron-garnet (YIG, $Y_3Fe_5O_{12}$). The garnet-structured ferrites can readily be modified by substitution of a range of REE ions for some of the Y ions, and substitution of Ga^{3+}, Al^{3+}, or other ions for some of the Fe^{3+} ions. In this way the material can be optimized to reduce the saturation magnetization and its temperature coefficient, and increase the bubble mobility and hence the speed of operation of the store.

Bubble memories utilizing synthetic REE garnets are beginning to be used on a limited scale for communication and computer systems as a replacement for small disc stores, particularly where it is advisable to avoid mechanical systems, such as applications in space.

12.8. Use development

A potentially large use for REE is in the safe storage of hydrogen. The use of lanthanum-nickel hydrides ($LaNi_5H_6$) permits the reversible storage of hydrogen at ambient temperatures and modest pressures. At normal temperature, hydrogen desorption occurs at a pressure of 2 bars; more hydrogen can be stored in a given volume of hydride than in the same volume of liquid hydrogen. Other metallic compounds (e.g., Mg or FeTi) have similar properties of absorption when hot and desorption when cool, but the high performance of $LaNi_5H_6$ is likely to ensure that it will find applications for hydrogen storage and in catalytic hydrogenation of organic compounds (Poirier, 1981).

A new light bulb has been developed which contains rare earth powders coating the interior wall of the bulb (Anonymous, 1980c). These convert any ultraviolet radiation into visible light and so give an approximately fourfold increase in efficiency.

There is an increasing interest being shown in the use of compacted graphite iron, which has properties mid-way between those of flake-graphite and nodular cast irons, especially for applications which require good resistance to repeated thermal stressing. Examples are ingot moulds, exhaust manifolds, and large-diameter brake discs. One way of producing this structure is by the use of mischmetal. At present the production of the compacted graphite iron is very small.

There are many other new uses. For example, since the mid 1970's, yttrium-stabilized, cubic zirconium oxide has been widely accepted as a diamond substitute. The index of refraction, dispersion and hardnesses are all similar and it is difficult to distinguish them apart especially in small mounted stones (Anonymous, 1980a). A new coating for gas-turbine buckets which triples their life has been developed. The Co-Cr-Al-Y alloy coating is plasma sprayed. The platinum alloys currently used are of higher cost (Anonymous, 1980b).

The effort devoted to the research and development of applications of the REE is clearly demonstrated by recent reviews of the patent literature, particularly that by Villani (1980), and of the future possibilities, e.g., Cannon (1980). New uses will undoubtedly be developed for this versatile group of elements, but as in the past, the success of the REE industry will depend on the relative merits of expensive REE technology when compared to cheaper alternative or substitute materials.

Acknowledgements

The authors would like to thank their colleagues in the Minerals Strategy and Economics Research Unit for their constructive advice and in particular Dr. D. Humphreys for statistical information, Mr. D. Bate for bibliographic services and Mrs. W. Khosla for graphical work. The helpful advice and information provided by many companies is also gratefully acknowledged.

This chapter is published by permission of the Director, Institute of Geological Sciences (N.E.R.C.).

References

Adams, J.W. and Staatz, M.H., 1973. Rare-earth elements. *U.S. Geol. Surv., Prof. Paper*, 820: 547—556.

Alexander, J.B., Harrel, G.M. and Flinter, B.H., 1964. Chemical analyses of Malayan rocks, commercial ores, alluvial mineral concentrates, 1903—1963. *Prof. Paper, Geol. Surv. Malaya*, E-64, 1-C: 295 pp.

Altschuler, Z.S., 1980. The geochemistry of trace elements in marine phosphorites, I. Characteristic abundances and enrichment. *Soc. Econ. Palaeontol. Mineral., Spec. Publ.*, 29: 19—30.

Altschuler, Z.S., Berman, S. and Cuttitta, F., 1967. Rare earths in phosphorites — geochemistry and potential recovery. *U.S. Geol. Surv., Prof. Paper*, 575-B: 1—9.

Anonymous, 1973. Cobalt/rare earth magnets. *Chem. Eng. News*, 51(31): 11.

Anonymous, 1979a. *Anuario Mineral Brasileiro, 1979, Ano VIII.* Departamento Nacional da Producao Mineral, Brasilia, 324 pp.

Anonymous, 1979b. Mining heavy mineral deposits in south Western Australia. *Quarry Mine Pit*, 18(9): 8—11.

Anonymous, 1979c. Rare earths: industry profile and market review. *Ind. Miner. (London)*, 138: 21—59.

Anonymous, 1980a. Caught in the act. *Rare-earth Inf. Center News*, 15(4): 4.

Anonymous, 1980b. Improved coating. *Rare-earth Inf. Center News*, 15(4): 3.

Anonymous, 1980c. Let there be (cheap) light. *Rare-earth Inf. Center News*, 15(3): 3.

Argall, G.O., 1980. Three iron ore bodies of Bayan Obo. *World Min.*, 33(1): 38—41.

Bamford, C.R., 1977. *Colour Generation and Control in Glass.* Elsevier, Amsterdam, 224 pp.

Baxter, J.L., 1977. Heavy mineral sand deposits of Western Australia. *Miner. Resour. Bull., Geol. Surv. West. Aust.*, 10: 148 pp.

Becker, J.J., 1970. Permanent magnets. *Sci. Am.*, 223(6): 92—100.

Bennett, H.W. and Sandell, L.P., 1974. Rare earth additions to electric furnace steels for sulfide shape control. *J. Metals*, 26(2): 21—24.

Blank, S.L., Nielsen, J.W., Biolshi, W.A., 1976. Preparation and properties of magnetic garnet films containing divalent and tetravalent ions. *J. Electrochem. Soc.*, 123: 856—863.

Bobeck, A.H. and Scovil, H.E.D., 1971. Magnetic bubbles. *Sci. Am.*, 224(6): 78—90.

Bril, K.J., 1964. Mass extraction and separation. In: L. Eyring (Editor), *Progress in the Science and Technology of the Rare Earths, 1.* Pergamon, Oxford, pp. 30—61.

Brown, C.S., 1973. Production and application of rare earth ferrimagnetic garnets, 2. Application in bubble domain stores. In: O.B. Michelsen (Editor), *Analysis and Application of Rare Earth Materials*, Universitetsforlaget, Oslo, pp. 315—330.

Butler, K.H., 1980. *Fluorescent Lamp Phosphors, Technology and Theory.* Pennsylvania State University Press, London, 351 pp.

Cannon, J.G., 1980. Rare earths — future applications. *Am. Chem. Soc. Meet., San Francisco, Calif., August 25, 1980* (preprint).

Coetzee, C.B., 1976. Rare earths. In: C.B. Coetzee (Editor), *Mineral Resources of the Republic of South Africa.* Department of Mines, Geological Survey, Pretoria, 5th ed., pp. 199—291.

Cranstone, D.A., 1981. Rare earths. *Can. Miner. Yearb.*, 1979: 365—371.

Deans, T., 1966. Economic mineralogy of African carbonatites. In: D.F. Tuttle and J. Gittins (Editors), *Carbonatites*, Interscience, New York, N.Y., pp. 385—413.

Deans, T., 1978. Mineral production from carbonatite complexes: a world review. In: *Proceedings of the First International Symposium on Carbonatites, Junho, 1976, Pocos de Caldas, MG, Brasil*, pp. 123—133.

De Keyser, F. and Cook, P.J., 1972. Geology of the Middle Cambrian phosphorites and associated sediments of northwestern Queensland. *Bull. Bur. Miner. Resour., Geol. Geophys. Aust.*, 138: 79 pp.

De Kun, N., 1965. *The Mineral Resources of Africa.* Elsevier, Amsterdam, 740 pp.

Eisele, J.A. and Bauer, D.J., 1974. Recovery and separation of rare-earth elements, barium and strontium from bastnaesite with sulfuric acid. *U.S. Bur. Mines, Rep. Invest.*, 7990: 10 pp.

Garnar, T.E., 1981. Heavy minerals industry of North America. In: *Proceedings of the 4th "Industrial Minerals" International Congress, Atlanta, Ga., 1980. Metal Bull. PLC (London)*, pp. 29—42.

Golovanov, G.A., Zhelnin, V.S., Kotilevski, V.I. and Makarov, A.M., 1969. Processing of apatite ores at the ore-dressing plants of "Apatite" Complex. In: *VIIIth International Mineral Processing Congress, Leningrad, 1968, 1.* Instituto Mekhanobr, Leningrad, pp. 489—502.

Greinacher, E., 1980. Overview — history of RE application, RE market today. *Am. Chem. Soc. Meet., San Francisco, Calif., August 25, 1980* (preprint).

Gulbrandsen, R.A., 1966. Chemical composition of phosphorites of the Phosphoria Formation. *Geochim. Cosmochim. Acta*, 30: 769—778.

Hammerbeck, E.C.I., 1976. Titanium. In: C.B. Coetzee (Editor), *Mineral Resources of the Republic of South Africa*. Department of Mines, Geological Survey, Pretoria, 5th ed., pp. 221—226.

Harrah, H.W., 1967. Rare earth concentration at Molbydenum Corporation of America, 2. Solvent extraction plant. *Deco Trefoil*, 31(5): 9—16.

Harris, P.M. and Jackson, D.V., 1966. Investigations into the recovery of niobium from Mrima Hill deposit. *Trans. Inst. Min. Metall., Sect. C*, 75: 95—111.

Hawkins, B.W., 1975. Mary Kathleen uranium deposit. In: C.L. Knight (Editor), *Economic Geology of Australia and Papua New Guinea, 1. Metals*. Australasian Institute of Mining and Metallurgy, Parkville, Vict., pp. 398—402.

Herath, M.M.J.W., 1977. Mineral based industries of Sri Lanka. *Liaison Rep. Commonw. Geol. Liaison Off.*, 131: 43 pp.

Herriott, D.R., 1968. Applications of laser light. *Sci. Am.*, 219(3): 141—156.

Hill, W.B., 1975. Rare earths — Western Australia. In: C.L. Knight (Editor), *Economic Geology of Australia and Papua New Guinea, 4. Industrial Minerals and Rocks*. Australasian Institute of Mining and Metallurgy, Parkville, Vict., p. 331.

Hirschorn, I.S., 1980. Trends in the use of mischmetal (mixed rare earth metals). *Am. Chem. Soc. Meet., San Francisco, Calif., August 25, 1980* (preprint).

Holt, D.N., 1965. The Kangankunde Hill rare earth prospect. *Bull. Geol. Surv., Dep. Malawi*, 20: 130 pp.

Ionescu, E., Tomescu, E. and Rachita, R., 1980. Contributions a la recuperation des lanthanides des phosphates naturels. In: *Proceedings 2nd International Congress on Phosphorus Compounds, April 21—25, 1980. Boston, Mass.* Institut Mondial du Phosphate, Paris, pp. 745—758.

Isokangas, P., 1978. Finland. In: S.H.U. Bowie, A. Kvalheim and H.W. Haslam (Editors), *Mineral Deposits of Europe, 1. Northwest Europe*. Institution of Mining and Metallurgy and the Mineralogical Society, London, pp. 39—92.

Johnson, N.L., 1966. Rare earth concentration at Molybdenum Corporation of America. *Deco Trefoil*, 30(4): 9—16.

Kaczmarek, J., 1980. Rare earths: discovery and commercial separations. *Am. Chem. Soc. Meet., San Francisco, Calif., August 26, 1980* (preprint).

Karve, V.M., Madhaven, T.R. and Somnay, J.Y. 1966. Mineral recovery from beach sands. *Min. Mag.*, 114: 10—15.

Kennard, F.L., 1980. Oxygen sensors. *Am. Chem. Soc. Meet., San Francisco, Calif., August 25, 1980* (preprint).

Kim, W.J., 1968. Outline of placer deposits in Korea. In: *Report of the Fourth Session of the Committee for Co-ordination of Joint Prospecting for Mineral Resources in Asian Offshore Areas*. Economic Commission for Asia and the Far East, Tai-pei, pp. 96—101.

Kovalenko, V.I., Vladykin, N.V., Konusova, V.V., Smirnova, Ye.V. and Goreglyad, A.V., 1976. Concentrated rare-earth mineralization in the southern part of the Gobi Desert, Mongolian People's Republic. *Dokl. Akad. Nauk SSSR*, 230: 213—216.

Lee, K.Y., 1970. Some rare-element mineral deposits in mainland China. *Bull. U.S. Geol. Surv.*, 1312-N: 34 pp.

Lee, R.W., 1979. The temperature dependence of demagnetization of $MM_{1-x}Sm_xCo_5$. *J. Appl. Phys.*, 50: 2337—2339.

Lempicki, A. and Samelson, H., 1967. Liquid lasers. *Sci. Am.* 216(6): 81—90.

Leonardos, O.H., 1974. Origin and provenance of fossil and recent monazite in Brazil. *Econ. Geol.*, 69: 1126—1128.

Lissiman, J.C. and Oxenford, R.J., 1975. Eneabba rutile-zircon-ilmenite sand deposit, W.A. In: C.L. Knight (Editor), *Economic Geology of Australia and Papua New Guinea, 1. Metals.* Australasian Institute of Mining and Metallurgy, Parkville, Vict., pp. 1062—1070.

Lounamaa, N., Mattila, T., Judin, V.P. and Sund, H.E., 1980. Recovery of rare earths from phosphate rock by solvent extraction. In: *Proceedings 2nd International Congress on Phosphorus Compounds, April 21—25, 1980, Boston, Mass.* Institut Mondial du Phosphate, Paris, pp. 759—768.

Lucas, B.H. and Ritcey, G.M., 1975. Examination of a rare earth solvent-extraction circuit for possible upgrading of the yttrium product. *Can. Inst. Min. Metall., Bull.*, 68: 124—130.

Luyckx, L.A., 1980. The rare earth metals in steel. *Am. Chem. Soc. Meet., San Francisco, Calif., August 25, 1980* (preprint).

Luyckx, L. and Jackman, J.R., 1978. Mischmetal in steelmaking — present and future. In: G.J. McCarthy and J.J. Rhyne (Editors), *The Rare Earths in Modern Science and Technology.* Plenum Press, New York, N.Y., pp. 339.

Maciel, A.C. and Cruz, P.R., 1973. Perfil analitico do torio e terras raras. *Bol. Dep. Nac. Prod. Miner. Bras.*, 28: 72 pp.

Mandle, H.H. and Mandle, R.M., 1966. Uses and applications. In: L. Eyring (Editor), *Progress in the Science and Technology of the Rare Earths, 2.* Pergamon, Oxford, pp. 190—297.

Mathers, J.E., 1973. Production of rare earth red phosphors for colour television and lighting application. In: O.B. Michelsen (Editor), *Analysis and Application of Rare Earth Materials.* Universitetsforlaget, Oslo, pp. 241—262.

Matzko, J.J. and Overstreet, W.C., 1977. Black monazite from Taiwan. *Proc. Geol. Soc. China*, 20: 16—35.

McKellar, J.B., 1975. The eastern Australian rutile province. In: C.L. Knight (Editor), *Economic Geology of Australia and Papua New Guinea, 1. Metals.* Australasian Institute of Mining and Metallurgy, Parkville, Vict., pp. 1055—1062.

McNeil, M., 1979. *Brazil's Uranium/Thorium Deposits: Geology, Reserves, Potential.* Miller Freeman, San Francisco, Calif., 126 pp.

Michelsen, O.D. (Editor), 1973. *Analysis and Application of Rare Earth Materials.* Universitetsforlaget, Oslo, 375 pp.

Moore, C.M., 1979. Rare earth elements and yttrium. *Miner. Commod. Profile, U.S. Bur. Mines.* 16 pp.

Morrice, E. and Wong, M.M., 1979. Fused-salt electrowinning and electrorefining of rare-earth and yttrium metals. *Miner. Sci. Eng.*, 11: 125—136.

Murthy, T.K.S. and Gupta, C.K., 1980. Rare earth resources, their extraction and application. In: E.C. Subbarao and W.E. Wallace (Editors), *Science and Technology of Rare Earth Materials.* Academic Press, New York, N.Y., pp. 3—23.

Neumeier, L.A. and Betts, B.A., 1978. A new look at nodulizing ductile iron with yttrium and mischmetal additives. *Trans. Am. Foundrymen's Soc.*, 86: 249—266.

Nielsen, J.W., 1979. Magnetic bubble materials. *Annu. Rev. Mater. Sci.*, 9: 87—121.

Noakes, L.C., 1968a. Reconnaissance of some placer deposits in the Republic of Korea. In: *Report of the Fourth Session of the Committee for Co-ordination of Joint Prospecting for Mineral Resources in Asian Offshore Areas.* Economic Commission for Asia and the Far East, pp. 102—107.

Noakes, L.C., 1968b. Reconnaissance of heavy mineral deposits in beach sands of Taiwan, China. In: *Report of the Fourth Session of the Committee for Co-ordination of Joint Prospecting for Mineral Resources in Asian Offshore Areas.* Economic Commission for Asia and the Far East, pp. 88—95.

Olson, J.C., Shawe, D.R., Pray, L.C. and Sharp, W.N., 1954. Rare-earth mineral deposits of the Mountains Pass district, San Bernadino County, California. *U.S. Geol. Surv., Prof. Paper*, 261: 75 pp.

Overstreet, W.C., 1967. The geologic occurrence of monazite. *U.S. Geol. Surv., Prof. Paper*, 530: 327 pp.

Parker, J.G. and Baroch, C.T., 1971. The rare-earth elements, yttrium, and thorium: a materials survey. *U.S. Bur. Mines, Inf. Circ.*, 8476: 92 pp.

Phillips, B.M. and Bassett, N., 1979. Development of the Eneabba mineral sands operation of Associated Minerals Consolidated Limited. In: *Western Australia Conference, 1979.* Australasian Institute of Mining and Metallurgy, Parkville, Vict., pp. 69—77.

Pings, W.B., 1969. The rare earths today. *Colo. Sch. Mines, Miner. Ind. Bull.*, 12(2): 19 pp.

Pirkle, E.C., Pirkle, W.A. and Yoho, W.H., 1974. The Green Cove Springs and Boulougne heavy-mineral sand deposits of Florida. *Econ. Geol.*, 69: 1129—1137.

Poirier, Ph., 1981. New developments in rare earths markets. In: *Proceedings of the 4th Industrial Minerals International Congress, Atlanta, Ga., 1980. Metal Bull. PLC (London)*, pp. 205—209.

Poulose, K.V., 1972. Heavy mineral and glass sand deposits of Kerala coast. *Indian Miner.*, 26: 118—124.

Puustinen, K., 1971. Geology of the Siilinjarvi carbonatite complex, eastern Finland. *Bull. Comm. Geol. Finl.*, 249: 5—43.

Rabatin, J.G., 1980. Rare earth X-ray phosphors for medical radiography. *Am. Chem. Soc. Meet., San Francisco, Calif., August 25, 1980* (preprint).

Raman, A., 1976. Uses of rare earth metals and alloys in metallurgy, I. Applications in ferrous materials. *Z. Metallkd.*, 67: 780—789.

Raman, A., 1977. Uses of rare earth metals and alloys in metallurgy, Part II. Applications in nonferrous materials. *Z. Metallkd.*, 68: 163—171.

Richardson, F.D. and Jeffes, J.H.E., 1980. Sidney Thomas's invention and its later impact. *Ironmaking Steelmaking*, 5: 222—226.

Roberts, D.E. and Hudson, G.R.T., 1983. The Olympic Dam copper-uranium-gold deposit, Roxby Downs, South Australia. *Econ. Geol.* (in press).

Robinson, W.O., 1948. The presence and determination of molybdenum and rare earths in phosphate rock. *Soil Sci.*, 66: 317—322.

Rose, E.R., 1979. Rare-earth prospects in Canada. *Can. Inst. Min. Metall., Bull.*, 72: 110—116.

Rosenblum, S., 1974. Analyses and economic potential of monazite in Liberia. *J. Res. U.S. Geol. Surv.*, 2: 689—692.

Roskill Information Services Ltd., 1980. *The Economics of Rare Earths and Yttrium.* Roskill, London, 4th ed., 103 pp.

Ross, I.N., 1975. The growing industrial role of the rare-earth metals. *Metals Mater.*, October, 42—45.

Russell, B.G., 1977. The possible recovery, during the manufacture of phosphoric acid, of rare earths from Foskor concentrate. *Bull. Miner. Bur., Dep. Mines, S. Afr.*, 1: 8 pp.

Schawlow, A.L., 1963. Advances in optical masers. *Sci. Am.*, 209(1): 34—45.

Semenov, E.I., 1974. Economic mineralogy of alkaline rocks. In: H. Sørensen (Editor), *The Alkaline Rocks.* John Wiley, New York, N.Y., pp. 543—552.

Tecotzky, M., 1973. Rare earth applications in luminescence. In: O.B. Michelsen (Editor), *Analysis and Application of Rare Earth Materials.* Universitetsforlaget, Oslo, pp. 359—372.

Van Wambeke, L., 1977. The Karonge rare earth deposits, Republic of Burundi new mineralogical-geochemical data and origin of the mineralization. *Miner. Deposita*, 12: 373—380.

Villani, F., 1980. *Rare Earth Technology and Applications.* Noyes Data Corporation, N.J., 367 pp.

Volkova, M.I. and Melentiev, B.N., 1939. Chemical composition of the Khibiny apatites. *Dokl. Adad. Nauk SSSR*, 35: 120—122.

von Backstrom, J.W., 1976. Thorium. In: C.B. Coetzee (Editor), *Mineral Resources of the Republic of South Africa*. Department of Mines, Geological Survey, Pretoria, 5th ed., pp. 209—212.

Ward, J., 1979. Rare earths and thorium. *Aust. Miner. Ind., Annu. Rev.*, 1977: 219—221.

Ward, J., 1980. Rare earths and thorium. *Aust. Miner. Ind. Annu. Rev.*, 1978: 196—197.

Ward, J., 1981. Rare earths and thorium. *Aust. Miner. Ind. Annu. Rev.*, 1979: 216—217.

Watson, M.D. and Snyman, C.P., 1975. The geology and mineralogy of the fluorite deposits at the Buffalo fluorspar mine on Buffelsfontein, 347 KR, Naboomspruit District. *Trans. Geol. Soc. S. Afr.*, 78: 137—175.

Wedow, H., 1967. The Morro do Ferro thorium and rare-earth ore deposit, Pocos de Caldas district, Brazil. *Bull. U.S. Geol. Surv.*, 1185-D: 34 pp.

Welch, B.K., Sofoulis, J. and Fitzgerald, A.C.F., 1975. Mineral sand deposits of the Capel area W.A. In: C.L. Knight (Editor), *Economic Geology of Australia and Papua New Guinea, 1. Metals*. Australasian Institute of Mining and Metallurgy, Parkville, Vict., pp. 1070—1088.

Wells, R.G., 1978. Nonmetallic and intermetallic phases containing rare earths in low alloy steels treated with mischmetal alloys. In: G. J. McCarthy and J.J. Rhyne (Editors), *The Rare Earths in Modern Science and Technology*. Plenum Press, New York, N.Y., pp. 341—346.

Wilson, W.G., 1979. The control of particles and other precipitates formed in cast iron and their effects on the structure. *83rd Casting Congress, Am. Foundrymen's Soc., Birmingham, Ala., May 1979*. Reactive Metals and Alloys Corp., Pittsburg, Ill., 29 pp.

Wilson, W.G., Kay, D.A.R. and Vahed, A., 1974. The use of thermodynamics and phase equilibria to predict behaviour of the rare earth elements in steel. *J. Metals*, 26(5): 14—23.

Wood, G.E., 1979. Rare earths. *Can. Miner. Yearb.*, 1977: 373—377.

Chapter 13

ANALYTICAL CHEMISTRY

PAUL HENDERSON and ROBERT J. PANKHURST

13.1. Introduction

Technical advances in analytical chemistry during recent years have enabled the accurate, routine measurement of individual REE concentrations in most geological materials even when the elements are in very low abundance and despite the fact that the REE form a coherent group in which the chemical properties of the members are very similar. Indeed, the improvement in analytical ability has come about mainly, but not entirely, from the use of the physical rather than the strictly chemical properties of the REE. The advances have been the prime cause of the rapid development of REE geochemistry in which measurement and interpretation of the chemical fractionations have proved so useful in helping to elucidate various petrogenetic problems.

Much of the geochemical study of the REE has involved the measurement of trace concentrations; to this end, certain methods have been more successful than others. Today, neutron activation analysis and mass spectrometric isotope dilution are the two most commonly used and best established techniques. Mass spectrometry is also used to determine variations in isotope ratios in natural specimens (Chapter 11). Inductively coupled plasma (ICP) emission spectrometry is a relative newcomer; it shows much promise for a wide range of elements, including REE, and is already being used routinely in a few geochemical laboratories. This chapter describes and assesses the application of these three techniques. Other methods in geochemistry include X-ray fluorescence, flameless atomic absorption spectroscopy, and electron-probe microanalysis. Some brief comments and literature citations for these techniques, and also for the use of beta-track mapping in certain experimental studies, are included.

The aim of this chapter is to provide an assessment of the scope and limitations of the principal analytical methods to give the geochemist some insight into the reliability and significance of published analytical data. It also aims to give those about to embark on analytical work some recommended procedures, as well as to help in choosing the best method for their particular requirements. The three principal methods discussed are sufficiently distinct from each other so that no overview of common principles is possible; each is discussed in turn.

13.2. Neutron activation analysis

Of the physical determinative methods, neutron activation analysis is one of the most sensitive and versatile. It is readily capable of the simultaneous determination of many elements at the parts per million (ppm), some at the parts per billion (ppb) levels, often without the need to destroy the specimen.

Samples, along with separate standards or comparators, are placed in a neutron flux (nuclear reactor or other neutron source) where nuclear reactions give rise to new, usually radioactive, isotopes of the elements present. After irradiation, the gamma radiations emitted by the radioactive isotopes are discriminated, measured, and compared with the radiations from the standards. Because each radioactive isotope has its own characteristic energy (or energies) of gamma radiation it is possible to identify the presence of a particular isotope, while the intensity of that radiation is directly proportional to the amount of that isotope present. The element La may be used as an example. Naturally occurring La consists of two isotopes: ^{139}La (99.91% abundance) and ^{138}La (0.09%). On irradiation with thermal neutrons (energy around 0.025 eV) a small proportion of each isotope will undergo a neutron-capture reaction to give an increase in nuclear mass of 1 amu. Hence for ^{139}La:

$$^{139}\mathrm{La} + \mathrm{n} \rightarrow {}^{140}\mathrm{La} + \gamma$$

(The reaction is normally written ^{139}La(n,γ)^{140}La.) ^{140}La is radioactive, with a half-life of 40.27 hours. Its decay involves the emission of gamma rays of various energies but those of 487, 816 and 1596 keV are the most intense. The gamma rays from the radioactive decay are distinct from those produced during the neutron-capture reaction given above.

Discrimination and intensity measurement of the gamma radiations are usually performed using lithium-drifted germanium (GeLi) or ultra-pure germanium (Ge) solid-state detectors linked to a suitable counting device such as a multi-channel analyser. The energy resolution of modern detectors is such that analysis for the REE in a wide range of matrices and at low concentration levels may be performed on small quantities of sample without any chemical treatment. Hence, this method is called "instrumental neutron activation analysis" (INAA). The concentration of a particular REE in a sample is directly proportional to the area of a photopeak associated with the appropriate isotope. The ratio of this area to that of the standard is used to calculate the concentration. The precision of the method is primarily dependent, therefore, on counting statistics.

Occasionally the matrix is of such a composition that there may be significant interference on the REE photopeaks by photopeaks from some other elements in the gamma-ray spectrum. Alternatively, the REE may be at extremely low concentrations. In these cases, chemical separation of the REE from some or all of the other elements may be necessary, usually after neutron irradiation, before the stage of gamma-ray detection and counting. This is called "radiochemical neutron activation analysis" (RNAA). Reviews of the principles and applications of neutron activation analysis can be found in a number of texts (e.g., De Soete et al., 1972; Muecke, 1980; Amiel, 1981).

This section is confined to the application of INAA to analysis for REE in rocks and minerals, with a brief review of some radiochemical methods. A short discussion on epithermal neutron activation analysis (ENAA) is also given. In epithermal neutron irradiation, the thermal neutron component of the neutron flux is prevented from reaching the samples by the use of a cadmium shield, often in the form of a cadmium-lined irradiation container or as a cadmium-lined core tube. The (n,γ) reactions are then induced by epithermal neutrons which can pass through the cadmium, but since the interactions of the various nuclei with epithermal neutrons are often very different from those with thermal neutrons the resultant activities will also be different. The advantages and disadvantages of this method are discussed below.

Instrumental neutron activation analysis

Selected data on the isotopes of the REE, which are relevant to INAA, are given in Table 13.1. Relative abundances of the naturally-occurring isotopes are given in Chapter 1. Study of Table 13.1 reveals that there is a wide variation in the ease with which individual REE may be determined. Pr is rarely determinable because ^{142m}Pr does not emit gamma rays, while the gamma rays from ^{142}Pr are very weak. Er is sometimes determinable but only with difficulty, principally because of its unsuitable half-life (7.52 hours), this being too long for a short irradiation time (as used in the determination of Dy) and too short for those irradiation and counting conditions suitable for most other REE. Difficulties may also be encountered in the determination of Gd and Ho, but these elements can sometimes be measured after epithermal neutron activation (see below). Good results can be obtained for the remaining REE.

The precise conditions of the analysis, such as sample weight, irradiation times, and decay and counting times, all depend on the sample to be analysed. The following account relates to a typical INAA procedure but many variants are possible.

Preparation of samples and standards. About 100 mg of powdered rock or mineral sample are suitable for most concentration levels. Considerably less

TABLE 13.1

Selected data on REE isotopes

Isotope	Natural abundance (%)	Thermal neutron capture cross section (barn)	Radio-nuclide	Half-life
^{139}La	99.91	9.1	^{140}La	40.27 hours
^{140}Ce	88.48	0.56	^{141}Ce	32.28 days
^{141}Pr	100	3.9	^{142m}Pr	14.6 minutes
		8	^{142}Pr	19.13 hours
^{146}Nd	17.18	1.4	^{147}Nd	11.06 days
^{152}Sm	26.74	208	^{153}Sm	1.948 days
^{151}Eu	47.82	4	^{152n}Eu	1.60 hours
		3000	^{152m}Eu	9.3 hours
		5300	^{152}Eu	12.7 years
^{152}Gd	0.20	~1000	^{153}Gd	241.6 days
^{158}Gd	24.87	2.4	^{159}Gd	18.56 hours
^{159}Tb	100	27	^{160}Tb	72.1 days
^{164}Dy	28.18	1820	^{165m}Dy	1.26 minutes
		780	^{165}Dy	2.334 hours
^{165}Ho	100	3.2	^{166m}Ho	1200 years
		62	^{166}Ho	1.117 days
^{170}Er	14.88	5.7	^{171}Er	7.52 hours
^{169}Tm	100	103	^{170}Tm	128.6 days
^{168}Yb	0.135	3200	^{169}Yb	30.7 days
^{174}Yb	31.84	46	^{175m}Yb	67×10^{-3} seconds
		19	^{175}Yb	4.19 days
^{176}Lu	2.59	7	^{177m}Lu	160.9 days
		315	^{177n}Lu	0.16×10^{-3} seconds
		1740	^{177}Lu	6.71 days

Sources: De Soete et al. (1972), Erdtmann and Soyka (1979).

material can be used if the sample is in short supply or the REE concentrations exceed trace amounts. The use of large sample weights can introduce errors relating to the geometry of counting and absorption of low-energy gamma rays as well as the additional hazard of a greater quantity of radioactive material. Care should be taken during sample crushing and preparation not to introduce interfering elements (e.g., W; see later). Mineral grains may also be analysed. The powders or grains are placed in small polythene vials which are heat sealed and numbered. Alternatively, the samples may be wrapped in pure, pre-cleaned Al foil.

Calibration standards (i.e., comparators) are prepared from stock solutions containing known quantities of the REE, produced by the dissolution of ultra-pure REE or rare earth compounds of known stoichiometry; they can also be obtained commercially (e.g., see Borley and Rogers, 1979). Dilution and mixing of the stock solutions gives a bulk solution from which an appropriate aliquot (100 μl is convenient) may be taken. This aliquot is placed on to a small disc of filter paper within a polythene vial or on to Al foil, and then evaporated to dryness. An example of the constituents of an aliquot that is suitable for analysis of a wide range of minerals or rocks, taking account of relative abundances and extent of activation, is given in Table 13.2. At least two sealed vials or wrapped foils of the standards should be placed with each batch of up to ten samples in a polythene or aluminium irradiation container.

The use of standard rocks as comparators cannot be recommended, not only because of the use of reference material which is in limited supply but because of uncertainties over the exact concentrations of the elements, the possibility of sample heterogeneity, and the existence of uncorrected interferences. For most rocks and minerals the matrix rarely creates the kind of problem (e.g., neutron self-shielding) which requires the use of standards similar in constitution to the samples, but it is advisable to incorporate one vial containing a reference rock in each sample batch to enable a check on the accuracy of the analysis.

Polythene degrades when subjected to a high neutron flux and thus, if long irradiations (i.e., equivalent to greater than 1.8×10^{17} n cm^{-2}) are envisaged, the sample and standard containers should be either sealed silica tubes or Al envelopes.

Irradiation conditions. Because of the different half-lives (Table 13.3), it is preferable to have two irradiation periods. The first should be of about 15 minutes' duration in a thermal neutron flux of 10^{12} n cm^{-2} s^{-1} or its equivalent (i.e., to give an integrated flux of about 10^{15} n cm^{-2}), for the determination of Dy. Sample insertion and removal is achieved by means of a pneumatic transfer system. The second irradiation, for the remaining REE,

TABLE 13.2

Elemental abundances in the aliquot used for standardization

Element	Abundance (μg)
La	12.5
Ce	20.0
Nd	10.0
Sm	1.0
Eu	0.3
Gd	3.0
Tb	1.0
Dy	1.0
Ho	1.0
Tm	0.5
Yb	1.0
Lu	0.3

can normally be of the order of 30 hours in a flux of 10^{12} n cm^{-2} s^{-1} (integrated flux $\simeq 10^{17}$ n cm^{-2}). If there is a gradient in the thermal neutron flux along the length of the irradiation container, a correction will be necessary. The flux variation can be measured by placing identical calibration standards at each end and in the middle of the irradiation container or by use of other monitors, such as Fe wire, wrapped around each ampoule (Gibson and Jagam, 1980).

TABLE 13.3

Principal gamma rays of rare earth radionuclides

Isotope	Half-life	Gamma-ray energy[a] (keV)	Relative intensity (%)
^{140}La	40.27 hours	*328.8*	19.4
		487.0	45.0
		815.8	23.4
		1596.2	100
^{141}Ce	32.28 days	*145.4*	100
^{142}Pr	19.13 hours	1575.9	100
^{147}Nd	11.06 days	*91.10*	100
		531.00	47.7
^{153}Sm	1.948 days	*69.67*	18.6
		103.18	100
^{152}Eu	12.7 years	*121.78*	100
		344.30	92.3
^{153}Gd	241.6 days	*97.50*	50
		103.20	36.3
^{159}Gd	18.56 hours	363.56	100
^{160}Tb	72.1 days	*86.80*	44.7
		298.57	91.3
		876.37	100
^{165m}Dy	1.26 minutes	108.20	58.9
		515.50	32.5
^{165}Dy	2.334 hours	*94.70*	83.6
^{166}Ho	1.117 days	*80.57*	100
^{171}Er	7.52 hours	*308.29*	100
^{170}Tm	128.6 days	*84.26*	100
^{169}Yb	30.7 days	*63.12*	57.6
		109.77	23.0
		177.18	28.2
		197.97	46.1
^{175}Yb	4.19 days	113.80	29.4
		396.32	100
^{177m}Lu	160.9 days	55.79	99.4
^{177}Lu	6.71 days	112.95	34.6
		208.36	100

[a]Italics indicate the energies commonly used in INAA.

Gamma-ray spectrometry. This is best done at three or four different time intervals after irradiation, principally to allow activity from other isotopes (e.g., ^{24}Na, half-life = 15 hours) to die away but also to make best use of the different half-lives of the various REE (Table 13.3). A suggested counting sequence is given in Table 13.4. Counting times should be sufficient to accrue 10,000 counts (for a counting precision of 1%) or more for each photopeak of interest.

A planar Ge(Li) or Ge detector with a good response (i.e., detection efficiency) to photons in the energy range 50—350 keV, and with a resolution of 580 eV (full width half maximum) or better at 122 keV, is the most suitable for detecting many of the REE indicated by the gamma-ray energies listed in Table 13.3 (Schock, 1977). Intrinsic Ge detectors of the required specification are available commercially and it is possible to determine routinely 12 (including Dy) of the 14 REE with their use (Henderson and Williams, 1981). Lanthanum can also be measured readily using a coaxial Ge(Li) detector with an energy response suitable for the 1596.2-keV photopeak (Table 13.3). Such a detector could be used also to determine those REE with relatively high-energy gamma peaks (e.g., Nd, Tb, Eu) but the resolution is unlikely to be as good as in the case of planar detectors and difficulties may be encountered from interference. If, however, many elements in the samples other than the REE are to be determined by INAA it would be appropriate to have two detectors: a planar one of high resolution for photopeaks up to 350 keV, and a coaxial Ge(Li) or Ge, or even a thick planar Ge detector with relatively good efficiency for detecting photons in the energy range 200 keV to 1.5 MeV.

The good resolution now available in modern planar detectors has done much to reduce the problem of the mutual interference of photopeaks and to make INAA a reliable method for the determination of REE in many rocks and minerals. In most cases it should no longer be necessary to take account of many interferences which hitherto could have been significant (e.g., the interference on the ^{140}La 328.8-keV peak by the ^{51}Cr 320-keV peak). Computer software also allows the theoretical resolution of partially

TABLE 13.4

INAA counting sequence

Counting session	Irradiation	Decay period[a]	Elements determined
1	short	0.5 hours	Dy
2	long	5—7 days	La, Sm, Ho
3	long	14 days	Ce, Nd, Eu, Tb, Tm, Yb, Lu
4	long	⩾1 month	Ce, Eu, Gd, Tb, Tm

[a]Period between end of irradiation and commencement of counting

interfering peaks. Interferences which require a direct means of correction do, however, still exist for the REE; a selection of the more important ones in INAA is given in Table 13.5. The correction may be made by measuring the intensity of a suitable reference peak (see Table 13.5) of the interfering isotope and applying a correction factor previously measured using a pure form of the irradiated interfering element (Henderson and Williams, 1981). The existence of possible interferences can be ascertained from listings of gamma-ray energies (e.g., Erdtmann and Soyka, 1979).

Special care over interferences is needed in the case of uranium-rich material because the neutron-induced fission of uranium gives products including REE isotopes (especially: ^{140}La, ^{141}Ce, ^{147}Nd, and ^{153}Sm). The use of tungsten carbide crushing equipment during sample preparation can lead to significant interference of the ^{153}Sm photopeak at 69.67 keV (half-life = 1.95 days) from the X-ray of ^{187}W at 69.20 keV (half-life = 23.9 hours).

The presence of beta emitters in the sample will give rise to a Bremsstrahlung spectrum which could add significantly to the background radiation and thereby reduce the recorded peak intensity. This could be an important effect in the analysis of phosphates (^{32}P is a beta emitter) but it can be prevented by placing a suitable beta absorber (e.g., perspex) between the sample and the detector assembly. Background that is produced from scattering of the gamma rays on any shielding material can be reduced by making the shields as large as possible. Lead shields should be lined with thin cadmium and copper sheets to convert the generated 86-keV X-ray of lead to the 12-keV X-ray of copper. Compton continuum can be reduced using an anti-coincidence counting system (De Soete et al., 1972).

TABLE 13.5

Important interfering photopeaks and reference correction peaks

Interfered isotope	Half-life (days)	Energy (keV)	Interfering isotope	Half-life (days)	Interfering energy (keV)	Reference energy (keV)
^{141}Ce	32.38	145.45	^{175}Yb	4.19	144.86	396.32
			$^{233}Pa(Th)$	27.4	145.40	94.66
^{153}Sm	1.948	69.67	^{153}Gd	241.6	69.60	97.50
			^{187}W	0.99	69.20	72.00
^{153}Sm	1.948	103.18	^{153}Gd	241.6	103.20	97.50
			$^{239}Np(U)$	2.355	103.70	106.13
			$^{233}Pa(Th)$	27.4	103.86	94.66
^{152}Eu	4640	121.78	$^{239}Np(U)$	2.355	120.70	106.13
^{160}Tb	72.1	86.80	$^{233}Pa(Th)$	27.4	86.61	94.66
^{170}Tm	128.6	84.26	^{182}Ta	115.0	84.68	100.11
^{169}Yb	30.7	63.12	^{176m}Lu	0.15	63.20	55.79
			^{177m}Lu	160.9	63.20	55.79
^{177}Lu	6.71	208.36	^{169}Yb	30.7	207.00	109.77

Data handling. Pulses from the detector proceed via an amplification stage to a multi-channel analyser which converts the energy of each pulse into digital form by recording the event as a pulse count in a particular channel of the analyser, each channel representing an energy increment. In order satisfactorily to record the gamma-ray spectrum incorporating all the measured photopeaks for the REE by a low-energy photon detector, and without losing the resolution (i.e., peak separation) provided by the detector, it is necessary to assign at least 4000 channels to the energy range. Each channel then has a width equivalent to about 100 eV or less.

The required gamma-ray spectrum may be processed in a number of ways. The objective is to obtain an accurate measure of the peak area associated with each isotope in the sample. This area is then related to that of the standard to obtain the concentration, any correction for radioactive decay (half-life) being made to compensate for the time interval between counting the standard and the sample. Peak-fitting methods are sometimes necessary when peaks are incompletely resolved from each other. Many computer programs are available for this purpose (e.g., the programs SAMPO — Routti and Pressin, 1969, and ALPHA — Schonfeld, 1966), and the manufacturers of computer-based multi-channel analysers often provide suitable software.

The reader is referred to general reviews of instrumentation and spectra handling in INAA (e.g., De Soete et al., 1972; Krugers, 1973; Jagam and Muecke, 1980) for further details.

Accuracy and precision. The analytical precision of INAA is strongly but not entirely related to counting statistics. Hence, on counting statistics alone a precision of about 1% should be attainable if peak areas contain 10,000 counts or more. Small variations in counting geometry, irradiation conditions, peak integration, and so on mean that the precision is rarely better than 2% (1 standard deviation) for many REE. At the kind of REE concentrations found in many basalts (e.g., BCR-1 standard rock), a precision of about 2—4% should be attainable for the elements La, Ce, Nd, Sm, Eu, and Yb; 3—6% for Tb and Lu; and 4—10% for Gd, Ho and Tm. Determination of Dy by the short-lived ^{165}Dy isotope is unlikely to yield a precision much better than 5%. An accuracy of about 5% or better is achievable for nine REE (La, Ce, Nd, Sm, Eu, Tb, Dy, Yb and Lu). Using a standard additions method for calibration, Potts et al. (1981) obtained accuracies better than 5% for nine REE, with precision of 2—4%, in their INAA analyses of 29 international rock standards.

Epithermal neutron activation analysis (ENAA). If during irradiation in a mixed-energy neutron flux (e.g., nuclear reactor) the samples and standards are enclosed in a thermal neutron absorber, such as cadmium or boron, which is relatively transparent to epithermal neutrons (energy from ~0.4 eV to 1 MeV), isotopes are produced in different proportions than when irradiation is without an absorber. This is because the ratio of the resonance activation integral to thermal neutron-capture cross section is different from one element to the next (De Soete et al., 1972), and this can be advantageous where the activation of a REE nuclide is enhanced relative to that of an interfering nuclide. It has been shown, however, that for the REE a significant advantage is obtained only with Sm, Gd, Tb and Tm (Steinnes, 1971; Baedecker et al., 1977; Watterson, 1978). There is also some advantage during the determination of Ho but in general, when analyses for many of the REE are required, thermal neutron activation is the preferred method, especially as the induced activity is so much greater under these conditions than with epithermal neutron activation.

Radiochemical neutron activation analysis (RNAA)

This method is selected when the required elements cannot be determined readily or with sufficient accuracy by INAA because of very low concentrations, small sample weight, or the presence of a significant background or other interference. The purpose is to separate chemically the REE as a group from other elements in the sample, usually after neutron irradiation, and to use the separated REE fraction for quantitative gamma-ray analysis. Low abundances of the elements to be determined require the addition of a known quantity of inactive REE to the irradiated sample in order to act as a chemical "carrier" for the trace quantities in the sample during the chemical separation processes. The carrier also enables the determination of any loss during the separation (see below).

A list of steps in RNAA procedure for REE might be as follows:

(1) Sample preparation and irradiation.
(2) Sample decomposition in presence of carriers.
(3) Chemical separation of REE as a group.
(4) Precipitation of known rare earth compound.
(5) Plating-out of precipitate on to planchet suitable for counting.
(6) Gamma-ray analysis.
(7) Determination of precipitate yield, i.e., the proportion of REE compound recovered to that theoretically obtainable from the known quantity of carrier added at step 2. Correction of the radiometric data for yield.

There are many published RNAA schemes for REE determination and it should be possible for the analyst to select that which best suits his needs. Recently, Smet et al. (1978) have described a procedure for group separation of 24 elements, including REE, and Duffield and Gilmore (1979) developed a rapid and straightforward scheme for the analysis of 14 REE.

Reported separation procedures (step 3 above) for the REE include the use of a fluoride-hydroxide precipitation cycle (e.g., Potts et al., 1973; Brunfelt et al., 1977; Duffield and Gilmore, 1979); oxalate precipitation (e.g., Elek et al., 1973; Duffield and Gilmore, 1979); ion exchange; and adsorption on preformed precipitates (e.g., Csajka, 1973). The presence of Fe and Sc can contribute much to the background of the gamma-ray spectrum; their removal can be achieved readily by solvent extraction using tributylphosphate (Brunfelt et al., 1977; Smet et al., 1978) although they (especially Sc) are also lost during repeated fluoride-hydroxide precipitation cycles. Iron can also be extracted with di-iso-propylether and scandium with diethylether.

Solutions containing the separated REE can be counted directly instead of precipitating and plating out a known compound (steps 4 and 5), provided that the counting geometry is the same for sample and standard solutions. Yields may be determined by a variety of standard analytical methods, although one of the best ways is to carry out a short re-irradiation and counting procedure on the final precipitate or solution. If the carrier contained more than one REE this method will indicate any differential loss of individual REE. Some authors (e.g., Csajka, 1973) claim that their recommended separation methods are quantitative so that determination of yield is unnecessary.

RNAA procedures for the separation of individual REE have been described (e.g., gas chromatography and high-voltage electrophoresis: Becker et al., 1975) but the technical capabilities of modern gamma-ray detectors should remove the need for such operations, at least in the analysis of most rocks and minerals.

Group separation of the REE *before* irradiation has some distinct advantages over post-irradiation separation although there is the risk (usually negligible in the case of REE) of contamination during chemical separation. These advantages include: lower activities after irradiation through the absence of elements of high activity (e.g., Na); ability to measure REE with short half-lives (e.g., Dy, Ho) immediately after irradiation; and the fact that yields may be monitored directly, and before neutron irradiation, by use of a radiotracer added at the stage of sample digestion. Pre-irradiation procedures have been described by a few authors including Voldet and Haerdi (1976, 1978), Randa (1976) and Croudace (1980). The method is particularly suitable for analysis of natural water (e.g., Hirose et al., 1978).

One of the principal advantages of RNAA over INAA is enhanced sensitivity. Chou (1980) has compiled a list of some of the lowest concentrations actually measured by RNAA as an approximate guide to sensitivities. Most of the REE have been determined at concentrations slightly below 0.05 ppm, and with Eu, Tm, Lu below 0.005 ppm. Ce and Nd have been determined at levels between 0.05 and 0.1 ppm.

In summary, neutron activation analysis is a very specific and sensitive technique which can be non-destructive. Simultaneous determination of many elements is possible, and the instrumental mode requires few manhours. For the REE it has proved to be an excellent method and is widely used in geochemical research laboratories. Its principal disadvantages are:

the need for special handling facilities for the radioactive materials; the cost of neutron irradiations which can sometimes be high if they were to be obtained commercially; and the unsuitability of the method for the analysis of some elements. The cost of suitable equipment for gamma-ray analysis varies considerably, depending on the degree of sophistication required. In general, it is significantly less than that of a mass-spectrometric facility.

13.3. Mass-spectrometric isotope dilution analysis

The basis of this method is the determination of the isotopic composition of an element by mass spectrometry. It is a very accurate method, applicable to ten of the REE, including all those which are critical in interpretations of geological processes. It is also sensitive in that it seldom requires more than 50—100 mg of material for a full analysis and so is suitable for separated minerals. For these reasons it has played a very prominent part in the development of REE geochemistry, especially the quantitative treatment of REE partitioning between crystals and magma during igneous fractionation. Some of the most notable contributions in this field using isotope dilution are those of Gast (1968), Schnetzler and Philpotts (1970), Kay and Gast (1973), O'Nions and Grønvold (1973), Arth and Hanson (1975) and Sun and Hanson (1975). In general, however, these papers give only minimal information on analytical procedure. The instrumentation of mass spectrometry is not well documented, so it will be appropriate to start with a brief summary of its theory and practice. A more extended introduction was given by Webster (1960) and this remains useful, although instrumentation has advanced considerably since it was written.

Mass spectrometry

Singly charged ions are generated from the element to be analysed in a suitable *ion source*, which usually operates either by electron bombardment of gaseous samples (gas source) or by thermal evaporation from a glowing filament in the case of solid samples (solid source). It is the latter with which we are concerned for REE analysis. The filament substrate is a high-melting-point metal such as W, Ta or Re and chemically-prepared samples are loaded directly on to it. The ions formed when the filament is heated under vacuum to temperatures above about 1000°C are accelerated through a potential difference of several thousand volts and focussed through a collimating slit as a slightly diverging beam. Most of the ions formed in this way are singly charged, regardless of the chemical behaviour of the parent element, and they therefore achieve a constant kinetic energy given by:

$$eV = 1/2\, mv^2$$

where e is the electronic charge, V the overall accelerating potential, m the ionic mass, and v its velocity.

In the next part of the mass spectrometer, called the *analyser*, the ion beam is fired into an evacuated curved tube between the pole pieces of an electro-magnet where the ions are deflected into curved trajectories by the electromagnetic force acting perpendicular to both the line of flight and the magnetic flux. The relationship:

$$eBv = mv^2/r$$

gives the radius of curvature in terms of the field strength (B) and the previously defined quantities, whence it follows that:

$$r = (2\,eVm)^{1/2}/eB$$

Thus a mass spectrum is produced in which the lighter ions are deflected with a smaller radius of curvature and the radial dispersion between consecutive mass numbers decreases according to $\sqrt{m}$.

By varying the magnetic field strength (or sometimes the accelerating voltage), ions of different masses may be tuned to the critical radius of curvature at which they alone pass through a slit at the *collector*. Here they are detected quantitatively, for example by allowing the ion current to drain to earth through a very large value resistor (e.g., 10^{10} ohms) and measuring the voltage developed across it. Comparison of the size of the signal at any two mass numbers allows the corresponding isotopic ratio in the original solid material to be calculated. Important facets of mass-spectrometer design include efficient ion optics (to refocus the largest possible proportion of ions as a coherent beam at the collector) and sensitive detection and measurement techniques. In solid-source mass spectrometry it is also sometimes necessary to control or monitor Rayleigh-type mass fractionation during evaporation from the filament. Modern instruments have direct digital recording, on-line data handling and, increasingly, fully automatic computer control of the analysis. Digital control of the field strength greatly simplifies the location and identification of complex spectra such as those of the REE. In the very near future it may also be expected that the use of multiple collectors to measure several isotopes simultaneously will prove of considerable value.

Isotope dilution

Most elements have a fixed isotopic composition in natural materials. For the relatively abundant light elements (up to about Si in the periodic table), mass fractionation during chemical exchange reactions can produce quite significant variations, but this is negligible for elements as heavy as the rare earths where proportional mass differences are small. Variations also occur where specific isotopes are produced naturally by radioactive decay. For example, ^{143}Nd is a decay product of ^{147}Sm, but the statistics of the decay are such that the ^{143}Nd/^{144}Nd ratio in rocks ranges only from 0.507 to about 0.515 (see Chapter 11). The isotopic composition of the naturally-occurring REE is given in Chapter 1, Table 1.4. Four elements — Pr, Tb, Ho and Tm — are mono-isotopic and therefore cannot be analysed by the method described here. La and Lu have two isotopes each, one major and one trace; Eu also has two but they are present in roughly equal proportions. Each of the remaining REE has several isotopes, some well represented, others present only in trace amounts, but with no one isotope greatly exceeding any other in abundance (except for ^{140}Ce). Isotopes may, however, be produced or enriched artificially by nuclear reactions or electromagnetic separation. Atomic research laboratories such as Harwell (U.K.) and Oak Ridge (U.S.A.) manufacture and sell a wide range of purified non-radioactive isotopes and this is an essential factor in isotope dilution analysis. If a known amount of such an isotopic "tracer" or "spike" is mixed with an unknown amount of the natural element, the latter quantity can be deduced from the composition of the mixture as determined by mass spectrometry (see Fig. 13.1). It can be shown that the amounts in moles of normal (x) and spike element (y) are related by the equation:

$$\frac{x}{y} = \frac{s_k}{n_k} \frac{(S_{i,k} - M_{i,k})}{(M_{i,k} - N_{i,k})} \tag{13.1}$$

where the subscripts i and k refer to *any* pair of isotopes, $S_{i,k}$ is the abundance ratio of these two isotopes in the spike, $N_{i,k}$ is the same ratio in the normal element, and $M_{i,k}$ that in the mixture. The quantities s_k and n_k are the abundances of the denominator isotope in the spike and normal element, respectively (expressed as a proportion or percentage). For the remainder of this section it will be useful to assume that the kth isotope is that most enriched in the spike, i.e., both the terms in brackets in equation (13.1) are negative. Clearly this relationship can be converted to a weight ratio by multiplying by the ratio of the atomic weights of the normal spike elements A_n/A_s:

$$A_s = \frac{\Sigma M_i \cdot s_i}{\Sigma s_i} \quad \text{or} \quad \frac{\Sigma M_i \cdot S_{i,j}}{\Sigma S_{i,j}} \tag{13.2}$$

where the terms are summed over all the isotopes of the element (M_i being the atomic mass of the ith isotope, s_i its abundance and $S_{i,j}$ its ratio to another fixed but arbitrary isotope).

Although it is only strictly necessary to determine a single isotope ratio, error magnification will clearly become large when $S_{i,k} \rightarrow N_{i,k}$ or as $M_{i,k}$

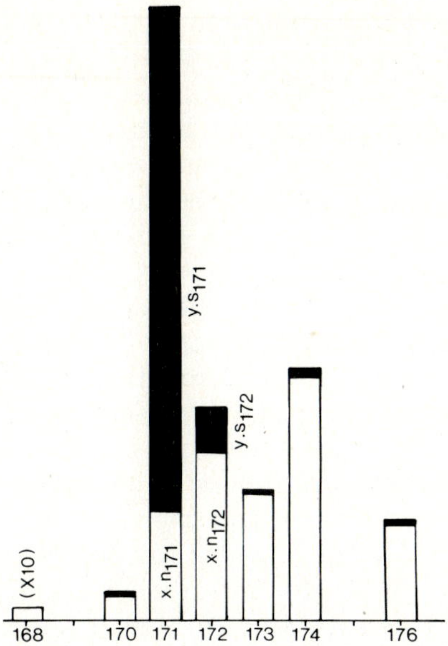

Fig. 13.1. Diagrammatic representation of the mass spectrum resulting from a mixture of x moles of natural Yb with y moles of spike Yb enriched in ^{171}Yb. If the proportional abundances of ^{171}Yb and ^{172}Yb in the normal and spike element are designated, respectively, n_{171}, n_{172}, s_{171} and s_{172} then the ^{171}Yb/^{172}Yb ratio of the mixture is given by:

$$M_{171,172} = (x \cdot n_{171} + y \cdot s_{171})/(x \cdot n_{172} + y \cdot s_{172})$$

Rearrangement of this gives an equation of the generalized type shown as equation (13.1).

approaches either end-member. So the choice of which pair of isotopes to use is important, as well as achieving a reasonably balanced mixture. Spikes are most useful if they represent a very high degree of enrichment of an isotope which is relatively rare in the normal element. This is not always possible; for example, it is prohibitively expensive to manufacture La which is more than about 10% enriched in ^{138}La. However, all the polybaric elements between La and Lu are available with enrichments of over 90% in a minor isotope. Measurement of a further isotope pair ($M_{j,k}$) enables an independent estimate of the x/y ratio, or it can be used to correct for any instrumental mass fractionation (see Boelrijk, 1967), although usually this is only significant for the REE if very precise results are required.

Efficiency of ionization from the mass-spectrometer filament is greatly reduced in the presence of other, more easily ionized elements so that prior chemical separation is essential. Moreover, with some fifty REE isotopes occurring in the mass range 138—176 *and* a tendency for the lighter elements to form complex oxide ions (e.g., MO^+) as well as metal ions it has been customary to sub-divide into smaller, more manageable, groups of elements to be analysed on separate filaments. A threefold division into light (La, Ce), middle (Nd—Gd) and heavy (Dy—Lu) REE is often made although if La is not analysed a twofold division between Gd and Dy is adequate. Thirlwall (1982) has recently described a technique for analysing the REE as a complete group on one filament, without the need for this extra separation. This is a very important advance in the practical application of the method. Ion exchange chromatography is invariably used as the separation technique, but with certain variations.

The crucial stage in isotope dilution is to homogenize the sample (e.g., 0.1—1 g of whole-rock powder) with the appropriate amount of the spike elements. A roughly equal weight of the normal and spike elements is usually optimum, so it helps to have a reasonable idea of the expected REE contents to avoid unnecessary error multiplication associated with gross over- or under-spiking, especially the latter. If homogenization is accomplished during initial decomposition then subsequent chemical treatment need not be quantitative in yield (cf., RNAA, section 13.2), since the isotope ratios will remain constant unless further natural REE are introduced via the reagents. The spikes are conveniently added as solutions (in which several or all the elements may be mixed) and evaporated carefully with the sample. Each individual spike has to be pre-calibrated by the same procedure — mixing with an accurately known (gravimetric) amount of the natural element and measuring the resultant isotope compositions. Calibration solutions are normally made from metals or ignited oxides of spectrographic purity (LREE oxides, especially La_2O_3, absorb H_2O and CO_2 very rapidly from the atmosphere). Sample decomposition may often be carried out by HF/HNO_3 or $HF/HClO_4$ attack in open F.E.P. beakers, but for whole-rock samples where high concentrations of REE may be present in resistant minor phases such as zircon, decomposition in a P.T.F.E. pressure vessel or fusion with Li metaborate is preferable.

Ion-exchange separation may be carried out on a standard cation-resin column such as Dowex-50, in a HCl or HNO_3 medium. One of the simplest methods used is to elute with 2.5 M HCl until Sr is removed and then collect two successive fractions in 6.2 M HCl, the first enriched in HREE and the second in LREE (Shimizu, 1974). Better separation of the REE from each other and from Ba is achieved by a two-stage method in which the REE are removed as a single group in 6 M HNO_3 after major elements, Ba and Sr have been eluted in 2 M HNO_3. This is then followed by a second column in which the REE are subdivided using 3 M HCl (Hanson, 1976). An alternative method exploiting mixed solvent complexes on anion exchange resin is described by Hooker et al. (1975) and is summarized in Fig. 13.2. This probably achieves the best separation from Ba — an important factor if La is to be determined. More sophisticated techniques are used to separate Sm and Nd for precise Nd isotope composition measurements (e.g., Eugster et al., 1970) but these are not necessary for multi-element REE isotope dilution

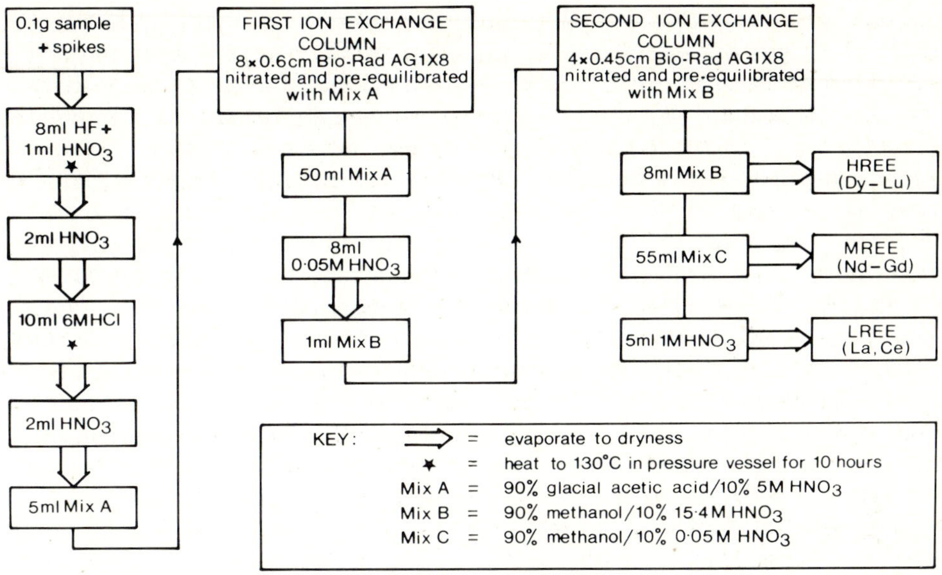

Fig. 13.2. Flow diagram showing a slightly modified version of the chemical separation scheme for REE given by Hooker et al. (1975). The ion-exchange columns are made up afresh for each sample in quartz tubes with quartz-wool plugs, otherwise P.T.F.E. vessels are used throughout. The resin is nitrated by passing dilute HNO_3 through the chloride form until Cl^- is no longer liberated (this may be done in bulk on a much larger column). The resin is transferred as a slurry in a few column volumes of the first eluting mixture, taking care to avoid air bubbles, prior to introduction of the sample solution.

analysis. The use of reagents purified by sub-boiling distillation in quartz keeps REE blanks to negligible levels for almost all practical purposes (below 5 ng for La and Ce, and well below 1 ng for the rest).

The separate REE fractions are loaded directly on to a tantalum or rhenium mass-spectrometer filament with a small drop of water or dilute HCl, which is then taken to dryness. In the mass-spectrometer source, evaporation and ionization of the sample may be induced in a single step by direct heating of the loaded filament to temperatures above 1000°C, requiring a DC current of 2—4 A. This is called "single-filament ionization". A more efficient though less convenient technique is to use a separate high-temperature ionization filament while the sample is evaporated at a lower temperature from one or two side filaments in close proximity (known as "double-" or "triple-" filament ionizations, respectively). Triple- (or double-) filament ionization is more sensitive and is better for the low-abundance involatile HREE (Dy, Er and Lu). It can produce LREE ions of both M^+ and MO^+ species according to operating conditions. The single-filament technique, which is very convenient to operate, results in a very high ratio of MO^+/M^+ ions for La, Ce, Pr, Gd and Tb, as well as Ba if present, a much lower ratio for Dy, but principally metal ions for the remainder. Disadvantages are the rapid evaporation of certain elements at certain stages, requiring timely measurement, more extreme mass fractionation, and low sensitivity for the HREE (except Yb) which sometimes necessitates the use of an electron multiplier or scintillation detector in the measurement circuit. Triple-filament ionization is essential for the analysis of all REE using one filament loading, as described by Thirlwall (1982).

Different elements are emitted from the filament at different temperatures so that most elements can be analysed sequentially in the presence of few interfering ions, the filament temperature being raised successively. Nevertheless, some interferences (e.g., ^{142}Nd on ^{142}Ce) are unavoidable (see Fig. 13.3). In such cases corrections may have to be made before equation (13.1) can be solved for concentration, as indicated in the following notes on the measurement of each element. The elements are discussed in the order in which they would normally be analysed using a three-fraction split (i.e., with La and Ce as a separate fraction). This order is slightly modified in the single loading method of Thirlwall (1982).

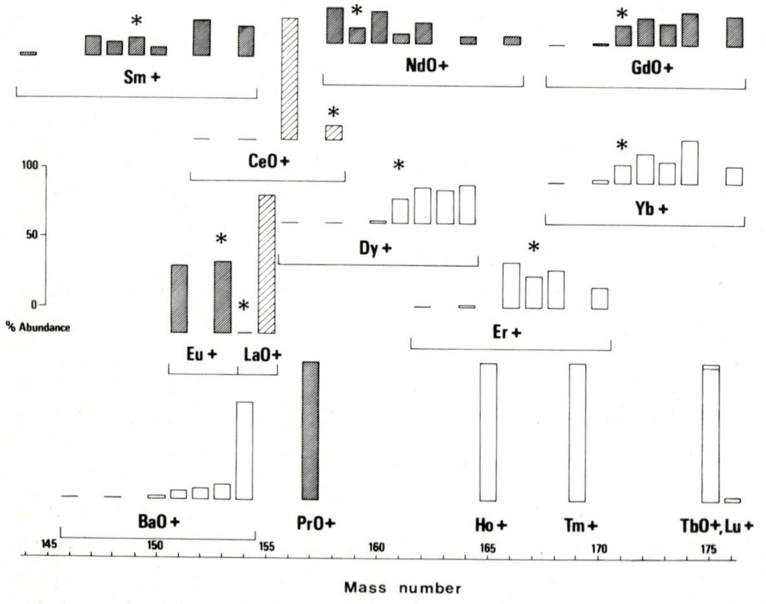

Fig. 13.3. Diagrammatic representation of observed REE mass spectra using single Ta filament thermal ionization. Isobaric interferences occur wherever two elements have isotopes in the same vertical line, but their effect is obviated or reduced by differences in emission temperature (see text) and by separation into two or three groups (the shading here indicates separation into two groups, La—Gd and Dy—Lu). The mono-isotopes are not used, but still occur in the observed spectra. An asterisk indicates the spike isotope in the mixture.

Europium. This is the first element to be ionized from the middle REE fraction (or the light fraction where only two have been collected). In the absence of Ba there are no interferences and only two isotopes — 151 and 153. Spikes are available highly enriched in either isotope.

Samarium. Emission of Sm^+ grows as that of Eu^+ peaks or begins to die off, often without any increase in filament current. There are no isobars of ^{149}Sm, which is therefore a good choice as spike. With triple-filament ionization there are potential interferences from LaO^+ (154), CeO^+ and Gd^+ (154 and 152) and Nd^+ (150 and 148) as well as BaO^+. These can be minimized by good chemical separation and by measuring before Ce and Nd start to emit, but there are no significant interferences on $^{147}Sm/^{149}Sm$. In the case of single-filament ionization, the interferences from Nd and Gd are virtually removed since these emit predominantly as oxide ions, so that $^{152}Sm/^{149}Sm$ may also be measured as a check or to enable fractionation corrections.

Neodymium. A significant increase in the temperature of the evaporation filament(s) is required to establish a strong stable NdO^+ ion beam, at which stage Sm and Eu may die off. ^{142}Ce interferes with the most abundant natural isotope and at the heavy end of the spectrum there is some overlap with Sm (triple-filament ionization) or the HREE (single filament). $^{144}Nd/^{143}Nd$ is probably the most reliable ratio but ^{146}Nd is often measured as well.

Gadolinium. Analysis of Gd^+ can potentially suffer interference from Sm^+ (154 and 152), LaO^+ (155), CeO^+ (156 and 158), PrO^+ (157) and Dy^+ (especially 160). On the other hand, with single-filament analysis of GdO^+ there is complete isobaric overlap with Yb^+. Fortunately in both cases the problem is minimized by gadolinium's low volatility, requiring filament temperatures well above those for most of the interferences, e.g., about 1700°C on single Ta. Thirlwall (1982) analysed the GdO^+ spectrum by triple-filament ionization with a deliberately reduced centre filament temperature. ^{155}Gd or ^{157}Gd are suitable as a spike isotope and ^{156}Gd and ^{158}Gd as the main natural isotopes.

Cerium. When analysed in the same chemical fraction as the above elements, Ce must be measured at the earliest possible stage, before the Nd beam grows too large, since there is no way of avoiding interference on the obvious choice of ratio: $^{142}Ce/^{140}Ce$ (142 being preferred as spike). A correction must be made by measuring ^{143}Nd (or ^{143}NdO with single-filament ionization) simultaneously with the Ce isotopes. The $^{142}Nd/^{143}Nd$ ratio in the sample can only be calculated after Nd has itself been analysed so that the data for Ce may have to be stored until then, although the Nd analysis can be carried out first if there are three chemical fractions. In the notation already defined, the following general relationship then applies:

$$\frac{(S_{i,k} - M_{i,k})}{(M_{i,k} - N_{i,k})} = \frac{(S_{j,k} - M_{j,k})}{(M_{j,k} - N_{j,k})} \tag{13.3}$$

whence $M_{j,k}$ can be calculated. In the present specific case, $i = 144$, $k = 143$, $j = 142$ and $M_{j,k}$ is the $^{142}Nd/^{143}Nd$ ratio to be calculated. The measured cerium data are then corrected thus:

$$\left(\frac{^{142}\text{Ce}}{^{140}\text{Ce}}\right)_{\text{true}} = \left(\frac{^{142}(\text{Ce} + \text{Nd})}{^{140}\text{Ce}}\right)_{\text{meas}} - \left(\frac{^{142}\text{Nd}}{^{143}\text{Nd}}\right)_{\text{calc}} \cdot \left(\frac{^{143}\text{Nd}}{^{140}\text{Ce}}\right)_{\text{meas}} \tag{13.4}$$

The magnitude of this correction is greatly reduced if Ce is collected, along with La, in a separate fraction from the second ion exchange column, but it should still be made since it can be as much as 5%. The correction is carried out identically in single-filament analysis of NdO^+, oxygen isotope splitting effects being negligible.

Lanthanum. One of the chief problems with La is the very low natural abundance of ^{138}La. Spikes prepared by isotope separation are extremely expensive beyond about 8% enrichment in this isotope. Error magnification in the term $(S_{i,k} - M_{i,k})$ is serious for overspiking, whereas if underspiked the peak is so small that it is very prone to interference. The main interferences are from ^{138}Ce and, with single-filament ionization, ^{154}Sm. If ^{140}Ce and ^{149}Sm are monitored during the measurement of $^{139}La/^{138}La$, then corrections can be applied in an analogous manner to that described above. Nevertheless, measurements should be made at various filament temperatures to check that the results are consistent with a changing proportion of interference (raising the temperature can often improve the ratio of La to Ce emission).

Ytterbium. Apart from the above mentioned overlap with GdO^+, Yb^+ is virtually free of isobars in the mass range 171—174, the least abundant ^{171}Yb making a useful spike. Ionization occurs at a rather low temperature (~1200—1300°C, similar to that for Sr on a single Ta filament) so that little interference is experienced.

Dysprosium. A significant increase in filament temperature is necessary to ionize the remaining HREE. The ion beams obtained are often small (ca. 10^{-13}A) and for very low abundances some form of signal magnification is useful in the case of single-filament analysis. There is some interference from NdO^+ (161, 162 and 164), but this usually burns off as the temperature is increased. Agreement between $^{163}Dy/^{161}Dy$ and $^{164}Dy/^{161}Dy$ is a good indication of this since there is no interference at mass 163.

Erbium. Er^+ emission tends to grow during the measurement of Dy^+, often without further increase in filament current. $^{166}Er/^{167}Er$ and $^{168}Er/^{167}Er$

should both be measured since there could be traces of interference from $^{150}NdO^+$ and $^{168}Yb^+$, respectively.

Lutetium. This is the most difficult element to analyse by this method. Because of its very low natural abundance in geological materials it provides only a weak short-lived ion beam at a high temperature (e.g., a single-filament current of about 4 A). It is prone to the smallest traces of interference on both its two isotopes, from $^{159}TbO^+$, and from $^{176}Yb^+$ and $^{160}DyO^+$, although the last two can be corrected for by monitoring ^{171}Yb and ^{161}DyO during the measurement of Lu, as explained above for Ce and La.

Accuracy and sensitivity

Given a reasonably stable ion beam, isotope ratios in the range 0.1—10.0 can be measured with a precision of 0.1% or better on modern mass spectrometers. Uncertainties concerning mass fractionation during ionization and occasionally unidentified isobaric interferences (e.g., from multiply-charged or complex ions, including MCl^+ and even hydrocarbons) sometimes degrade the accuracy of this measurement to 0.25—0.50%. Thirlwall (1982) shows that with sufficient attention to these factors Sm/Nd concentration ratios are reproducible to 0.2%. Allowing for the propogation of errors from the spike calibration procedure (which will include the precision of gravimetric, volumetric and mass-spectrometric steps) it is still reasonable to predict an overall accuracy for other REE of around 1% in the determined concentration — a figure which is supported by within-laboratory replication. This order of accuracy is not easily matched for the REE by other analytical methods, but is not always achieved even by isotope dilution. In unfavourable circumstances, where the ion beam is too short-lived to reach normal precision, or where very large (>20%) isobaric interference corrections have to be made (as is sometimes the case for La or Lu), overall accuracy may deteriorate to about 5%. Comparison of results obtained on standard samples by different laboratories generally supports these estimates, although occasionally a rather more discrepant result may emerge for just one or two elements. Such discrepancies are usually systematic in that they appear in all analysed standards. Possible causes are inadequate decomposition or errors in the preparation of good stoichiometric salts for spike calibrations, so that normalization between laboratories on the basis of such comparisons may well be justifiable. For most purposes there is more concern with intralaboratory consistency.

In absolute terms, the efficiency of thermal ionization is quite low — say only one ion per 10^2—10^4 atoms of the element, depending on chemical purity. Some elements (e.g., Eu) are inherently more easily ionized than others (HREE generally being less so). Even though only a few percent of the ions produced may actually be registered at the collector, it is nevertheless possible to obtain good ion beams of 10^{-12} A for 10 minutes or more from as little as 1 ng of the parent REE, half of which can be spike. Thus it is clear that sensitivity is controlled mostly by blank levels. Allowing for a maximum reasonable sample weight of 1 g for chemical processing and 50% recovery, determination of 10 ng each of La and Ce and rather less of the other elements should be possible, corresponding to concentrations of 0.01 ppm. Of course, at such low levels it is necessary either to accept low accuracy or else make special efforts to reduce blanks further. Using suitable pre-concentration steps, REE at concentrations down to about 10^{-6} ppm may be determined successfully (e.g. in seawater).

Summary

Mass-spectrometric isotope dilution is a determinative method for the polybaric REE which offers a combination of high selectivity, high accuracy and high sensitivity at present unmatched by any other single method. It is particularly suitable for determining chondrite-normalized REE patterns in geological materials since both light and heavy ends are well established (Ce, Nd and Yb among the most reliable elements) and there is optimum control on Eu anomalies (Sm-Eu-Gd). The lack of capability for Pr and hence reduced control over possible Ce anomalies is perhaps its one deficiency in this respect.

The chief disadvantages of the method are not only the need for expensive equipment (a suitable solid-source mass spectrometer currently costs over U.K. £100,000) and skilled operators (although computer-control is gradually replacing these), but also the time and effort required. The chemical separation described in Fig. 13.2 takes 4—5 working days, though the time spent is greatly reduced if only the primary group separation is made. The time required for mass spectrometry averages around 1—2 hours per fraction analysed so that the limit here is closer to two full REE analyses per day. Automation and multiple collection could improve the throughput but it is as well to remember that mass-spectrometry laboratories generally have other major commitments (e.g., geochronology). For these reasons it is probable that as new, rapid and accurate analytical methods are developed (especially, for example, ICP emission spectrometry) isotope dilution will be reserved for cases especially requiring its high accuracy and sensitivity, as well as for the ultimate calibration of other methods.

13.4. Inductively coupled plasma emission spectrometry

The development of plasma sources for atomic emission spectroscopy (Greenfield et al., 1964; Wendt and Fassel, 1965) has opened up a new and important avenue in analytical geochemistry. Inductively coupled plasma

(ICP) emission spectrometry shows great promise of becoming a routine method for simultaneous multi-element analysis which is both relatively cheap and rapid.

As presently developed the method uses a solution of the sample which is passed, as an aerosol from a nebulizer, into an electrodeless argon plasma. The plasma is sustained by inductive coupling to radio frequency magnetic fields, and is connected with a multi-channel spectrometer. The injected aerosol remains in the plasma for a relatively long period (~2 ms) at a temperature of between 6000 and 8000 K; these conditions leading to a very high degree of atomization and the excitation of numerous atomic and ionic spectral lines. The spectrometer consists of a diffraction grating and exit slits for particular spectral lines, each with its own photomultiplier. A roving detector can also be used for scanning or diagnostic purposes. The signals from the photomultipliers may be read directly (as intensities) or fed into an appropriate data handling assembly which will relate the spectrum from the sample to stored linear calibration lines and thereby calculate the concentration levels. Some spectrometers have only one detector which is moved automatically to pre-set positions in order to perform a sequential analysis. Alternatively, photographic recording can be employed (e.g., Brenner et al., 1981). Fassel (1978) gives a concise review of the method and the history of its development.

ICP spectrometry has been shown (e.g., Walsh and Howie, 1980) to be a technique of good sensitivity and precision for analysis of major and trace constituents of silicate rocks or minerals. Although the method requires the dissolution of the sample it is none the less relatively rapid since, once the sample solution is prepared, only two minutes or less are required for the simultaneous determination of many elements. Sample preparation may be carried out by many of the standard techniques (e.g., fusion with $LiBO_2$, or attack with HF acid). A comprehensive scheme for the determination of the REE in rocks and minerals has been published by Walsh et al. (1981). Attack by HF and $HClO_4$ acids on about 0.5 g of powdered samples, followed by dissolution in HCl, was used. Any resistant minerals can be fused with NaOH or KHF_2. Another scheme (Brenner et al., 1981), uses various preparation techniques, including fusion with Na_2O_2.

To ensure good sensitivity in the analysis for trace quantities of REE it is recommended that the elements are separated as a group from most other trace elements and major elements. Walsh et al. (1981) used the exchange resin Dowex AG 50W-8, which quantitatively holds the REE. The elements are eluted using 500 ml of 4 M HCl, and the resultant solution is evaporated to dryness, followed by redissolution in 5 ml of 10% HCl. This separation procedure is simple and rapid.

The operating conditions of the spectrometer, which include power setting, gas flow and sample-solution uptake rates, are set at positions that are optimum for the determination of all elements to be analysed with each injection of the sample solution, whether by simultaneous or sequential recording methods. It is not possible to make one list of recommended settings because of constructional variations shown by different instruments; the effects of different settings on REE spectral line intensity have been discussed by Brenner et al. (1981). The measurement of REE in a wide range of rocks and minerals can be achieved by using the suggested spectral lines listed in Table 13.6, although it should be remembered that other lines might also be suitable. Listings of spectral lines may be found in Broekaert et al. (1979), Parsons et al. (1980) and Boumans (1981). Spectral interferences are absent or slight if the REE are separated as a group (see Table 13.6). Church (1981) performed multi-element analysis, including the elements La, Ce, Nd, Sm, Eu, Gd and Yb, on 54 geochemical reference samples using ICP spectrometry. He gives details of lines used (some of them different from those in Table 13.6), detection limits and some interferences.

ICP spectrometry is a sensitive technique especially, in the case of the REE, if the elements are separated from other elements as described above. The method is particularly sensitive in the determination of Nd, Eu, Dy, Ho, Yb and Lu where levels of about 0.5 ppm may be readily and accurately measured. Most of the remaining REE may be determined down to about 1 or 2 ppm except Ce, Tb and Tm which do not have very sensitive spectral lines. Tm is determinable only with difficulty by ICP spectrometry.

ICP spectrometry is being rapidly developed as an analytical method in

TABLE 13.6

Suggested spectral lines for use in analysis of silicates by ICP spectrometry

Element	Line (nm)	Comments
La	398.85	the more sensitive 394.91 nm line is subject to interference
Ce	418.66	
Pr	422.30	small interferences from Ce, Sr, and Ca; correction necessary
Nd	406.11	
Sm	359.26	small interferences from Nd and Gd; correction necessary.
	373.92	less sensitive than 359.26; small interference from Pr; correction necessary
Eu	381.97	trace interference from Nd; correction necessary
Gd	335.05	small interference from Ca, correction may be required
Tb	350.92	
Dy	353.17	
Ho	345.60	
Er	390.63	
	369.27	
Tm	313.13	
Yb	328.94	
Lu	261.54	

Based principally on Walsh et al. (1981). Other lines are also likely to be suitable; see text.

geochemistry, and the best conditions for analysis are being actively investigated. It is inappropriate, therefore, for firm recommendations on operating conditions to be made at this stage. The intending analyst is advised to maintain a steady surveillance of the analytical literature and to be prepared to search for the optimum conditions for his particular analytical task. He will soon appreciate that the method is rapid and sensitive, and that correction procedures for any spectral interferences are usually straightforward. These factors lead to low-cost analysis. The disadvantages are that dissolution of the sample is required, element separations are needed, and relatively large sample weights (i.e., about 0.5 g) are used. This last disadvantage may be overcome in the future with the development of controlled volatilization of the rock or mineral sample by using a laser beam.

13.5. Other methods

Other analytical techniques which should be mentioned are: atomic absorption spectrophotometry, spark source mass spectrometry, X-ray fluorescence, and electron probe microanalysis. These are discussed briefly in turn. Mention is also made of the use of radioisotopes in the experimental study of element distribution.

In relation to the methods described earlier in this chapter, *atomic absorption spectrophotometry* (AAS) has not proved very successful in the analysis for REE occurring at trace levels, especially since the reliable determination of Ce is virtually impossible. A solution of the rock or mineral sample is aspirated into a nitrous oxide/acetylene flame which, at the prevailing temperature, produces significant ionization of the REE with resultant loss of response. It is necessary to add an easily ionizable element such as Na or K (usually as a chloride to a concentration of about 0.1% alkali metal) to suppress this effect. Furthermore, the best instrumental conditions are different from one element to another.

There are few papers on the application of AAS to REE geochemistry. Van Loon et al. (1971) and Ooghe and Verbeek (1974) have assessed the method as applied to rare earth ores and REE-rich silicates. It is clear that it is more sensitive for the determination of HREE (but excluding Lu) than for the LREE, sensitivities down to about 1 ppm being achieved for some elements. Operating conditions are given by the above authors. Price (1979) gives a modern account of the method and tabulates operating conditions, absorption, emission lines for many elements including the REE.

Spark-source mass spectrometry combines some of the principles of emission spectroscopy with those of mass spectrometry. The powdered sample is carefully mixed with graphite containing Lu as an internal standard, and formed by pressure into two electrodes. These are sparked at high voltage (about 25 kV) under a high vacuum (10^{-7} torr). The ions from the sample are focussed and passed through a mass spectrometer, the mass spectrum (usually from Be to U) being recorded on a photographic plate (also under vacuum). The density of each line on the developed plate is read by means of a microphotometer. The method requires a skilled operator and great care in sample preparation. In the right hands, an accuracy of ±5% and typical sensitivities of 0.1—0.5 ppm for the REE can be achieved. Taylor and Gorton (1977) have listed additional advantages of the method as follows: (a) multi-element capability (30 elements simultaneously), (b) small sample weights, down to about 10 mg, and (c) rapid acquisition of data (4 photoplates per day). Disadvantages include the high capital cost of the equipment and the need for constant surveillance during analytical runs. Despite the advantages of the technique there are very few spark source mass spectrometers operating in geochemical laboratories today; with developments in ICP spectrometry as well as activation analysis and isotope dilution, there is unlikely to be a growth in the use of this technique. A good account of the method is given by Nicholls and Wood (1975), there are useful papers by Taylor (1965, 1971) and Taylor and Gorton (1977), and there is a review on the application to analysis for REE by Taylor (1979). Ure and Bacon (1978) used the method to determine concentrations of 50 elements, including some REE, in soils and rocks, with a precision of ±10—15%.

X-ray fluorescence spectrometry is a widely-used technique for the rapid determination of both major and trace elements. It is ideal for high throughput requirements, but since trace elements are usually best analysed in a pressed pellet made from about 10 g of the powdered sample, its use for separated minerals or other small samples is extremely limited. The method depends on the quantitative measurements of specific X-rays emitted during electronic transitions within atoms excited by irradiation with primary X-rays. The latter are generated within a source such as a Coolidge tube with a metal target (Au is best suited for the REE). A very thorough introduction to the theory and practice of the method, together with useful notes on procedure, are given by Norrish and Chappell (1977) and in view of the limitations of the method applied to REE, only a brief summary is given here.

Fluorescent X-rays are emitted as lines of discrete wavelength controlled by the energy of each internal transition, so that each element has its own characteristic spectrum of lines. The most intense are the short-wavelength K-series lines (shorter than 0.1 nm for Kr and heavier elements), associated with high-energy transitions into the electronic shell closest to the nucleus. The series consists of several lines of decreasing intensity, labelled $K\alpha_1$, $K\alpha_2$, $K\beta_1$, etc., according to the origin and ground-state energy of the most probable electrons involved. A second series of less intense but more numerous lines occurs for transitions into the second shell (L-series) and so on. Excitation potentials for the K-series are the highest and for the REE they are too high to be achieved by a significant proportion of the primary

X-rays, so that analysis must be based on the L-series lines. Unfortunately these longer wavelength lines (0.15—0.25 nm for the REE) are compressed into a spectral region where they are interfered by minor L-lines from Cs, Ba and each other, as well as by the intense K-lines of some lighter elements such as Ca, Ti and V. Tables of wavelengths for all the elements are given by Norrish and Chappell, but it must be remembered that the resolution of the detector (the fluorescent spectrum is dispersed by diffraction from the surface of an artificial crystal such as LiF) is inadequate to separate lines such as $K\alpha_{1,2}$ and $K\beta_{1,3}$. In practice one is virtually forced to use $L\alpha_1$ for La and Pr and $L\beta_1$ lines for Ce and Nd (see Norrish and Chappell for other analytical conditions), and limits of detection are correspondingly poor — ca. 2 ppm and 4 ppm, respectively. There is little possibility for analysis of any heavier REE than these without prior chemical separation or concentration. One separation scheme devised by Eby (1972) uses 0.1—1 g sample weight, and an ion-exchange procedure. A precision of ±10% to ±20% is claimed for REE concentrations down to about 2 ppm.

X-ray fluorescence is a comparative method, calibrated against known standards, and even here correction must be made for the variable self-absorption of fluorescent X-rays by the samples and standards. Since REE concentrations in whole rocks are generally only a few times the limit of detection (except for REE-rich alkaline rocks), accuracy is often poor, say 10—20%. However, if routine analyses are available, La and Ce determined by this method can be a very useful guide to the use of more sophisticated techniques (e.g., in the estimation of necessary irradiation and other conditions for INAA or spike/sample ratios for isotope dilution).

Electron probe microanalysis (Reed, 1975) offers the capability of quantitatively analysing small areas (~10 μm diameter) of cut and polished mineral, metal or glass surfaces. Provided that a good accuracy is not required it should be possible to determine concentrations down to about 0.1 wt% of each REE, although difficulties will be encountered with Eu and some of the HREE. For accuracy of about ±5% or better it is necessary to have a concentration of around 1% or more of each REE. Interferences can be a serious problem and it is necessary to make a careful choice of peaks using wavelength dispersion. Exley (1980) in analysing allanites and some other REE-rich minerals recommends that $L\beta_1$ peaks are used for the analysis of Pr, Nd, Sm, Gd and Dy, and $L\alpha_1$ for La, Ce, Er and Yb. Interference from Fe will prevent the accurate analysis of Er in iron-rich specimens, while interference with the determination of Ce concentrations from high Ti concentrations can be avoided by using the $L\beta_1$ peak and making a correction for slight interference from Nd. Åmli and Griffin (1975) have determined empirical correction factors for use in the analysis of REE minerals, while Smith and Reed (1981) have developed a general procedure for the calculation of background corrections in wavelength-dispersive analysis, which gives improved detection limits. Their method was applied successfully to part of the spectrum from a specimen of monazite.

The advantages of the electron microprobe method include the ability to detect compositional zoning in mineral grains, and the ease and speed of analysis. The disadvantages for geochemical studies include the poor limit of detection, the difficulty of determining Eu (results for which can be critical in many geochemical interpretations), and the scarcity of good REE microprobe standards. This last disadvantage may soon be overcome since there are now several laboratories developing the application of the microprobe to REE geochemistry. Drake and Weill (1972) have produced some synthetic glasses suitable for determination of Y and REE. However, analyses of some REE-rich mineral grains by both microprobe and INAA show good agreement for La, Ce, Nd and Sm (Henderson, 1980; Brooks et al., 1981).

The study of REE distributions, at trace levels, in experimental mineralogy can be made simple, from an analytical viewpoint, by the use of radiotracers and beta-track *autoradiography* (Mysen and Seitz, 1975). Radioactive (beta-emitting) REE compounds (usually as chlorides or nitrates) are added, usually in very small quantities, to the materials to be used in the experiment, and homogenized by careful grinding and mixing. After the experiment, the charge is mounted, cut and polished, and a nuclear emulsion (e.g., Kodak 4463 electron image film, or Ilford K5 plates) placed in close contact with the surface where it is left for an appropriate time (the duration depends on the concentration and half-life of the added radiotracer). The passages of the beta particles are registered in the emulsion of the film and, after development, appear as small dots, the density of which is proportional to the concentration of the radionuclide.

A general review of autoradiography has been given by Bowie (1977). Useful papers dealing with the application of the method to problems in experimental mineralogy include Mysen and Seitz (1975), Harrison and Wood (1980), and Harrison (1981). The method can be applied readily to all REE except Pr, Nd, Gd and Yb.

13.6. Concluding statement

The ease of analysis for individual REE in a wide range of rocks, minerals and waters has increased considerably during the last decade through the advent of new techniques, such as ICP spectrometry, and the development of more established methods. Emphasis has been placed in this chapter on those techniques which have been shown capable of producing reliable results on a routine basis for many matrices and many REE at an accuracy of about 6% or better, namely neutron activation analysis, isotope dilution mass spectrometry, and ICP spectrometry. Each of these three methods requires its own special laboratory and trained staff; such facilities might not be available to all researchers, but much can be achieved with the other methods.

It is not appropriate to put the various methods in some order of merit — any facility is likely to be used for more than REE analysis, and many researchers will use the existing facilities rather than attempt to establish new methods and laboratories. It can be said, however, that where the highest accuracy is required, then mass-spectrometric isotope dilution analysis should be used. Instrumental neutron activation analysis can be particularly useful for giving a well-defined, chondrite-normalized REE pattern, and ICP spectrometry is probably the best technique when analyses for many elements, including the REE, are needed. X-ray fluorescence is useful for reconnaissance work provided that data on the HREE are not a prerequisite.

Future and recently-developed techniques (e.g., particle-induced X-ray emission, PIXE) are bound to complement and overlap the methods described in this chapter but we are confident that the three principal methods given here will remain the standard techniques for many years to come.

Acknowledgements

We are grateful to Chris Hawkesworth, Phil Potts, Nicholas Walsh, John Whitley and Terry Williams for helpful reviews of this chapter.

References

Amiel, S. (Editor), 1981. *Studies in Analytical Chemistry, 3. Nondestructive Activation Analysis.*, Elsevier, Amsterdam, 369 pp.

Åmli, R. and Griffin, W.L., 1975. Microprobe analysis of REE minerals using empirical correction factors. *Am. Mineral.*, 60: 599—606.

Arth, J.G. and Hanson, G.N., 1975. Geochemistry and origin of the early Precambrian crust of northeastern Minnesota. *Geochim. Cosmochim. Acta*, 39: 325—362.

Baedecker, P.A., Rowe, J.J. and Steinnes, E., 1977. Application of epithermal neutron activation in multielement analysis of silicate rocks employing both coaxial Ge(Li) and low energy photon detector systems. *J. Radioanal. Chem.*, 40: 115—146.

Becker, R., Buchtela, K., Grass, F., Kittl, R. and Miller, G., 1975. Abtrennung und radiochemische Bestimmung geringer Mengen seltener Erden in extraterrestrischen Material. *Z. Anal. Chem.*, 274: 1—7.

Boelrijk, N.A.I.M., 1967. A general formula for "double" isotope dilution analysis. *Chem. Geol.*, 3: 323—325.

Borley, G.D. and Rogers, N., 1979. Comparison of rare-earth element data obtained by neutron activation analysis using international rock and multielement standards. *Geostand. Newsl.*, 3: 89—92.

Boumans, P.W.J.M., 1981. *Line Coincidence Tables for Inductively Coupled Plasma Atomic Emission Spectrometry.* Pergamon, Oxford, 951 pp.

Bowie, S.H.U., 1977. Radiographic techniques. In: J. Zussman (Editor) *Physical Methods in Determinative Mineralogy.* Academic Press, New York, N.Y., 2nd ed., pp. 677—687.

Brenner, I.B., Watson, A.E., Steele, T.W., Jones, E.A. and Goncalves, M., 1981. Application of argon-nitrogen inductively-coupled radiofrequency plasma (ICP) to the analysis of geological and related materials for their rare earth contents. *Spectrochim. Acta*, 36B: 785—797.

Broekaert, J.A.C., Leis, F. and Laqua, K., 1979. Application of an inductively coupled plasma to the emission spectroscopic determination of rare earths in mineralogical samples. *Spectrochim. Acta.*, 34B: 73—84.

Brooks, C.K., Henderson, P. and Rønsbo, J.G., 1981. Rare-earth partition between allanite and glass in the obsidian of Sandy Braes, Northern Ireland. *Mineral. Mag.*, 44: 157—160.

Brunfelt, A.O., Roelandts, I. and Steinnes, E., 1977. Some new methods for determination of rare-earth elements in geological materials using thermal- and epithermal-neutron activation. *J. Radioanal. Chem.* 38: 451—459.

Chou, C.-L., 1980. Radiochemical neutron activation analysis. In: G.K. Muecke (Editor), Neutron Activation Analysis in the Geosciences. *Mineral. Assoc. Can., Short Course Handb.*, 5: 133—166.

Church, C.E., 1981. Multi-element analysis of fifty-four geochemical reference samples using inductively coupled plasma — atomic emission spectrometry. *Geostand. Newsl.*, 5: 133—160.

Croudace, I.W., 1980. The use of pre-irradiation group separations with neutron activation analysis for the determination of the rare-earths in silicate rocks. *J. Radioanal. Chem.*, 59: 323—330.

Csajka, M., 1973. Fast radiochemical separation of rare-earths from rock samples by retention on lanthanum oxalate. *Radiochem. Radioanalyt. Lett.*, 13: 151—159.

De Soete, D., Gijbels, R. and Hoste, J., 1972. *Neutron Activation Analysis.* Wiley-Interscience, New York, N.Y., 836 pp.

Drake, M.J. and Weill, D.F., 1972. New rare-earth element standards for electron microprobe analysis. *Chem. Geol.*, 10: 179—181.

Duffield, J. and Gilmore, G.R., 1979. An optimum method for the determination of rare earth elements by neutron activation analysis. *J. Radioanal. Chem.*, 48: 135—145.

Eby, G.N., 1972. Determination of rare-earth, yttrium and scandium abundances in rocks and minerals by an ion-exchange—X-ray fluorescence procedure. *Anal. Chem.*, 44: 2137—2143.

Elek, A., Pemeczki, G., Szabo, E. and Dogadkin, N.N., 1973. Determination of twelve rare-earth metals in the Luna-16 lunar sample by neutron-activation analysis. *Radiochem. Radioanalyt. Lett.*, 15: 123—133.

Erdtmann, G. and Soyka, W., 1979. *The Gamma Rays of the Radionuclides.* Verlag Chemie, Weinheim, 862 pp.

Eugster, O., Tera, F., Burnett, D.S. and Wasserburg, G.J., 1970. The isotopic composition of gadolinium and neutron capture effects in some meteorites. *J. Geophys. Res.*, 75: 2753—2768.

Exley, R.A., 1980. Microprobe studies of REE-rich accessory minerals: implications for Skye granite petrogenesis and REE mobility in hydrothermal systems. *Earth Planet. Sci. Lett.*, 48: 97—110.

Fassel, V.A., 1978. Quantitative elemental analyses by plasma emission spectroscopy. *Science*, 202: 183—191.

Gast, P.W., 1968. Trace element fractionation and the origin of tholeiitic and alkaline magma types. *Geochim. Cosmochim. Acta*, 32: 1057—1086.

Gibson, I.L. and Jagam, P., 1980. Instrumental neutron activation analysis of rocks and minerals. In: G.K. Muecke (Editor) *Neutron Activation Analysis in the Geosciences. Mineral. Assoc. Can., Short Course Handb.*, 5: 109—131.

Greenfield, S., Jones, I.L. and Berry, C.T., 1964. High pressure plasmas as spectroscopic emission sources. *Analyst*, 89: 713—720.

Hanson, G.N., 1976. Rare earth element analysis by isotope dilution. In: *Proceedings 7th*

IMR Symposium, Accuracy in Trace Analysis: Sampling, Sample Handling, and Analysis. U.S. Natl. Bur. Stand., Spec. Publ., 422: 937—949.

Harrison, W.J., 1981. Partition coefficients for REE between garnets and liquids: implications of non-Henry's Law behaviour for models of basalt origin and evolution. *Geochim. Cosmochim. Acta*, 45: 1529—1544.

Harrison, W.J. and Wood, B.J., 1980. An experimental investigation of the partitioning of REE between garnet and liquid with reference to the role of defect equilibria. *Contrib. Mineral. Petrol.*, 72: 145—155.

Henderson, P., 1980. Rare earth partition between sphene, apatite, and other coexisting minerals of the Kangerdlugssuaq intrusion, E. Greenland. *Contrib. Mineral. Petrol.*, 27: 81—85.

Henderson, P. and Williams, C.T., 1981. Application of intrinsic Ge detectors to the instrumental neutron activation analysis for rare earth elements in rocks and minerals. *J. Radioanal. Chem.*, 67: 445—452.

Hirose, A., Kobori, K. and Ishii, D., 1978. Determination of rare earth elements and heavy metals in river water by preconcentration in Chelax 100 and neutron activation. *Anal. Chim. Acta*, 97: 303—310.

Hooker, P.J., O'Nions, R.K. and Pankhurst, R.J., 1975. Determination of rare-earth elements in U.S.G.S. standard rocks by mixed-solvent exchange and mass-spectrometric isotope dilution. *Chem. Geol.*, 16: 189—196.

Jagam, P. and Muecke, G.K., 1980. Instrumentation in neutron activation analysis. In: G.K. Muecke (Editor) Neutron Activation Analysis in the Geosciences. *Mineral. Assoc. Can., Short Course Handb.*, 5: 73—108.

Kay, R.W. and Gast, P.W., 1973. The rare earth content and origin of alkali-rich basalts. *J. Geol.*, 81: 653—682.

Krugers, J. (Editor), 1973. *Instrumentation in Applied Nuclear Chemistry*. Plenum Press, New York, N.Y., 383 pp.

Muecke, G.K. (Editor), 1980. Neutron Activation Analysis in the Geosciences. *Mineral. Assoc. Can., Short Course Handb.*, 5: 279 pp.

Mysen, B.O. and Seitz, M.G., 1975. Trace element partitioning determined by beta track mapping: an experimental study using carbon and samarium as examples. *J. Geophys. Res.*, 80: 2627—2635.

Nicholls, G.D. and Wood, M., 1975. Elemental analysis using mass spectrographic techniques. In: A.W. Nicol (Editor), *Physiochemical Methods of Mineral Analysis*. Plenum Press, New York, N.Y., pp. 195—229.

Norrish, K. and Chappell, B.W., 1977. X-ray fluorescence spectrometry. In: J. Zussman (Editor), *Physical Methods in Determinative Mineralogy*. Academic Press, New York, N.Y., 2nd ed., pp. 201—272.

O'Nions, R.K. and Grønvold, K., 1973. Petrogenetic relationships of acid and basic rocks in Iceland: Sr isotopes and rare-earth elements in late and postglacial volcanics. *Earth Planet Sci. Lett.*, 19: 397—409.

Ooghe, W. and Verbeek, F., 1974. Atomic absorption spectrometry of the lanthanides in minerals and ores. *Anal. Chim. Acta*, 73: 87—95.

Parsons, M.L., Forster, A. and Anderson, D., 1980. *An Atlas of Spectral Interferences in ICP Spectroscopy*. Plenum Press, New York, N.Y., 654 pp.

Potts, M.J., Early, T.O. and Hermann, A.G., 1973. Determination of rare earth element distribution patterns in rocks and minerals by neutron activation analysis. *Z. Anal. Chem.*, 263: 97—100.

Potts, P.J., Thorpe, O.W. and Watson, J.S., 1981. Determination of the rare-earth element abundances in 29 international rock standards by instrumental neutron activation analysis: a critical appraisal of calibration errors. *Chem. Geol.*, 34: 331—352.

Price, W.J., 1979. *Spectrochemical Analysis by Atomic Absorption*. Heyden, London, 392 pp.

Randa, Z., 1976. Routine neutron-activation determination of rare-earth metals in rocks and similar materials by pre-activation group separation. *Radiochem. Radioanal. Lett.*, 24: 177—188.

Reed, S.J.B., 1975. *Electron Microprobe Analysis.* Cambridge University Press, Cambridge, 400 pp.

Routti, J.T. and Pressin, S.G., 1969. Photopeak method for the computer analysis of gamma-ray spectra from semiconductor detectors. *Nucl. Instrum. Meth.*, 72: 125—142.

Schnetzler, C.C. and Philpotts, J.A., 1970. Partition coefficients of some rare-earth elements between igneous matrix material and rock-forming mineral phenocrysts, II. *Geochim. Cosmochim. Acta*, 34: 331—340.

Schock, H.H., 1977. Comparison of a coaxial Ge(Li) and a planar Ge detector in instrumental neutron activation analysis of geologic materials. *J. Radioanal. Chem.*, 36: 557—564.

Schonfeld, E., 1966. ALPHA — a computer program for the determination of radioisotopes by least-squares resolution of the gamma-ray spectra. *Nucl. Instrum. Meth.*, 42: 213—218.

Shimizu, N., 1974. An isotope dilution technique for analysis of the rare earth elements. *Carnegie Inst. Washington, Yearb.*, 73: 941—943.

Smet, T., Hertogen, J., Gijbels, R. and Hoste, J., 1978. A group separation scheme for radiochemical neutron activation analysis for 24 trace elements in rocks and minerals. *Anal. Chim. Acta*, 101: 45—62.

Smith, D.G.W. and Reed, S.J.B., 1981. The calculation of background in wavelength-dispersive electron microprobe analysis. *X-ray Spectrom.*, 10: 198—202.

Steinnes, E., 1971. Epithermal neutron activation analysis of geological material. In: A.D. Brunfelt and E. Steinnes (Editors), *Activation Analysis in Geochemistry and Cosmochemistry*. Universitetsforlaget, Oslo, pp. 113—128.

Sun, S.S. and Hanson, G.N., 1975. Origin of Ross Island basanitoids and limitations upon the heterogeneity of mantle sources for alkali basalts and nephelinites. *Contrib. Mineral. Petrol.*, 52: 77—106.

Taylor, S.R., 1965. Geochemical analysis by spark source mass-spectrography. *Geochim. Cosmochim. Acta*, 29: 1243—1262.

Taylor, S.R., 1971. Geochemical application of spark source mass spectrography, II. Photoplate data processing. *Geochim. Cosmochim. Acta*, 35: 1187—1196.

Taylor, S.R., 1979. Trace element analysis of rare earth elements by spark source mass spectrometry. In: K.A. Gschneidner, Jr. and L. Eyring (Editors), *Handbook on the Physics and Chemistry of Rare Earths, 4.* North-Holland, Amsterdam, pp. 359—376.

Taylor, S.R. and Gorton, M.P., 1977. Geochemical application of spark source mass spectrography, III. Element sensitivity, precision and accuracy. *Geochim. Cosmochim. Acta*, 41: 1375—1380.

Thirlwall, M.F., 1982. A triple-filament method for rapid and precise analysis of rare-earth elements by isotope dilution. *Chem. Geol.*, 35: 155—166.

Ure, A.M. and Bacon, J.R., 1978. Comprehensive analysis of soils and rocks by spark-source mass spectrometry. *Analyst*, 103: 807—822.

Van Loon, J.C., Galbraith, J.H. and Aarden, H.M., 1971. The determination of yttrium, europium, terbium, dysprosium, holmium, erbium, thulium, ytterbium and lutetium in minerals by atomic absorption spectrometry. *Analyst*, 96: 47—50.

Voldet, P. and Haerdi, W., 1976. Determination of europium and dysprosium in rocks by neutron activation and high-resolution X-ray spectrometry. *Anal. Chim. Acta*, 87: 227—231.

Voldet, P. and Haerdi, W., 1978. Determination of rare-earth elements in rocks by neutron activation followed by high-resolution X-ray spectrometry or γ-spectrometry. *Anal. Chim. Acta*, 97: 185—189.

Walsh, J.N. and Howie, R.A., 1980. An evaluation of the performance of an inductively coupled plasma source spectrometer for the determination of the major and trace constituents of silicate rocks and minerals. *Mineral. Mag.*, 43: 967—974.

Walsh, J.N., Buckley, F. and Barker, J., 1981. The simultaneous determination of the rare-earth elements in rocks using inductively coupled plasma source spectrometry. *Chem. Geol.*, 33: 141—153.

Watterson, J.I.W., 1978. The instrumental neutron-activation analysis of granites from the Bushveld complex. *Natl. Inst. Metall. (S. Afr.) Rep.*, 1917: 67 pp.

Webster, R.K., 1960. Mass spectrometric isotope dilution analysis. In: A.A. Smales and L.R. Wager (Editors), *Methods in Geochemistry*. Interscience, New York, N.Y., pp. 202—246.

Wendt, R.H. and Fassel, V.A., 1965. Induction-coupled plasma spectrometric excitation source. *Anal. Chem.*, 37: 920—922.

SUBJECT INDEX

Abee meteorite, 89, 93
Abukumalite, 46
Achondrites, 94–103
Acid washing of minerals, 185
Activity coefficient, 66
Adams Mine, Ontario, Canada, 303
Aeschynite, 36, 41
Aeschynite-priorite series, 41
Afar Rift, Ethiopia, 261
African ore deposits, 427–428, 429, 434, 435
—, production of, 444, 445, 448
Agardite, 36, 54
Aggregates, 71
Agrellite, 36, 51
Alkali basalt group, 238, 256, 257–258
Alkali feldspar, 27
Alkaline igneous rocks, 238, 245, 254–262
—, as rare earth source, 427–429
Allan Hills meteorite, 102, 103
Allanite, 14, 18, 19, 24, 26, 27, 36, 45
Allende meteorite, 65, 71–80, 83–86
Alloys, 452–453
Alps, 156–158
Alpine peridotites, 153–166, 195
Altervalent substitution, 17, 22
Aluminium-26, 74
Amphibole, 24, 26, 27
— in inclusions, 182–183
Analytical standards, 470–471
Analytical techniques, 414, 467–495
—, accuracy of, 475, 487
—, precision of, 475
—, sensitivity of, 477, 487–488
Ancylite, 36, 39
Andesites, 276–281
— in Chile, 404
Angra dos Reis meteorite, 99–100, 379, 382
Anhydrite, 303
"Anomalous" ridge segments, 210, 213
Anorthosites, 277, 281–285
Apatite, 18, 24, 25, 26, 27, 51
—, biogenic, 355–356
— deposits, 435
—, marine, 360
— ore, 424, 425, 435
Apatite group, 51
Apatite-rich inclusions, 183
Archaean rocks, dating of, 380–382
Ashcroftine, 36, 46
Assab, Ethiopia, 177–181, 184
Ataq, South Yemen, 182, 398
Atlanta meteorite, 93
Atlantic fracture zones, 172–173
Atlantic Ocean
— basalts, 209–214, 225–227
— floor, 170–175
— water, 353, 354
— —, isotopes in, 407
Atomic absorption spectrophotometry, 491
Atomic weights, 3
Australian ore deposits, 431–432, 435, 436
—, production of, 444, 445, 448
Autoradiography, 494
Azores Islands, 215, 217–218

Barberton greenstone belt, Africa, 253–254, 290
Barite, 303, 358, 359
Basalt, 126–127, 131, 238, 385
—, back-arc basin, 262–267
—, continental, 262–267
— heterogeneity, 146
—, island arc, 262–267
—, ocean floor, 205–230, 403
—, ocean island, 214–220, 384
—, ocean plateau, 220–222
—, ocean ridge, 208–214, 386
—, submarine weathering of, 348–349
Bastnäsite, 36, 40, 41
— ore, 424, 425, 427

Bastnäsite (*contd.*)
— production, 424, 444, 445, 447
Bauxite, 347
Bay of Islands, Newfoundland, 167, 170, 383, 387
Belovite, 36, 52
Ben Ghnema batholith, North Africa, 298
Beneficiation, 438–442
Beni Bouchera, Morocco, 156–161, 165
Betafite — *see* Pyrochlore group
Biogenic apatite, 355–356
Bishop Tuff, California, U.S.A., 147, 288, 300
Blomstrandine, 42
Borralan complex, Scotland, 249, 333–334
Bovedy meteorite, 86
Braitschite, 36, 45
Brannerite, 36, 42
— ore, 424
Brazilian ore deposits, 428, 432
—, production of, 444, 445, 448
Britholite, 36, 46, 428
Brocenite, 43
Brockite, 36, 52
Brokeoff volcano, California, U.S.A., 279
Bubble memories, 459

Ca,Al-rich inclusions, 65, 71–86
Calcite, 301–302
Calkinsite, 36, 39
Callander Bay, Canada, 245, 247, 248
Campo del Cielo meteorite, 104–105
Canadian ore deposits, 434
—, production of, 444
Cañon Diablo meteorite, 105
Cappelenite, 36, 46
Captains Bay pluton, Alaska, U.S.A., 291
Carbocernaite, 36, 39
Carbonatites, 238, 243–249
—, petrogenesis of, 245
— as rare earth source, 427–429
Carlsberg Ridge, Indian Ocean, 172
Caryocerite — *see* Melanocerite
Catalysts, 449–450
Caysichite, 36, 51
Celadonite, 335
Cenosite — *see* Kainosite
Ceramic manufacture, 455–457
Cerianite, 36, 42
Cerite, 36, 41, 46
Cerium
— anomaly, 9
— — in metalliferous sediments, 360
— — in meteorites, 81, 84–86, 87–88
—, atomic weight, 3
— in chondrites, 8, 10
— in Earth, 4–6
— in electrodes, 456
—, electronic configuration, 3
— in glass manufacture, 456
—, ionic radius, 16
— isotopes, 7
— in manganese nodules, 364
—, mass spectrometry for, 485
—, oxidation states, 12, 352
— in seawater, 352
— in solar system, 4, 5
— in Sun, 5
—, *see also* Rare earth elements
Cerotungstite, 36, 45
Chassignites, 101–102
Chassigny meteorite, 102
Chert, 353, 368
Chevkinite, 36, 46
Chinese ore deposits, 428, 434
Chlorite, 335, 336–337
Chondrite meteorites, 64, 71–94
—, analyses of, 89–94
—, anomalous REE patterns in, 92–94
—, carbonaceous, 71, 92
— classification, 64
Chondrite normalization, 8–11, 90–91
Chondritic uniform reservoir, 381–382
Chondrules, 70, 71, 86–89
Chukhrovite, 36, 54
CHUR, *see* Chondritic uniform reservoir
Churchite, 36, 52
Clays, 349, 357, 366
Colorado mineral belt, U.S.A., 302
Colouring agents, 455–457
Complexes, 15, 304–305, 324, 328, 331–333
—, effect of pH on, 328–330
Condensate minerals, 67
Condensation theory, 65–70
Contamination
—, crustal, 390–396
— in kimberlite genesis, 243
Continental tholeiites, 238, 262–267
Cordylite, 36, 40, 41
Core, composition of, 154
Craters of the Moon National Monument, U.S.A., lavas, 131
Crust
—, composition of, 4–6, 154
—, evolution of, 408, 411
Crustal contamination, 390–396
Crystal zoning, 133–138

Data presentation, 6
Davidite, 36, 42
Decay constant, 376
Deep Sea Drilling Project, 207, 220
Destructive plate margins, 400–405
Diagenesis, 367–369
Diogenites, 97–99
Diorites, 277, 285–293
Dispersed element, 3
Distribution coefficient, 13, 18–28, 34, 118–122, 135
—, olivine/melt, 168
—, solid/gas, 66–69
Dolomite, 301–302
Dome Mine, Ontario, Canada, 302, 303
Donnayite, 36, 39
Doverite — *see* Synchysite
Dreiser Weiher, Germany, 177–179, 181, 183
Dunite, 154, 163–164, 166, 193
Dysanalyte, 36, 43
—, *see also* Perovskite group
Dysprosium
—, atomic weight, 3
— in chondrites, 8, 10
— in Earth, 4–6
—, electronic configuration of, 3
—, ionic radius, 16
— isotopes, 7
—, mass spectrometry for, 486
— in solar system, 4, 5
— in Sun, 5
—, *see also* Rare earth elements

E-type basalts, 210
Earth
—, core of, 154
—, crust of, 4–6, 154
—, evolution of, 408–411
—, isotope ratios in, 381–382, 388–390
—, mantle of, 154
—, rare earth abundances in, 126
East Pacific Rise, 210–212, 228
—, sediments, 361
Eclogites, 35, 137
Economic deposits, 424–438
—, locations of, 426
Electron probe microanalysis, 493–494
Electronic configuration, 2–3, 12, 33
Element
— condensation, 65–70, 71
— co-ordination, 14–17
Emperor Seamounts, 219–221
Equilibrium melting and crystallization, 122–128, 225
Erbium
—, atomic weight, 3
— in chondrites, 8, 10
— as colouring agent, 456
— in Earth, 4–6
—, electronic configuration, 3
—, ionic radius, 16
— isotopes, 7
—, mass spectrometry of, 486
— in solar system, 4, 5
— in Sun, 5
—, *see also* Rare earth elements
Eucrite meteorites, 94–97, 412
—, isotope studies of, 412
Eudialyte, 36, 46
Europium
— anomaly, 9, 130
— — in meteorites, 73–75, 81, 87–88
— — in sulphides, 305
—, atomic weight, 3
— in chondrites, 8, 10
— in Earth, 4–6
—, electronic configuration, 3
—, ionic radius, 16
— isotopes, 7
—, mass spectrometry of, 485
—, oxidation states, 12, 304–305
—, oxide behaviour, 12
— in solar system, 4, 5
— in Sun, 5
—, *see also* Rare earth elements
Euxenite, 36, 42
— ore, 424
Euxenite-polycrase series, 42
Ewaldite, 36, 39, 40
Extraction of elements, modelling, 142–144

FAMOUS area, North Atlantic, 147, 211, 212, 214, 225, 361–362
Fen, Norway, 245, 247, 248
Fenitization, 249, 333
Fergusonite, 36, 43
Fergusonite-formanite series, 43
Ferrous metallurgy, 450–452
Fersmite, 36, 43
Finnish ore deposits, 430, 435
Fish debris, 355–356
Florencite, 36, 52
Fluocerite, 36, 38, 41
Fluorite, 28, 38
Foraminifera, 355–356

Forest City meteorite, 87–88
Formanite, 36, 43
Fractional crystallization, 128–138, 222–224
— in alkaline igneous rock genesis, 261–262
— in andesite genesis, 280–281
— in anorthosite genesis, 284
— in carbonatite genesis, 246–247
— in granitic rock genesis, 288, 291–293, 297–300
— in kimberlite genesis, 242
— in komatiite genesis, 254
— in tholeiite genesis, 266

Gadolinite, 36, 40, 46
Gadolinium
—, atomic weight, 3
— in chondrites, 8, 10
— in Earth, 4–6
— in electrode manufacture, 456
—, electronic configuration, 3
—, ionic radius, 16
— isotopes, 7
—, mass spectrometry of, 485
— in solar system, 4, 5
— in Sun, 5
—, *see also* Rare earth elements
Gagarinite, 36, 38, 41
Galapagos spreading centre, 361–362
Galena, 301
Gamma rays, 472, 474
Gamma-ray detection, 468, 473–475
Garnet, 22, 24, 26, 27
—, synthetic, 459, 460
Garnet lherzolites, 162
—, inclusions, 186–187
Garonne–Dordogne river system, 328, 351
Gas mantles, 443
Geochronology, 379–383
Glass alteration, 318, 320, 322, 325
Glass manufacture, 455–457
Glass polishing, 455
Globigerina, 355–356
Gneiss, 347, 348
Gorgona Island, Columbia, 254
Gough Island, Atlantic, 219
Goyazite, 428
Granites, 277, 285–301
—, Caledonian, Scotland, 387
—, isotope variation, 384
—, Namibian, 387
Granodiorite, 285–293
—, weathering of, 329–330, 347
Greenschist metamorphism, 336
Greenstones, 327, 336–337, 348
Grenville Province, U.S.A., 284
Greywacke, 267, 368
Groundwater, 328–331
Group I inclusions, 71–74
Group II inclusions, 72, 75–80
Group III inclusions, 72, 73, 75–80
Group IV inclusions, 80

Hafnium isotopes, 375
— in bulk Earth, 412
Half-lives, 7
Halmyrolysis, 346, 356, 357
Harzburgite, 154, 161, 163–164, 166–167, 176–186, 193–195
Havertö meteorite, 100
Hawaiian Islands, 215, 219–221
Heavy rare earth elements, 2
Hellandite, 36, 47
Henry's law, 18, 21–23, 121
Hibonite, 36, 43, 79
Highly incompatible elements, 208
Holmium
—, atomic weight, 3
— in chondrites, 8, 10
— as colouring agent, 456
— in Earth, 4–6
—, electronic configuration, 3
—, ionic radius, 16
— isotope, 7
— in solar system, 4, 5
— in Sun, 5
—, *see also* Rare earth elements
Howardites, 98–99
Huanghoite, 36, 40
Hydrogen storage, 460
Hydrogenous minerals, 357–360
Hydrothermal fluid, 323, 331–334
Hydrothermal systems, 301–306, 331–334

Iceland, 131, 215–217
Iimoriite, 36, 47
Ijolites, 238, 244, 246
Ilimaussite, 36, 47
Ilmajokite, 36, 47
Immiscible melts, 21
— in carbonatite genesis, 247–248
Inclusions
—, amphibole-bearing, 183
—, apatite-rich, 183
—, garnet lherzolite, 186–187
—, harzburgite, 176–186
— in kimberlites, 186–188

Inclusions (*contd.*)
— in meteorites, Group I, 71–74
— —, Group II, 72, 75–80
— —, Group III, 72, 73, 75–80
— —, Group IV, 80
—, pyroxenite, 187–192
—, spinel lherzolite, 176–186
—, ultramafic, in basalts, 176–186
Incompatible element, 130, 160
—, definition of, 207
— in upper mantle, 396–400
Indian Ocean
— basalts, 209–210
— floor, 173
— water, 353
Indian Ocean Ridge, 210
Indian ore deposits, 432–433
—, production of, 444, 445, 448
Indonesian ore deposits, 433
Inductively coupled plasma emission spectrometry, 488, 491
Ion exchange separation, 482, 489
Ionic radius, 10, 14–18, 34
Ionization potential, 78
Iraqite, 36, 47
Iron-manganese deposits, 360–364, 367
— in soils, 347
Iron meteorites, 101–106
Ishikawaite, 44
Island arc volcanic rocks, 238, 262–267
—, isotope studies, 401–402
Isochron, 378
— diagram, 378
— equation, 377, 411
Isotope, 7, 469, 470
— correlations, 389
— dilution analysis, 475–488
— evolution diagram, 378
— tracers, 388–405
Itinome-gata, Japan, 178–179, 181, 184

Joaquinite, 37, 47
Johnstown meteorite, 98, 99
Johnstrupite — *see* Mosandrike
Juvinas meteorite, 379, 382
—, isotope ratios, 413

Kainosite, 37, 48
Kapfenstein, Austria, 178–179, 181, 189
Karnasurtite, 37, 48
Karonage, Burundi republic, 427
Kazakhstan massif, U.S.S.R., 332
Kenna meteorite, 100, 101
Keweenawan lavas, U.S.A., 132
Kiama, Australia, 183
Kiglapait intrusion, Canada, 132
Kilbourne Hole, New Mexico, 179–181
Kimberlites, 238, 239–243
—, South African, 187, 243
Knopite — *see* Perovskite group
Kohar meteorite, 93
Kiovinite — *see* Florencite
Komatiites, 238, 251–254
Korean ore deposits, 433

Lafayette meteorite, 102
Lamprophyres, 238, 249–251
Lance meteorite, 82
Lanthanide contraction, 15
Lanthanides, 2
Lanthanite, 37, 40
Lanthanons, 2
Lanthanum
—, atomic weight, 3
— in chondrites, 8, 10
— in Earth, 4–6
—, electronic configuration, 3
—, ionic radius, 16
— isotopes, 7
—, mass spectrometry for, 486
— in optical glasses, 456
— in solar system, 4, 5
— in Sun, 5
—, *see also* Rare earth elements
Laplandite, 37, 48
Lasers, 458–459
Lashaine volcano, Tanzania, 187
Layered intrusions, 132
Lead-zinc deposits, 301–302, 305
Leedey chondrite, 10, 91
Lewisian rocks, Scotland, 380, 381, 393
Lherz, France, 156–158
Lherzolite, 154–162, 172–173, 176–187, 193–195
Light rare earth elements, 2
Liguria, Italy, 156–161, 170
Limestones, 367, 368
Lithophile element, 2
Lizard, Cornwall, England, 155–158, 169
Loparite — *see* Perovskite group
Lovchorrite — *see* Mosandrite
Loveringite, 37, 43
Lutetium
—, atomic weight, 3
— in chondrites, 8, 10
— in Earth, 4–6
—, electronic configuration, 3

Lutetium (*contd.*)
—, ionic radius, 16
—, isotope decay, 375
— isotopes, 7, 375
—, mass spectrometry for, 487
— in solar system, 4, 5
— in Sun, 5
—, *see also* Rare earth elements
Lutetium-hafnium isotope method, 411–413

Mckelveyite, 37, 39, 40
McMurdo volcanics, Antarctica, 261
Madagascar ore deposits, 434, 435
Magnetite, 24, 26, 27
Magnets, 453–455
Malaysian ore deposits, 433
—, production of, 444
Malvern meteorite, 98
Manganese nodules, 362–364, 367
—, isotopes in, 403, 406, 407
Mantle
— evolution, 408–411
— heterogeneity, 192, 196, 207, 225–228
— isochrons, 399
— metasomatism, 182, 227–228
Marianas island arc, Pacific, 265
Marine minerals, 346, 357–361
Mass fractionation, 415
Mass spectrometry, 414, 478–488
—, spark-source, 491–492
Masuda-Coryell diagram, 8
Medicine Lake, California, 280
Malanocerite, 37, 48
Mesosiderites, 103–104
Meta-igneous rocks, 298
Metalliferous sediments, 360–364
Metamorphic minerals, 35
Metamorphic redistribution, 28
Metamorphic rocks, 28
Metamorphism
—, hydrous, 323
—, zeolite facies, 320–322, 323
Metasomatic fluid, 28, 323, 331–334
Metasomatism, 331–334, 335
— in mantle, 227–228, 398–399
Meteorite inclusions
—, Ca,Al-rich, 71–86
—, coarse-grained, 71–74
—, fine-grained, 71–80
—, FUN, 83–86
—, ultra-refractory condensates, 81–83
Meteorites, 64, 71–107
—, achondrites, 94–103
—, chondrites, 64, 71–94
—, mesosiderites, 103–104
—, pallasites, 104
—, shergottites, 102–103
—, ureilites, 100–101
Micas, 347, 348
Microlite — *see* Pyrochlore group
Mid-Atlantic Ridge, 173, 209–214, 336, 337, 338, 349
Mid-ocean ridge basalt, 159, 171, 211–214
—, E-type, 213
—, isotope variation of, 384, 401
—, N-type, 210–212
—, origin of, 222–229
—, T-type, 213–214
Mid-ocean ridges, 205–206, 208–214
Middle rare earth elements, 2
Mineral alteration, 319–322
Mineral processing, 438–442
Minettes, 249–250
Mining, 438–442
Mixing models, 145
Mobility, 16, 17, 28, 169
— in crust of Earth, 317–339
—, factors affecting, 318
— through weathering, 319–322, 346–350
Model ages, 383–388
—, calculation of, 416
Mokoia meteorite, 88
Monazite, 14, 37, 52
— as ore, 424, 425
— production, 424, 444, 445, 446, 448
Montmorillonite, 349
Monzogranites, 293–301
Moore County meteorite, 94, 95, 96
MORB — *see* Mid-ocean ridge basalt
Morristown meteorite, 103
Mosandrite, 37, 48
Mount Albert, Quebec, Canada, 155–159
Mountain Pass district, California, U.S.A., 427
Mull, Isle of, Scotland, 320, 393
Munro Township, Ontario, Canada, 253–254
Murchison meteorite, 81–83
Mylonites, 170–172

Nakhlites, 102
Namaqualand, South Africa, 298
Nauru Basin, Pacific, 220–222
Navajo volcanics, New Mexico, 187
Neodymium
—, atomic weight, 3
— in chondrites, 8, 10

Neodymium (*contd.*)
— as colouring agent, 456
— in Earth, 4–6
—, electronic configuration, 3
—, ionic radius, 16
— isotopes, 7, 375, 414, 415
— — in bulk Earth, 381–382
— — in seawater, 406
—, mass spectrometry for, 485
— in solar system, 4, 5
— in Sun, 5
—, *see also* Rare earth elements
Nernst distribution coefficient, 118–120
Neutron activation analysis, 468–478
—, epithermal, 469, 475–476
—, instrumental, 468, 469–476
—, radiochemical, 476–478
Nigerian ore deposits, 434
—, production of, 444
Non-ferrous metallurgy, 452–453
Nontronite, 349
"Normal" ridge segments, 210–212
Normalization, 8–11
— to average rock, 8
— to chondrites, 8–11
— to North American shales composite, 9, 345
North American shales composite, 8–10, 344, 366
Nucleosynthesis, 90
Nuevo Laredo meteorite, 95, 96
Nunivak Island, Alaska, U.S.A., 183

Ocean basin rocks, 205–230, 403
Ocean floor peridotites, 170–176
Ocean island basalts, 214–220
—, isotope variation in, 384
Ocean plateau volcanism, 220–222
Ocean ridge basalts, 208–214
Oceanic oozes, 355
Oceans, 353
—, supply of REE to, 350
—, mass balances in, 364–365
Odessa meteorite, 105
Oka, Quebec, Canada, 245, 246, 248
Okanoganite, 37, 51
Olivine, 24, 26
Ontong-Java plateau, Pacific, 220–221
Ophiolites, 166–170, 326
Ore deposits, 301–306, 424–436
—, demand for, 443–447
— extraction, 438–442
— reserves, 436–438
Ore extraction, 438–442
Ore genesis, 304–306
Ornans meteorite, 82–83, 86
Orthite — *see* Allanite
Othris, Greece, 167
Oxidation state, 10, 12–14, 304–305
Oxygen fugacity, 13, 20–21

Pacific Ocean
— basalts, 209–212
— water, 353
— —, isotopes in, 407
Palagonite, 326
Palagonitization, 318, 325
Pallasites, 104
Parisite, 37, 40, 41
Parnallee meteorite, 87–88
Partial melting, 122–128
— in alkaline igneous rock genesis, 258–259
— in andesite genesis, 279–280
— in anorthosite genesis, 284
— in carbonatite genesis, 246
— in granitic rock genesis, 289–291, 296–297
—, incremental, 138–140
— in kimberlite genesis, 241–242
— in komatiite genesis, 253–254
—, non-modal, 160
— in tholeiite genesis, 265–266
Partition, 17–28, 118–122
—, temperature effect on, 18–20
Partition coefficient — *see* Distribution coefficient
Pasamonte meteorite, 94, 95
Pegmatites, 35
Peridotites, 126, 153–196
Periodic table, 1
Perovskite, 14, 67, 69, 82
— group, 43
Perrierite, 37, 46, 48
Petrogenetic modelling, 115–148
Phillipsite, 357, 358
Phlogopite, 24, 26
Phosinaite, 37, 51
Phosphorites, 359–360
Phosphors, 457–458
Photopeak interferences, 474
Pillow basalts, 325–326, 336
Pindos, Greece, 167
Piñon meteorite, 105
Placer deposits, 430–435, 438
Plagioclase feldspar, 19, 20, 21, 24, 25, 27
Plankton, 355
Plume model, 226–228

Point Sal ophiolite, California, U.S.A., 326–327
Polycrase, 37, 42
Polymignite, 37, 44
Potassium-rich basalts, 238, 256, 258
Praseodymium
—, atomic weight, 3
— in chondrites, 8, 10
— as colouring agent, 456
— in Earth, 4–6
—, electronic configuration, 3
—, ionic radius, 16
— isotope, 7
— in solar system, 4, 5
— in Sun, 5
—, *see also* Rare earth elements
Primary mantle melts, 129
Priorite, 37, 41, 42
Production, 443–448
Promethium
—, atomic weight, 3
—, electronic configuration, 3
Puerto-Rico trench, West Indies, 173–175
Pyrochlore, 37, 39, 44
— group, 44
Pyrophoric alloys, 453
Pyroxene, 19, 21, 24, 26, 27
— in inclusions, 182–185
Pyroxenite, 154, 164–166, 195–196
— inclusions, 187–192

Radioactive decay, 376–379
Rare earth elements
—, atomic weights, 3
— in chondrules, 86–89
—, condensation, 65–71, 77
— in core of Earth, 154
— in crust of Earth, 4–6, 154
—, definition of, 2
—, demand for, 443–447
—, deposits of, 424–436
—, distribution coefficients, 23–28
— in Earth, 4–6, 126
—, electronic configurations, 3
—, heavy, 2
—, ionic radii, 14–17
—, ionization potentials, 78
—, light, 2
—, in mantle, 154, 192–193
—, middle, 2
— in nebular condensate, 68
—, oxidation states, 10–14
—, partition, 17–28
—, production, 443–447
—, reserves, 436–438
— in seawater, 323–325, 351–354
—, sedimentation rates, 359
—, solar abundances, 3–6, 79
— in solar system, 3–5
Residence times, 351–352, 406
—, definition of, 405–406
Resources, 436–438
Retzian, 37, 54
Reunion Island, Indian Ocean, 218–219
Reydarfjordur, Iceland, 320
Reykjanes Peninsula, Iceland, 131, 215, 320
Reykjanes Ridge, 211, 214, 215, 227
Rhabdophane, 37, 53
Richardton meteorite, 87–88
Rinkite — *see* Mosandrite
Rinkolite — *see* Mosandrite
Rio Grand rift, U.S.A., volcanic rocks, 132
River water, 328, 351
Roccamonfina, Italy, 394–395, 400
Roda meteorite, 97–98
Rogaland, Norway, 284, 285, 297
Ronda, Spain, 156–162, 164, 165, 166, 169
Röntgenite, 37, 40, 41
Ross Island, Antarctica, 259
Rowlandite — *see* Thalenite

Sahamalite, 37, 40
St. Paul's Rocks, Atlantic, 170–172
Salt Lake Crater, Hawaii, U.S.A., 178–179, 181, 189–190
Samail, Oman, 167, 168
Samarium,
—, atomic weight, 3
— in chondrites, 8, 10
— in Earth, 4–6
—, electronic configuration, 3
—, ionic radius, 16
—, isotope decay constant, 379
—, isotope half-life, 379
— isotopes, 7, 375, 414, 415
—, mass spectrometry for, 485
—, oxidation states, 12
— in solar system, 4, 5
— in Sun, 5
—, *see also* Rare earth elements
Samarium-neodymium isotope method, 376–411
Samarskite, 37, 44
San Carlos, Arizona, U.S.A., 179–181, 191
Sandstones, 367, 368
Santa Clara meteorite, 105

Saryarkite, 37, 51
Sazhinite, 37, 48
Scandium, 2
— in Sun, 5
— in solar system, 5
Scheteligite, 37, 45
Scourion complex, Scotland, 289
Seabrook Lake, Canada, 245, 246, 248, 249
Seawater, 323–327, 350, 351–354
—, isotope variations in, 405–408
—, removal of REE from, 351, 354–364
Sediment sorting, 348
Sedimentary rocks, 365–367
—, model ages of, 387
Sedimentation rates, 359
Sediments, 365–367
Semenovite, 37, 49
Separation processes, modelling, 117–118
Serpentinization, 169
Serra de Magé meteorite, 94, 95, 96
Shales, 344, 365–366
—, definition of, 343
Shergottites, 102–103
Shergotty meteorite, 102, 103
Shoshonites, 238, 256, 258
Sierra Nevada Batholith, U.S.A., 293
Sioux County meteorite, 94, 95
Skaergaard intrusion, Greenland, 132, 146–147, 266
Skye, Isle of, Scotland, 393
Smectites, 349, 357, 358
Sokoman iron formation, Canada, 303
Solar abundances, 5, 89–92
Solar nebula, 78
Solar system,
—, origin of, 63
Solubility products, 323–324
Soret effect, 147, 300
Sources of rare earths, 423–438
Spanish Peaks igneous complex, Colorado, U.S.A., 250, 280, 294, 295, 297
Spark-source mass spectrometry, 491–492
Sparta complex, Oregon, U.S.A., 290, 291
Speciation, 324, 352
Spencite — *see* Tritomite
Sphene, 14, 18, 26, 27, 49
Spilites, 318
Sri Lankan ore deposits, 434
—, production of, 444
Standard rocks, 163–164
Stannern meteorite, 95
Steenkampskraal, South Africa, 429, 446
Steenstrupine, 37, 49
Stillwellite, 37, 49
Strontium, 13
Substitution, 17–18
Summer Coon volcano, Colorado, U.S.A., 280
Syenogranites, 293–301
Synchysite, 37, 38, 40, 41

Tadzhikite — *see* Hellandite
TAG area, North Atlantic, 361–362
Tatahouine meteorite, 97–98
Television tubes, 457
Tengerite, 37, 38, 40
Terbium
—, atomic weight, 3
— in chondrites, 8, 10
— in Earth, 4–6
—, electronic configuration, 3
—, ionic radius, 16
— isotope, 7
—, oxidation states, 12
— in solar system, 4, 5
— in Sun, 5
— *see also* Rare earth elements
Thailand ore deposits, 433
—, production of, 444
Thalenite, 37, 49
Thermodynamic activity, 66
Thermodynamic properties, 29
Thermogravitational diffusion, 147, 300–301
Tholeiitic basalts
—, back-arc basin, 262–267
—, continental, 262–267
—, island arc, 262–267
Tholeiitic intrusive complexes, 146–147, 238, 266
Thorbastnäsite, 40, 41
Thorite, 49
Thortveitite, 37, 50
Thulium
— anomaly, 75–77, 81, 85
—, atomic weight, 3
— in chondrites, 8, 10
— in Earth, 4–6
—, electronic configuration, 3
— ionic radius, 16
— isotope, 7
— in solar system, 4, 5
— in Sun, 5
—, *see also* Rare earth elements
Tinaquillo, Venezuela, 155–159
Titanite — *see* Sphene
Tombarthite, 37, 50
Tonalites, 285–293

Törnebohmite, 37, 50
Tourmaline, 335
Tranquillitite, 37, 50
"Transitional" ridge segments, 211, 213–214
Trapped liquid, 147
Tritomite, 37, 50
Trondhjemites, 285–293
Troodos, Cyprus, 167
Tundrite, 38, 50
Tveitite, 38, 39
Tysonite — *see* Fluocerite

Ultramafic inclusions, 176–192
Ultramafic rocks, 153–196
—, inclusions, 176–192
—, nomenclature, 155
U.S.A. ore deposits, 427, 433
—, production of, 444, 445, 448
U.S.G.S. standard rocks, 163–164
U.S.S.R. ore deposits, 435
Uptake by minerals, 334–337, 354
Ureilites, 100–101
Use development, 460
Uses of rare earths, 447–460

Van't Hoff equation, 18
Vein deposits, 429–430
Victoria, Australia, 177–179, 181
Vigezzite, 42
Vitusite, 38, 53
Volatile transport
— in carbonatite genesis, 249
— of HREE, 288
— in kimberlite genesis, 242
Vourinos, Greece, 167

Wabigoon belt, Ontario, Canada, 309
Weathering, 318, 319–322, 345, 346–350
—, continental, 346–348
—, submarine, 348–350
Weekeroo Station meteorite, 105
Weinschenkite — *see* Churchite
Welagonite, 40
Wolf River Batholith, Wisconsin, U.S.A., 285, 296, 298–299
World reserves, 436–438

Xenotime, 38, 54
— ore, 424, 425
— — production, 424, 444, 447
X-ray fluorescence spectrometry, 492–493

Yardymly meteorite, 105
Ytterbium,
— anomaly, 75, 81
—, atomic weight, 3
— in chondrites, 8, 10
— in Earth, 4–6
—, electronic configuration, 3
—, ionic radius, 16
— isotopes, 7
—, mass spectrometry for, 486
—, oxidation states
— in solar system, 4, 5
— in Sun, 5
—, *see also* Rare earth elements
Yttrialite — *see* Thalenite
Yttrium, 2
— in electrodes, 456
— in refractories, 456
— in solar system, 5
— in Sun, 5
Yttrocerite, 38
Yttrocrasite, 38, 45
Yttrofluorite, 38
Yttrotantalite, 38, 45
Yttrotungstite, 38, 45

Zabargad Island, Red Sea, 172
Zagami meteorite, 102
Zaire ore deposits, 434
Zeolite facies metamorphism, 320–321, 326–327, 335
Zhonghuacerite, 38, 41
Zircon, 14, 18, 19, 24, 27, 50
Zone melting, 140–142
— in kimberlite genesis, 242